INTERACTIVE MATHEMATICS

Intermediate Algebra

**Personal
Academic
Notebook**

ACADEMIC
SYSTEMS

Interactive Mathematics – Intermediate Algebra
©1997-1999 Academic Systems Corporation
All rights reserved.

Academic Systems Corporation
444 Castro Street, Suite 1200
Mountain View, CA 94041

800.694.6850
www.academic.com

ISBN   1-928962-02-5

06/99

TABLE OF CONTENTS

TOPIC EII ESSENTIALS OF ALGEBRA

LESSON EII.A REAL NUMBERS AND EXPONENTS 1

Overview 2

Explain   Real Numbers and Notation 3

                           Integer Exponents 9

Homework   Homework Problems 11

Apply   Practice Problems 12

Evaluate   Practice Test 14

LESSON EII.B POLYNOMIALS 15

Overview 16

Explain   Polynomial Operations 17

                           Factoring Polynomials 26

Homework   Homework Problems 33

Apply   Practice Problems 34

Evaluate   Practice Test 36

LESSON EII.C EQUATIONS AND INEQUALITIES 37

Overview 38

Explain   Linear 39

Homework   Homework Problems 46

Apply   Practice Problems 47

Evaluate   Practice Test 48

LESSON EII.D RATIONAL EXPRESSIONS 49

Overview 50

Explain   Rational Expressions 51

                           Rational Equations 60

Homework   Homework Problems 65

Apply   Practice Problems 66

Evaluate   Practice Test 68

LESSON EII.E GRAPHING LINES 69

Overview 70

Explain   Graphing Lines 71

                           Finding Equations 78

Homework   Homework Problems 84

Apply   Practice Problems 85

Evaluate   Practice Test 87

LESSON EII.F ABSOLUTE VALUE 89
Overview 90
Explain   Solving Equations 91
                           Solving Inequalities 97
Homework   Homework Problems 102
Apply   Practice Problems 103
Evaluate   Practice Test 105
Cumulative Activities   Cumulative Review 107

TOPIC 6 EXPONENTS AND POLYNOMIALS
LESSON 6.1 EXPONENTS 109
Overview 110
Explain   Properties of Exponents 111
Homework   Homework Problems 115
Apply   Practice Problems 116
Evaluate   Practice Test 117

LESSON 6.2 POLYNOMIAL OPERATIONS I 119
Overview 120
Explain   Adding and Subtracting 121
                           Multiplying and Dividing 125
Homework   Homework Problems 128
Apply   Practice Problems 130
Evaluate   Practice Test 132

LESSON 6.3 POLYNOMIAL OPERATIONS II 133
Overview 134
Explain   Multiplying Binomials 135
                           Multiplying and Dividing 138
Explore   Sample Problems 141
Homework   Homework Problems 143
Apply   Practice Problems 145
Evaluate   Practice Test 146
Cumulative Activities   Cumulative Review 147

TOPIC 7 FACTORING
LESSON 7.1 FACTORING POLYNOMIALS I 149
Overview 150
Explain   Greatest Common Factor 151
                           Grouping 153
Homework   Homework Problems 156
Apply   Practice Problems 157
Evaluate   Practice Test 159

LESSON 7.2 FACTORING POLYNOMIALS II 161
    Overview 162
    Explain   Trinomials I 163
                     Trinomials II 167
    Explore   Sample Problems 176
    Homework   Homework Problems 178
    Apply   Practice Problems 179
    Evaluate   Practice Test 180

LESSON 7.3 FACTORING BY PATTERNS 181
    Overview 182
    Explain   Recognizing Patterns 183
    Homework   Homework Problems 187
    Apply   Practice Problems 188
    Evaluate   Practice Test 189
    Cumulative Activities   Cumulative Review 191

TOPIC 8 RATIONAL EXPRESSIONS
    LESSON 8.1 RATIONAL EXPRESSIONS I 193
        Overview 194
        Explain   Multiplying and Dividing 195
                     Adding and Subtracting 200
        Homework   Homework Problems 203
        Apply   Practice Problems 204
        Evaluate   Practice Test 206

    LESSON 8.2 RATIONAL EXPRESSIONS II 207
        Overview 208
        Explain   Negative Exponents 209
                     Multiplying and Dividing 215
                     Adding and Subtracting 219
        Homework   Homework Problems 225
        Apply   Practice Problems 227
        Evaluate   Practice Test 230

    LESSON 8.3 EQUATIONS WITH FRACTIONS 231
        Overview 232
        Explain   Solving Equations 233
        Homework   Homework Problems 236
        Apply   Practice Problems 237
        Evaluate   Practice Test 238

LESSON 8.4 PROBLEM SOLVING 239
 Overview 240
 Explain   Rational Expressions 241
 Homework   Homework Problems 245
 Apply   Practice Problems 246
 Evaluate   Practice Test 248
 Cumulative Activities Cumulative Review 249

TOPIC 9 RATIONAL EXPONENTS AND RADICALS

 LESSON 9.1 RATIONAL EXPONENTS 251
  Overview 252
  Explain   Roots and Exponents 253
          Simplifying Radicals 257
          Operations on Radicals 261
  Homework   Homework Problems 266
  Apply   Practice Problems 268
  Evaluate   Practice Test 271
  Cumulative Activities Cumulative Review 273

TOPIC 10 QUADRATIC EQUATIONS

 LESSON 10.1 QUADRATIC EQUATIONS I 275
  Overview 276
  Explain   Solving by Factoring...277
          Solving by Square Roots 280
  Homework   Homework Problems 282
  Apply   Practice Problems 284
  Evaluate   Practice Test 285

 LESSON 10.2 QUADRATIC EQUATIONS II 287
  Overview 288
  Explain   Completing the Square......................................289
          The Quadratic Formula.....................................294
  Explore   Sample Problems 298
  Homework   Homework Problems 300
  Apply   Practice Problems 302
  Evaluate   Practice Test 304

 LESSON 10.3 COMPLEX NUMBERS 305
  Overview 306
  Explain   Complex Number System.................................307
  Homework   Homework Problems 315
  Apply   Practice Problems 316
  Evaluate   Practice Test 317
  Cumulative Activities Cumulative Review 319

TOPIC 11 FUNCTIONS AND GRAPHING

LESSON 11.1 FUNCTIONS...321
Overview		322
Explain	Functions and Graphs	323
	Linear Functions	329
	Quadratic Functions	333
Explore	Sample Problems	339
Homework	Homework Problems	340
Apply	Practice Problems	346
Evaluate	Practice Test	355

LESSON 11.2 THE ALGEBRA OF FUNCTIONS 359
Overview		360
Explain	The Algebra of Functions	361
	Inverse Functions	369
Explore	Sample Problems	374
Homework	Homework Problems	377
Apply	Practice Problems	381
Evaluate	Practice Test	385
Cumulative Activities	Cumulative Review	387

TOPIC 12 THE EXPONENTIAL AND LOGARITHMIC FUNCTIONS

LESSON 12.1 EXPONENTIAL FUNCTIONS 389
Overview		390
Explain	The Exponential Function	391
Homework	Homework Problems	402
Apply	Practice Problems	403
Evaluate	Practice Test	404

LESSON 12.2 LOGS AND THEIR PROPERTIES 405
Overview		406
Explain	The Logarithmic Function	407
	Logarithmic Properties	412
Explore	Sample Problems	414
Homework	Homework Problems	418
Apply	Practice Problems	420
Evaluate	Practice Test	422

LESSON 12.3 APPLICATIONS OF LOGS 423
Overview		424
Explain	Natural and Common Logs	425
	Solving Equations	432
Homework	Homework Problems	439
Apply	Practice Problems	441
Evaluate	Practice Test	443
Cumulative Activities	Cumulative Review	445

TOPIC 13 MORE NONLINEAR EQUATIONS AND INEQUALITIES

    LESSON 13.1 NONLINEAR EQUATIONS..449

Overview	....	450
Explain	Solving Equations	451
	Radical Equations	454
Explore	Sample Problems	458
Homework	Homework Problems	460
Apply	Practice Problems	461
Evaluate	Practice Test	462

    LESSON 13.2 NONLINEAR SYSTEMS...463

Overview	....	464
Explain	Solving Systems	465
Explore	Sample Problems	474
Homework	Homework Problems	476
Apply	Practice Problems	478
Evaluate	Practice Test	480

    LESSON 13.3 INEQUALITIES...483

Overview	....	484
Explain	Quadratic Inequalites	485
	Rational Inequalities	490
Explore	Sample Problems	498
Homework	Homework Problems	500
Apply	Practice Problems	504
Evaluate	Practice Test	507
Cumulative Activities	Cumulative Review	509

  ANSWERS 513

  INDEX 543

LESSON EII.A – REAL NUMBERS AND EXPONENTS

OVERVIEW

Here's what you'll learn in this lesson:

Real Numbers and Notation

a. Number line and notation

b. Operations on signed numbers

c. Properties of real numbers

Integer Exponents

a. Nonnegative exponents

b. Properties of exponents

In algebra you use many numbers, symbols, operations and properties to simplify expressions. Once you are familiar with these basic tools, you can use algebra to solve many different types of problems.

 **EXPLAIN**

REAL NUMBERS AND NOTATION

Summary

An Introduction to Real Numbers and the Number Line

The first numbers were used for counting things, like the number of sheep in a flock or the number of members in a family. This type of number is now called a counting number or natural number. Examples are: 1, 2, 3, 4, 5, …

As trade and the use of money developed, the idea of zero and negative numbers became necessary to record losses or gains. For this sort of work the integers are the most useful. Examples are: …, −4, −3, −2, −1, 0, 1, 2, 3, …

However, there are many everyday quantities which are not measured in whole numbers. You might go to the store and buy a half gallon of juice, or three and a half pounds of cheese, or three quarters of a yard of fabric. This uses yet another type of number called a rational number or fraction. Fractions are written using two whole numbers. For example, $\frac{3}{8}$ and $\frac{11}{29}$ are both fractions.

Here are some examples of sets of numbers:

Counting numbers	1, 2, 3, 4, 5, 6, 7, …
Whole numbers	0, 1, 2, 3, 4, 5, 6, 7, …
Integers	…, −4, −3, −2, −1, 0, 1, 2, 3, 4, 5, …
Rational numbers	Numbers which can be written as a fraction $\frac{a}{b}$, where a and b are integers and $b \neq 0$. Counting numbers, whole numbers, and integers are all rational numbers. Some examples are: $\frac{7}{15}$, $-\frac{4}{27}$, $\frac{13}{1}$
Irrational numbers	Numbers which cannot be written as a fraction $\frac{a}{b}$, where a and b are integers and $b \neq 0$. Some examples are π, $\sqrt{3}$, $\sqrt{11}$

The rational numbers and irrational numbers taken together are the real numbers. Real numbers can be represented as points on a number line.

Comparison Symbols

The signs =, ≠, <, >, ≤ or ≥ are used to compare and order numbers.

Symbol	Meaning
=	equal to
≠	not equal to
<	less than
>	greater than
≤	less than or equal to
≥	greater than or equal to

For example, $-4 < 6$ means that the number -4 is less than the number 6. That is, -4 lies to the left of 6 on the number line.

Similarly, $x \geq 3$ means that the variable x can be any real number greater than 3 or equal to 3.

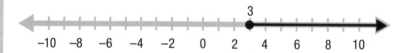

Absolute Value

The absolute value of a number gives the distance of that number from zero on the number line. For example, $|-7|$ denotes that -7 is 7 units from zero on the number line.

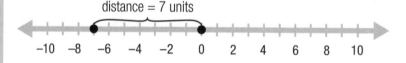

Vertical bars enclosing a number are used to denote the absolute value of that number. Absolute value bars leave positive or zero numbers alone but change the sign of negative numbers. So the absolute value of a number is never negative.

For example:

$$|15| = 15$$

$$|0| = 0$$

$$|-7| = 7$$

Grouping Symbols

Parentheses or brackets are often used to group symbols together. Grouping symbols determine the order in which operations are performed.

Here's an example: $[7 - 2 \cdot (5 + 13)]$

Operations on Signed Numbers

You can combine positive numbers using the operations of addition, subtraction, multiplication and division. For example:

$$2 + 18 = 20 \qquad\qquad \frac{5}{6} - \frac{1}{4} = \frac{7}{12}$$

$$13 \cdot 5 = 65 \qquad\qquad 36 \div 3 = 12$$

You can also add, subtract, multiply and divide positive **and** negative numbers using operations on signed numbers.

Adding Signed Numbers

To add signed numbers:

If both numbers are positive, their sum is positive.	$2 + 7 = 9$
If both numbers are negative, their sum is negative.	$-3 + (-4) = -7$
If the numbers have different signs, ignore their signs and subtract the smaller number from the larger number. The sum has the same sign as the larger number.	$-3 + 5 = 2$ $3 + (-5) = -2$

Subtracting Signed Numbers

To subtract signed numbers:

1. Change the subtraction sign (–) to an addition sign (+) and change the sign of the number being subtracted.

2. Add according to the rules for addition of signed numbers.

For example, to find $3 - 8$:

1. Change – to + and change the sign of the number being subtracted.
    $$3 + (-8)$$

2. Add according to the rules for addition of signed numbers.
    $$= -5$$

So, $3 - 8 = -5$.

Multiplying Signed Numbers

To multiply signed numbers:

If the numbers have the same sign, their product is positive.	positive · positive = positive negative · negative = positive	$3 \cdot 5 = 15$ $(-3) \cdot (-5) = 15$
If the numbers have different signs, their product is negative.	positive · negative = negative negative · positive = negative	$3 \cdot (-5) = -15$ $(-3) \cdot 5 = -15$

Dividing Signed Numbers

To divide signed numbers, use the same rules for signs as you used for multiplication:

If the numbers have the same sign, their quotient is positive.	positive ÷ positive = positive negative ÷ negative = positive	$20 \div 4 = 5$ $(-20) \div (-4) = 5$
If the numbers have different signs, their quotient is negative.	positive ÷ negative = negative negative ÷ positive = negative	$20 \div (-4) = -5$ $(-20) \div 4 = -5$

Exponents

Exponents are used to indicate repeated multiplication of the same number.

For example:

$$2^6 = \underbrace{2 \cdot 2 \cdot 2 \cdot 2 \cdot 2 \cdot 2}_{6 \text{ factors}} = 64$$

In this case, the number 2 is called the base and the number 6 is called the exponent. The exponent 6 indicates that there are 6 factors of 2.

Properties of Real Numbers

On the following page are some properties that will help you simplify calculations with real numbers.

Property	Examples	General Rule
		(a, b, and c are real numbers)
Commutative Property of Addition	$2 + 8 = 8 + 2$	$a + b = b + a$
Commutative Property of Multiplication	$4 \cdot (-8) = (-8) \cdot 4$	$a \cdot b = b \cdot a$
Associative Property of Addition	$(2 + 8) + 3 = 2 + (8 + 3)$	$(a + b) + c = a + (b + c)$
Associative Property of Multiplication	$(2 \cdot 7) \cdot 3 = 2 \cdot (7 \cdot 3)$	$(a \cdot b) \cdot c = a \cdot (b \cdot c)$
Distributive Property	$6 \cdot (3 + 9) = 6 \cdot 3 + 6 \cdot 9$	$a \cdot (b + c) = a \cdot b + a \cdot c$
Additive Identity	$33 + 0 = 33$	$a + 0 = a$
Multiplicative Identity	$39 \cdot 1 = 39$	$a \cdot 1 = a$
Additive Inverse	$27 + (-27) = 0$	$a + (-a) = 0$
Multiplicative Inverse	$15 \cdot \frac{1}{15} = 1$	$a \cdot \frac{1}{a} = 1$

Order of Operations

When you are simplifying expressions that involve several operations, you must do them in the correct order. Here is the order you should use:

1. Perform all operations inside parentheses or brackets.

2. Simplify terms with exponents.

3. Multiply or divide, working from left to right.

4. Add or subtract, working from left to right.

Here is a simple example. To find $6 + 2 \cdot 4$:

    1. Multiply. $= 6 + \mathbf{8}$

    2. Add. $= \mathbf{14}$

Here's another example. To simplify the expression $2 - 3^2 \div 3 - (8 - 5) \cdot 4$:

    1. Work within the parentheses. $= 2 - 3^2 \div 3 - \mathbf{3} \cdot 4$

    2. Simplify the term with the exponent. $= 2 - \mathbf{9} \div 3 - 3 \cdot 4$

    3. Multiply and divide from left to right. $= 2 - \mathbf{3} - \mathbf{12}$

    4. Add and subtract from left to right. $= \mathbf{-1} - 12$

                                               $= \mathbf{-13}$

*One way to recall the order of operations is to remember the phrase "Please Excuse My Dear Aunt Sally." The first letter of each word corresponds to an operation – Parentheses, Exponents, Multiply or Divide, Add or Subtract. However, be careful. This does **not** tell you to multiply before you divide or to add before you subtract.*

Sample Problems

1. From the list below, pick the numbers that are rational numbers.

    $-5, 0, \frac{1}{13}, \sqrt{7}, 2, \sqrt{9}, -\frac{1}{2}$

    ☑ a. Integers are rational numbers. List all of the integers. $-5, 0, 2, \sqrt{9}$

               (Since $\sqrt{9} = 3$, it too is an integer.)

b. $\frac{1}{13}, -\frac{1}{2}$

     ☐ b. Fractions where the numerator and denominator are integers are also rational numbers. List all such fractions.   _____

c. $-5, 0, \frac{1}{13}, 2, \sqrt{9}, -\frac{1}{2}$

     ☐ c. List all of the rational numbers.   _____

2. Find: $-11 + (-13)$

a. 24

     ☐ a. Ignore the signs and add the two numbers. $11 + 13 =$ _____

b. -24

     ☐ b. The answer has a negative sign. $=$ _____

3. Find: $(-7) \cdot (8)$

a. 56

     ☐ a. Ignore the signs and multiply the numbers. $7 \cdot 8 =$ _____

b. -56

     ☐ b. Write the answer with its correct sign. $=$ _____

4. Use the Distributive Property to calculate: $5 \cdot (7 - 4)$

     ☑ a. Use the Distributive Property. $= 5 \cdot 7 - 5 \cdot 4$

b. $35 - 20$
    15

     ☐ b. Simplify. $=$ _____

                         $=$ _____

5. Find: $4 - 2 \cdot (6 - 9)$

     ☑ a. Subtract inside the parentheses. $= 4 - 2 \cdot (-3)$

b. 6

     ☐ b. Multiply. $= 4 +$ _____

c. 10

     ☐ c. Add. $=$ _____

INTEGER EXPONENTS

Summary

Nonnegative Integer Exponents

Exponential notation is a shorthand way of writing repeated multiplication.

For example: $3 \cdot 3 \cdot 3 \cdot 3 \cdot 3 \cdot 3 \cdot 3 = 3^7$

Here, 3 is called the base and 7 is called the exponent or power.

Here is a similar example that uses a variable: $x^5 = x \cdot x \cdot x \cdot x \cdot x$

Here, x is the base and 5 is the exponent or power.

More generally: $x^n = \underbrace{x \cdot x \cdot x \cdot \ldots \cdot x \cdot x}_{n \text{ copies of } x}$

Here, n is a positive integer.

Properties of Exponents

There are several properties which make it easier for you to simplify, multiply, and divide expressions that contain exponents. These are listed below.

Property	Example
Multiplication: $x^m \cdot x^n = x^{m+n}$	$4^3 \cdot 4^7$ $= 4^{3+7}$ $= 4^{10}$
Division: $\frac{x^m}{x^n} = x^{m-n}$ (Here, $x \neq 0$)	$\frac{12^8}{12^5}$ $= 12^{8-5}$ $= 12^3$
Power of a Power: $(x^m)^n = x^{m \cdot n}$	$(7^4)^3$ $= 7^{4 \cdot 3}$ $= 7^{12}$
Power of a Product: $(x \cdot y)^n = x^n \cdot y^n$	$(5 \cdot 8)^9$ $= 5^9 \cdot 8^9$
Power of a Quotient: $\left(\frac{x}{y}\right)^n = \frac{x^n}{y^n}$ (Here, $y \neq 0$)	$\left(\frac{5}{11}\right)^4$ $= \frac{5^4}{11^4}$
Zero Exponent: $x^0 = 1$ (Here, $x \neq 0$)	$9^0 = 1$

To multiply expressions with the same base, add the exponents.

To divide expressions with the same base, subtract the bottom exponent from the top exponent.

To raise a power to a power, multiply the exponents.

To raise a product to a power, raise each factor to that power.

To raise a fraction to a power, raise the numerator and denominator to that power.

Remember, when you see a zero exponent (and the base is not zero), the answer is one.

Be careful, though. There are some cases where the properties of exponents do not apply.

For example, to simplify $5^4 \cdot 7^2$ you cannot use the Multiplication Property, because the two expressions have different bases. Instead, you have to multiply everything out:

$$5^4 \cdot 7^2$$
$$= (5 \cdot 5 \cdot 5 \cdot 5) \cdot (7 \cdot 7)$$
$$= 30{,}625$$

Here is another example. To simplify $3^4 - 3^5$ you can multiply everything out and then add. This is because even though the base is the same, there is no exponent property for addition.

$$3^4 - 3^5$$
$$= 81 - 243$$
$$= -162$$

You can also use the Distributive Property.

$$3^4 - 3^5$$
$$= 3^4(1 - 3)$$
$$= 3^4(-2)$$
$$= 81(-2)$$
$$= -162$$

Answers to Sample Problems

Sample Problems

1. Calculate: 3^4

   ✓ a. Write the base.   The base is 3.

b. The exponent is 4.

   ☐ b. Write the exponent.   _____

c. $3 \cdot 3 \cdot 3 \cdot 3 = 81$

   ☐ c. Write the answer.   _____

2. Simplify: $(2xy)^4 \cdot x^{12} \cdot y^8$

   ✓ a. First use the Power of a Product rule. $= 2^4 \cdot x^4 \cdot y^4 \cdot x^{12} \cdot y^8$

b. $2^4 \cdot x^{16} \cdot y^4 \cdot y^8$

   ☐ b. Use the Multiplication Property for the factors with base x. $=$ _____

c. $2^4 \cdot x^{16} \cdot y^{12}$ or $16x^{16}y^{12}$

   ☐ c. Use the Multiplication Property for the factors with base y. $=$ _____

3. Simplify: $\left(\dfrac{3}{5}\right)^7 \cdot 3^4 \cdot 5^2$

   ✓ a. First use the Power of a Quotient property. $= \dfrac{3^7}{5^7} \cdot 3^4 \cdot 5^2$

                                                      $= \dfrac{3^7 \cdot 3^4 \cdot 5^2}{5^7}$

b. $\dfrac{3^{11} \cdot 5^2}{5^7}$

   ☐ b. Use the Multiplication Property for the factors with base 3. $=$ _____

c. $\dfrac{3^{11}}{5^5}$ or $3^{11}5^{-5}$

   ☐ c. Use the Division Property for the factors with base 5. $=$ _____

 **HOMEWORK**

Homework Problems

Circle the homework problems assigned to you by the computer, then complete them below.

 Explain

Real Numbers and Notation

1. Circle all the inequality symbols that could replace the question mark to make this statement true: $-4 \ ? \ 3$

    $<$ $\leq$ $=$

    $>$ $\geq$ $\neq$

2. Find: $\dfrac{-45}{9}$

3. Rewrite using the Commutative Property: $17 + 8$

4. Circle the number that is **not** an integer:

    $0, -\dfrac{1}{2}, \sqrt{16}, \dfrac{4}{2}, -77$

5. Find: $-7 - (-3)$

6. Rewrite using the Associative Property: $5 \cdot (11 \cdot 23)$

7. Find: $|-9|$

8. Find: $2 - 5 - (-8)$

9. Find the additive inverse of 7 and the multiplicative inverse of 7.

10. Circle all the inequality symbols that could replace the question mark to make this statement true: $|23| \ ? \ -|-23|$

    $<$ $\leq$ $=$

    $>$ $\geq$ $\neq$

11. Find: $9 + \dfrac{45}{-9}$

12. Find: $8 + 2[3 - 4(7 - 2)]$

Integer Exponents

13. Identify the base and the exponent, then calculate: 4^3

14. Simplify using the properties of exponent: $4^{13} \cdot 4^6$

15. Find: 23^0

16. Write using an exponent:

    $7 \cdot 7 \cdot 7 \cdot 7 \cdot 7 \cdot 7 \cdot 7 \cdot 7 \cdot 7 \cdot 7 \cdot 7 \cdot 7$

17. Simplify using the properties of exponents: $\dfrac{7^{13} \cdot 12^6}{7^8 \cdot 12^2}$

18. Simplify: $2^3 + 2^5$

19. Write using an exponent: $y \cdot y \cdot y \cdot y \cdot y \cdot y$

20. Simplify using the properties of exponents: $\left(x^3 \cdot y^6\right)^4$

21. Simplify : $\left(2^3 \cdot 3^2\right)$

22. Write using exponents: $3 \cdot 3 \cdot y \cdot y \cdot y \cdot y \cdot y \cdot z \cdot z \cdot z \cdot z$

23. Simplify using the properties of exponents: $\left(\dfrac{1}{2}x\right)^3 \cdot x^2 \cdot (2y)^3$

24. Simplify using the properties of exponents: $\left[\dfrac{(2x)^3}{(2y)^2}\right]^3 \cdot y^6 \cdot x^4$

 APPLY

Practice Problems

Here are some additional practice problems for you to try.

Real Numbers and Notation

1. Circle the true statements.

    $7 = 12$ $3 \geq 0$ $2 \neq 8$

    $2 \leq 2$ $15 < 2$

2. Circle the true statements.

    $4 \neq 6$ $7 \leq 7$ $2 > 5$

    $6 \geq 0$ $13 < 1$

3. Find the absolute values:

    a. $|3.5|$ c. $|-0.005|$ e. $|36.1|$

    b. $|36|$ d. $|-14.33|$

4. Find the absolute values:

    a. $|0.26|$ c. $|-0.13|$ e. $|-2.69|$

    b. $|-28|$ d. $|15.22|$

5. Find using the Distributive Property: $6 \cdot (11 + 5)$

6. Find using the Distributive Property: $(10 + 5) \cdot 7$

7. Find using the Distributive Property: $7 \cdot (9 - 3)$

8. Which of the following numbers are not integers?

    $-4, 5.2, 7, -\sqrt{8}, \frac{7}{8}, \sqrt{4}$

9. Which of the following numbers are not integers?

    $-2, \sqrt{7}, 4, -\frac{1}{3}, 0, -\sqrt{36}$

10. Which of the following numbers are integers?

    $-21, \sqrt{100}, 3.75, \frac{5}{1}, \frac{1}{5}, \sqrt{24}$

11. Find the additive inverse of 8 and find the multiplicative inverse of 8.

12. Find the additive inverse of 5 and find the multiplicative inverse of 5.

13. Find the additive inverse of $\frac{2}{3}$ and find the multiplicative inverse of $\frac{2}{3}$.

14. Which of the following is not a rational number?

    $\pi, -65, 2.257, \sqrt{7}, -3$

15. Which of the following is not a rational number?

    $-1, 0, 0.14, \sqrt{19}$

16. Which of the following is a rational number?

    $4, \sqrt{2}, \frac{3}{5}, -\sqrt{5}$

17. Find: $8 - (-3) - 10$

18. Find: $9 - 17 - (-26)$

19. Find: $6 - (-2) - 7$

20. Find: $10 - \left(3 - \frac{16}{4}\right)$

21. Find: $7 + \left(\frac{9}{3} - 5\right)$

22. Find: $6 + \left(2 - \frac{27}{3}\right)$

23. Find: $-33 - (-5) - (-27)$

24. Find: $16 - (-4) \cdot (-3)$

25. Find: $-15 - (-4) - (-9)$

26. Find: $-\sqrt{45} + \sqrt{45}$

27. Find: $\left(\sqrt{17}\right) + \left(-\sqrt{17}\right)$

28. Find: $\sqrt{34} - \sqrt{34}$

Integer Exponents

29. Identify the base and the exponent, then calculate: 3^5

30. Identify the base and the exponent, then calculate: 2^5

31. Find: $(-5)^0$

32. Find: $-(5^0)$

33. Write using exponents: $m \cdot m \cdot m \cdot m \cdot m \cdot m \cdot n \cdot n \cdot n \cdot n$

34. Write using exponents: $x \cdot x \cdot x \cdot x \cdot y \cdot y \cdot y \cdot z$

35. Write using exponents: $a \cdot a \cdot a \cdot b \cdot b \cdot b \cdot b \cdot b \cdot b \cdot c \cdot c$

36. Simplify: $3^2 + 4^3$

37. Simplify: $4^2 + 2^3$

38. Simplify: $3^4 - 2^6$

39. Simplify using the properties of exponents: $7^5 - 7^4 \cdot 7^6$

40. Simplify using the properties of exponents: $8^3 \cdot 8^5 + 8^7$

41. Simplify using the properties of exponents: $\frac{5^{13}}{5^6} + 5^4$

42. Simplify using the properties of exponents: $x^3 \cdot y^2 \cdot z^4 \cdot x$

43. Simplify using the properties of exponents: $m^2 \cdot n^7 \cdot m^4 \cdot n^3$

44. Simplify using the properties of exponents: $a^3 \cdot b^9 \cdot c \cdot b^4$

45. Simplify using the properties of exponents: $(a^4 \cdot b^2)^5$

46. Simplify using the properties of exponents: $(x^3 \cdot y^5)^7$

47. Simplify using the properties of exponents: $(x^5 \cdot y^7 \cdot z^8)^9$

48. Simplify using the properties of exponents: $\frac{a^8 \cdot b^9}{a \cdot b^3}$

49. Simplify using the properties of exponents: $\frac{x^{12} \cdot y^8}{x^4 \cdot y^6}$

50. Simplify using the properties of exponents: $\frac{a^5 \cdot b^{12} \cdot c^7}{a^5 \cdot b^6 \cdot c^5}$

51. Simplify using the properties of exponents: $(2a)^3 \cdot a^5 \cdot (3b)^4$

52. Simplify using the properties of exponents: $\left(\frac{1}{3}x\right)^2 \cdot x^4 \cdot (2y)^5$

53. Simplify using the properties of exponents: $\left(\frac{3a}{b^2}\right)^3 \cdot b^7 \cdot \left(\frac{2}{3}a\right)^2$

54. Simplify using the properties of exponents: $\frac{15a^{10}}{5a^5}$

55. Simplify using the properties of exponents: $\frac{4x^6}{2x^8}$

56. Simplify using the properties of exponents: $\frac{3b^3}{18b^{11}}$

EVALUATE

Practice Test

Take this practice test to be sure that you are prepared for the final quiz in Evaluate.

1. Find: $|-7|$

2. Find: $7 - 4 - (-12)$

3. Fill in the missing value in each equation below:

    $13 + ? = 0$

    $13 \cdot ? = 1$

4. Find: $6 - 2\,[3 - 5(4 - 2)]$

5. Write using exponents: $y \cdot y \cdot y \cdot z \cdot z \cdot z \cdot z \cdot z$

6. Simplify using the properties of exponents: $\left(x^5 \cdot y^2\right)^7$

7. Simplify: $\left(2^4 \cdot 3^2\right)$

8. Simplify using the properties of exponents: $\left[\dfrac{(2x)^3}{(3y)^2}\right]^2 \cdot y^8 \cdot x^5$

LESSON EII.B — POLYNOMIALS

OVERVIEW

Here's what you'll learn in this lesson:

Polynomial Operations

a. Algebra building blocks

b. Evaluating polynomials

c. Adding and subtracting polynomials

d. Multiplying and dividing polynomials

Factoring Polynomials

a. Factoring out the greatest common factor

b. Factoring by grouping

c. Factoring trinomials

d. Factoring a difference of two squares

e. Factoring sums and differences of two cubes

f. Factoring using a combination of methods

In this lesson, you will review how to evaluate, add, subtract, multiply, and divide polynomials. Then you will use these techniques to solve equations. In addition, you will review different methods for factoring polynomials.

 EXPLAIN

POLYNOMIAL OPERATIONS

Summary

Algebra Building Blocks

You have reviewed some of the basic properties of real numbers and how to work with them. Now you will review some of the basic definitions of algebra.

Here are some algebraic expressions:

x $\qquad$ $5x^4 - 11 - 7xy + 2y^2$ $\qquad$ $8z - \dfrac{2}{3}$

$\dfrac{1}{x}$ $\qquad$ $4zx^7 - 5xy + 113$ $\qquad$ $a^4 - 11a^3b - 19a^2b^2$

An algebraic expression is made up of different parts.

Here's an example:

$3x^7 - 18 - 11xy + 3y^2$

This expression is made up of four terms: $3x^7$, -18, $-11xy$, and $3y^2$

Three of the terms have variables: x or y

$3x^7 - 18 - 11xy + 3y^2$

variables

The numbers in front of the variables are called coefficients: 3, −11, and 3

$3x^7 - 18 - 11xy + 3y^2$

coefficients

The terms have exponents: 7 and 2

exponents

$3x^7 - 18 - 11xy + 3y^2$

The term without any variables is called a constant term: −18

$3x^7 - 18 - 11xy + 3y^2$

constant term

The term −11xy also has the exponent 1 associated with the x and the y, but we usually don't write it.

Polynomials

A polynomial is a special kind of algebraic expression that has one or more variables and one or more terms.

For a polynomial in one variable x, each term has the form ax^r, where the coefficient, a, is any real number, and the exponent, r, is a nonnegative integer.

For example: $2x^3 - x^2 + 7x - 1$

$$a = 2 \quad a = -1 \quad a = 7 \quad a = -1$$
$$r = 3 \quad r = 2 \quad r = 1 \quad r = 0$$

For a polynomial in two variables x and y, each term has the form ax^ry^s, where a is any real number, and r and s are non-negative integers.

For example:

$$4y^2 + 23x^4y^5 - 18x^4y^1 + 3y^2 - 11x^1 + 17$$

$$a = 4 \quad a = 23 \quad a = -18 \quad a = 3 \quad a = -11 \quad a = 17$$
$$r = 0 \quad r = 4 \quad r = 4 \quad r = 0 \quad r = 1 \quad r = 0$$
$$s = 2 \quad s = 5 \quad s = 1 \quad s = 2 \quad s = 0 \quad s = 0$$

Polynomials with one, two, or three terms have special names:

A monomial has one term: $\qquad 2z^5x^8$

A binomial has two terms: $\qquad 15xy + 3y^2$

A trinomial has three terms: $\qquad 23x^2 - 17x + 9$

An algebraic expression is not a polynomial if any of its terms cannot be written in the form ax^r. For example, these algebraic expressions are **not** polynomials:

$$\frac{6}{x} - 3y \qquad \sqrt{9x^3y^2} + 10 \qquad 3 - \frac{6}{\sqrt{x}}$$

The Degree of a Polynomial

The degree of a term of a polynomial is the sum of the exponents of the variables in that term. The degree of a polynomial is the degree of the term with the highest degree.

For example, to find the degree of this polynomial:
$$4y^2 + 23x^4y^5 - 18x^4y^1 + 3y^2 - 11x^1 + 17$$

• find the degree of each term.

$$4y^2 + 23x^4y^5 - 18x^4y^1 + 3y^2 - 11x^1 + 17$$
$$\lor \qquad \lor$$
$$4+5 \quad 4+1$$

$$2 \qquad 9 \qquad 5 \qquad 2 \qquad 1 \qquad 0$$

The term $23x^4y^5$ has degree 9, the highest degree of all the terms. Therefore, the degree of the polynomial is 9.

Evaluating Polynomials

Sometimes the variables in a polynomial are assigned specific numerical values. In these cases you can evaluate the polynomial by replacing the variables with those values.

To evaluate a polynomial:

1. Substitute the given values for the variables in place of each variable.

2. Use the order of operations to simplify.

For example, to evaluate the polynomial $3x^7y - 4x - 2y^2 + 11$ when $x = -1$ and $y = 4$:

$$3x^7y - 4x - 2y^2 + 11$$

1. Substitute −1 in place of x and 4 in place of y.

$$= 3(-1)^7(4) - 4(-1) - 2(4)^2 + 11$$

2. Simplify.

$$= 3(-1)(4) - 4(-1) - 2(16) + 11$$
$$= -12 + 4 - 32 + 11$$
$$= -29$$

So when $x = -1$ and $y = 4$, $3x^7y - 4x - 2y^2 + 11 = -29$.

Adding and Subtracting Polynomials

Before you add or subtract polynomials, you need to review what is meant by like, or similar, terms.

Like terms have the same variables with the same exponents.

These are like terms: $13x$ and $7x$ $32xy$ and $45yx$ $2yx^2$ and $-7yx^2$

These are **not** like terms: 8 and $8x$ x and x^2 x and xy

When you combine like terms, you leave the variables alone and add or subtract their coefficients.

Here's an example of how to combine like terms: $13x + 5x = (13 + 5)x = 18x$

Here's another example: $3x^2y - 7x^2y = (3 - 7)x^2y = -4x^2y$

To add or subtract polynomials:

1. Remove parentheses. Use the distributive property if necessary.

2. Write like terms next to each other.

3. Combine like terms. Use the rules for signed numbers.

For example, to find $(3x^3 - 5x^2 + 6x - 11) - (7x^3 - 5x + 15)$:

1. Remove the parentheses. Use the distributive property on the second polynomial.

    $= 3x^3 - 5x^2 + 6x - 11 - 7x^3 + 5x - 15$

2. Write like terms next to each other.

    $= 3x^3 - 7x^3 - 5x^2 + 6x + 5x - 11 - 15$

3. Combine like terms.

    $= -4x^3 - 5x^2 + 11x - 26$

Multiplying Polynomials

To multiply two polynomials, you first need to learn how to multiply two monomials.

To multiply two monomials:

1. Multiply the coefficients and multiply the variables using the properties of exponents.

For example, to multiply the monomials $8x$ and $3x^4$:

1. Multiply the coefficients and variables.

    $(8x) \cdot (3x^4)$
    $= 8 \cdot 3 \cdot x \cdot x^4$
    $= 24x^5$

So, $8x \cdot 3x^4 = 24x^5$.

Now you can multiply a monomial by a polynomial with more than one term. You will use the distributive property.

To multiply a monomial by a polynomial:

1. Distribute the monomial to each term in the other polynomial.

2. Simplify.

For example, to find $5x \cdot (x^2 + 6x - 8)$:

1. Distribute the $5x$.

    $= 5x \cdot x^2 + 5x \cdot 6x - 5x \cdot 8$

2. Simplify.

    $= 5x^3 + 30x^2 - 40x$

So, $5x \cdot (x^2 + 6x - 8) = 5x^3 + 30x^2 - 40x$.

Finally, you can multiply any polynomial by any other polynomial.

To multiply a polynomial by another polynomial:

Another way to recall how to do this is to remember that each term in the first polynomial is multiplied by each term in the second polynomial.

1. Distribute each term in the first polynomial to the second polynomial.

2. Distribute again to remove the parentheses.

3. Multiply the resulting pairs of monomials.

4. Combine like terms if necessary.

For example, to find $(3z - 5)(3z^2 + 7z - 4)$:

1. Distribute each term in the first polynomial to the second polynomial.

$$= 3z(3z^2 + 7z - 4) - 5(3z^2 + 7z - 4)$$

2. Distribute again to remove the parentheses.

$$= 3z \cdot 3z^2 + 3z \cdot 7z + 3z \cdot (-4) - 5 \cdot 3z^2 - 5 \cdot 7z - 5 \cdot (-4)$$

3. Multiply the resulting pairs of monomials.

$$= 9z^3 + 21z^2 - 12z - 15z^2 - 35z + 20$$

4. Combine like terms.

$$= 9z^3 + 6z^2 - 47z + 20$$

So, $(3z - 5)(3z^2 + 7z - 4) = 9z^3 + 6z^2 - 47z + 20$.

Using the FOIL Method to Multiply Two Binomials

When you multiply two binomials there is a standard pattern that can help you. It is called the FOIL method. The letters in the word "FOIL" show you the order in which to multiply.

The general format is:

$$(a + b)(c + d) = a \cdot c + a \cdot d + b \cdot c + b \cdot d$$

$$ F + O + I + L$$

$$ \text{First} + \text{Outer} + \text{Inner} + \text{Last}$$

To multiply two binomials using the **FOIL** method:

1. Multiply the **F**irst terms of the binomials.

2. Multiply the **O**uter terms (the terms next to the outer parentheses).

3. Multiply the **I**nner terms (the terms next to the inner parentheses).

4. Multiply the **L**ast terms.

5. Add the terms. Be sure to combine like terms if necessary.

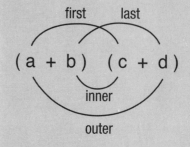

For example, to find $(y - 7)(y + 2)$:

1. Multiply the **F**irst terms of the binomials.

$$y \cdot y = y^2$$

2. Multiply the **O**uter terms.

$$y \cdot 2 = 2y$$

3. Multiply the **I**nner terms.

$$-7 \cdot y = -7y$$

4. Multiply the **L**ast terms.

$$-7 \cdot 2 = -14$$

5. Add the terms and combine like terms.

$$y^2 + 2y - 7y - 14$$
$$= y^2 - 5y - 14$$

So, $(y - 7)(y + 2) = y^2 - 5y - 14$.

This picture may help you remember how to use the FOIL method. Notice how the connecting lines form a face: F and L make the eyebrows, O makes the smile and I the nose.

Using Patterns to Multiply Two Binomials

Certain patterns are common when multiplying two binomials. These patterns can be used to simplify such calculations.

Here are the patterns:

Perfect square: $(a + b)^2 = (a + b)(a + b) = a^2 + 2ba + b^2$

Perfect square: $(a - b)^2 = (a - b)(a - b) = a^2 - 2ba + b^2$

Difference of two squares: $(a - b)(a + b) = a^2 - b^2$

To use these patterns to multiply two binomials:

1. Identify the correct pattern.

2. Identify the values to substitute for a and b.

3. Substitute these values.

4. Simplify.

For example, to use a pattern to find $(5x - 4)(5x - 4)$:

1. Identify the correct pattern.	Perfect square: $(a - b)(a - b) = a^2 - 2ba + b^2$
2. Identify the values for a and b.	$a = 5x$ $b = 4$
3. Substitute these values.	$(5x)^2 - 2(4)(5x) + (4)^2$
4. Simplify.	$= 25x^2 - 40x + 16$

So, $(5x - 4)(5x - 4) = 25x^2 - 40x + 16$.

As another example, to use a pattern to find $(2x^2 - y)(2x^2 + y)$:

1. Identify the correct pattern.	Difference of two squares: $(a - b)(a + b) = a^2 - b^2$
2. Identify the values for a and b.	$a = 2x^2$ $b = y$
3. Substitute these values.	$(2x^2)^2 - y^2$
4. Simplify.	$= 4x^4 - y^2$

So, $(2x^2 - y)(2x^2 + y) = 4x^4 - y^2$.

Dividing Polynomials

To divide two polynomials you first need to learn how to divide two monomials.

To divide two monomials, divide the coefficients and divide the variables using the properties of exponents.

For example, to divide the monomial $15y^{10}$ by the monomial $3y^4$:

1. Divide the coefficients and variables.

$$\frac{15y^{10}}{3y^4}$$
$$= 5y^{10-4}$$
$$= 5y^6$$

So, $\frac{15y^{10}}{3y^4} = 5y^6$.

Now you can divide a polynomial with more than one term by a monomial.

To divide a polynomial by a monomial:

1. Divide each term in the numerator by the denominator.

2. Do each division with the rules for dividing two monomials.

For example, to find $\frac{18y^7 - 6y^5}{3y^4}$:

1. Divide each term in the numerator by the denominator.

$$= \frac{18y^7}{3y^4} - \frac{6y^5}{3y^4}$$

2. Do each division.

$$= 6y^{7-4} - 2y^{5-4}$$
$$= 6y^3 - 2y$$

So, $\frac{18y^7 - 6y^5}{3y^4} = 6y^3 - 2y$.

Finally, you can divide any polynomial by any other polynomial.

To divide a polynomial by a polynomial where each has more than one term:

1. Arrange the terms of each polynomial in descending order. In the dividend, write missing terms as $0x^r$ where r is the exponent of the missing term.

2. Write the problem in long division form.

3. Divide the first term of the dividend by the first term of the divisor.

4. Multiply the divisor by the term you found in step (3).

5. Subtract the expression in step (4) from the dividend.

6. Continue dividing until the degree of the remainder is less than the degree of the divisor.

7. The answer is the expression that appears above the division sign plus the fraction $\frac{\text{remainder}}{\text{divisor}}$.

Just like with numbers, with polynomials division by zero is not allowed.

Don't confuse $\frac{a-b}{c}$ with $\frac{c}{a-b}$.

Remember, $\frac{a-b}{c} = \frac{a}{c} - \frac{b}{c}$. But you can't do this with $\frac{c}{a-b}$.

For example, to find $\dfrac{2x^2 + 7x + 8}{x + 3}$:

1. The terms are already in descending order.

2. Write the problem in long division form.

$$x + 3 \overline{)2x^2 + 7x + 8}$$

3. Divide the first term of the dividend by the first term of the divisor.

$$\begin{array}{r} 2x \\ x + 3 \overline{)2x^2 + 7x + 8} \end{array}$$

4. Multiply the divisor by the term in (3).

$$2x^2 + 6x$$

5. Subtract the expression you found in step (4) from the dividend.

$$\begin{array}{r} 2x \\ x + 3 \overline{)2x^2 + 7x + 8} \\ -(2x^2 + 6x) \\ \hline x + 8 \end{array}$$

6. Continue dividing until the degree of the remainder is less than the degree of the divisor.

$$\begin{array}{r} 2x \ + 1 \\ x + 3 \overline{)2x^2 + 7x + 8} \\ -(2x^2 + 6x) \\ \hline x + 8 \\ -(x + 3) \\ \hline 5 \end{array}$$

7. Write your answer.

$$2x + 1 + \dfrac{5}{x + 3}$$

Answers to Sample Problems

b. *1 + 3 = 4*

c. *−113*

d. *8*

Sample Problems

1. For this polynomial: $14yx^4 - 5x^5z^3 + 3zx^3 + 11xy - 113$

    a. Identify the third term.
    b. Find the degree of the third term.
    c. Identify the constant term.
    d. Find the degree of the polynomial.

    ☑ a. Identify the third term. $3zx^3$

    ☐ b. Find the degree of the third term. _____

    ☐ c. Identify the constant term. _____

    ☐ d. Find the degree of the polynomial. _____

2. Evaluate this polynomial when $x = 2$ and $z = -3$: $8zx^3 - 5x^5z^2 + 3zx^3 + 11x$

Answers to Sample Problems

☑ a. Substitute $x = 2$ and
$z = -3$ into the polynomial. $8(-3)(2)^3 - 5(2)^5(-3)^2 + 3(-3)(2)^3 + 11(2)$

☐ b. Use the order of
operations to simplify.

= _____

b. -1682

3. Simplify: $(2x^2)^4 + x^2 \cdot x^3 + (5x^5 - 4x^8)$

☑ a. First use the power of $= 2^4x^{2 \cdot 4} + x^2 \cdot x^3 + (5x^5 - 4x^8)$
a power property. $= 16x^8 + x^2 \cdot x^3 + (5x^5 - 4x^8)$

☑ b. Then use the multiplication $= 16x^8 + x^5 + (5x^5 - 4x^8)$
property of exponents.

☐ c. Now combine like terms.   = _____

c. $12x^8 + 6x^5$

4. Find: $(3x - 5y)(2x^2 + 7xy - 4y)$

☑ a. First distribute each term $= 3x(2x^2 + 7xy - 4y) - 5y(2x^2 + 7xy - 4y)$
in the first polynomial to
the second polynomial.

☑ b. Distribute again. $= 3x \cdot 2x^2 + 3x \cdot 7xy - 3x \cdot 4y - 5y \cdot 2x^2 - 5y \cdot 7xy - 5y \cdot (-4y)$

☐ c. Multiply the resulting
pairs of monomials.   = _____

☐ d. Combine like terms.   = _____

c. $6x^3 + 21x^2y - 12xy - 10x^2y - 35xy^2 + 20y^2$

d. $6x^3 + 11x^2y - 12xy - 35xy^2 + 20y^2$

5. Find: $\dfrac{24xy^9 - 8x^4y^5}{4xy^4}$

☑ a. Divide each term in the $\dfrac{24xy^9 - 8x^4y^5}{4xy^4}$
numerator by the
denominator. $= \dfrac{24xy^9}{4xy^4} - \dfrac{8x^4y^5}{4xy^4}$

☐ b. Do each division using
the rules for dividing
two monomials.

= _____

b. $6y^5 - 2x^3y$

FACTORING POLYNOMIALS

Summary

Factoring by Finding the Greatest Common Factor (GCF)

Here is a quick review of how to find the GCF of a collection of monomials. You will need it in order to factor a polynomial by factoring out the GCF.

As an example, here's how to find the GCF of the monomials $10x^3y^2$ and $8y^2$:

1. Factor each monomial into its prime factors.

$$10x^3y^2 = 2 \cdot 5 \cdot x \cdot x \cdot x \cdot y \cdot y$$
$$8y^2 = 2 \cdot 2 \cdot 2 \cdot y \cdot y$$

2. List each common prime factor the smallest number of times it appears in any factorization.

$2, y, y$

3. Multiply the prime factors in the list.

$2 \cdot y \cdot y = 2y^2$

So, the GCF of the monomials $10x^3y^2$ and $8y^2$ is $2y^2$.

Now you can factor a polynomial by factoring out the GCF of its monomial terms. Here are the steps:

1. Identify the monomial terms of the polynomial.

2. Factor each monomial term.

3. Find the GCF of the monomial terms.

4. Rewrite each term of the polynomial using the GCF.

5. Factor out the GCF.

For example, to factor the polynomial $10x^3y^2 + 8y^2$:

1. Identify the monomial terms.

$10x^3y^2, 8y^2$

2. Factor each monomial term.

$$10x^3y^2 = 2 \cdot 5 \cdot x \cdot x \cdot x \cdot y \cdot y$$
$$8y^2 = 2 \cdot 2 \cdot 2 \cdot y \cdot y$$

3. Find the GCF of the monomial terms.

$GCF = 2 \cdot y \cdot y = 2y^2$

4. Rewrite each term of the polynomial using the GCF.

$$10x^3y^2 = 2y^2 \cdot 5x^3$$
$$8y^2 = 2y^2 \cdot 4$$

5. Factor out the GCF.

$$10x^3y^2 + 8y^2 = (2y^2)(5x^3 + 4)$$

Sometimes you can just look at the terms and guess the GCF. Here, 2 is the largest integer which is a factor of 10 and 8, x is not a factor of each term, and y^2 is the largest y-factor of each term. So $2y^2$ is the GCF.

To check your factoring, you can multiply using the distributive property:

Is $10x^3y^2 + 8y^2 = (2y^2)(5x^3 + 4)$?

Is $10x^3y^2 + 8y^2 = (2y^2)(5x^3) + (2y^2)(4)$?

Is $10x^3y^2 + 8y^2 = 10x^3y^2 + 8y^2$? Yes.

Factoring by Grouping

Sometimes the only factor common to all the terms is 1. But there may be groups of terms with a larger common factor. Here's how to factor by grouping:

1. Factor each term.

2. Group terms with common factors.

3. Factor out the GCF in each grouping.

4. Factor out the binomial GCF of the polynomial.

For example, to factor the polynomial $x^2 + xy + x + y$:

1.	Factor each term.	$x^2 = x \cdot x$
		$xy = x \cdot y$
		$x = x$
		$y = y$
2.	Group terms with common factors.	$(x^2 + xy) + (x + y)$
		$(x \cdot x + x \cdot y) + (x + y)$
3.	Factor out the GCF in each grouping.	$x(x + y) + 1(x + y)$
4.	Factor out the binomial GCF of the polynomial.	$(x + y)(x + 1)$

So, $x^2 + xy + x + y = (x + y)(x + 1)$.

Factoring Trinomials

There are two ways to factor a trinomial of the form $ax^2 + bx + c$.

The first way is by grouping and the second way is by trial and error.

To factor a trinomial of the form $ax^2 + bx + c$ by grouping:

1. Find two numbers whose sum is b and whose product is $a \cdot c$.

2. Use these two numbers to rewrite bx as a sum.

3. Then factor by grouping.

For example, to factor $3x^2 + 7x + 4$:

1.	Here, $a = 3$, $b = 7$, and $c = 4$, so find two numbers whose sum is 7 and whose product is 12.	3 and 4
2.	Rewrite $7x$ as a sum.	$3x + 4x$

3. Factor by grouping. $3x^2 + 7x + 4 = 3x^2 + 3x + 4x + 4$

    • Factor each term. $\qquad\qquad = (3 \cdot x \cdot x) + (3 \cdot x) + (2 \cdot 2 \cdot x) + (2 \cdot 2)$

    • Group terms with $\qquad\qquad = [(3 \cdot x \cdot x) + (3 \cdot x)] + [(2 \cdot 2 \cdot x) + (2 \cdot 2)]$
      common factors.

    • Factor out the GCF $\qquad\qquad = 3x(x + 1) + 4(x + 1)$
      in each grouping.

    • Factor out the binomial $\qquad = (x + 1)(3x + 4)$
      GCF of the polynomial.

So $3x^2 + 7x + 4 = (x + 1)(3x + 4)$.

You can also factor the trinomial $3x^2 + 7x + 4$ by trial and error using these steps:

1. Try different x-terms whose product is $3x^2$ and different constants whose product is 4.

2. Then use the FOIL method to find factors whose "inner" and "outer" products add together to make $7x$. Here are three possibilities:

$$3x^2 + 7x + 4 \qquad\qquad (3x + 1)(x + 4)$$
$$(3x + 2)(x + 2)$$
$$(3x + 4)(x + 1)$$

The first possibility, $(3x + 1)(x + 4)$, is incorrect. The product of the x-terms, $3x$ and x, is $3x^2$, which is correct. The product of the constant terms, 1 and 4, is 4, which is also correct. But the "inner" and "outer" products are $(1)(x)$ and $(3x)(4)$ which, when added, give $13x$ not $7x$.

The second possibility, $(3x + 2)(x + 2)$, is incorrect. The product of the x-terms, $3x$ and x, is $3x^2$, which is correct. The product of the constant terms, 2 and 2, is 4, which is also correct. But the "inner" and "outer" products are $(2)(x)$ and $(3x)(2)$ which, when added, give $8x$ not $7x$.

The correct factorization is the third possibility, $(3x + 4)(x + 1)$. The product of the x-terms, $3x$ and x, is $3x^2$, which is correct. The product of the constant terms, 4 and 1, is 4, which is also correct. And the "inner" and "outer" products are $(4)(x)$ and $(3x)(1)$ which, when added, give $7x$.

Factoring a Perfect Square Trinomial

A perfect square trinomial is a polynomial that can be written so that it:
    • has three terms
    • has a first term that is a perfect square, a^2
    • has a third term that is a perfect square, b^2
    • has a second term that is twice the product of a and b, $2ba$

The patterns for factoring perfect square trinomials are:
$$a^2 + 2ba + b^2 = (a + b)^2$$
$$a^2 - 2ba + b^2 = (a - b)^2$$

For example, to factor $4x^2 + 12xz + 9z^2$:

1. Identify the correct pattern. $a^2 + 2ba + b^2 = (a + b)^2$

2. Substitute $2x$ for a and $3z$ for b. $(2x)^2 + 2(3z)(2x) + (3z)^2 = (2x + 3z)^2$

So, $4x^2 + 12xz + 9z^2 = (2x + 3z)^2$.

As another example, to factor $9x^2 - 24x + 16$:

1. Identify the correct pattern. $a^2 - 2ba + b^2 = (a - b)^2$
2. Substitute $3x$ for a and 4 for b. $(3x)^2 - 2(4)(3x) + (4)^2 = (3x - 4)^2$

So, $9x^2 - 24x + 16 = (3x - 4)^2$.

Factoring a Difference of Two Squares

A difference of two squares is a polynomial that can be written so that it:
- has two terms
- has a first term that is a perfect square, a^2
- has a second term that is a perfect square, b^2
- has a minus sign between the terms

The pattern for factoring a difference of two squares is: $a^2 - b^2 = (a + b)(a - b)$

For example, to factor $25x^2 - y^6$:

1. Use this pattern. $a^2 - b^2 = (a + b)(a - b)$

2. Substitute $5x$ for a and y^3 for b. $(5x)^2 - \left(y^3\right)^2 = \left(5x + y^3\right)\left(5x - y^3\right)$

So, $25x^2 - y^6 = \left(5x + y^3\right)\left(5x - y^3\right)$.

Factoring Sums and Differences of Two Cubes

A difference of two cubes or a sum of two cubes are polynomials that:
- have two terms
- have a first term that is a perfect cube, a^3
- have a second term that is a perfect cube, b^3

The patterns for factoring a difference or sum of two cubes are:
$$a^3 + b^3 = (a + b)\left(a^2 - ab + b^2\right)$$

$$a^3 - b^3 = (a - b)\left(a^2 + ab + b^2\right)$$

For example, to factor $z^6 + 27w^3$:

1. Identify the correct pattern. $a^3 + b^3 = (a + b)(a^2 - ab + b^2)$
2. Substitute z^2 for a and $(z^2)^3 + (3w)^3 = (z^2 + 3w)[(z^2)^2 - (z^2)(3w) + (3w)^2]$
   $3w$ for b. $= (z^2 + 3w)(z^4 - 3z^2w + 9w^2)$

So, $z^6 + 27w^3 = (z^2 + 3w)(z^4 - 3z^2w + 9w^2)$.

As another example, to factor $125x^3 - 1$:

1. Identify the correct pattern. $a^3 - b^3 = (a - b)(a^2 + ab + b^2)$

2. Substitute $5x$ for a and 1 for b. $(5x)^3 - (1)^3 = (5x - 1)[(5x)^2 + (5x)(1) + (1)^2]$
                                              $= (5x - 1)(25x^2 + 5x + 1)$

So, $125x^3 - 1 = (5x - 1)(25x^2 + 5x + 1)$.

Factoring Using a Combination of Methods

Often, a polynomial looks as if it will not factor. Sometimes you need to use more than one of the techniques you have learned. For example, you may be able to first factor out the GCF of the terms, and then it may fit into one of the patterns.

A good plan is to:

1. Check for common factors.

2. Count the number of terms to help you identify a way to factor:
   - If there are 2 terms, try the pattern for a difference of two squares or a difference/sum of two cubes.
   - If there are 3 terms, try one of the patterns for trinomial factoring.
   - If there are 4 or more terms, try grouping.

For example, to factor the polynomial $6x^3y - 84x^2y + 294xy$:

1. Factor out the GCF of the terms. $= 6xy(x^2 - 14x + 49)$

2. Identify a pattern for factoring $(a^2 - 2ba + b^2) = (a - b)^2$
   $(x^2 - 14x + 49)$.

3. Substitute x for a and 7 for b. $= 6xy(x - 7)^2$

So, $6x^3y - 84x^2y + 294xy = 6xy(x - 7)^2$.

Sample Problems

1. Factor this polynomial: $15y^2x^4 - 5x^5y^3 + 10y^2x^3$.

   ☑ a. Identify the terms of the polynomial. $15y^2x^4, 5x^5y^3, 10y^2x^3$

   ☐ b. Factor each monomial term.
   $15y^2x^4 = $ _____
   $5x^5y^3 = $ _____
   $10y^2x^3 = $ _____

   ☐ c. Find the GCF of the monomial terms.   GCF = _____

   ☐ d. Rewrite each term of the polynomial using the GCF.
   $15y^2x^4 = $ _____
   $5x^5y^3 = $ _____
   $10y^2x^3 = $ _____

   ☐ e. Factor out the GCF.   _____

2. Factor by grouping: $3x^2y - 6x + 4xy^2 - 8y$

   ☑ a. Factor each term.

   $$3x^2y = 3 \cdot x \cdot x \cdot y$$

   $$6x = 2 \cdot 3 \cdot x$$

   $$4xy^2 = 2 \cdot 2 \cdot x \cdot y \cdot y$$

   $$8y = 2 \cdot 2 \cdot 2 \cdot y$$

   ☑ b. Group terms with common factors. $= (3x^2y - 6x) + (4xy^2 - 8y)$

   ☐ c. Factor out the GCF in each grouping.   ____(___ − ___) + ____(___ − ___)

   ☐ d. Factor out the binomial GCF of the polynomial.   = (_____)(_____)

3. Factor this trinomial: $6x^2 - 5x - 4$

   ☑ a. Try different x-terms whose product is $6x^2$ and different constants whose product is -4.

   $6x^2 = 6x \cdot x$ or $3x \cdot 2x$

   $-4 = -1 \cdot 4$
   or $1 \cdot -4$
   or $-2 \cdot 2$
   or $2 \cdot -2$

   ☐ b. Use the FOIL method to find factors whose "inner" and "outer" products add together to make $-5x$.   (_____)(_____)

   ☐ c. Check your answer by multiplying.

Answers to Sample Problems

b. $3 \cdot 5 \cdot y \cdot y \cdot x \cdot x \cdot x \cdot x$
$5 \cdot x \cdot x \cdot x \cdot x \cdot x \cdot y \cdot y \cdot y$
$2 \cdot 5 \cdot y \cdot y \cdot x \cdot x \cdot x$

c. $5 \cdot x \cdot x \cdot x \cdot y \cdot y$ or $5x^3y^2$

d. $5x^3y^2 \cdot 3x$
$5x^3y^2 \cdot x^2y$
$5x^3y^2 \cdot 2$

e. $5x^3y^2 \cdot (3x - x^2y + 2)$

c. $3x, xy, 2, 4y, xy, 2$

d. $3x + 4y, xy - 2$

b. $3x - 4, 2x + 1$

c. $(3x - 4)(2x + 1)$
$= 6x^2 + 3x - 8x - 4$
$= 6x^2 - 5x - 4$

4. Factor: $8w^3 - x^6y^{12}$

    ☑ a. There are no common factors. So check for a factoring pattern.

        Difference of two cubes:
$$a^3 - b^3 = (a - b)(a^2 + ab + b^2)$$

    ☐ b. Identify what to substitute for a and b.

        $a = \underline{\hspace{1.5cm}}$

        $b = \underline{\hspace{1.5cm}}$

    ☐ c. Substitute into the factoring pattern.

        $= (\underline{\hspace{1.5cm}})(\underline{\hspace{2.5cm}})$

5. Factor: $8x^3y - 18xy$

    ☑ a. First factor out the GCF of the two terms.

        $8x^3y - 18xy = 2xy(4x^2 - 9)$

    ☐ b. Check for a recognizable factoring pattern for $(4x^2 - 9)$.

        $\underline{\hspace{3.5cm}}$

    ☐ c. Identify what to substitute for a and b.

        $a = \underline{\hspace{2cm}}$

        $b = \underline{\hspace{2cm}}$

    ☐ d. Substitute into the factoring pattern.

        $= 2xy(\underline{\hspace{1.5cm}})(\underline{\hspace{1.5cm}})$

HOMEWORK

Homework Problems

Circle the homework problems assigned to you by the computer, then complete them below.

☀ Explain

Polynomial Operations

1. Find the degree of the third term of this polynomial:
   $13x^4y^5 + 7y^2 - 18x^4y^2 - 11x + 17$

2. Find: $13xz - 5xz + 8xz$

3. Find: $(2y - 3)(5y + 4)$

4. Find the degree of this polynomial:
   $13x^3y^5 + 7y^2 - 12xy + 18x^4y^6 - 11x + 2$

5. Find: $(5x^3 - 2x^2 + 6x - 13) - (7x^3 - 8x + 10)$

6. Find: $(2ab - 5c^2)^2$

7. Evaluate when $x = -2$: $3x^2 - 5x - 7$

8. Find: $5x^2 \cdot (x^3 - 4x + 8)$

9. Find: $\frac{32y^7 - 16y^5}{4y^3}$

10. Evaluate when $x = -1$ and $y = 3$: $3x^7y - 4x - 2y^2 + 11$

11. Find: $(2x - 7)(x^2 - 4x + 1)$

12. Find: $(3x^2 + 4x - 7) \div (x - 2)$

Factoring Polynomials

13. Factor: $12x^7z^2 - 18x^4z^4 - 24x^5z^6$

14. Factor: $x^4 - 36$

15. Factor: $2x + 4 + xy + 2y$

16. Factor: $8x^3y^2 - 18xy^4$

17. Factor: $27 + a^6$

18. Factor: $54b + 2a^3b$

19. Factor: $x^2 + 4x - 21$

20. Factor: $4x^2 + 12xy + 9y^2$

21. Factor: $2x^3 - 20x^2 + 32x$

22. Factor: $28y^2 - 13y - 6$

23. Factor: $8p^3 - q^9$

24. Factor: $x^6 - y^6$

 **APPLY**

Practice Problems

Here are some additional practice problems for you to try.

Polynomial Operations

1. Find the degree of the second term of this polynomial:
   $9x^4y^3 + 23x^3y^2z^3 - 2z^2 + 5x$

2. Find the degree of the third term of this polynomial:
   $7a^3b^2 - 5a^3b + 16ab^4c^3 - 6b$

3. Find the degree of this polynomial:
   $4a^3b^4c^5 + 7b^8c^3 - 25a^3b^3 - 13$

4. Find the degree of this polynomial:
   $12m^7n^6 - 7m^3n^7p^5 - 8m^{10} + 7$

5. Find: $(12x^4 + 7x^3 - 6x + 10) + (3x^3 - 2x - 8)$

6. Find: $(5x^3 + 8x^2 - 7x - 15) - (2x^3 + 9x - 20)$

7. Find: $(11x^3 - 9x^2 + 6x - 4) - (x^3 - 5x^2 - 12)$

8. Find: $(6a^3b + 4a^2b^2 - 3ab^3) + (-4a^3b - 13a^2b^2 - 6ab^3)$

9. Find: $(5xy^3 - 19x^2y + 8xy) - (-2x^2y - 16x^3y + 7x)$

10. Find:
    $(15x^3y^2z - 10x^2y - 25xz^2) - (-3xy^2 - 14x^2z + 14x^3y^2z)$

11. Find: $(4x + 9)(2x + 3)$

12. Find: $(2x - 5)(3x - 4)$

13. Find: $(3y - 2)(4y + 7)$

14. Find: $(a^2 + 2b)^2$

15. Find: $(4xy - 2z^2)^2$

16. Find: $(3mn^2 - 5n^3)(3mn^2 + 5n^3)$

17. Find: $6a^2 \cdot (3a^3 + 4a - 5)$

18. Find: $5x^3 \cdot (x^4 - 6x + 11)$

19. Find: $-3xy^2 \cdot (7x^3y^2 - 10x^2y - 9x)$

20. Evaluate when $a = 3$ and $b = -2$: $a^2b^2 + 3a^2 - 5b^2$

21. Evaluate when $x = -4$ and $y = 1$:
    $2xy^7 - 5x^2 + xy^3 - 8x + 9$

22. Evaluate when $x = -1$, $y = 3$ and $z = -2$:
    $x^2y - y^2z + 2xyz + 8xy - 13$

23. Find: $\dfrac{10x^5 - 5x^3 + 25x^2}{5x^2}$

24. Find: $\dfrac{16x^5 + 5x^2y + 12y^2}{4x^2y^2}$

25. Find: $\dfrac{21a^6 - 49a^3b^3 - 35b^4}{7a^3b^3}$

26. Find: $(x + 2)(3x^2 + 7x + 6)$

27. Find: $(x - 3)(2x^2 - 5x + 4)$

28. Find: $(x + 2y)(x^2 - 2xy + 4y^2)$

Factoring Polynomials

29. Factor: $m^2 - 64$

30. Factor: $w^2 - 49$

31. Factor: $4a^2 - 9$

32. Factor: $35x^5y^3 - 14x^3y^2 + 21x^2y$

33. Factor: $24a^4b^6 + 8a^5b^2$

34. Factor: $3a^2 - 6a + 4ab - 8b$

35. Factor: $5xy - 15x + 2y - 6$

36. Factor: $12a^4 + 6a^3 - 90a^2$

37. Factor: $6x^5 + 33x^4 - 63x^3$

38. Factor: $6w^3 - 22w^2 - 8w$

39. Factor: $a^2 + 14a + 49$

40. Factor: $x^2 - 12x + 36$

41. Factor: $4m^2 - 20m + 25$

42. Factor: $16a^2 - 72a + 81$

43. Factor: $9x^2 + 24x + 16$

44. Factor: $49m^2 + 70mn + 25n^2$

45. Factor: $a^3 - 64$

46. Factor: $8m^3 + 1$

47. Factor: $x^3 + 27$

48. Factor: $x^6 + 27y^3$

49. Factor: $64p^3 - q^9$

50. Factor: $125a^9 + 64b^9$

51. Factor: $25m^4 - 16n^2$

52. Factor: $16x^4 - 625y^8$

53. Factor: $16a^4 - 81b^4$

54. Factor: $x^2 + 9x - 36$

55. Factor: $a^2 - 6a - 16$

56. Factor: $4m^2 + 45m - 36$

 **EVALUATE**

Practice Test

Take this practice test to be sure that you are prepared for the final quiz in Evaluate.

1. Find the degree of this polynomial:

    $11zx - 5x^2 z^3 - 4xz^2 + 8z - 13$

2. Find: $\dfrac{40y^7 - 15y^9 + 5y^5}{5y^5}$

3. Evaluate when $x = -1$ and $y = -4$: $3x^9y - 2x - 2y^2 + 3y$.

4. Find: $(2x - 5y)(x^2 - 4xy + 3)$

5. Factor: $12x^7z^4 - 30x^3z^9 - 42x^5z^6$

6. Factor: $3x^2 + 13x - 30$

7. Factor: $4x^2 - 20xy + 25y^2$

8. Factor: $3x^3 - 30x^2 + 48x$

LESSON EII.C – EQUATIONS AND INEQUALITIES

OVERVIEW

Once you know ways to simplify algebraic expressions, you can use these techniques as you solve equations and inequalities. Solving equations and inequalities is an important part of algebra.

In this lesson, you will solve linear equations and inequalities.

EXPLAIN

LINEAR

Summary

Equations and Inequalities

You have already learned how to simplify expressions such as $7(x + 5)$ and $3x - 2 + 5x$. When you relate expressions to each other with an equals sign or an inequality sign, you create an equation or an inequality.

An example of an equation is: $7(x + 5) = 3x - 2 + 5x$.

An example of an inequality is: $7(y + 5) > 3y - 2 + 5y$.

In this lesson you will review how to solve linear equations and linear inequalities.

Solving Linear Equations

To solve a linear equation, you must isolate the variable—that is, you must get the variable by itself on one side of the equation. Some equations are simple enough that you can find the solution just by looking at the equation. Other equations are more complicated and you need a systematic approach to find the solution.

Here's a way to solve a linear equation:

1. Remove any parentheses using the distributive property.

2. Combine like terms on each side of the equation.

3. Then do the following, as necessary:

   • Add the same quantity to both sides of the equation.

   • Subtract the same quantity from both sides of the equation.

   • Multiply or divide both sides of the equation by the same nonzero quantity.

4. Check the solution.

For example, to solve the equation $3x + 7 - 6x = -2$:

1. Combine the like terms, $3x$ and $-6x$.　　　　$-3x + 7 = -2$

2. Subtract 7 from both sides.　　　　$-3x + 7 - 7 = -2 - 7$

$$-3x = -9$$

3. Divide both sides by -3.　　　　$\dfrac{-3x}{-3} = \dfrac{-9}{-3}$

$$x = 3$$

The distributive property states that for all real numbers a, b, and c:

$$a(b + c) = a \cdot b + a \cdot c$$

For example:

$$3(4 + 5) = 3 \cdot 4 + 3 \cdot 5$$
$$= 12 + 15$$
$$= 27$$

4. Check the solution, $x = 3$.

Is $3(3) + 7 - 6(3) = -2$?

Is $9 + 7 - 18 = -2$?

Is $-2 = -2$? Yes.

So, the solution of the equation $3x + 7 - 6x = -2$ is $x = 3$.

Here is another example. To solve the equation $5(x + 1) - 2 = 18$:

1. Distribute the 5.

$5 \cdot x + 5 \cdot 1 - 2 = 18$

$5x + 5 - 2 = 18$

2. Combine like terms on the left side.

$5x + 3 = 18$

3. Subtract 3 from both sides.

$5x + 3 - 3 = 18 - 3$

$5x = 15$

4. Divide both sides by 5.

$\dfrac{5x}{5} = \dfrac{15}{5}$

$x = 3$

5. Check the solution, $x = 3$.

Is $5(3 + 1) - 2 = 18$?

Is $5(4) - 2 = 18$?

Is $20 - 2 = 18$?

Is $18 = 18$? Yes.

So, the solution of the equation $5(x + 1) - 2 = 18$ is $x = 3$.

Solving Linear Equations that Contain Fractions

When a linear equation contains fractions, it is often easier to solve the equation if you first clear the fractions. Then you can use the steps you just learned to solve equations without fractions.

To clear the fractions, multiply both sides of the equation by the least common denominator (LCD) of the fractions.

Sometimes you can figure out the LCD just by looking at the denominators. Try this method on the first example below.

To find the LCD using a formal method:

1. Factor each denominator into its prime factors.

2. List each prime factor the greatest number of times it appears in any one of the denominators.

3. Multiply the prime factors in the list.

For example, to solve $\dfrac{1}{2}x = \dfrac{1}{3}(x - 4)$:

1. Find the LCD of the fractions.

   - Factor the denominators.

     $2 = 2$

     $3 = 3$

   - List each factor the greatest
     number of times it appears
     in any one of the denominators.

     $2, 3$

   - Multiply the factors in the list.

     $LCD = 2 \cdot 3 = 6$

2. Multiply by 6 to clear the
   fractions.

   $6 \cdot \frac{1}{2}x = 6 \cdot \frac{1}{3}(x - 4)$

   $3x = 2(x - 4)$

3. Distribute the 2.

   $3x = 2 \cdot x - 2 \cdot 4$

   $3x = 2x - 8$

4. Subtract $2x$ from both sides.

   $3x - 2x = 2x - 8 - 2x$

   $x = -8$

5. Check the solution.

   Is $\frac{1}{2}(-8) = \frac{1}{3}(-8 - 4)$?

   Is $\quad -4 = \frac{1}{3}(-12) \quad$?

   Is $\quad -4 = -4 \quad\quad$? Yes.

So, $x = -8$ is the solution of the equation $\frac{1}{2}x = \frac{1}{3}(x - 4)$.

Consider another example. To solve $\frac{x}{8} = \frac{1}{20}x + \frac{1}{10}$:

1. Find the LCD of the fractions.

   - Factor the denominators.

     $8 = 2 \cdot 2 \cdot 2$

     $10 = 2 \cdot 5$

     $20 = 2 \cdot 2 \cdot 5$

   - List each factor the greatest
     number of times it appears in
     any one of the denominators.

     $2, 2, 2, 5$

   - Multiply the factors in the list.

     $LCD = 2 \cdot 2 \cdot 2 \cdot 5 = 40$

2. Multiply by 40 to clear the
   fractions.

   $40 \cdot \frac{x}{8} = 40 \cdot \left(\frac{1}{20}x + \frac{1}{10}\right)$

   $5x = 40 \cdot \frac{1}{20}x + 40 \cdot \frac{1}{10}$

   $5x = 2x + 4$

3. Subtract $2x$ from both sides.

   $5x - 2x = 2x + 4 - 2x$

   $3x = 4$

4. Divide both sides by 3.

$$\frac{3x}{3} = \frac{4}{3}$$

$$x = \frac{4}{3}$$

5. Check the solution.

Is $\dfrac{\frac{4}{3}}{8} = \dfrac{1}{20}\left(\dfrac{4}{3}\right) + \dfrac{1}{10}$?

Is $\dfrac{4}{24} = \dfrac{1}{15} + \dfrac{1}{10}$?

Is $\dfrac{1}{6} = \dfrac{2+3}{30}$?

Is $\dfrac{1}{6} = \dfrac{1}{6}$? Yes.

So, $x = \dfrac{4}{3}$ is the solution of the equation $\dfrac{x}{8} = \dfrac{1}{20}x + \dfrac{1}{10}$.

Solving Equations with Multiple Variables

Often formulas or equations have more than one variable. You can solve for one of the variables in terms of the others.

This is the formula for the circumference of a circle. C is the circumference, r is the radius, and π is a number approximately equal to 3.14.

For example, to solve the formula $C = 2\pi r$ for r:

1. Divide both sides by 2.

$$\frac{C}{2} = \frac{2\pi r}{2}$$

$$\frac{C}{2} = \pi r$$

2. Divide both sides by π.

$$\frac{\frac{C}{2}}{\pi} = \frac{\pi r}{\pi}$$

$$\frac{C}{2\pi} = r$$

So, $r = \dfrac{C}{2\pi}$.

As another example, to solve $w = xy + 5z$ for x:

1. Subtract $5z$ from both sides.

$$w - 5z = xy + 5z - 5z$$

$$w - 5z = xy$$

2. Divide both sides by y.

$$\frac{w-5z}{y} = \frac{xy}{y}$$

$$\frac{w-5z}{y} = x$$

So, $x = \dfrac{w-5z}{y}$.

Solving Linear Inequalities

Remember there are equations with no solutions. There are also equations, called identities, where every value of the variable is a solution.

While the linear equations you have seen have only one solution, linear inequalities usually have an infinite number of solutions. But you use the same strategy to solve both. However, it is important to remember one key difference: when you multiply or divide an inequality by a negative number, you must reverse the direction of the inequality.

For example, to solve $6(3 - x) > 7$:

1. Distribute the 6.

$$6 \cdot 3 - 6 \cdot x > 7$$
$$18 - 6x > 7$$

2. Subtract 18 from both sides.

$$18 - 6x - 18 > 7 - 18$$
$$-6x > -11$$

3. Divide both sides by -6.

$$\frac{-6x}{-6} < \frac{-11}{-6}$$
$$x < \frac{11}{6}$$

So all real numbers less than $\frac{11}{6}$ satisfy this inequality. You can graph the solution on a number line.

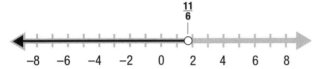

As another example, to solve $-7(3 - x) \geq 5 + 7x$:

1. Distribute the -7.

$$-7 \cdot 3 - (-7) \cdot x \geq 5 + 7x$$
$$-21 + 7x \geq 5 + 7x$$

2. Add 21 to both sides.

$$-21 + 7x + 21 \geq 5 + 7x + 21$$
$$7x \geq 7x + 26$$

3. Subtract $7x$ from both sides.

$$7x - 7x \geq 7x + 26 - 7x$$
$$0 \geq 26 \ \text{ No!}$$

Since 0 is not greater than or equal to 26, this inequality has no solution.

Solving Compound Linear Inequalities

A compound inequality is a shorthand way of writing two inequalities. When you solve a compound inequality and you add, subtract, multiply or divide, you must do so to all sides of the inequality.

For example, to solve $9 \geq 1 - 4y > 1$:

1. Subtract 1 from all sides.

$$9 - 1 \geq 1 - 4y - 1 > 1 - 1$$
$$8 \geq \quad -4y \quad > 0$$

2. Divide all sides by -4.

$$\frac{8}{-4} \leq \quad \frac{-4y}{-4} \quad < \frac{0}{-4}$$
$$-2 \leq \quad y \quad < 0$$

The graph of this solution is shown below.

If you're having trouble remembering when to reverse the direction of a sign, try this: **always** reverse the direction of the sign when you multiply or divide both sides by a negative number, even with an equality sign. Just notice that a flipped equality sign doesn't change the solution.

Remember, you do not reverse the inequality sign when you distribute a negative number. Only flip the sign when you multiply or divide both sides of the inequality by a negative number.

Another way to solve a compound inequality is to break up the compound inequality and solve the two simple inequalities separately.

Another way to solve a compound inequality is to break up the compound inequality and solve the two simple inequalities separately.

Here is another example. To solve $3 + x < 4x - 6 < 9 + x$:

1. Subtract x from all sides. $\qquad$ $3 + x - x < 4x - 6 - x < 9 + x - x$

$$3 < \quad 3x - 6 \quad < 9$$

2. Add 6 to all sides. $\qquad$ $3 + 6 < 3x - 6 + 6 < 9 + 6$

$$9 < \quad 3x \quad < 15$$

3. Divide all sides by 3. $\qquad$ $\dfrac{9}{3} < \quad \dfrac{3x}{3} \quad < \dfrac{15}{3}$

$$3 < \quad x \quad < 5$$

The graph of this solution is shown below.

Sample Problems

1. Solve for x: $10 - \dfrac{3}{7}x = 4 + \dfrac{1}{3}x$

   ☐ a. Multiply by the LCD to clear the fractions.

$\underline{\quad}(10 - \dfrac{3}{7}x) = \underline{\quad}(4 + \dfrac{1}{3}x)$

$210 - \underline{\quad} = 84 + \underline{\quad}$

   ☑ b. Subtract 210 from both sides.

$210 - 9x - 210 = 84 + 7x - 210$

$-9x = 7x - 126$

   ☐ c. Subtract $7x$ from both sides.

$-9x - 7x = 7x - 126 - 7x$

$\underline{\quad} = -126$

   ☐ d. Divide both sides by −16.

$\dfrac{-16x}{-16} = \underline{\quad}$

$x = \underline{\quad}$

2. Solve the formula $E = K + mgh$ for h.

   ☑ a. Subtract K from both sides.

$E - K = K + mgh - K$

$E - K = mgh$

   ☐ b. Divide both sides by m.

$\underline{\quad} = \dfrac{mgh}{m}$

$\underline{\quad} = \underline{\quad}$

   ☐ c. Divide both sides by g.

$\underline{\quad} = \dfrac{gh}{g}$

$\underline{\quad} = h$

3. Solve for x: $-12\left(x + \frac{1}{2}\right) < 30$

   ☐ a. Distribute the -12.

   $-12 \cdot x + -12 \cdot \frac{1}{2}$ ____ 30

   $-12x - 6$ ____ 30

   ☐ b. Add 6 to both sides.

   $-12x - 6 + 6$ ____ $30 + 6$

   $-12x$ ____ 36

   ☐ c. Divide both sides by -12.

   $\frac{-12x}{-12}$ ____ $\frac{36}{-12}$

   x ____ -3

4. Solve for y: $1 \le \frac{2}{5}y + \frac{9}{5} < 2$

   ☐ a. Multiply by the LCD to clear

   ____ $\cdot 1 \le$ ____ $\cdot \left(\frac{2}{5}y + \frac{9}{5}\right) <$ ____ $\cdot 2$

   the fractions.

   ____ $\le$ ____ $\cdot \left(\frac{2}{5}y + \frac{9}{5}\right) <$ ____

   ☐ b. Distribute and simplify.

   $5 \le 5 \cdot \frac{2}{5}y + 5 \cdot \frac{9}{5} < 10$

   $5 \le$ ____ $+$ ____ < 10

   ☑ c. Subtract 9 from all sides.

   $5 - 9 \le 2y + 9 - 9 < 10 - 9$

   $-4 \le 2y < 1$

   ☐ d. Divide all sides by 2.

   $\frac{-4}{2}$ ____ $\frac{2y}{2}$ ____ $\frac{1}{2}$

   -2 ____ y ____ $\frac{1}{2}$

HOMEWORK

Homework Problems

Circle the homework problems assigned to you by the computer, then complete them below.

 Explain

Linear Equations and Inequalities

1. Solve for x: $5(x + 3) = 40$

2. Solve for m: $F = ma$

3. Solve for x: $\frac{2}{3}x + 1 \geq 7$

4. Solve for y: $2y + 3 = 2y + 5$

5. Solve for x: $y(2 - x) = 11$

6. Solve for x: $-5(x + 4) > 3$

7. Solve for x: $3(3 - x) = 5 + x$

8. Solve for y: $0 < 3y + 2 - 4y < 7$

9. Solve for z: $-2 < 2(z + 2) \leq 6$

10. Solve for x: $\frac{1}{4}(x + 3) = \frac{1}{6}x + 1$

11. Solve for x: $\frac{3x + 5}{10} - \frac{1}{4}x = \frac{1}{3}(x + 1)$

12. Solve for y: $\frac{18 - 2y}{4} > \frac{3}{2}\left(y + \frac{1}{3}\right) \geq \frac{3}{2} - \frac{1}{2}y$

 **APPLY**

Practice Problems

Here are some additional practice problems for you to try.

Linear Equations and Inequalities

1. Solve for r: $C = 2\pi r$

2. Solve for m: $Fd = \frac{1}{2}mv^2$

3. Solve for y: $3x - 5y = 15$

4. Solve for x: $ax + b = c$

5. Solve for m: $y = mx + b$

6. Solve for x: $7x + 3 < 24$

7. Solve for x: $-\frac{3}{5}x - 2 \le 10$

8. Solve for y: $\frac{4}{3}y + 1 \ge -7$

9. Solve for y: $3x(y - 5) = -24$

10. Solve for y: $7x(4 - y) = 56$

11. Solve for y: $12(y - 7) < 36$

12. Solve for x: $-5(x + 6) > 14$

13. Solve for y: $-3(y + 2) + 2y \ge 17$

14. Solve for y: $3(4 - 2y) = 4 + 2y$

15. Solve for y: $2(6 + y) = 9 - y$

16. Solve for y: $17 - 3(5 - 2y) = 8 + 4y$

17. Solve for x: $-6 < 5x - 12 - 2x < 9$

18. Solve for x: $-2 \le 2x - 6 - 4x \le 8$

19. Solve for x: $16 - x \le 5x + 12 - 2x \le 24 - x$

20. Solve for y: $\frac{1}{5}(y - 5) = \frac{2}{5} - \frac{1}{4}y$

21. Solve for x: $\frac{1}{4}\left(x - \frac{5}{3}\right) = \frac{2}{3}(x + 5)$

22. Solve for x: $\frac{2}{3}\left(x + \frac{1}{5}\right) = \frac{1}{5}(x - 5)$

23. Solve for x: $\frac{7}{3}(y + 2) = \frac{3}{4}(x + 3)$

24. Solve for y: $\frac{3y - 1}{2} = \frac{4}{5}(x - 10)$

25. Solve for y: $\frac{4y - 8}{3} = \frac{2}{5}(x + 10)$

26. Solve for x: $\frac{3}{5}(x + 9) = \frac{1}{10}(x - 14)$

27. Solve for y: $\frac{2}{3}(y + 8) = \frac{4}{9}(y + 6)$

28. Solve for x: $\frac{3}{4}(x + 6) = \frac{5}{12}\left(x - \frac{2}{5}\right)$

Practice Test

Take this practice test to be sure that you are prepared for the final quiz in Evaluate.

1. Solve for x: $2(x + 4) - 3x = 0$

2. Solve for y: $15y - 20 = 14y - 9$

3. Circle the value of z that is the solution of the equation
$12\left(z + \dfrac{1}{3}\right) = -20.$

   $z = 3$

   $z = 0$

   $z = -2$

4. Solve for x: $-4 - 4(2x + 3) = 16$

5. Solve for x: $\dfrac{1}{2}x + 4 = \dfrac{1}{5}x + 7$

6. Solve for x: $z = 3x - 5y$

7. Circle the values of y that are solutions of the inequality $3y + 5 < 14.$

   $y = -1$

   $y = 1$

   $y = 3$

   $y = 5$

8. Solve the following inequality, then graph the solution on the number line.

   $0 \le 5(x + 2) < 25$

LESSON EII.D – RATIONAL EXPRESSIONS

OVERVIEW

Here's what you'll learn in this lesson:

Rational Expressions

a. Negative integer exponents

b. Writing rational expressions in lowest terms

c. Multiplying and dividing rational expressions

d. Adding and subtracting rational expressions

e. Simplifying complex fractions

Rational Equations

a. Solving equations that contain rational expressions

Your work with fractions in arithmetic has prepared you to work with algebraic fractions, which are also called rational expressions. For example, since you know how to add, subtract, multiply, and divide fractions, you are ready to do the same operations with rational expressions. Similarly, your knowledge of how to solve equations that contain fractions has prepared you to solve equations that contain rational expressions.

In this lesson, you will work with rational expressions and you will solve rational equations. In working with rational expressions, you will also extend your knowledge of positive exponents to include exponents that are negative integers.

 **EXPLAIN**

RATIONAL EXPRESSIONS

Summary

Negative Integer Exponents

You know that 7^3 is a shorthand notation for $7 \cdot 7 \cdot 7$.

Similarly, 7^{-3} is a shorthand notation for $\dfrac{1}{7 \cdot 7 \cdot 7}$.

That is, $7^{-3} = \dfrac{1}{7^3}$.

In general, a negative integer exponent is defined like this:

$$x^{-n} = \frac{1}{x^n}$$

(Here n is a positive integer, and $x \neq 0$.)

In other words, you can rewrite x^{-n} without using a negative exponent by writing a fraction like this:

1. Put a 1 in the numerator.

2. Put the base, x, in the denominator, and change the sign of the exponent.

3. Simplify.

For example, you can rewrite 5^{-4} like this:

$$5^{-4} = \frac{1}{5^4} = \frac{1}{5 \cdot 5 \cdot 5 \cdot 5} = \frac{1}{625}.$$

Properties of Negative Integer Exponents

You are already familiar with several properties of positive exponents. You can use these same properties when you work with negative exponents. The following table summarizes these properties, and gives examples with positive integer exponents and negative integer exponents.

Property of Exponents	Example with Positive Integer Exponents	Example with Negative Integer Exponents
Multiplication Property $x^m \cdot x^n = x^{m+n}$	$3^2 \cdot 3^4$ $= 3^{2+4}$ $= 3^6$	$3^{-2} \cdot 3^{-4}$ $= 3^{-2+(-4)}$ $= 3^{-6}$
Division Property $\dfrac{x^m}{x^n} = x^{m-n}$ (Here, $x \neq 0$.)	$\dfrac{4^7}{4^5}$ $= 4^{7-5}$ $= 4^2$	$\dfrac{4^{-7}}{4^{-5}}$ $= 4^{-7-(-5)}$ $= 4^{-7+5}$ $= 4^{-2}$
Power of a Power Property $(x^m)^n = x^{m \cdot n}$	$(5^2)^3$ $= 5^{2 \cdot 3}$ $= 5^6$	$(5^{-2})^{-3}$ $= 5^{(-2)(-3)}$ $= 5^6$
Power of a Product Property $(x \cdot y)^n = x^n \cdot y^n$	$(5 \cdot 7)^3$ $= 5^3 \cdot 7^3$	$(5 \cdot 7)^{-3}$ $= 5^{-3} \cdot 7^{-3}$
Power of a Quotient Property $\left(\dfrac{x}{y}\right)^n = \dfrac{x^n}{y^n}$ (Here, $y \neq 0$.)	$\left(\dfrac{3}{5}\right)^4$ $= \dfrac{3^4}{5^4}$	$\left(\dfrac{3}{5}\right)^{-4}$ $= \dfrac{3^{-4}}{5^{-4}}$

In addition, here are two other useful properties of negative exponents.

This property helps you rewrite a fraction where both the numerator and denominator are raised to a negative power:

$$\frac{x^{-m}}{y^{-n}} = \frac{y^n}{x^m} \qquad \text{(Here, } x \neq 0 \text{ and } y \neq 0.)$$

For example: $\dfrac{5^{-3}}{7^{-9}} = \dfrac{7^9}{5^3}$

This property helps you simplify a fraction raised to a negative power:

$$\left(\frac{x}{y}\right)^{-n} = \left(\frac{y}{x}\right)^n \qquad \text{(Again, } x \neq 0 \text{ and } y \neq 0.)$$

For example: $\left(\dfrac{3}{8}\right)^{-11} = \left(\dfrac{8}{3}\right)^{11}$

Finally, here is the property of zero exponents:

$$x^0 = 1 \qquad \text{(Here, } x \neq 0.)$$

Writing Rational Expressions in Lowest Terms

Working with negative exponents leads to working with rational expressions. A rational expression, which is also called an algebraic fraction, is a quotient of two polynomials, where the polynomial in the denominator is not equal to zero. Here, you'll review how to reduce rational expressions to lowest terms, and then you'll learn how to multiply, divide, add and subtract them.

Here's how to reduce a rational expression to lowest terms:

1. Factor the numerator.

2. Factor the denominator.

3. Cancel pairs of factors that are common to both the numerator and the denominator.

4. Simplify.

For example, to reduce $\dfrac{x^2 - 5x - 14}{x^2 + 3x + 2}$ to lowest terms:

1. Factor the numerator. $\qquad = \dfrac{(x-7)(x+2)}{x^2 + 3x + 2}$

2. Factor the denominator. $\qquad = \dfrac{(x-7)(x+2)}{(x+1)(x+2)}$

3. Cancel common factors. $\qquad = \dfrac{(x-7)\cancel{(x+2)}^{1}}{(x+1)\cancel{(x+2)}_{1}}$

4. Simplify. $\qquad = \dfrac{x-7}{x+1}$

So $\dfrac{x^2 - 5x - 14}{x^2 + 3x + 2}$ reduced to lowest terms is $\dfrac{x-7}{x+1}$.

Multiplying and Dividing Rational Expressions

You can multiply and divide rational expressions in much the same way that you multiply and divide fractions.

Here's how to multiply rational expressions:

1. Factor the numerators and denominators.

2. Cancel all pairs of factors common to the numerators and denominators.

3. Multiply the numerators; multiply the denominators.

For example, to find $\dfrac{3x^2}{7y} \cdot \dfrac{21y^6}{x^3}$:

1. Factor the numerators and denominators. $\qquad = \dfrac{3 \cdot x^2}{7 \cdot y} \cdot \dfrac{7 \cdot 3 \cdot y^5 \cdot y}{x^2 \cdot x}$

2. Cancel all pairs of factors common to the numerators and denominators. $\qquad = \dfrac{3 \cdot \cancel{x^2}^{1}}{\cancel{7}_{1} \cdot \cancel{y}_{1}} \cdot \dfrac{\cancel{7}^{1} \cdot 3 \cdot y^5 \cdot \cancel{y}^{1}}{\cancel{x^2}_{1} \cdot x}$

3. Multiply the numerators. Multiply the denominators. $\qquad = \dfrac{9y^5}{x}$

So $\dfrac{3x^2}{7y} \cdot \dfrac{21y^6}{x^3} = \dfrac{9y^5}{x}$.

You want to try to find common factors when you factor the numerators and denominators. So you factor x^3 as $x^2 \cdot x$ in the denominator since $3x^2 = 3 \cdot x^2$ in the numerator.

Here's how to divide one rational expression by a second rational expression:

1. Invert the second rational expression and change $\div$ to $\cdot$.

2. Follow the steps for multiplying.

For example, to find $\dfrac{7x^2}{8y} \div \dfrac{21x^3}{y^4}$:

1. Invert the second rational expression and change $\div$ to $\cdot$.

$$= \frac{7x^2}{8y} \cdot \frac{y^4}{21x^3}$$

2. Follow the steps for multiplying.

• Factor the numerators and denominators.

$$= \frac{7 \cdot x^2}{8 \cdot y} \cdot \frac{y^3 \cdot y}{21 \cdot x^2 \cdot x}$$

• Cancel pairs of factors common to the numerator and denominator.

$$= \frac{\overset{1}{\cancel{7}} \cdot \overset{1}{\cancel{x^2}}}{8 \cdot \cancel{y}} \cdot \frac{y^3 \cdot \overset{1}{\cancel{y}}}{\underset{3}{\cancel{21}}\underset{1}{\cancel{x^2}} \cdot x}$$

• Multiply the numerators. Multiply the denominators.

$$= \frac{y^3}{24x}$$

So $\dfrac{7x^2}{8y} \div \dfrac{21x^3}{y^4} = \dfrac{y^3}{24x}$.

Adding and Subtracting Rational Expressions

Here's how you add or subtract rational expressions.

To add or subtract two rational expressions with the same denominator:

1. Add (or subtract) the numerators. The denominator stays the same.

2. Reduce the resulting rational expression to lowest terms.

For example, to find $\dfrac{7x}{24xy^3} + \dfrac{11x - 3}{24xy^3}$:

1. Add the numerators. The denominator stays the same.

$$= \frac{7x + 11x - 3}{24xy^3}$$

$$= \frac{18x - 3}{24xy^3}$$

2. Reduce to lowest terms:

• Factor the numerator and denominator.

$$= \frac{3(6x - 1)}{3 \cdot 8 \cdot x \cdot y^3}$$

• Cancel pairs of factors common to the numerator and denominator.

$$= \frac{\overset{1}{\cancel{3}}(6x - 1)}{\underset{1}{\cancel{3}} \cdot 8 \cdot x \cdot y^3}$$

• Simplify.

$$= \frac{6x - 1}{8xy^3}$$

So $\dfrac{7x}{24xy^3} + \dfrac{11x - 3}{24xy^3} = \dfrac{6x - 1}{8xy^3}$.

To add or subtract rational expressions with different denominators, it is useful to find the least common denominator (LCD) of the rational expressions.

Sometimes you can just look at the denominators and quickly guess the LCD. If not, you can use the following more formal method.

To find the LCD of a collection of rational expressions:

1. Factor each denominator.

2. List each factor the greatest number of times that it appears in any one of the factorizations.

3. Multiply the factors in the list. The result is the least common denominator of the rational expressions.

For example, to find the LCD of $\frac{3}{4xy}$ and $\frac{y}{5x}$:

1. Factor each denominator.

$$4xy = 4 \cdot x \cdot y$$

$$5x = 5 \cdot x$$

2. List each factor the greatest number of times that it appears in any one of the factorizations.

$$4, 5, x, y$$

3. Multiply the factors in the list. The result is the least common denominator of the rational expressions.

$$4 \cdot 5 \cdot x \cdot y = 20xy$$

So the LCD of $\frac{3}{4xy}$ and $\frac{y}{5x}$ is $20xy$.

To add (or subtract) two rational expressions with different denominators:

1. Find the LCD of the rational expressions.

2. Rewrite each rational expression with this LCD.

3. Add (or subtract) the numerators. The denominator stays the same.

4. Reduce the resulting rational expression to lowest terms.

For example, to find $\frac{3}{4xy} - \frac{y}{5x}$:

1. Find the LCD of the rational expressions.

$$LCD = 20xy$$

2. Rewrite each rational expression with this LCD.

$$\frac{3}{4xy} - \frac{y}{5x}$$

$$= \frac{3}{4xy} \cdot \frac{5}{5} - \frac{y}{5x} \cdot \frac{4y}{4y}$$

$$= \frac{15}{20xy} - \frac{4y^2}{20xy}$$

3. Subtract the numerators. The denominator stays the same.

$$= \frac{15 - 4y^2}{20xy}$$

4. This expression is already in lowest terms.

So $\frac{3}{4xy} - \frac{y}{5x} = \frac{15 - 4y^2}{20xy}$.

Complex Fractions

Fractions that contains other fractions in the numerator or denominator are called complex fractions.

Here are some examples:

$$\dfrac{\dfrac{x^2}{11}}{\dfrac{7x}{5}} \qquad\qquad \dfrac{3y + \dfrac{1}{x}}{7} \qquad\qquad \dfrac{\dfrac{5x - 1}{x^3 - 2x}}{\dfrac{3x}{x - 2} + \dfrac{7}{x}}$$

There are two methods for simplifying complex fractions.

Method 1:

1. If there is more than one term in the numerator of the complex fraction, combine the terms by adding or subtracting.

2. If there is more than one term in the denominator of the complex fraction, combine the terms by adding or subtracting.

3. Rewrite the complex fraction using the division symbol, ÷.

4. Invert the second rational expression and change ÷ to ·.

5. Follow the steps for multiplying.

For example, to simplify the complex fraction $\dfrac{\dfrac{x^3}{25}}{\dfrac{x}{5}}$ using Method 1:

1. There is only one term in the numerator.

2. There is only one term in the denominator.

3. Rewrite using the division symbol, ÷. $\qquad = \dfrac{x^3}{25} \div \dfrac{x}{5}$

4. Invert the second rational expression and change ÷ to ·. $\qquad = \dfrac{x^3}{25} \cdot \dfrac{5}{x}$

5. Multiply.

   • Factor the numerators and denominators. $\qquad = \dfrac{x^2 \cdot x}{5 \cdot 5} \cdot \dfrac{5}{x}$

   • Cancel pairs of factors common to the numerators and denominators. $\qquad = \dfrac{x^2 \cdot \overset{1}{\cancel{x}}}{\cancel{5} \cdot 5} \cdot \dfrac{\overset{1}{\cancel{5}}}{\underset{1}{\cancel{x}}}$

   • Multiply the numerators. Multiply the denominators. $\qquad = \dfrac{x^2}{5}$

So $\dfrac{\dfrac{x^3}{25}}{\dfrac{x}{5}} = \dfrac{x^2}{5}$.

Method 2:

1. Find the LCD of the fractions contained in the complex fraction.

2. Multiply the numerator and denominator of the complex fraction by this LCD.

3. Simplify.

For example, to simplify the complex fraction $\dfrac{\frac{x^3}{25}}{\frac{x}{5}}$ using Method 2:

1. Find the LCD of the fractions contained in this complex fraction.

    - Factor each denominator.

        $25 = 5 \cdot 5$

        $5 = 5 \cdot 1$

    - List each factor the greatest number of times it appears in any one of the factorizations.

        $5, 5$

    - Multiply the factors in the list. The result is the LCD.

        $LCD = 5 \cdot 5 = 25$

2. Multiply the numerator and denominator by 25.

    $= \dfrac{\frac{x^3}{25} \cdot 25}{\frac{x}{5} \cdot 25}$

3. Simplify.

    - Cancel the 25 from the numerator and denominator of the top fraction.

        $= \dfrac{\frac{x^3}{\cancel{25}} \cdot \overset{1}{\cancel{25}}}{\frac{x}{5} \cdot 25}$

        $= \dfrac{\frac{x^3}{x}}{\frac{x}{5} \cdot 25}$

    - Cancel the 5 from the numerator and denominator of the bottom fraction.

        $= \dfrac{x^3}{\frac{x}{\underset{1}{\cancel{5}}} \cdot \overset{5}{\cancel{25}}}$

        $= \dfrac{x^3}{5x}$

    - Cancel the x from the numerator and denominator.

        $= \dfrac{x^2 \cdot \overset{1}{\cancel{x}}}{5\underset{1}{\cancel{x}}}$

        $= \dfrac{x^2}{5}$

So, $\dfrac{\frac{x^3}{25}}{\frac{x}{5}} = \dfrac{x^2}{5}$.

Sample Problems

1. Simplify and write without using negative exponents: $\dfrac{18x^5 y^{-2}}{6x^{-3} y^4}$

   ☑ a. Use the property $\dfrac{y^{-m}}{x^{-n}} = \dfrac{x^n}{y^m}$ $\qquad\qquad = \dfrac{18x^5 x^3}{6y^2 y^4}$

   to rewrite $\dfrac{y^{-2}}{x^{-3}} = \dfrac{x^3}{y^2}$.

b. $\dfrac{18x^8}{6y^6}$

   ☐ b. Use the Multiplication Property. $\qquad =$ _____

c. $\dfrac{3x^8}{y^6}$

   ☐ c. Reduce to lowest terms. $\qquad =$ _____

2. Reduce to lowest terms: $\dfrac{x^2 - 9}{x - 2} \cdot \dfrac{x^2 - 5x + 6}{x - 3}$

   ☑ a. Factor the numerators and $\qquad = \dfrac{(x-3)(x+3)}{x-2} \cdot \dfrac{(x-3)(x-2)}{x-3}$

   denominators.

b. $(x + 3)(x - 3)$

   ☐ b. Cancel all pairs of factors
   common to the numerators
   and denominators and multiply. $\quad =$ _____

3. Find: $\dfrac{x^2 + 5x + 4}{x^2 + x - 2} \div \dfrac{x^2 - x - 20}{x^2 - 3x - 10}$

   ☑ a. Invert the second rational $\qquad = \dfrac{x^2 + 5x + 4}{x^2 + x - 2} \cdot \dfrac{x^2 - 3x - 10}{x^2 - x - 20}$

   expression and change ÷ to ·.

b. $\dfrac{(x + 4)(x + 1)}{(x + 2)(x - 1)} \cdot \dfrac{(x - 5)(x + 2)}{(x - 5)(x + 4)}$

   ☐ b. Factor the numerators and
   denominators. $\qquad\qquad\qquad =$ _____

   ☐ c. Cancel all pairs of factors
   common to the numerators

c. $\dfrac{x + 1}{x - 1}$

   and denominators and multiply. $\quad =$ _____

4. Find: $\dfrac{x}{3x^2 - 8x - 3} + \dfrac{4}{3x^2 - 10x + 3}$

   ☐ a. Find the LCD.

   • Factor each denominator. $\qquad 3x^2 - 8x - 3 = (3x + 1)(x - 3)$

   $3x^2 - 10x + 3 = (3x - 1)(x - 3)$

a. $(3x + 1), (x - 3), (3x - 1)$

   • List each factor the greatest $\qquad\qquad$ _____
   number of times it appears
   in any one of the factorizations.

$(3x + 1)(x - 3)(3x - 1)$

   • Write the LCD as the product $\qquad$ LCD = _____
   of the factors in the list.

☑ b. Rewrite the first rational
   expression with this LCD.

$$\frac{x}{3x^2 - 8x - 3}$$

$$= \frac{x}{(3x + 1)(x - 3)}$$

$$= \frac{x(3x - 1)}{(3x + 1)(x - 3)(3x - 1)}$$

☐ c. Rewrite the second rational
   expression with this LCD.

$$\frac{4}{3x^2 - 10x + 3}$$

$$= \frac{4}{(3x - 1)(x - 3)}$$

c. $\dfrac{4(3x + 1)}{(3x - 1)(x - 3)(3x + 1)}$

$$= \underline{\hspace{4cm}}$$

☐ d. Add the rational expressions
   by adding the numerators and
   keeping the denominator the same.

d. $\dfrac{x(3x - 1) + 4(3x + 1)}{(3x - 1)(x - 3)(3x + 1)}$

$$= \underline{\hspace{4cm}}$$

☐ e. Distribute and combine like
   terms in the numerator.

e. $\dfrac{3x^2 + 11x + 4}{(3x - 1)(x - 3)(3x + 1)}$

$$= \underline{\hspace{4cm}}$$

5. Simplify this complex fraction: $\dfrac{3 - \dfrac{1}{x}}{\dfrac{5}{x^2}}$

☑ a. Combine the two terms in
   the numerator.

$$3 - \frac{1}{x}$$

$$= \frac{3x}{x} - \frac{1}{x}$$

$$= \frac{3x - 1}{x}$$

☐ b. Rewrite the complex fraction
   using ÷.

$$\frac{\dfrac{3x - 1}{x}}{\dfrac{5}{x^2}} = \underline{\hspace{2cm}} \div \underline{\hspace{2cm}}$$

b. $\left(\dfrac{3x - 1}{x}\right), \left(\dfrac{5}{x^2}\right)$

☐ c. Invert the second rational
   expression and change ÷ to · .

$$= \underline{\hspace{2cm}} \cdot \underline{\hspace{2cm}}$$

c. $\left(\dfrac{3x - 1}{x}\right), \left(\dfrac{x^2}{5}\right)$

☐ d. Cancel factors common to
   the numerators and denominators
   and multiply.

$$= \underline{\hspace{2cm}}$$

d. $\dfrac{x(3x - 1)}{5}$ or $\dfrac{3x^2 - x}{5}$

RATIONAL EQUATIONS

Summary

Solving Equations that Contain Rational Expressions

Here are some examples of equations that contain rational expressions:

$$\frac{x}{3} + 2 = \frac{3}{4} \qquad 3x - \frac{4}{5} = \frac{x}{4} + \frac{1}{20} \qquad \frac{3}{x^2 - 9} - \frac{5}{x^2 - 4} = \frac{x}{x^2 - 5x + 6}$$

The first step in solving such equations is to find the least common denominator (LCD) of the rational expressions in the equation. Multiplying both sides of the equation by this LCD will clear the fractions and make the equation easier to solve.

Often you can just look at the denominators and guess the LCD.

Or you can use this formal method:

1. Factor each denominator.

2. List each factor the greatest number of times it appears in any one of the factorizations.

3. Multiply the factors in the list. The result is the LCD.

For example, to find the LCD of these rational expressions:

$$\frac{2}{x + 5}, \frac{1}{x - 5}, \text{ and } \frac{16}{x^2 - 25}$$

1. Factor each denominator. $\qquad\qquad\qquad\qquad\qquad\qquad x + 5 = x + 5$

$$x - 5 = x - 5$$

$$x^2 - 25 = (x + 5)(x - 5)$$

2. List each factor the greatest number of $\qquad (x + 5), (x - 5)$
times it appears in any one of the factorizations.

3. Multiply the factors in the list. The result is the LCD. $\quad (x + 5)(x - 5)$

So the LCD of $\frac{2}{x + 5}, \frac{1}{x - 5},$ and $\frac{16}{x^2 - 25}$ is $(x + 5)(x - 5)$.

Once you have determined the LCD of a collection of rational expressions, you are ready to solve an equation that contains rational expressions.

To solve an equation that contains rational expressions:

1. Find the LCD of the fractions in the equation.

2. Multiply both sides of the equation by this LCD.

3. Use the distributive property if necessary.

4. Cancel any pairs of factors that are common to both the numerators and denominators.

5. Solve the remaining equation.

6. Check the solution by substituting it in the original equation.

The last step, the check step, is important. When you multiply both sides of an equation by an expression that contains a variable, it is possible to introduce "solutions" which do not actually satisfy the original equation. These are called extraneous, or false, solutions. To make sure that you do not have an extraneous solution, you should substitute the answer into the original equation.

For example, to solve the equation $\frac{2}{x+5} + \frac{1}{x-5} = \frac{16}{x^2-25}$:

1. Find the LCD of the fractions in the equation (see example above). $\qquad$ $LCD = (x+5)(x-5)$

2. Multiply both sides of the equation by this LCD.
$$(x+5)(x-5) \cdot \left(\frac{2}{x+5} + \frac{1}{x-5} \right) = (x+5)(x-5) \cdot \frac{16}{x^2-25}$$

3. Use the Distributive Property.
$$(x+5)(x-5) \cdot \frac{2}{x+5} + (x+5)(x-5) \cdot \frac{1}{x-5} = (x+5)(x-5) \cdot \frac{16}{x^2-25}$$

4. Cancel any pairs of factors that are common to both the numerators and denominators.
$$(x+5)(x-5) \cdot \frac{2}{x+5} + (x+5)(x-5) \cdot \frac{1}{x-5} = (x+5)(x-5) \cdot \frac{16}{x^2-25}$$
$$2 \cdot (x-5) + 1 \cdot (x+5) = 16$$

5. Solve the remaining equation.
$$2x - 10 + x + 5 = 16$$
$$3x = 21$$
$$x = 7$$

6. Check the solution by substituting it in the original equation.

Is $\frac{2}{7+5} + \frac{1}{7-5} = \frac{16}{7^2-25}$?

Is $\frac{2}{12} + \frac{1}{2} = \frac{16}{24}$?

Is $\frac{1}{6} + \frac{3}{6} = \frac{4}{6}$?

Is $\frac{4}{6} = \frac{4}{6}$? Yes.

So $x = 7$ is the solution of the equation $\frac{2}{x+5} + \frac{1}{x-5} = \frac{16}{x^2-25}$.

Here's another example.

To solve the equation $\frac{y-1}{y-6} = \frac{5}{y-6}$:

1. Find the LCD of the fractions in the equation. $\qquad$ $LCD = y - 6$

2. Multiply both sides of the equation by this LCD. $\qquad$ $(y-6) \cdot \frac{y-1}{y-6} = (y-6) \cdot \frac{5}{y-6}$

3. The distributive property is not necessary.

4. Cancel any pairs of factors that are common to both the numerators and denominators.

$$(\cancel{y-6})^1 \cdot \frac{y-1}{\cancel{y-6}_1} = (\cancel{y-6})^1 \cdot \frac{5}{\cancel{y-6}_1}$$

$$y - 1 = 5$$

5. Solve the remaining equation.

$$y = 6$$

6. Check the solution by substituting it in the original equation.

Is $\dfrac{6-1}{6-6} = \dfrac{5}{6-6}$?

Is $\dfrac{6-1}{0} = \dfrac{5}{0}$?

Since division by zero is undefined, $x = 6$ is an extraneous solution.

So the equation $\dfrac{y-1}{y-6} = \dfrac{5}{y-6}$ has no solutions.

Sample Problems

1. Find the LCD of the rational expressions in this equation:

$$\frac{7-x}{x^2-9} - \frac{15}{x^2-4} = \frac{x+11}{x^2-5x+6}$$

☑ a. Factor each denominator.

$$x^2 - 9 = (x-3)(x+3)$$

$$x^2 - 4 = (x-2)(x+2)$$

$$x^2 - 5x + 6 = (x-2)(x-3)$$

☐ b. List each factor the greatest number of times it appears in any one of the factorizations.

_____

☐ c. Write the LCD as the product of the factors in the list.

LCD = _____

2. Solve this equation: $\frac{3x}{8} = \frac{5}{12}$

☑ a. Find the LCD of the rational expressions.

LCD = 24

☐ b. Multiply both sides of the equation by the LCD.

_____ = _____

☐ c. Cancel any pairs of factors that are common to both the numerators and denominators, and simplify.

_____ = _____

☐ d. Solve the remaining equation.

$x = $ _____

3. Solve this equation: $\frac{x}{5} + 2 = \frac{3}{4} - x$

☑ a. Find the LCD of the rational expressions.

LCD = 20

☐ b. Multiply both sides of the equation by 20.

$\frac{x}{5} + 2 = \frac{3}{4} - x$

_____ = _____

☐ c. Use the distributive property.

_____ = _____

☐ d. Cancel factors common to the numerators and denominators.

_____ = _____

☐ e. Solve the remaining equation.

$x = $ _____

Answers to Sample Problems

b. $24 \cdot \frac{3x}{8}$, $24 \cdot \frac{5}{12}$

c. $9x$, 10

d. $\frac{10}{9}$

b. $20 \cdot \left(\frac{x}{5} + 2\right)$, $20 \cdot \left(\frac{3}{4} - x\right)$

c. $20 \cdot \frac{x}{5} + 40$, $20 \cdot \frac{3}{4} - 20x$

d. $4x + 40$, $15 - 20x$

e. $-\frac{25}{24}$

4. Solve this equation: $\dfrac{3y}{y^2-4} + \dfrac{3}{y-2} = \dfrac{1}{y+2}$

☑ a. Find the LCD of the rational expressions.

    • Factor the denominators.

$$y^2 - 4 = (y+2)(y-2)$$
$$y - 2 = y - 2$$
$$y + 2 = y + 2$$

    • List each factor the greatest number of times it appears in any one of the factorizations.

$$(y+2),\ (y-2)$$

    • Write the LCD as the product of the factors in the list.

$$\text{LCD} = (y+2)(y-2)$$

☐ b. Multiply both sides of the equation by this LCD.

_____ = _____

☐ c. Use the distributive property.

_____ = _____

☐ d. Cancel any pairs of factors that are common to both the numerators and denominators.

_____ = _____

☐ e. Solve the remaining equation.

$y = $ _____

☑ f. Substitute $y = -\dfrac{8}{5}$ into the original equation to check that it is a solution.

Is $\dfrac{3 \cdot \left(-\dfrac{8}{5}\right)}{\left(-\dfrac{8}{5}\right)^2 - 4} + \dfrac{3}{-\dfrac{8}{5} - 2} = \dfrac{1}{-\dfrac{8}{5} + 2}$?

Is $\dfrac{-\dfrac{24}{5}}{\dfrac{64}{25} - \dfrac{100}{25}} + \dfrac{3}{-\dfrac{8}{5} - \dfrac{10}{5}} = \dfrac{1}{-\dfrac{8}{5} + \dfrac{10}{5}}$?

Is $\dfrac{-\dfrac{24}{5}}{-\dfrac{36}{25}} + \dfrac{3}{-\dfrac{18}{5}} = \dfrac{1}{\dfrac{2}{5}}$?

Is $-\dfrac{24}{5} \cdot \left(-\dfrac{25}{36}\right) + 3 \cdot \left(-\dfrac{5}{18}\right) = 1 \cdot \dfrac{5}{2}$?

Is $\dfrac{10}{3} + \left(-\dfrac{5}{6}\right) = \dfrac{5}{2}$?

Is $\dfrac{15}{6} = \dfrac{5}{2}$? Yes.

So $y = -\dfrac{8}{5}$ is the solution.

Answers (left column):

b. $(y+2)(y-2) \cdot \left(\dfrac{3y}{y^2-4} + \dfrac{3}{y-2}\right)$,

$(y+2)(y-2) \cdot \dfrac{1}{y+2}$

c. $(y+2)(y-2) \cdot \dfrac{3y}{y^2-4}$

$+ (y+2)(y-2) \cdot \dfrac{3}{y-2}$,

$(y+2)(y-2) \cdot \dfrac{1}{y+2}$

d. $3y + 3 \cdot (y+2),\ y-2$

e. $-\dfrac{8}{5}$

HOMEWORK

Homework Problems

Circle the homework problems assigned to you by the computer, then complete them below.

Explain

Rational Expressions

1. Write without using a negative exponent: 13^{-5}

2. Find: $\dfrac{4x^2}{7y} \cdot \dfrac{28y^9}{x^4}$

3. Find: $\dfrac{1-5x}{23xy^5} + \dfrac{11x-3}{23xy^5}$

4. Reduce to lowest terms: $\dfrac{x^2 + 3x - 10}{x^2 - 5x + 6}$

5. Find: $\dfrac{4x^2y^5}{7} \div \dfrac{28y^9}{x^4}$

6. Find: $\dfrac{2+x}{3x} - \dfrac{1+y}{5y}$

7. Reduce to lowest terms and write without using a negative exponent: $\dfrac{28x^{-4}y^{-9}}{7x^6y^{-11}}$

8. Find: $\dfrac{x^2-9}{x^2-16} \div \dfrac{2x-6}{3x+12}$

9. Simplify: $\dfrac{\dfrac{x^3+12}{18}}{\dfrac{x^2}{12}}$

10. Simplify: $\dfrac{a+2}{3ab} \cdot \dfrac{6b}{a^2+a-2} \div \dfrac{5}{a-1}$

11. Simplify: $\dfrac{6b}{a^2+a-2} + \dfrac{3}{a+2} - \dfrac{5}{a-1}$

12. Simplify: $\dfrac{3-\dfrac{1}{2x}}{2+\dfrac{1}{3x}}$

Rational Equations

13. Find the LCD of the fractions in this equation: $\dfrac{x}{6} + 2 = \dfrac{3}{4}$

14. Solve for x: $\dfrac{x}{6} + 2 = \dfrac{3}{4}$

15. Solve for y: $\dfrac{5+y}{2y} = \dfrac{7}{6}$

16. Find the LCD of the fractions in this equation:
$$\dfrac{2}{8x^6} - 3x = \dfrac{7}{12x^8}$$

17. Solve for x: $\dfrac{6}{x} - 3 = \dfrac{3}{2x}$

18. Solve for x: $\dfrac{2}{x} + 1 = \dfrac{2+2x}{x}$

19. Find the LCD of the fractions in this equation:
$$\dfrac{2}{3a-6} - 5a = \dfrac{7}{4a+8}$$

20. Solve for x: $\dfrac{3}{x^2-2x} = \dfrac{2}{x^2+x}$

21. Solve for x: $\dfrac{2}{x^2+8x+16} = \dfrac{5}{x^2-16}$

22. Find the LCD of the fractions in this equation:
$$\dfrac{2}{2y-3} - \dfrac{7y}{2y^2-y-3} = \dfrac{13}{y+1}$$

23. Solve for y: $\dfrac{2}{y^2-4y+3} = \dfrac{5}{y^2+y-12}$

24. Solve for x: $\dfrac{1}{x^2-5x+6} = \dfrac{2}{2x^2-6x}$

APPLY

Practice Problems

Here are some additional practice problems for you to try.

Rational Expressions

1. Write without using a negative exponent: 3^{-10}

2. Write without using a negative exponent: 7^{-11}

3. Write without using a negative exponent: y^{-5}

4. Write without using negative exponents: $3x^{-4}$

5. Simplify and write without using negative exponents: $2^3 y^{-7}$

6. Reduce to lowest terms and write without using negative exponents: $\dfrac{15m^{-5}n}{3m^{-3}n^{-6}}$

7. Reduce to lowest terms and write without using negative exponents: $\dfrac{12a^{-5}b^{-7}}{24a^{-10}b^{-3}}$

8. Reduce to lowest terms and write without using negative exponents: $\dfrac{21a^{-4}b^{-3}}{7a^{-6}b}$

9. Reduce to lowest terms: $\dfrac{x^2 - x - 12}{x^2 + 5x + 6}$

10. Reduce to lowest terms: $\dfrac{x^2 - 4x - 5}{x^2 - 2x - 15}$

11. Reduce to lowest terms: $\dfrac{x^2 + 3x - 10}{x^2 + 6x + 5}$

12. Reduce to lowest terms: $\dfrac{x^3 - 16x}{x^3 - 8x^2 + 16x}$

13. Reduce to lowest terms: $\dfrac{x^4 + 10x^3 + 25x^2}{x^3 + x^2 - 20x}$

14. Simplify: $\dfrac{\dfrac{z^2 + 3}{16z}}{\dfrac{5}{8z^3}}$

15. Simplify: $\dfrac{\dfrac{y^2 - 4}{10y^3}}{\dfrac{y + 2}{5y^2}}$

16. Simplify: $\dfrac{\dfrac{y^2 + 2}{12y}}{\dfrac{3}{4y^2}}$

17. Simplify: $\dfrac{3 + \dfrac{1}{3b}}{2 - \dfrac{1}{6b}}$

18. Simplify: $\dfrac{6 - \dfrac{1}{2c}}{3 - \dfrac{1}{4c}}$

19. Simplify: $\dfrac{5 - \dfrac{1}{4a}}{3 + \dfrac{1}{2a}}$

20. Simplify: $\dfrac{15x^3}{4y^2} \div \dfrac{5x^2 y}{6}$

21. Simplify: $\dfrac{x^2 - 25}{x^2 - 64} \div \dfrac{-3x + 15}{3x + 24}$

22. Simplify: $\dfrac{x + 5}{3xy^2} \cdot \dfrac{9x}{x^2 + 3x - 10} \div \dfrac{2y}{x - 2}$

23. Simplify: $\dfrac{y^2 - 9}{7x^3 y} \div \dfrac{y^2 + 6y + 9}{14xy} \cdot \dfrac{y + 3}{2x}$

24. Find: $\dfrac{y + 6}{2x^2 y} \cdot \dfrac{8y}{y^2 + 2y - 24} \div \dfrac{2x}{y - 4}$

25. Find: $\dfrac{y + 1}{2y} - \dfrac{2x - 3}{3x}$

26. Simplify: $\dfrac{2y}{x^2 + 2x - 35} - \dfrac{3}{x + 7}$

27. Simplify: $\dfrac{z + 3}{2z} - \dfrac{2x + 3}{3x}$

28. Simplify: $\dfrac{2a}{a^2 + 5a - 24} - \dfrac{3}{a + 8}$

Rational Equations

29. Find the LCD of the fractions in this equation: $\dfrac{3}{7x^5} = x - \dfrac{9}{14x^2}$

30. Find the LCD of the fractions in this equation: $\dfrac{7}{x^3} + \dfrac{x}{6} = \dfrac{10}{3x^2}$

31. Find the LCD of the fractions in this equation: $\dfrac{2}{5x^4} - x = \dfrac{9}{10x^3}$

32. Find the LCD of the fractions in this equation: $\dfrac{2y}{y + 3} = y + \dfrac{7}{6y}$

33. Find the LCD of the fractions in this equation:
$\dfrac{2a}{a + 5} - a = \dfrac{3}{a^2 - 25}$

34. Find the LCD of the fractions in this equation:
$\dfrac{1}{4x - 12} + \dfrac{14x}{x^2 + 3x} = \dfrac{x}{8}$

35. Find the LCD of the fractions in this equation:
$$\frac{2b-5}{b+2} - \frac{5b}{2b^2-b-10} = \frac{1}{2b-5}$$

36. Find the LCD of the fractions in this equation:
$$\frac{x+7}{x-3} + \frac{2}{x^2-9} = \frac{3x-1}{x^2+6x+9}$$

37. Find the LCD of the fractions in this equation:
$$\frac{3a+2}{a+5} - \frac{7a}{2a^2+3a-35} = \frac{12}{2a-7}$$

38. Solve for x: $\dfrac{3x}{4} - 2 = \dfrac{5}{8}$

39. Solve for y: $\dfrac{5y}{4} + 3 = \dfrac{13}{16}$

40. Solve for x: $\dfrac{2x}{3} + 1 = \dfrac{4}{5}$

41. Solve for a: $\dfrac{a+1}{5a} = \dfrac{4}{7}$

42. Solve for b: $\dfrac{2-b}{3b} = \dfrac{3}{5}$

43. Solve for y: $\dfrac{1-y}{3y} = \dfrac{2}{5}$

44. Solve for x: $\dfrac{3}{x} + 7 = -\dfrac{2}{5x}$

45. Solve for y: $\dfrac{7}{y} - 6 = \dfrac{5}{8y}$

46. Solve for x: $\dfrac{2}{x} - 5 = -\dfrac{4}{3x}$

47. Solve for x: $\dfrac{8}{3x+2} = \dfrac{4}{5x}$

48. Solve for y: $\dfrac{6}{5y} = \dfrac{3}{4y-5}$

49. Solve for x: $\dfrac{6}{4x-5} = \dfrac{2}{3x}$

50. Solve for a: $\dfrac{2}{a-1} = \dfrac{3}{a-2}$

51. Solve for b: $\dfrac{6}{b+5} = \dfrac{-4}{b-10}$

52. Solve for a: $\dfrac{3}{a-1} = \dfrac{5}{a-3}$

53. Solve for y: $2 - \dfrac{3x+1}{x} = \dfrac{5}{2x}$

54. Solve for z: $8 + \dfrac{4}{z-3} = -\dfrac{12}{3z-9}$

55. Solve for r: $\dfrac{4}{2r+8} - 6 = \dfrac{8}{r+4}$

56. Solve for q: $4 - \dfrac{3}{q-6} = \dfrac{2}{2q-12}$

EVALUATE

Practice Test

Take this practice test to be sure that you are prepared for the final quiz in Evaluate.

1. Write without using negative exponents and reduce to lowest terms: $\dfrac{35x^{-5}y^{-9}}{5x^6y^{-12}}$

2. Simplify: $\dfrac{a+2}{2ab} \cdot \dfrac{6bc}{a^2-a-6} \div \dfrac{7}{a-3}$

3. Simplify: $\dfrac{5a}{2a^2-5a-3} + \dfrac{2}{a-3} - \dfrac{1}{2a+1}$

4. Simplify: $\dfrac{1-\dfrac{1}{2x}}{2+\dfrac{1}{3x^2}}$

5. Find the LCD of the fractions in this equation: $\dfrac{2}{3a-9} - 5a = \dfrac{7}{5a+20} + 7$

6. Solve for x: $\dfrac{3x}{4} - 2 = \dfrac{5}{6}$

7. Solve for x: $\dfrac{2-x}{x} = \dfrac{2}{x}$

8. Solve for x: $\dfrac{3}{x^2-2x} = \dfrac{2}{x^2+x} + \dfrac{4}{x^2-x-2}$

LESSON EII.E – GRAPHING LINES

Here's what you'll learn in this lesson:

Graphing Lines

a. The Cartesian coordinate system

b. The distance formula

c. Graphing linear equations

d. Finding the x- and y-intercepts

e. Finding the slope of a line

Finding Equations

a. The point-slope form of a line

b. The standard form of a line

c. The slope-intercept form of a line

d. Horizontal lines

e. Vertical lines

f. Parallel and perpendicular lines

OVERVIEW

In this lesson you will review how to graph linear equations and how to find the equation of a line. In the process, you will reacquaint yourself with the Cartesian coordinate system.

 EXPLAIN

GRAPHING LINES

Summary

The Cartesian Coordinate System

You have learned how to graph real numbers on a number line. Now you'll review how to plot ordered pairs of numbers. In the 17th century, the French philosopher and mathematician René Descartes invented the Cartesian coordinate system as a means to represent ordered pairs of numbers. The Cartesian coordinate system, or xy-plane, consists of two number lines placed at right angles to each other that break the plane into four quadrants. The horizontal number line is called the x-axis and the vertical number line is called the y-axis. These axes intersect at the point (0, 0), called the origin.

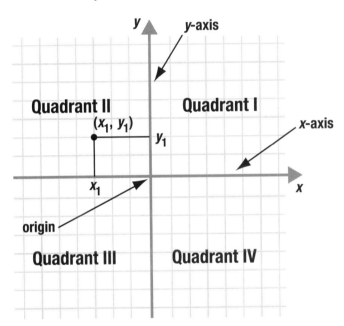

Given an ordered pair of real numbers (x_1, y_1) where x_1 is the x-coordinate (abscissa) and y_1 is the y-coordinate (ordinate), you can plot this ordered pair as a point in the xy-plane.

To plot a point:

1. Draw a vertical line through the x-coordinate of the point.

2. Draw a horizontal line through the y-coordinate of the point.

3. Plot the point where these two lines intersect.

As an example, points P (1, 3) and Q (−2, −2) are plotted in Figure EII.E.1.

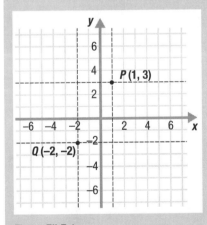

Figure EII.E.1

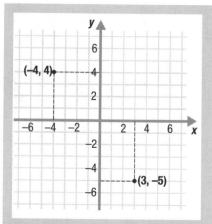

Figure EII.E.2

Drawing horizontal and vertical lines is a very formal method of plotting points or finding coordinates. With experience, you will be able to do both without actually drawing the lines.

If a line is horizontal, the y-coordinate will be the same for every point on the line. That is, $y_1 = y_2$. Substituting this into the distance formula gives the horizontal distance formula.

Similarly, if a line is vertical, the x-coordinate will be the same for every point on the line. That is, $x_1 = x_2$. Once again, if you substitute this into the distance formula and simplify, you will get the vertical distance formula.

It doesn't matter which point you call (x_1, y_1) or which point you call (x_2, y_2). Either way, you'll get the same answer.

Similarly, given a point in the *xy*-plane, you can find its coordinates. To find the coordinates of a point:

1. Draw a vertical line from the point to the *x*-axis. The line intersects the *x*-axis at the *x*-coordinate.

2. Draw a horizontal line from the point to the *y*-axis. The line intersects the *y*-axis at the *y*-coordinate.

For an example, look at Figure EII.E.2, where the points $(-4, 4)$ and $(3, -5)$ are given.

The Distance Formula

Once you can plot points, you can find the distance between points. In general, you can always find the distance between any two points (x_1, y_1) and (x_2, y_2) in the plane. This distance, *d*, is given by the distance formula:

$$d = \sqrt{(x_2 - x_1)^2 + (y_2 - y_1)^2}$$

If the points lie on a vertical or horizontal line, this formula reduces to simpler expressions to calculate the distance.

On a horizontal line, the distance, *d*, between two points (x_1, y_1) and (x_2, y_2) is given by:

$$d = |x_2 - x_1|$$

On a vertical line, the distance, *d*, between two points (x_1, y_1) and (x_2, y_2) is given by:

$$d = |y_2 - y_1|$$

To find the distance, *d*, between two points:

1. Label one of the points (x_1, y_1) and the other (x_2, y_2).

2. Determine if the points lie on a horizontal or vertical line.

    • If the points lie on a horizontal line, their *y*-values are equal.

    • If the points lie on a vertical line, their *x*-values are equal.

3. Apply the appropriate distance formula.

For example, to find the distance, *d*, between the points $(-1, 7)$ and $(5, 7)$:

1. Label one of the points (x_1, y_1) and the other (x_2, y_2).

    Let $(x_1, y_1) = (-1, 7)$
    Let $(x_2, y_2) = (5, 7)$

2. Determine if the points lie on a horizontal or vertical line. See the grid in Figure EII.E.3.

    Are the *x*-values equal? No.
    Are the *y*-values equal? Yes.
    The points lie on a horizontal line.

3. Apply the horizontal distance formula.

$$d = |x_2 - x_1|$$
$$= |5 - (-1)|$$
$$= |5 + 1|$$
$$= |6|$$
$$= 6$$

So the distance between the points $(-1, 7)$ and $(5, 7)$ is 6.

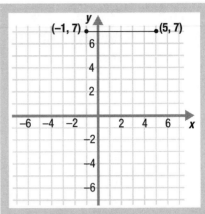

Figure EII.E.3

As another example, to find the distance, d, between the points $(-1, 7)$ and $(5, 1)$:

1. Label one of the points (x_1, y_1) and the other (x_2, y_2).

Let $(x_1, y_1) = (-1, 7)$
Let $(x_2, y_2) = (5, 1)$

2. Determine if the points lie on a horizontal or vertical line. See the grid in Figure EII.E.4.

Are the x-values equal? No.
Are the y-values equal? No.
The points do not lie on a vertical or a horizontal line.

3. Apply the distance formula.

$$d = \sqrt{(x_2 - x_1)^2 + (y_2 - y_1)^2}$$
$$= \sqrt{[(5 - (-1)]^2 + (1 - 7)^2}$$
$$= \sqrt{(6)^2 + (-6)^2}$$
$$= \sqrt{36 + 36}$$
$$= \sqrt{72}$$
$$= 6\sqrt{2}$$

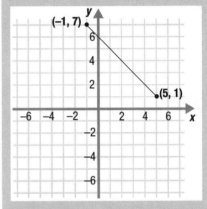

Figure EII.E.4

So the distance between the points $(-1, 7)$ and $(5, 1)$ is $6\sqrt{2}$.

Graphing Linear Equations

The graph of a linear equation is a line. To graph a linear equation:

1. Make a table of ordered pairs that satisfy the equation.

2. Plot these ordered pairs.

3. Draw a line through the plotted points.

For example, to graph the linear equation $x + 3y = 6$:

1. Make a table of ordered pairs that satisfy the equation.

x	y
6	0
3	1
0	2
-3	3

2. Plot these ordered pairs.

3. Draw a line through the plotted points. See the grid in Figure EII.E.5

Here's another example. To graph the linear equation $4x - 2y = 6$:

Since a line is determined by two points, you don't need more than two points to graph a line. However, plotting extra points will help you avoid errors.

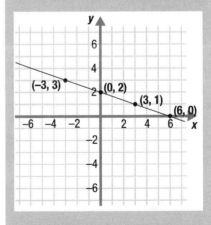

Figure EII.E.5

The points in this example are two special points called the x-intercept and the y-intercept. You'll learn more about these in the next section.

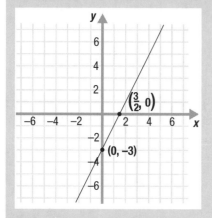

Figure EII.E.6

1. Make a table of ordered pairs that satisfy the equation.

x	y
$\frac{3}{2}$	0
0	−3

Choose $y = 0$:

$$4x - 2(0) = 6$$

$$4x = 6$$

$$x = \frac{6}{4}$$

$$= \frac{3}{2}$$

Choose $x = 0$:

$$4(0) - 2y = 6$$

$$-2y = 6$$

$$y = \frac{6}{-2}$$

$$= -3$$

2. Plot the ordered pairs.

3. Draw a line through the plotted points. See the grid in Figure EII.E.6.

Finding the x- and y-Intercepts of a Line

When graphing a line, two points on the line that are often easy to find are the x-intercept and the y-intercept. The x-intercept is the point where the line crosses the x-axis. Similarly, the y-intercept is the point where the line crosses the y-axis.

To find the intercepts of a line:

1. Set $y = 0$ and solve for x. This gives the x-intercept, $(x, 0)$.

2. Set $x = 0$ and solve for y. This gives the y-intercept, $(0, y)$.

For example, to find the intercepts of the line $5x + 2y = -1$:

1. Set $y = 0$ and solve for x.

$$5x + 2(0) = -1$$

$$5x = -1$$

$$x = -\frac{1}{5}$$

The x-intercept is $\left(-\frac{1}{5}, 0\right)$.

2. Set $x = 0$ and solve for y.

$$5(0) + 2y = -1$$

$$2y = -1$$

$$y = -\frac{1}{2}$$

The y-intercept is $\left(0, -\frac{1}{2}\right)$.

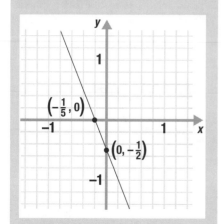

Figure EII.E.7

The intercepts of the line $5x + 2y = -1$ are plotted in Figure EII.E.7.

Here's another example. To find the intercepts of the line $y = 3x + 6$:

1. Set $y = 0$ and solve for x.

$$0 = 3x + 6$$

$$-3x = 6$$

$$x = \frac{6}{-3}$$

$$= -2$$

The x-intercept is $(-2, 0)$.

2. Set $x = 0$ and solve for y.

$$y = 3(0) + 6$$

$$y = 6$$

The y-intercept is $(0, 6)$.

The intercepts of the line $y = 3x + 6$ are plotted in Figure EII.E.8.

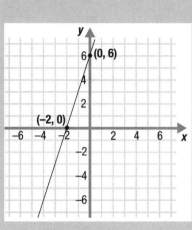

Figure EII.E.8

Finding the Slope of a Line

The slope of a line is a number that describes the steepness of the line. It is the ratio of the rise to the run in moving from one point on the line to another point on the line.

The slope, m, of the line containing the points (x_1, y_1) and (x_2, y_2) is given by:

$$m = \frac{\text{rise}}{\text{run}} = \frac{\text{change in } y}{\text{change in } x} = \frac{y_2 - y_1}{x_2 - x_1}$$

To find the slope of a line through two points (x_1, y_1) and (x_2, y_2):

1. Label one of the points (x_1, y_1) and the other (x_2, y_2).

2. Substitute the coordinates of the points into the definition of slope, $m = \frac{y_2 - y_1}{x_2 - x_1}$.

3. Simplify.

Again, it doesn't matter which point you call (x_1, y_1) or which point you call (x_2, y_2). Either way, you get the same slope.

For example, to find the slope of the line joining the points $(8, 2)$ and $(6, 5)$:

1. Label one of the points (x_1, y_1) and the other (x_2, y_2).

Let $(x_1, y_1) = (8, 2)$
Let $(x_2, y_2) = (6, 5)$

2. Substitute the coordinates of the points into the definition of slope.

$$m = \frac{5 - 2}{6 - 8}$$

3. Simplify.

$$m = \frac{3}{-2}$$

$$m = -\frac{3}{2}$$

So the slope of the line through the points $(8, 2)$ and $(6, 5)$ is $-\frac{3}{2}$.

Here's another example. To find the slope of the line passing through the points (3, −1) and (3, 14):

1. Label one of the points (x_1, y_1) and the other (x_2, y_2).

Let $(x_1, y_1) = (3, -1)$
Let $(x_2, y_2) = (3, 14)$

2. Substitute the coordinates of the points into the definition of slope.

$$m = \frac{14 - (-1)}{3 - 3}$$

3. Simplify.

$$m = \frac{14 + 1}{3 - 3}$$

 $m = $ ~~$\frac{15}{0}$~~

Division by 0 is undefined, so the slope of the line through the points (3, −1) and (3, 14) is undefined.

The slope of a vertical line is undefined.

The slope of a horizontal line is zero.

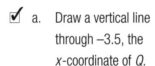
Sample Problems

1. On the grid below, plot the point Q (−3.5, 4).

   ☑ a. Draw a vertical line through −3.5, the x-coordinate of Q.

   ☐ b. Draw a horizontal line through 4, the y-coordinate of Q.

   ☐ c. Plot a point where the lines intersect. Label the point Q.

   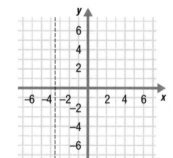

2. Find the distance, d, between the point (−1, −2) and the point (7, 10).

   ☑ a. Label one of the points (x_1, y_1) and the other (x_2, y_2).

   Let $(x_1, y_1) = (-1, -2)$
   Let $(x_2, y_2) = (7, 10)$

   ☐ b. Determine if the points lie on a horizontal or vertical line.

   Are the x-values equal? _____
   Are the y-values equal? _____
   The points do not lie on a horizontal or vertical line.

   ☑ c. Substitute the coordinates of the points into the distance formula.

   $$d = \sqrt{[7 - (-1)]^2 + [10 - (-2)]^2}$$

   ☐ d. Simplify.

   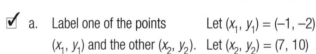 $d = $ _____

3. Graph the linear equation $5x + 4y = 20$.

☑ a. Make a table of ordered   Choose $y = 0$.
       pairs that satisfy the equation.

                                     $x =$ _____

                                     Choose $x = 0$.

                                     $y =$ _____

x	y
___	0
0	___

☐ b. Plot the ordered pairs.

☐ c. Draw a line through
       the plotted points.

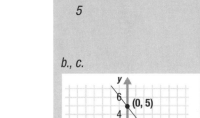

4. Find the x- and y-intercepts of the line $y - 7 = -9(x + 2)$.

☑ a   Set $y = 0$ and solve for x. $0 - 7 = -9(x + 2)$

                                            $-7 = -9x - 18$

                                            $11 = -9x$

                                            $x = -\dfrac{11}{9}$

☐ b. Write the x-intercept.   The x-intercept is (____, ____).

☐ c. Set $x = 0$ and solve for y. $y - 7 = -9(0 + 2)$

                                            $y =$ ___

☐ d. Write the y-intercept.   The y-intercept is (____, ____).

5. Find the slope of the line containing the points $(4, 8)$ and $(12, 6)$.

☐ a. Label one of the points   Let $(x_1, y_1) = (4, 8)$
       (x_1, y_1) and the other (x_2, y_2).   Let $(x_2, y_2) = ($ ____, ____ $)$

☐ b. Substitute the coordinates of the $m = \dfrac{6 - 8}{\text{___} - \text{___}}$
       points into the definition of slope.

☐ c. Simplify. $m =$ ____

FINDING EQUATIONS

Summary

The Point-Slope Form of the Equation of a Line

You can find the equation of a line if you know the slope, m, of the line and one point, (x_1, y_1), on the line.

To find the point-slope form for the equation of the line through (x_1, y_1) with slope m:

1. Pick any point on the line. Call it (x, y).

2. Substitute the slope, m, and the points (x_1, y_1) and (x, y) into the definition of slope: $m = \frac{y - y_1}{x - x_1}$

3. Multiply both sides of the equation by the denominator, $(x - x_1)$.

4. Rewrite the equation as $(y - y_1) = m(x - x_1)$.

The equation $(y - y_1) = m(x - x_1)$ is the point-slope form of the equation of a line.

For example, to find the equation of the line through the point $(1, 4)$ with slope -7.

1. Pick any point on the line. (x, y)

2. Substitute the slope, $m = -7$, $\frac{y - 4}{x - 1} = -7$
    the point $(1, 4)$, and the point
    (x, y) into the definition of slope.

3. Multiply both sides by $x - 1$. $(x - 1) \cdot \frac{y - 4}{x - 1} = (x - 1) \cdot (-7)$

4. Rewrite the equation. $y - 4 = -7(x - 1)$

So the equation of the line through the point $(1, 4)$ with slope -7 is $y - 4 = -7(x - 1)$.

It is not necessary to develop the equation from scratch every time. Having derived the point-slope form, you can just substitute the given slope, m, and the point (x_1, y_1) into the final form of the equation $(y - y_1) = m(x - x_1)$.

Here's an example. To find the equation of the line through the point $(-3, 8)$ with slope 14:

1. Substitute the slope $m = 14$ $y - 8 = 14[x - (-3)]$
    and the point $(-3, 8)$ into the
    point-slope form.

2. Simplify. $y - 8 = 14(x + 3)$

So the equation of the line through the point $(-3, 8)$ with slope 14 is $y - 8 = 14(x + 3)$.

If you are given two points, you can still use the point-slope form to find the equation of the line.

To find the point-slope form of the equation of a line through two points:

1. Find the slope of the line by substituting the coordinates of the two points into the definition of slope.

2. Use the slope and either one of the points to write the equation in point-slope form.

For example, to find the equation of the line through the points (–2, –3) and (0, 5):

1. Find the slope of the line.

$$m = \frac{5 - (-3)}{0 - (-2)}$$
$$= \frac{5 + 3}{2}$$
$$= \frac{8}{2}$$
$$= 4$$

2. Use the slope, $m = 4$, and one of the points, say (–2, –3), to write the equation in point-slope form.

$$y - (-3) = 4[x - (-2)]$$
$$y + 3 = 4(x + 2)$$

So the equation of the line through the points (–2, –3) and (0, 5) is $y + 3 = 4(x + 2)$.

The Standard Form of the Equation of a Line

The equation of a line in point-slope form can be rewritten in several different forms. One of these is the standard form. The equation of a line in standard form is $Ax + By = C$. Here, A and B are not both equal to zero.

To change a linear equation in another form to standard form:

1. Distribute (if necessary) to remove parentheses.

2. Move all the x- and y-terms to the left side of the equation.

3. Move all of the constant terms to the right side of the equation.

4. If necessary, rearrange the terms on the left so the equation is in the form $Ax + By = C$.

For example, to rewrite the equation of the line $y - 8 = 11(x - 2)$ in standard form:

1. Distribute the 11. $y - 8 = 11x - 22$

2. Subtract $11x$ from both sides. $y - 8 - 11x = -22$

3. Add 8 to both sides. $y - 11x = -14$

4. Switch the x- and y-terms. $-11x + y = -14$

So the equation of the line $y - 8 = 11(x - 2)$ in standard form is $-11x + y = -14$.

Here's a different type of example. To write the equation of the line through the point (7, 6) with slope $m = -5$ in standard form:

1. Substitute the slope, $m = -5$, and the point (7, 6) into the point-slope form.

$$y - 6 = -5(x - 7)$$

2. Distribute the -5.

$$y - 6 = -5x + 35$$

3. Add $5x$ to both sides.

$$y - 6 + 5x = 35$$

4. Add 6 to both sides.

$$y + 5x = 41$$

5. Switch the x- and y-terms.

$$5x + y = 41$$

So the standard form of the equation of the line through (7, 6) with slope -5 is $5x + y = 41$.

The Slope-Intercept Form of the Equation of a Line

The slope-intercept form of the equation of a line is $y = mx + b$. Here, m is the slope of the line and (0, b) is the y-intercept.

To change a linear equation in another form to the slope-intercept form:

1. Solve the equation for y.

2. If necessary, rearrange the terms on the right so the equation is in the form $y = mx + b$.

For example, to find the slope-intercept form of the equation $3x + 12y = 6$:

1. Solve the equation for y.

$$3x + 12y = 6$$

$$3x + 12y - 3x = 6 - 3x$$

$$12y = 6 - 3x$$

$$\frac{12y}{12} = \frac{6}{12} - \frac{3}{12}x$$

$$y = \frac{1}{2} - \frac{1}{4}x$$

2. Rearrange the terms on the right.

$$y = -\frac{1}{4}x + \frac{1}{2}$$

So the slope-intercept form of the equation $3x + 12y = 6$ is $y = -\frac{1}{4}x + \frac{1}{2}$.

Here's another example. To write the equation of the line with slope 2 through the point $\left(\frac{1}{2}, 7\right)$ in slope-intercept form:

1. Write the point-slope form of the equation of the line.

$$y - 7 = 2\left(x - \frac{1}{2}\right)$$

2. Solve the equation for y.

$$y = 2\left(x - \frac{1}{2}\right) + 7$$

$$y = 2x - 1 + 7$$

$$y = 2x + 6$$

So the slope-intercept form of the line through the point $\left(\frac{1}{2}, 7\right)$ with slope 2 is $y = 2x + 6$.

To find the slope-intercept form of a linear equation when you are given the slope, m, and the y-intercept, $(0, b)$:

Note that when $x = 0$, $y = m(0) + b$ or $y = b$.

1. Start with the equation $y = mx + b$.

2. Substitute the slope for m.

3. Substitute the y-coordinate of the y-intercept for b.

Here's an example. To find the slope-intercept form of the line with slope $m = \frac{6}{11}$ and y-intercept $(0, 2)$:

1.	Start with the equation $y = mx + b$.	$y = mx + b$
2.	Substitute the slope for m.	$y = \frac{6}{11}x + b$
3.	Substitute the y-coordinate of the y-intercept for b.	$y = \frac{6}{11}x + 2$

So the slope-intercept form of the line with slope $m = \frac{6}{11}$ and y-intercept $(0, 2)$ is $y = \frac{6}{11}x + 2$.

Horizontal Lines

The equation of a horizontal line is $y = c$. Here, c is a constant.

To find the equation of a horizontal line:

1. Find the y-coordinate of any point on the line.

2. Substitute the value of the y-coordinate for c in the equation $y = c$.

For example, to find the equation of the horizontal line through the point $(-18, 4)$:

1.	Find the y-coordinate of any point on the line.	4
2.	Substitute 4 for c.	$y = 4$

So the equation of the horizontal line through $(-18, 4)$ is $y = 4$.

Vertical Lines

The equation of a vertical line is $x = c$. Here, c is a constant.

To find the equation of a vertical line:

1. Find the x-coordinate of any point on the line.

2. Substitute the value of the x-coordinate for c in the equation $x = c$.

For example, to find the equation of the vertical line through the point $(7, -7)$:

1.	Find the x-coordinate of any point on the line.	7
2.	Substitute 7 for c.	$x = 7$

So the equation of the vertical line through $(7, -7)$ is $x = 7$.

Parallel and Perpendicular Lines

You know how to find the equation of a line given a point on the line and the slope of the line. Sometimes you may not be given the slope of a line when you need it. However, in some cases, these facts may help you find the slope:

- Parallel lines have the same slope.

- Perpendicular lines have slopes that are negative reciprocals of each other.

For example, to find the equation of the line that is parallel to $2x + y = -9$ and passes through the point $(0, -6)$:

1. Find the slope of the line
   $2x + y = -9$.

   Convert to slope-intercept form:
   $$2x + y = -9$$
   $$y = -2x - 9$$
   So the slope, m, is -2.

2. Find the slope of the parallel line.

   Parallel lines have the same slope, so $m = -2$.

3. Find the equation of the parallel line. Use the slope-intercept form, since you have the slope and the y-intercept.

   $$y = mx + b$$
   $$y = -2x - 6$$

So the equation of the line that is parallel to $2x + y = -9$ and passes through the point $(0, -6)$ is $y = -2x - 6$.

Here's another example. To find the equation of the line through $\left(\frac{3}{5}, \frac{4}{5}\right)$ that is perpendicular to $y = -\frac{1}{5}x + 8$:

1. Find the slope of the line
   $y = -\frac{1}{5}x + 8$.

   This equation is in slope-intercept form, so the slope, m, is $-\frac{1}{5}$.

2. Find the slope of the perpendicular line.

   Perpendicular lines have slopes that are negative reciprocals of each other, so:
   $$m = -\frac{1}{-\frac{1}{5}}$$
   $$= -1 \cdot -\frac{5}{1}$$
   $$= 5$$

3. Find the equation of the perpendicular line. Use the point-slope form, since you have a point and the slope.

   $$y - y_1 = m(x - x_1)$$
   $$y - \frac{4}{5} = 5\left(x - \frac{3}{5}\right)$$

So the equation of the line perpendicular to $y = -\frac{1}{5}x + 8$ that passes through $\left(\frac{3}{5}, \frac{4}{5}\right)$ is $y - \frac{4}{5} = 5\left(x - \frac{3}{5}\right)$.

To find the negative reciprocal of a fraction, just switch the numerator and the denominator and change the sign. For example, the negative recipocal of $\frac{3}{7}$ is $-\frac{7}{3}$. The negative recipocal of 5 is $-\frac{1}{5}$.

Sample Problems

1. Find the standard form of the equation of the line with slope $-\frac{1}{4}$ that passes through the point (3, 1).

   ☐ a. Write the point-slope form of the linear equation. $y - \underline{\quad} = -\frac{1}{4}(x - \underline{\quad})$ *a. 1, 3*

   ☑ b. Distribute. $y - 1 = -\frac{1}{4}x + \frac{3}{4}$

   ☑ c. Move the x-term to the left side. $y - 1 + \frac{1}{4}x = \frac{3}{4}$

   ☐ d. Move the constant terms to the right side. $y + \frac{1}{4}x = \underline{\quad}$ *d. $\frac{7}{4}$*

   ☐ e. Rearrange the left side of the equation. $\underline{\qquad\qquad} = \frac{7}{4}$ *e. $\frac{1}{4}x + y$*

2. Rewrite $12x + 7y = 49$ in slope-intercept form.

   ☐ a. Solve the equation for y. If necessary, rearrange the terms on the right. $y = \underline{\qquad\qquad}$ *a. $-\frac{12}{7}x + 7$*

3. Find the equation of the line parallel to the y-axis that passes through the point (12, 20).

   ☑ a. Determine if the line is vertical or horizontal.   The line parallel to the y-axis is a vertical line.

   ☐ b. Write the equation of a vertical line. $\underline{\quad}$ *b. $x = c$*

   ☐ c. Find the x-coordinate of any point on the line. $\underline{\quad}$ *c. 12*

   ☐ d. Substitute the value of the x-coordinate for c. $x = \underline{\quad}$ *d. 12*

4. Find the equation of the line through the point (4, −9) that is parallel to the line through the points (1, 3) and (5, 7).

   ☐ a. Find the slope of the line through the points (1, 3) and (5, 7). $m = \underline{\quad}$ *a. $\frac{7-3}{5-1}$ or 1*

   ☐ b. Find the slope of the parallel line. $m = \underline{\quad}$ *b. 1*

   ☐ c. Find the equation of the parallel line. Use point-slope form. $y - \underline{\quad} = \underline{\quad}(x - \underline{\quad})$ *c. (−9), 1, 4*
   $y + \underline{\quad} = x - \underline{\quad}$ *9, 4*

5. Find the equation of the line through (8, 7) that is perpendicular to the line $2x + 3y = 1$.

   ☐ a. Find the slope of the line $2x + 3y = 1$. $3y = -2x + 1$

   $y = -\frac{2}{3}x + \frac{1}{3}$

   $m = \underline{\quad}$ *a. $-\frac{2}{3}$*

   ☐ b. Find the slope of the perpendicular line. $m = \underline{\quad}$ *b. $\frac{3}{2}$*

   ☐ c. Find the equation of the perpendicular line. Use point-slope form. $\underline{\qquad\qquad}$ *c. $y - 7 = \frac{3}{2}(x - 8)$*

HOMEWORK

Homework Problems

Circle the homework problems assigned to you by the computer, then complete them below.

 **Explain**

Graphing Lines

1. Find the coordinates of the points P, Q, and R as shown in Figure EII.E.9.

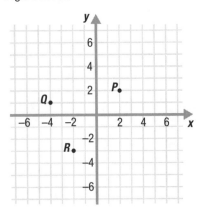

Figure EII.E.9

2. Find the distance between the points (0, 0) and (0, 7).

3. Plot the points in the table below, and then draw a line through the points.

x	y
−2	−2
0	−1
2	0
4	1

4. Find the x-intercept of the line $6x + 9y = 54$.

5. Find the distance between the points (−2, −3) and (4, −3).

6. Graph the line $y - 2 = -3(x + 1)$.

7. Find the slope, m, of the line through the points (12, 7) and (5, 4).

8. Find the distance between the points (−2, 9) and (−5, 5).

9. Graph the line $6x - 2y = -3$.

10. Find the x- and y-intercepts of the line $y + 3 = 16(x - 2)$.

11. Of the points A (3, 3), B (−2, 2), and C (1, 1), which two are closest together?

12. Graph the line with slope $m = -\frac{5}{2}$ that passes through the point (3, −2).

Finding Equations

13. Find the point-slope form of the equation of the line through the point (1, 2) with slope 5.

14. Write the linear equation $y = -16x + 9$ in standard form.

15. Find the equation of the horizontal line through the points (1, 12) and (6, 12).

16. Find the point-slope form of the equation of the line through the points (−7, 3) and (−4, 2).

17. Write the linear equation $y + 8 = 2(x + 6)$ in standard form.

18. Find the equation of the vertical line that crosses the x-axis 3 units to the left of the y-axis.

19. Find the slope-intercept form of the equation of the line through (0, 9) that is parallel to the line $y - 7 = 18(x + 14)$.

20. Find the standard form of the equation of the line passing through the points (2, 11) and (3, 12).

21. Find the equation of the line through $\left(\frac{3}{7}, \frac{1}{8}\right)$ that is perpendicular to the y-axis.

22. Find the point-slope form of the equation of the line through (4, −6) that is parallel to the line $36x + 18y = -42$.

23. Find the standard form of the equation of the line through (−1, 1) that is perpendicular to the line $6x - y = 11$.

24. Find the equation of the line perpendicular to the line $-5x - 40y = 10$ that has the same y-intercept as this line.

APPLY

Practice Problems

Here are some additional practice problems for you to try.

Graphing Lines

1. Find the coordinates of the points *P*, *Q*, and *R* below.

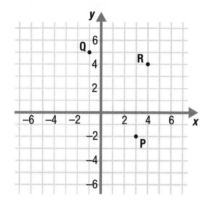

2. Find the coordinates of the points *S*, *T*, and *U* below.

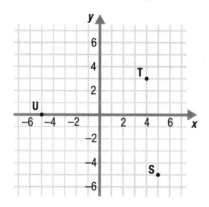

3. Circle the point below that is in Quadrant II.

   (4, 6)   (−3, 4)

   (0, −2)   (5, −7)

4. Circle the point below that is in Quadrant III.

   (−3, −2)   (4, −4)

   (7, 0)   (−4, 6)

5. Find the distance between the points (8, 4) and (3, 4).

6. Find the distance between the points (−13, 12) and (5, 12).

7. Find the distance between the points (−3, 7) and (−3, −2).

8. Find the distance between the points (9, 2) and (3, 10).

9. Find the distance between the points (8, −5) and (−4, 0).

10. Find the distance between the points (−4, 1) and (−1, −3).

11. Graph the line $y + 3 = -2x$.

12. Graph the line $y - 5 = 3x$.

13. Graph the line $y + 4 = -\frac{1}{2}x$.

14. Find the slope *m* of the line that passes through the points (4, 7) and (8, 13).

15. Find the slope *m* of the line that passes through the points (3, 5) and (−5, 7).

16. Find the slope *m* of the line passing through the points (5, −3) and (−3, −3).

17. Find the *x*- and *y*-intercepts of the line $2x - y = 10$.

18. Find the *x*- and *y*-intercepts of the line $3x - 5y = -15$.

19. Find the *x*- and *y*-intercepts of the line $4x + 3y = 21$.

20. Graph the line $y + 3 = -2(x - 4)$.

21. Graph the line $y - 2 = \frac{1}{2}(x + 6)$.

22. Graph the line $y + 4 = -\frac{2}{3}(x - 3)$.

23. Circle each equation below that represents a line that has slope 2.

   $y + 2 = -2(x - 7)$ $y = \frac{1}{2}x + 10$

   $y - 2x = 15$ $2x + 2y = 2$

24. Circle each equation below that represents a line that has slope −3.

   $y = -\frac{1}{3}x + 5$ $y - 6 = 3(x + 3)$

   $x + y = -3$ $3x + y = 9$

25. Circle each equation below that represents a line that has slope −1.

   $y = x - 1$ $2x + 2y = 14$

   $y - 5 = x + 3$ $y = -x + 2$

26. Graph the line that passes through the point (0, 3) with slope $m = 2$.

27. Graph the line with slope $m = -\dfrac{1}{2}$ that passes through the point (0, 1).

28. Graph the line that passes through the point (3, −1) with slope $m = \dfrac{4}{3}$.

Finding Equations

29. Find the point-slope form of the equation of the line that passes through the point (−3, 5) with slope 2.

30. Find the point-slope form of the equation of the line that passes through the point (2, −4) with slope $\dfrac{1}{4}$.

31. Find the equation of the horizontal line and the equation of the vertical line that pass through the point (−2, 6).

32. Find the equation of the horizontal line and the equation of the vertical line that pass through the point (0, 8).

33. Find the equation of the horizontal line and the equation of the vertical line that pass through the point (−3, 0).

34. Find the equation of the vertical line that crosses the x-axis 5 units to the right of the y-axis.

35. Find the equation of the horizontal line that crosses the y-axis 4 units below the x-axis.

36. Find the slope-intercept form of the equation of the line that passes through the point (3, 1) with slope −5.

37. Find the slope-intercept form of the equation of the line that passes through the point (−2, 2) with slope 4.

38. Find the slope-intercept form of the equation of the line that passes through the point (−2, 7) with slope $\dfrac{1}{2}$.

39. Find the slope-intercept form of the equation of the line that passes through the points (3, −2) and (1, 4).

40. Find the slope-intercept form of the equation of the line that passes through the points (3, −3) and (5, 1).

41. Find the slope-intercept form of the equation of the line that passes through the points (−6, 3) and (−3, 2).

42. Find the x-intercept of the line that passes through the point (2, 4) with slope −2.

43. Find the y-intercept of the line that passes through the point (3, 6) with slope $-\dfrac{1}{3}$.

44. Find the y-intercept of the line that passes through the point (−5, 2) with slope $-\dfrac{1}{5}$.

45. Find the standard form of the equation of the line that passes through the points (0, −2) and (4, −6).

46. Find the standard form of the equation of the line that passes through the points (0, 1) and (−5, −9).

47. Find the standard form of the equation of the line that passes through the points (3, 5) and (−6, 8).

48. Find the equation in slope-intercept form of the line that passes through (4, 2) and is perpendicular to $y = 2x$.

49. Find the equation in slope-intercept form of the line that passes through (5, −2) and is perpendicular to $y = \dfrac{1}{3}x + 7$.

50. Find the equation in slope-intercept form of the line that passes through (−4, −9) and is perpendicular to $y = -\dfrac{2}{5}x - 17$.

51. Find the equation in slope-intercept form of the line that passes through (−2, 2) and is parallel to $y = -3x$.

52. Find the equation in slope-intercept form of the line that passes through (6, 7) and is parallel to $y = \dfrac{2}{3}x + 10$.

53. Find the equation in slope-intercept form of the line that passes through (8, −3) and is parallel to $y = -\dfrac{1}{4}x - 15$.

54. Find the equation in slope-intercept form of the line that has x-intercept (4, 0) and y-intercept (0, −8).

55. Find the equation in slope-intercept form of the line that has x-intercept (−2, 0) and y-intercept (0, 6).

56. Find the equation in slope-intercept form of the line that has x-intercept (−6, 0) and y-intercept (0, −4).

 **EVALUATE**

Practice Test

Take this practice test to be sure that you are prepared for the final quiz in Evaluate.

1. Look at Figure EII.E.10. Determine in which quadrant the points P, Q, and R lie.

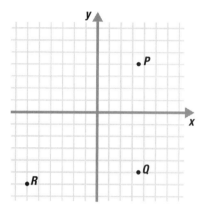

Figure EII.E.10

2. Find the distance, d, between the point (6, 11) and the point (−3, −1).

3. Find both the rise (change in y) and the run (change in x) between the points (2, 3) and (5, 1).

4. A line passes through the points (0, 6) and (−3, −3). Find its slope.

5. Circle the equation of the line through the point (20, −7) that has slope 9.

$$y - 7 = 9(x - 20)$$

$$y - 20 = 9(x + 7)$$

$$x + 7 = 9(y - 20)$$

$$y + 7 = 9(x - 20)$$

6. Find the y-intercept of the line that has slope $m = \frac{1}{2}$ and passes through the point $\left(-3, \frac{1}{2}\right)$.

7. Find the equation of the vertical line and the equation of the horizontal line that pass through the point (−11, 17).

8. Find the standard form of the equation of the line that passes through the point (0, 5) and is perpendicular to $y = -\frac{1}{8}x + 30$.

LESSON E II.F – ABSOLUTE VALUE

Here's what you'll learn in this lesson:

Solving Equations

a. Solving $|x| = a$

b. Solving $|Ax + B| = a$

c. Solving $|Ax + B| = |Cx + D|$

Solving Inequalities

a. Solving absolute value inequalities

Your favorite brand of candy is on sale, so you buy three bags. The labels on the bags say: "Contents: Approximately 25 pieces." When you count the pieces in each bag, you notice that every bag contains a different numbers of pieces, but the average number of pieces is 25.

To indicate how the number of pieces in each bag varies from the average, you can use absolute value.

In this lesson you will learn how to solve equations and inequalities involving absolute value.

 **EXPLAIN**

SOLVING EQUATIONS

Summary

Absolute Value

You have learned that the absolute value of a number is the distance of that number from 0 on the number line. Since distance is always a nonnegative number, the absolute value of a number is always nonnegative.

For example, the absolute value of 7, denoted by $|7|$, is 7, since the number 7 is a distance of 7 from 0 on the number line. Similarly, the absolute value of -7, denoted by $|-7|$, is also 7, since the number -7 is also a distance of 7 from 0 on the number line.

Solving Equations of the Form $|x| = a$

You can use what you know about absolute value to solve an equation that can be written in the form $|x| = a$. Here are the steps:

1. Write the equation in the form $|x| = a$.

2. Find the solutions based on the following:

    • If $a > 0$, the equation has two solutions, $x = -a$ or $x = a$.

    • If $a < 0$, the equation has no solutions.

    • If $a = 0$, the equation has one solution, $x = 0$.

For example, solve the equation $|x| = 6$:

1. The equation is in the form $|x| = a$. $|x| = 6$

2. Find the solutions. Since a is 6, $x = -6$ or $x = 6$
    $a > 0$, and the equation has two solutions.

So the solutions of $|x| = 6$ are $x = -6$ or $x = 6$.

You can check these solutions by substituting them into the original equation.

Check $x = -6$:   Check $x = 6$:

Is $|-6| = 6$?   Is $|6| = 6$?

Is $6 = 6$? Yes.   Is $6 = 6$? Yes.

As another example, to solve the equation $2|x| + 5 = 5$:

1. Write the equation $2|x| + 5 = 5$
   in the form $|x| = a$. $2|x| + 5 - 5 = 5 - 5$
                              $2|x| = 0$
                              $\dfrac{2|x|}{2} = \dfrac{0}{2}$
                              $|x| = 0$

2. Find the solutions. Since $a = 0$, $\qquad\qquad x = 0$
the equation has one solution.

So the solution of $2|x| + 5 = 5$ is $x = 0$.

You can check this solution by substituting it into the original equation.

Check $x = 0$:

Is $2|0| + 5 = 5$?

Is $2(0) + 5 = 5$?

Is $\quad 0 + 5 = 5$?

Is $\qquad 5 = 5$? Yes.

Similarly, solve the equation $|x| = -21$:

1. The equation is in the form $|x| = a$. $\qquad\qquad\qquad |x| = -21$

2. Find the solutions. Since a is -21, $a < 0$, $\qquad\qquad$ no solutions
and the equation has no solutions.

So there are no solutions of the equation $|x| = -21$.

Solving Equations of the Form $|ax + b| = c$

Here are the steps for solving an equation that can be written in the form $|ax + b| = c$, where $c \geq 0$:

1. Write the equation in the form $|ax + b| = c$.

You can use any variable, not just z, as the value to substitute for ax + b.

2. Substitute z for $ax + b$.

3. Solve the equation $|z| = c$ to get $z = -c$ or $z = c$.

4. Replace z with $ax + b$.

5. Solve for x.

For example, solve the equation $|3x| = 15$:

1. The equation is in the form $\qquad\qquad |3x| = 15$
$|ax + b| = c$. (Here, b is 0.)

2. Substitute z for $3x$. $\qquad\qquad\qquad\qquad |z| = 15$

3. Solve for z. $\qquad\qquad\qquad\qquad z = -15 \qquad$ or $\qquad z = 15$

4. Replace z with $3x$. $\qquad\qquad\qquad 3x = -15 \qquad\qquad\qquad 3x = 15$

5. Solve for x. $\qquad\qquad\qquad\qquad \dfrac{3x}{3} = \dfrac{-15}{3} \qquad\qquad \dfrac{3x}{3} = \dfrac{15}{3}$

$\qquad\qquad\qquad\qquad\qquad\qquad\qquad x = -5 \qquad$ or $\qquad x = 5$

So the solutions of the equation $|3x| = 15$ are $x = -5$ or $x = 5$.

You can check these solutions by substituting them into the original equation.

Check $x = -5$:

Is $|3(-5)| = 15$?

Is $|-15| = 15$?

Is $15 = 15$? Yes.

Check $x = 5$:

Is $|3(5)| = 15$?

Is $|15| = 15$?

Is $15 = 15$? Yes.

As a second example, solve $|2x - 3| = 9$:

1. The equation is in the form $|ax + b| = c$.

   $|2x - 3| = 9$

2. Substitute z for $2x - 3$.

   $|z| = 9$

3. Solve for z.

   $z = -9$ or $z = 9$

4. Replace z with $2x - 3$.

   $2x - 3 = -9$ or $2x - 3 = 9$

5. Solve for x.

   $2x - 3 + 3 = -9 + 3$ or $2x - 3 + 3 = 9 + 3$

   $2x = -6$ $2x = 12$

   $\dfrac{2x}{2} = \dfrac{-6}{2}$ $\dfrac{2x}{2} = \dfrac{12}{2}$

   $x = -3$ or $x = 6$

So the solutions of the equation $|2x - 3| = 9$ are $x = -3$ or $x = 6$.

You can check these solutions by substituting them into the original equation.

Check $x = -3$:

Is $|2(-3) - 3| = 9$?

Is $|-6 - 3| = 9$?

Is $|-9| = 9$?

Is $9 = 9$? Yes.

Check $x = 6$:

Is $|2(6) - 3| = 9$?

Is $|12 - 3| = 9$?

Is $|9| = 9$?

Is $9 = 9$? Yes.

Here's another example. To solve the equation $7|2x + 5| - 3 = 18$:

1. Write the equation in the form $|ax + b| = c$.

   $7|2x + 5| - 3 = 18$

   $7|2x + 5| - 3 + 3 = 18 + 3$

   $7|2x + 5| = 21$

   $\dfrac{7|2x + 5|}{7} = \dfrac{21}{7}$

   $|2x + 5| = 3$

2. Substitute z for $2x + 5$.

   $|z| = 3$

3. Solve for z.

   $z = -3$ or $z = 3$

4. Replace z with $2x + 5$. $2x + 5 = -3$ or $2x + 5 = 3$

5. Solve for x. $2x + 5 - 5 = -3 - 5$ or $2x + 5 - 5 = 3 - 5$

$$2x = -8 \qquad\qquad 2x = -2$$

$$\frac{2x}{2} = \frac{-8}{2} \qquad\qquad \frac{2x}{2} = \frac{-2}{2}$$

$$x = -4 \qquad \text{or} \qquad x = -1$$

So the solutions of the equation $7|2x + 5| - 3 = 18$ are $x = -4$ or $x = -1$.

You can check these solutions by substituting them into the original equation.

Check $x = -4$:

Is $7|2(-4) + 5| - 3 = 18$?

Is $7|-8 + 5| - 3 = 18$?

Is $7|-3| - 3 = 18$?

Is $7(3) - 3 = 18$?

Is $21 - 3 = 18$?

Is $18 = 18$? Yes.

Check $x = -1$:

Is $7|2(-1) + 5| - 3 = 18$?

Is $7|-2 + 5| - 3 = 18$?

Is $7|3| - 3 = 18$?

Is $7(3) - 3 = 18$?

Is $21 - 3 = 18$?

Is $18 = 18$? Yes.

Solving Equations of the Form $|ax + b| = |cx + d|$

Here are the steps for solving an equation that can be written in the form $|ax + b| = |cx + d|$:

1. Write the equation in the form $|ax + b| = |cx + d|$.

2. Substitute z for $ax + b$ and w for $cx + d$.

3. Solve the equation $|z| = |w|$ to get $z = w$ or $z = -w$.

4. Replace z with $ax + b$ and w with $cx + d$.

5. Solve for x.

Again, you can use any variables you like…you don't have to use z and w.

For example, solve the equation $|3x - 4| = |x + 8|$:

1. The equation is $|3x - 4| = |x + 8|$
   in the form
   $|ax + b| = |cx + d|$.

2. Substitute z for $3x - 4$ $|z| = |w|$
   and w for $x + 8$.

3. Solve for z. $z = w$ or $z = -w$

4. Replace z with $3x - 4$ $3x - 4 = x + 8$ or $3x - 4 = -(x + 8)$
   and w with $x + 8$.

5. Solve for x:

$$3x - 4 - x = x + 8 - x \quad \text{or} \quad 3x - 4 = -x - 8$$

$$2x - 4 = 8 \qquad\qquad 3x - 4 + x = -x - 8 + x$$

$$2x - 4 + 4 = 8 + 4 \qquad\qquad 4x - 4 = -8$$

$$2x = 12 \qquad\qquad 4x - 4 + 4 = -8 + 4$$

$$\frac{2x}{2} = \frac{12}{2} \qquad\qquad 4x = -4$$

$$\qquad\qquad \frac{4x}{4} = \frac{-4}{4}$$

$$x = 6 \qquad \text{or} \qquad x = -1$$

So the solutions of the equation $|3x - 4| = |x + 8|$ are $x = 6$ or $x = -1$.

You can check these solutions by substituting them into the original equation.

Check $x = 6$:

Is $|3(6) - 4| = |6 + 8|$?

Is $|18 - 4| = |14|$?

Is $|14| = 14$?

Is $14 = 14$? Yes.

Check $x = -1$:

Is $|3(-1) - 4| = |-1 + 8|$?

Is $|-3 - 4| = |7|$?

Is $|-7| = 7$?

Is $7 = 7$? Yes.

Sample Problems

1. Solve this equation: $|x| - 5 = 23 + 5$

☑ a. Write the equation in the form $|x| = a$.

$$|x| - 5 = 23 + 5$$
$$|x| - 5 = 28$$
$$|x| - 5 + 5 = 28 + 5$$
$$|x| = 33$$

☐ b. Find the solutions. Since a is 33, $a > 0$, and the equation has two solutions. $x = \underline{\quad}$ or $x = \underline{\quad}$

☐ c. Check the solutions in the original equation.

2. Solve this equation: $3|x + 9| - 7 = 24 - 28$

☑ a. Write the equation in the form $|ax + b| = c$.

$$3|x + 9| - 7 = 24 - 28$$
$$3|x + 9| - 7 = -4$$
$$3|x + 9| - 7 + 7 = -4 + 7$$
$$3|x + 9| = 3$$
$$\frac{3|x + 9|}{3} = \frac{3}{3}$$
$$|x + 9| = 1$$

b. 1

c. −1, 1

d. −1, 1

e. −10, −8

f. Here's one way to check.

Check x = −10:

Is 3 |−10 + 9| − 7 = 24 − 28?

Is   3|−1| − 7 = −4   ?

Is   3(1) − 7 = −4   ?

Is   3 − 7 = −4   ?

Is   −4 = −4   ? Yes.

Check x = −8:

Is 3 |−8 + 9| − 7 = 24 − 28?

Is   3|1| − 7 =   −4?

Is   3(1) − 7 = −4   ?

Is   3 − 7 = −4   ?

Is   −4 = −4   ? Yes.

c. w, −w

d. 2x − 5

f. Here's one way to check.

Check x = 0:

Is |2(0) + 5| = |2(0) − 5|?

Is |0 + 5| = |0 − 5| ?

Is |5| = |− 5| ?

Is   5 = 5   ? Yes.

☐ b. Substitute z for $x + 9$. $|z| = $ ____

☐ c. Solve for z. $z = $ ____ or $z = $ ____

☐ d. Replace z with $x + 9$. $x + 9 = $ ____ or $x + 9 = $ ____

☐ e. Solve for x. $x = $ _____ or $x = $ _____

☐ f. Check the solutions in
the original equation.

3. Solve this equation: $|2x + 5| = |2x − 5|$

☑ a. The equation is in the $|2x + 5| = |2x − 5|$
form $|ax + b| = |cx + d|$.

☑ b. Substitute z for $2x + 5$ $|z| = |w|$
and w for $2x − 5$.

☐ c. Solve for z. $z = $ ____   or $z = $ ____

☐ d. Replace z with $2x + 5 = 2x − 5$ or $2x + 5 = −($____$)$
$2x + 5$ and w
with $2x − 5$.

☑ e. Solve for x.

$$2x + 5 = 2x − 5 \quad \text{or} \quad 2x + 5 = −(2x − 5)$$
$$2x + 5 − 2x = 2x − 5 − 2x \qquad 2x + 5 = −2x + 5$$
$$5 = −5 \qquad 2x + 5 − 5 = −2x + 5 − 5$$
$$2x = −2x$$
$$2x + 2x = −2x + 2x$$
$$4x = 0$$
$$\frac{4x}{4} = \frac{0}{4}$$
$$x = 0$$

Since 5 ≠ −5, there is only one solution, $x = 0$.

☐ f. Check the solution
in the original equation.

SOLVING INEQUALITIES

Summary

Solving Inequalities of the Form $|x| < a$ or $|x| \leq a$

Recall that the absolute value of x, denoted by $|x|$, is the distance of x from 0 on the number line. You can use this fact to solve inequalities of the form $|x| < a$ (where $a > 0$) or $|x| \leq a$ (where $a \geq 0$).

To solve an inequality that can be written in the form $|x| < a$ or $|x| \leq a$.

1. Write the inequality of the form $|x| < a$ or $|x| \leq a$.

2. Find the solution based on the following:

    • If $|x| < a$, then the solution is all x such that $-a < x < a$.

    • If $|x| \leq a$, then the solution is all x such that $-a \leq x \leq a$.

For example, solve the inequality $|x| < 8$:

1. The inequality is in the form $|x| < a$. $|x| < 8$

2. Find the solution. Here, $a = 8$. $-8 < x < 8$

So the solution of $|x| < 8$ is all x such that $-8 < x < 8$. This solution consists of all numbers whose distance from 0 is less than 8 on the number line.

Many numbers are part of the solution of this inequality. Here, two of these numbers have been checked.

Check $x = -3.5$:   Check $x = 5$:

Is $|-3.5| < 8$?   Is $|5| < 8$?

Is $3.5 < 8$? Yes.   Is $5 < 8$? Yes.

Solving Inequalities of the Form $|x| > a$ or $|x| \geq a$

Here are the steps to solve inequalities that can be written in the form $|x| > a$ or $|x| \geq a$, where $a \geq 0$:

1. Write the inequality in the form $|x| > a$ or $|x| \geq a$.

2. Find the solution based on the following:

    • If $|x| > a$, then the solution is all x such that $x < -a$ or $x > a$.

    • If $|x| \geq a$, then the solution is all x such that $x \leq -a$ or $x \geq a$.

*When you write $-3 < x < 3$, read this $-3 < x$ **and** $x < 3$.*

*Remember, the open circles on this number line are used to indicate that the numbers -8 and 8 are **not** included in the solution.*

For example, solve the inequality $|x| \geq 3$:

1. The inequality is in the form $|x| \geq a$. $|x| \geq 3$

2. Find the solution. Here, $a = 3$. $x \leq -3$ or $x \geq 3$

So the solution of $|x| \geq 3$ is all x such that $x \leq -3$ or $x \geq 3$. This solution consists of all numbers whose distance from 0 is greater than 3 on the number line.

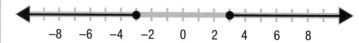

Many numbers are part of the solution of this inequality. Here, two of these numbers have been checked.

Check $x = -3$:   Check $x = 7$:

Is $|-3| \geq 3$?   Is $|7| \geq 3$?

Is $3 \geq 3$? Yes.   Is $7 \geq 3$? Yes.

Solving Inequalities of the Form $|ax + b| < c$ or $|ax + b| \leq c$

Just as you do when you solve equations involving absolute value, you can use substitution to solve inequalities involving absolute value.

To solve an inequality that can be written in the form $|ax + b| < c$ (where $c > 0$) or $|ax + b| \leq c$ (where $c \geq 0$):

1. Write the inequality in the form $|ax + b| < c$ or $|ax + b| \leq c$.

2. Substitute w for $ax + b$.

3. Solve the inequality $|w| < c$ or solve the inequality $|w| \leq c$.

4. Replace w with $ax + b$.

5. Solve for x.

For example, solve the inequality $|x + 5| < 10$:

1. The inequality is in the form $|ax + b| < c$. $|x + 5| < 10$

2. Substitute w for $x + 5$. $|w| < 10$

3. Solve the inequality $|w| < 10$. $-10 < w < 10$

4. Replace w with $x + 5$. $-10 < x + 5 < 10$

5. Solve for x. $-10 - 5 < x + 5 - 5 < 10 - 5$

     $-15 < x < 5$

So the solution of the inequality $|x + 5| < 10$ is all x such that $-15 < x < 5$. This solution can be graphed on a number line:

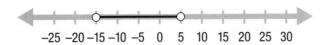

Many numbers are part of the solution of this inequality. Here, two of these numbers have been checked.

Check $x = 2.5$:

Is $|2.5 + 5| < 10$?

Is $|7.5| < 10$?

Is $7.5 < 10$? Yes.

Check $x = -7.5$:

Is $|-7.5 + 5| < 10$?

Is $|-2.5| < 10$?

Is $2.5 < 10$? Yes.

Solving Inequalities of the Form $|ax + b| > c$ or $|ax + b| \geq c$

To solve an inequality that can be written in the form $|ax + b| > c$ or $|ax + b| \geq c$, where $c \geq 0$:

1. Write the inequality in the form $|ax + b| > c$ or $|ax + b| \geq c$.

2. Substitute w for $ax + b$.

3. Solve the inequality $|w| > c$ or solve the inequality $|w| \geq c$.

4. Replace w with $ax + b$.

5. Solve for x.

For example, solve the inequality $|2x + 3| \geq 11$:

1. The inequality is in the form $|ax + b| \geq c$. $|2x + 3| \geq 11$

2. Substitute w for $2x + 3$. $|w| \geq 11$

3. Solve the inequality $|w| \geq 11$. $w \leq -11$ or $w \geq 11$

4. Replace w with $2x + 3$. $2x + 3 \leq -11$ or $2x + 3 \geq 11$

5. Solve for x.
$$2x + 3 - 3 \leq -11 - 3 \quad \text{or} \quad 2x + 3 - 3 \geq 11 - 3$$
$$2x \leq -14 \qquad\qquad\qquad 2x \geq 8$$
$$\frac{2x}{2} \leq \frac{-14}{2} \qquad\qquad\qquad \frac{2x}{2} \geq \frac{8}{2}$$
$$x \leq -7 \quad \text{or} \quad x \geq 4$$

So the solution of the inequality $|2x + 3| \geq 11$ is all x such that $x \leq -7$ or $x \geq 4$. This solution can be graphed on a number line:

Many numbers are part of the solution of this inequality. Here, two of these numbers have been checked.

Check $x = -8.3$:

Is $|2(-8.3) + 3| \geq 11$?

Is $|-16.6 + 3| \geq 11$?

Is $\quad |-13.6| \geq 11$?

Is $\qquad 13.6 \geq 11$? Yes.

Check $x = 5$

Is $|2(5) + 3| \geq 11$?

Is $|10 + 3| \geq 11$?

Is $\quad |13| \geq 11$?

Is $\qquad 13 \geq 11$? Yes.

Sample Problems

1. Solve this inequality: $|x| \leq 2.5$

   ☑ a. The inequality is in the form $|x| \leq a$. $\qquad |x| \leq 2.5$

   ☑ b. Find the solution. Here, $a = 2.5$. The $\qquad -2.5 \leq x \leq 2.5$
   solution is all numbers x whose distance
   from 0 is less than or equal to 2.5.

   ☐ c. Check two of the numbers that are
   part of the solution, $x = -1$ and $x = 2.5$.

2. Solve this inequality: $2|x| - 5 > 3$

   ☐ a. Write the inequality in $\qquad 2|x| - 5 > 3$
   the form $|x| > a$. $\qquad 2|x| - 5 + 5 > 3 + 5$
   $\qquad\qquad 2|x| > 8$

   $\qquad\qquad \dfrac{2|x|}{2} > \dfrac{8}{2}$

   $\qquad\qquad |x| > \underline{\qquad}$

   ☐ b. Find the solutions. $\underline{\qquad\qquad\qquad}$

   ☑ c. Check two of the Check $x = -5$:
   numbers that are Is $2|-5| - 5 > 3$?
   part of the solution, Is $\quad 2(5) - 5 > 3$?
   $x = -5$ and $x = 7$. Is $\quad\; 10 - 5 > 3$?
   Is $\qquad 5 > 3$? Yes.

   Check $x = 7$:
   Is $2|7| - 5 > 3$?
   Is $\quad 2(7) - 5 > 3$?
   Is $\quad\; 14 - 5 > 3$?
   Is $\qquad 9 > 3$? Yes.

3. Solve this inequality: $|7 - 3x| \leq 28$

   ☑ a. The inequality is in the form $|ax + b| \leq c$. $|7 - 3x| \leq 28$

✓ b. Substitute w for $7 - 3x$. $|w| \le 28$

☐ c. Solve the inequality $|w| \le 28$. _____

☐ d. Replace w with $7 - 3x$. _____

☐ e. Solve for x. (Remember to reverse the _____
direction of the inequality sign when you divide
both sidesof an inequality by a negative number.)

e. Here's a way to solve for x:
$$-28 \le 7 - 3x \le 28$$
$$-28 - 7 \le 7 - 3x - 7 \le 28 - 7$$
$$-35 \le -3x \le 21$$
$$\frac{-35}{-3} \ge \frac{-3x}{-3} \ge \frac{21}{-3}$$
$$\frac{35}{3} \ge x \ge -7$$

✓ f. Check two of the
numbers that are
part of the solution,
$x = -2$ and $x = 10$.

Check $x = -2$:
Is $|7 - 3(-2)| \le 28$?
Is $|7 + 6| \le 28$?
Is $|13| \le 28$?
Is $13 \le 28$? Yes.

Check $x = -10$:
Is $|7 - 3(10)| \le 28$?
Is $|7 - 30| \le 28$?
Is $|-23| \le 28$?
Is $23 \le 28$? Yes.

4. Solve this inequality: $2|3x + 7| - 10 \ge 74$

✓ a. Write the inequality
in the form $|ax + b| \ge c$.

$$2|3x + 7| - 10 \ge 74$$
$$2|3x + 7| - 10 + 10 \ge 74 + 10$$
$$2|3x + 7| \ge 84$$
$$\frac{2|3x + 7|}{2} \ge \frac{84}{2}$$
$$|3x + 7| \ge 42$$

d. $3x + 7 \le -42$ or $3x + 7 \ge 42$

e. Here's a way to solve for x:
$$3x + 7 \le -42 \quad or \quad 3x + 7 \ge 42$$
$$3x + 7 - 7 \le -42 - 7 \quad 3x + 7 - 7 \ge 42 - 7$$
$$3x \le -49 \qquad 3x \ge 35$$
$$\frac{3x}{3} \le \frac{49}{3} \qquad \frac{3x}{3} \ge \frac{35}{3}$$
$$x \le -\frac{49}{3} \quad or \quad x \ge \frac{35}{3}$$

✓ b. Substitute w for $3x + 7$. $|w| \ge 42$

☐ c. Solve the inequality $|w| \ge 42$. _____

☐ d. Replace w with $3x + 7$. _____

☐ e. Solve for x.

_____

f. Here's one way to check.
Check $x = -17$:
Is $2|3(-17) + 7| - 10 \ge 74$?
Is $2|-51 + 7| - 10 \ge 74$?
Is $2|-44| - 10 \ge 74$?
Is $2(44) - 10 \ge 74$?
Is $88 - 10 \ge 74$?
Is $78 \ge 74$? Yes.

☐ f. Check two of the _____ or _____
numbers that are
part of the solution,
$x = -17$ and $x = 12$.

Check $x = 12$:
Is $2|3(12) + 7| - 10 \ge 74$?
Is $2|36 + 7| - 10 \ge 74$?
Is $2|43| - 10 \ge 74$?
Is $2(43) - 10 \ge 74$?
Is $86 - 10 \ge 74$?
Is $76 \ge 74$? Yes.

 **HOMEWORK**

Homework Problems

Circle the homework problems assigned to you by the computer, then complete them below.

 Explain

Solving Equations

Solve the equations in problems (1) – (8) for x:

1. $|x| = 100$

2. $3|x| = 51$

3. $|x| + 10 = 7$

4. $|x| - 23 = 5 - 28$

5. $|x - 5| = 27$

6. $2|3x - 7| = 42$

7. $|5x - 6| + 71 = 72$

8. $4|3x + 8| - 18 = 14$

9. To win a prize in a contest, a contestant must guess how many jelly beans are in a jar. The jar contains 457 jelly beans. If a contestant's guess is within 15 of the actual number of jelly beans, the contestant wins a prize. Write an absolute value equation to represent the highest and lowest guesses that will receive a prize. Solve your equation to find the highest and lowest possible guesses.

10. The following formula is used to calculate percent error in a scientific experiment: $E = \frac{|a - e|}{e}$. Use this formula to find a if e is 0.156 and E is 0.1.

11. Solve for x: $|x - 8| = |x|$

12. Solve for x: $|2x + 7| = |3x - 5|$

Solving Inequalities

Solve the inequalities in problems (13) – (20) for x:

13. $|x| \geq 24$

14. $|x| < 8.27$

15. $|x| - 5 \leq -4$

16. $2|x| > 15$

17. $|x - 7.2| < 18.7$

18. $|2x + 5| \leq 21$

19. $3|x - 8| \geq 48$

20. $|5x + 9| \leq 100.5$

21. The absolute value of 5 less than 3 times a number is greater than 23. Find all possible numbers that satisfy this statement.

22. A sawmill produces 8 ft. long wall studs. The maximum desirable percent error for the lengths of the studs is 0.005. What range of lengths of studs is allowable.

    To answer this question, use the formula $E \geq \frac{|a - e|}{e}$. Let $E = 0.005$ and $e = 8$. Solve the inequality for a.

23. $3|4x + 2.4| - 9 < 23 - 26$

24. $|7 - 6x| > 42$

APPLY

Practice Problems

Here are some additional practice problems for you to try.

Solving Equations

1. Solve for y: $|y| = 128$

2. Solve for x: $|x| = -4$

3. Solve for x: $|x| = 250$

4. Solve for y: $4|y| = 56$

5. Solve for x: $-3|x| = -27$

6. Solve for x: $3|x| = 33$

7. Solve for x: $|x| + 5 = 26$

8. Solve for y: $|y| + 21 = 20$

9. Solve for x: $2|x| - 16 = 14$

10. Solve for y: $|y + 7| = 34$

11. Solve for x: $|3x + 6| = 21$

12. Solve for x: $|x - 6| = 48$

13. Solve for y: $3|2y - 5| = 39$

14. Solve for x: $4|3x - 3| = 60$

15. Solve for y: $-2|5y + 5| = 50$

16. Solve for y: $5|3y - 9| + 8 = 53$

17. Solve for x: $-3|4x + 8| + 38 = 2$

18. Solve for x: $4|2x + 5| - 6 = 22$

19. Solve for y: $\frac{1}{3}|3y - 3| + 13 = 25$

20. Solve for x: $\frac{2}{5}|5x + 15| - 24 = 36$

21. Solve for x: $\frac{1}{2}|2x - 4| - 19 = 5$

22. Solve for x: $|x - 8| = |x|$

23. Solve for y: $|y| = |12 - y|$

24. Solve for y: $|3y - 2| = |4y + 9|$

25. Solve for x: $|5x + 1| = |7x - 1|$

26. Solve for x: $|6x - 1| = |9 - 4x|$

27. Solve for y: $5|4y - 4| = 120$

28. Solve for x: $4|2x + 3| = 60$

Solving Inequalities

29. Solve for y: $|y| < 7$

30. Solve for x: $|x| \le 3$

31. Solve for x: $|3x| < 12$

32. Solve for y: $|y| \ge 23$

33. Solve for x: $|x| > 1$

34. Solve for x: $|5x| > 95$

35. Solve for y: $4|y| \le 36$

36. Solve for x: $3|x| \ge 45$

37. Solve for x: $5|3x| > 45$

38. Circle the inequalities below which have no solution.

$$|y| - 18 < -18$$
$$|y + 7| \ge -4$$
$$8|y| - 36 \le 27$$
$$-3|y| - 10 < 8$$

39. Circle the inequalities below which have no solution.

$$|x| + 9 < 9$$
$$|x + 2| < -3$$
$$7|x| < 99$$
$$-|2x + 9| > 0$$

40. Solve for y: $5 - 4|2y - 3| \le -39$

41. Solve for x: $4 + 3|4x - 2| > 22$

42. Solve for x: $-7 + 3|5x - 10| > 38$

43. Solve for y: $12 + 3|y - 6| \le 48$

44. Solve for x: $-3 + 2|x + 4| \le 7$

45. Solve for x: $-8 + 4|3x + 6| < 28$

46. Solve for y: $|12 - 6y| \le 48$

47. Solve for x: $|8 - 4x| < 16$

48. Solve for x: $2|14 - 7x| < 56$

49. Solve for y: $|8 - y| > 8$

50. Solve for y: $|5 - 2y| \ge 9$

51. Solve for x: $5|18 - 6x| > 30$

52. Find the inequality whose solution is graphed below.

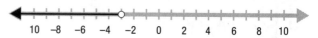

53. Find the inequality whose solution is graphed below.

54. Find the inequality whose solution is graphed below.

55. Find the inequality whose solution is graphed below.

56. Find the inequality whose solution is graphed below.

EVALUATE

Practice Test

Take this practice test to be sure that you are prepared for the final quiz in Evaluate.

1. Find the solutions of the following equations.

    a. $|x - 5| = 9$ b. $|8x| = 24$

2. Solve for x: $|x + 3| - 8 = 19$

3. Solve for x: $|2x + 5| = |x + 7|$

4. Circle the solution of this equation: $5|4x - 7| + 12 = 7$

    $x = 2$ or $x = 3$

    $x = -2$ or $x = -3$

    $x = 2$ or $x = -3$

    $x = -2$ or $x = 3$

    the equation has no solution

5. Circle the graph that represents the solution of this inequality: $|x - 5| \le 7$

6. Solve for x: $|4x - 6| > 18$

7. Circle the inequality whose solution is graphed below.

    $|x| \ge 4$

    $|x| \le 4$

    $|x| > 4$

    $|x| \le 4$

8. Solve for x: $|-4x - 4| < 16$

CUMULATIVE REVIEW PROBLEMS

These problems combine all of the material you have covered so far in this course. You may want to test your understanding of this material before you move on to the next topic. Or you may wish to do these problems to review for a test.

1. Which of the following is **not** a real number:

   $-\frac{3}{8}, 0, \sqrt{(-3) \cdot 5}, 42, \sqrt{6}$

2. Factor: $3x^2 + 21xy - x - 7y$

3. Find: $\frac{5x^2y^8}{2z} \cdot \frac{12x^2z}{y^6}$

4. Solve for x: $\frac{x-7}{x-2} - \frac{x}{x+7} = \frac{x+12}{x^2+5x-14}$

5. Solve for x: $\frac{4+3x}{2} > \frac{3}{4}(x - \frac{1}{3}) > \frac{5}{12} + \frac{3}{2}x$

6. Solve for x: $|3x + 7| = 12$

7. Find: $3 + 3[2^2 - 2(8+1)]$

8. Find the equation of the line through the point $(-3, -3)$ that is perpendicular to the line $y = 4x + 12$. Write your answer in slope-intercept form.

9. Find the reciprocal of $3 \cdot \left(\frac{x+y}{2x^2+4y^3}\right)$.

10. Find the slope m of the line through the points $(11, -4)$ and $(-3, 6)$.

11. Find the coordinates of the points in Figure EII.1.

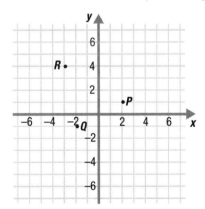

Figure EII.1

12. Solve for x: $\frac{1}{x^2 + x - 12} = \frac{3}{3x^2 + 12x}$

13. Solve for x: $\frac{x}{5} = \frac{x}{25} - \frac{1}{30}$

14. Solve for x: $|x| < 11$

15. Find the degree of this polynomial:

    $17x^6y^2 + 3y^4 - 8x^2y + 11x^3y^6 - 4x + 1$

16. Find the point-slope form of the equation of the line through the point $(-1, 2)$ with slope 3.

17. Simplify using the properties of exponents: $\left[\frac{(2x)^4}{(3y)^2}\right]^3 \cdot x^5y^6$

18. Reduce to lowest terms and write using positive exponents:
    $\frac{45x^{-3}y^{-3}z^{-4}}{9x^5y^{-2}z}$

19. Factor: $y^4 - 49$

20. Find: $\frac{48x^9 - 3x^4}{12x^3}$

21. Solve for x: $y = mx + b$

22. Find: $\frac{x^3 + 8}{32xyz^2} + \frac{x^2 - 4}{32xyz^2}$

23. Solve for x: $|x + 2| = |11x - 4|$

24. Factor: $28x^5y^2 - 12x^5y^4 + 20x^4y^3$

25. Solve for z: $-5(4z - 3) > 24 - 9z$

26. Simplify using the properties of exponent: $\frac{x^{17}}{x^{15}}$

27. Find: $\frac{6x^3}{2y^2} + \frac{13y^9}{x^2}$

28. Solve for x: $3|x - 2| \geq 12$

29. Reduce to lowest terms: $\frac{6xy^2 + 4y - 15xy - 10}{3xy^2 + 2y + 24xy + 16}$

30. Find:

$(6a^2 + 7ab + 6ab^2 + b^2 - 3) - (5b^2 - 17 - 4a^2b + 5ab)$

31. Solve for z: $-4 < 2(z - 5) \le 10$

32. Find: $\dfrac{x^2 + 2x - 15}{x^2 - 4} \div \dfrac{4x + 20}{3x - 6}$

33. Factor: $8x^2 + 2x - 21$

34. Solve for y: $\dfrac{2}{3}(y + 2) = \dfrac{1}{8}y + \dfrac{5}{6}$

35. Find the distance d between the points $(1, 3)$ and $(-4, 7)$.

36. Graph the solution to the inequality $-2|3x - 3| > -6$ on a number line.

37. Simplify: $\dfrac{\frac{x^4}{10}}{\frac{x^2}{5}}$

38. Evaluate when $y = -3$: $-2y^2 - 9y - 9$

39. Rewrite using the distributive property: $7 \cdot (12 + 33)$

40. Factor: $90x^3 - 25x^2 - 35x$

41. Simplify: $\dfrac{-3 + \frac{1}{x}}{-2 - \frac{1}{2x}}$

42. Simplify: $\dfrac{8b}{2a^2 + 5a - 3} - \dfrac{4}{a + 3} + \dfrac{9}{2a - 1}$

43. Find the distance d between the points $(1, 15)$ and $(1, -8)$.

44. Graph the line $4x + y = 12$.

45. Solve for x: $\dfrac{2x}{5} + 7 = \dfrac{8}{3}$

46. Factor: $16a^2 - b^8$

47. Find the equations of the vertical line and the horizontal line that pass through the point $(7, 1)$.

48. Solve for x: $|x| = 99$

49. Find: $(6x - 9)(3x + 5)$

50. Find the x- and y-intercepts of the line $y - 2 = 5(x + 1)$.

LESSON 6.1 – EXPONENTS

OVERVIEW

Here's what you'll learn in this lesson:

Properties of Exponents

a. *Definition of exponent, power, and base*

b. *Multiplication Property*

c. *Division Property*

d. *Powers raised to powers*

e. *Products raised to powers*

f. *Quotients raised to powers*

g. *The zero exponent*

If a friend agrees to give you one penny today, two pennies tomorrow, four pennies the next day, eight after that, and every day thereafter gives you double the amount he gave you the day before, by the end of a week he will have given you $1.27, and by the end of a month, $10,737,418.23!

While doubling your money every day may not be something you can relate to, it is an example of a situation that can be described using exponents. Another, perhaps more relevant example is the growth of your money in an account earning compound interest.

In this lesson, you will learn some general properties of exponents. These properties will help you later when you learn to simplify expressions involving exponents.

☀ EXPLAIN

PROPERTIES OF EXPONENTS

Summary

You have seen how to use exponential notation as a shortcut for writing multiplication of repeated factors. Now you'll learn some basic properties involving exponents.

Multiplication Property

There is a rule for multiplying exponential expressions with the same base.

For example, find $3^2 \cdot 3^3$:

$$3^2 \cdot 3^3 = 3 \cdot 3 \cdot 3 \cdot 3 \cdot 3 = 3^{2+3} = 3^5$$

In general:

$$x^m \cdot x^n = x^{m+n}$$

Here, m and n are positive integers.

This property is called the multiplication property of exponents.

Division Property

There are also rules for dividing two expressions with the same base.

For example, find $\frac{5^7}{5^4}$:

$$\frac{5^7}{5^4} = \frac{\overset{1}{\cancel{5}} \cdot \overset{1}{\cancel{5}} \cdot \overset{1}{\cancel{5}} \cdot \overset{1}{\cancel{5}} \cdot 5 \cdot 5 \cdot 5}{\underset{1}{\cancel{5}} \cdot \underset{1}{\cancel{5}} \cdot \underset{1}{\cancel{5}} \cdot \underset{1}{\cancel{5}}} = 5^{7-4} = 5^3$$

In general:

$$\frac{x^m}{x^n} = x^{m-n}$$

Here, $x \neq 0$, m and n are positive integers, and $m > n$.

You can also divide expressions where the exponent in the denominator is greater than the exponent in the numerator.

For example, find $\frac{5^4}{5^7}$:

$$\frac{5^4}{5^7} = \frac{\overset{1}{\cancel{5}} \cdot \overset{1}{\cancel{5}} \cdot \overset{1}{\cancel{5}} \cdot \overset{1}{\cancel{5}}}{\underset{1}{\cancel{5}} \cdot \underset{1}{\cancel{5}} \cdot \underset{1}{\cancel{5}} \cdot \underset{1}{\cancel{5}} \cdot 5 \cdot 5 \cdot 5} = \frac{1}{5^{7-4}} = \frac{1}{5^3}$$

To multiply two expressions with the same base, add their exponents.

To divide two expressions with the same base, subtract their exponents.

In general:

$$\frac{x^m}{x^n} = \frac{1}{x^{n-m}}$$

Here, $x \neq 0$, m and n are positive integers, and $m < n$.

These are called the division properties of exponents.

Power of a Power Property

There is a rule for raising an exponential expression to a power.

For example, find $(2^4)^3$:

$$(2^4)^3 = 2 \cdot 2 \cdot 2 \cdot 2 \quad \cdot \quad 2 \cdot 2 \cdot 2 \cdot 2 \quad \cdot \quad 2 \cdot 2 \cdot 2 \cdot 2 = 2^{4 \cdot 3} = 2^{12}$$

When an expression raised to a power is itself raised to a power, multiply the exponents.

In general:

$$(x^m)^n = x^{m \cdot n}$$

This is called the power of a power property of exponents.

Power of a Product Property

There is a rule for raising a product to a power.

Here you could have done the multiplication first and then applied the exponent: $(4 \cdot 5)^2 = (20)^2 = 400$. This may seem easier, but when you have an expression that includes variables as well as numbers, the other way is more useful.

For example, find $(4 \cdot 5)^2$:

$$(4 \cdot 5)^2 = (4 \cdot 5) \cdot (4 \cdot 5)$$
$$= 4 \cdot 5 \cdot 4 \cdot 5$$
$$= 4 \cdot 4 \cdot 5 \cdot 5$$
$$= 4^2 \cdot 5^2$$
$$= 16 \cdot 25$$
$$= 400$$

In general:

$$(xy)^n = x^n \cdot y^n$$

To raise a product to a power, raise each factor to the power.

This is called the power of a product property of exponents.

Power of a Quotient Property

There is a rule for raising a quotient to a power.

For example, find $\left(\frac{2}{5}\right)^3$:

$$\left(\frac{2}{5}\right)^3 = \frac{2}{5} \cdot \frac{2}{5} \cdot \frac{2}{5}$$

$$= \frac{2 \cdot 2 \cdot 2}{5 \cdot 5 \cdot 5}$$

$$= \frac{2^3}{5^3}$$

$$= \frac{8}{125}$$

In general:

$$\left(\frac{x}{y}\right)^n = \frac{x^n}{y^n}$$

Here, $y \neq 0$.

This is called the power of a quotient property of exponents.

Zero Power Property

There is a rule for raising a nonzero quantity to the zero power.

You know that:

$$\frac{x^m}{x^n} = x^{m-n} \text{ when } x \neq 0 \text{ and } m > n$$

and

$$\frac{x^m}{x^n} = \frac{1}{x^{n-m}} \text{ when } x \neq 0 \text{ and } m < n$$

But what happens when $m = n$?

For example, find $\frac{3^2}{3^2}$ in two ways:

1. Calculate the value of the numerator and denominator.
$$\frac{3^2}{3^2} = \frac{3 \cdot 3}{3 \cdot 3} = \frac{9}{9} = 1$$

2. Use the division property of exponents.
$$\frac{3^2}{3^2} = 3^{2-2} = 3^0$$

So, $3^0 = 1$.

In general:

$$x^0 = 1$$

Here, $x \neq 0$.

This is called the zero power property of exponents.

To raise a quotient to a power, raise the numerator and the denominator to the power.

Any nonzero number raised to the zero power is equal to 1.

Sample Problems

1. Find: $b^3 \cdot b^8 \cdot b^6$

    ☑ a. Add the exponents.
            The base stays the same. $= b^{3+8+6}$

                                                            $= b^{17}$

2. Find: $\dfrac{x^9}{x^5}$

    ☐ a. Subtract the exponent in the
            denominator from the exponent
            in the numerator. The base

            stays the same. $=$ _____

3. Find: $(3^4)^6$

    ☐ a. Use the power of a power
            property to multiply $= 3^{4 \cdot 6}$

            the exponents. $=$ _____

4. Find: $(b^2 \cdot c)^5$

    ☐ a. Use the power of a product
            property to raise each factor

            to the 5th power. $=$ _____

    ☐ b. Use the power of a power
            property to multiply the

            exponents. $=$ _____

5. Find: $\left(\dfrac{x}{y^3}\right)^2$

    ☐ a. Use the power of a quotient
            property to raise the numerator

            and denominator to the 2nd power. $=$ _____

    ☐ b. Use the power of a power

            property to multiply the exponents. $=$ _____

6. Find: $(3x^2y^5)^0$

    ☐ a. Use the zero power property

            to evaluate this expression. $=$ _____

 **HOMEWORK**

Homework Problems

Circle the homework problems assigned to you by the computer, then complete them below.

 Explain
Properties of Exponents

Use the appropriate properties of exponents to simplify the expressions in problems 1 through 12. (Keep your answers in exponential form where possible.)

1. Find:

    a. $3^2 \cdot 3^5$ b. $5^2 \cdot 5^5$

    c. $7^2 \cdot 7^5$

2. Find:

    a. $\dfrac{3^9}{3^5}$ b. $\dfrac{3^5}{3^9}$

    c. $\dfrac{3^9}{3^9}$

3. Find:

    a. $(7^3)^2$ b. $(7^2)^3$

4. Find:

    a. $(5 \cdot x)^3$ b. $(3 \cdot y)^2$

    c. $(a^2 \cdot b)^4$

5. Find:

    a. $\left(\dfrac{x^3 \cdot x^5}{x^4}\right)^2$ b. $\left(\dfrac{a^{12} \cdot a^6}{a^9 \cdot a^7}\right)^4$

    c. $\left(\dfrac{b^6 \cdot b^5}{b^3 \cdot b^8}\right)^3$ d. $\dfrac{2^3 \cdot x^5}{2^5 \cdot x^2}$

6. Find:

    a. $(a^2 \cdot a^3)^2 + (a^2 \cdot a^3)^2$

    b. $\dfrac{y^4 \cdot 3y^2}{y^8}$

    c. $x^4 \cdot x^9 \cdot x \cdot y^5 \cdot y^{11}$

7. Find:

    a. $(b^3)^2 \cdot (b^4)^3$

    b. $\dfrac{y^6}{y^{17}} \cdot (y^5)^2 \cdot (y^3)^4$

    c. $\dfrac{a^4 \cdot b^6}{a^{11} \cdot b^3}$

8. Find:

    a. $\dfrac{(xy)^4}{y^9 \cdot x^7}$ b. $\dfrac{(3b)^6}{(3b^2)^4}$

9. As animals grow, they get taller faster than they get stronger. In general, this proportion of increase in height to increase in strength can be written as $\dfrac{x^2}{x^3}$. Simplify this fraction.

10. An animal is proportionally stronger the smaller it is. If a person is 200 times as tall as an ant, figure out how much stronger a person is, pound for pound, by simplifying the expression $\dfrac{200^2}{200^3}$.

11. Find:

    a. $\left(\dfrac{4xy^2z}{5x^2yz^3}\right)^0$ b. $\dfrac{y^7 \cdot y}{y^9 \cdot y^2}$

    c. $\left(\dfrac{b^3 \cdot b^5}{b^6 \cdot b^3}\right)^4$ d. $-2x^0 + 4y^0$

12. Find:

    a. $\left(\dfrac{(x^3 \cdot x^4)^2}{x^7}\right)^5$ b. $\dfrac{(4a^2)^0 - 3b^0}{2}$

    c. $\left(\dfrac{(3x \cdot 3x^2)^2}{3^{11} \cdot x^7}\right)^3$ d. $\left(\dfrac{b^8}{(b^2 \cdot b^7)^3}\right)^4$

 APPLY

Practice Problems

Here are some additional practice problems for you to try.

Properties of Exponents

1. Find: $7^5 \cdot 7^3$. Leave your answer in exponential notation.

2. Find: $6^3 \cdot 6^4$. Leave your answer in exponential notation.

3. Find: $b^{12} \cdot b^3$

4. Find: $c^9 \cdot c^4$

5. Find: $a^6 \cdot a^5$

6. Find: $5^7 \div 5^3$. Leave your answer in exponential notation.

7. Find: $9^{10} \div 9^4$. Leave your answer in exponential notation.

8. Find: $\dfrac{m^{10}}{m^4}$

9. Find: $\dfrac{n^{20}}{n^{15}}$

10. Find: $\dfrac{b^{12}}{b^5}$

11. Find: $(5^3)^4$. Leave your answer in exponential notation.

12. Find: $(8^2)^5$. Leave your answer in exponential notation.

13. Find: $(13^5)^6$. Leave your answer in exponential notation.

14. Find: $(y^8)^3$

15. Find: $(z^{12})^4$

16. Find: $(x^9)^4$

17. Find: $(3 \cdot a)^4$

18. Find: $(4 \cdot b)^2$

19. Find: $(2 \cdot y)^3$

20. Find: $\dfrac{a^6 b^5}{a^8 b^2}$

21. Find: $\dfrac{m^7 n^4}{m^3 n^{10}}$

22. Find: $\dfrac{x^3 y^7 z^{12}}{x y^8 z^5}$

23. Find: 5^0

24. Find: 348^0

25. Find: x^0

26. Find: $5^1 + (4z)^0$

27. Find: $a^0 - (xyz)^0 + 3^1$

28. Find: $2^1 - (3x)^0 + y^0$

EVALUATE

Practice Test

Take this practice test to be sure that you are prepared for the final quiz in Evaluate.

1. Rewrite each expression below. Keep your answer in exponential form where possible.

   a. $11 \cdot 11 \cdot 11 \cdot 11$

   b. $3 \cdot 3 \cdot y \cdot y \cdot y \cdot y \cdot y$

   c. $5^{12} \cdot 5^8 \cdot 5^{23}$

   d. $x^7 \cdot y \cdot y^{19} \cdot x^{14} \cdot y^6$

   e. $7^8 \cdot b^5 \cdot b^8 \cdot 7^{10} \cdot b$

2. Rewrite each expression below in simplest form using exponents.

   a. $\dfrac{2 \cdot 2 \cdot 2 \cdot 2 \cdot 2 \cdot 2}{2 \cdot 2 \cdot 2}$

   b. $\dfrac{b^{20}}{b^{14}}$

   c. $\dfrac{3^{12} \cdot x^7}{3^9 \cdot x^{16}}$

   d. $\dfrac{y^{17}}{y^{14} \cdot y^3 \cdot y^4}$

3. Circle the expressions below that simplify to $\dfrac{x^3}{y^5}$.

   $\dfrac{x^6 y^2}{x^3 y^7}$ $\quad\quad\quad$ $\dfrac{y^{11} x^5}{y^2 x^4}$

   $\dfrac{x y^9}{x^6 y^4}$ $\quad\quad\quad$ $\dfrac{x^7 y}{x^4 y^6}$

4. Circle the expressions below that simplify to $5y$.

   $(31x^8)^0 \cdot 5y$

   $-(-5y)^0$

   $\dfrac{5y^2}{y}$

   $\dfrac{(5y)^2}{5y}$

   $\dfrac{5 \cdot 5 \cdot 5 \cdot y \cdot y \cdot y \cdot y}{5 \cdot 5 \cdot y \cdot y}$

5. Simplify each expression below.

   a. $(b^4 \cdot b^2)^8$

   b. $(3^5 \cdot a^6)^2$

   c. $(2^9 \cdot x^4 \cdot y^6)^{11}$

6. Simplify each expression below.

   a. $\left(\dfrac{5y^{10}}{3x^8}\right)^4$

   b. $\left(\dfrac{7a^3b^4}{5a^2}\right)^6$

7. Calculate the value of each expression below.

   a. $(4x)^0 - 2y^0$

   b. $(5xy^2 \cdot 4x^3)^0$

   c. $-2x^0 - y^0$

   d. $\dfrac{(4x)^0}{2} + \dfrac{3x^0}{2} + \dfrac{-2x^0}{2}$

8. Rewrite each expression below using a single exponent.

   a. $\left(\dfrac{a^4 \cdot a^5}{a \cdot a^3}\right)^7$

   b. $\left(\dfrac{a \cdot a^3}{a^4 \cdot a^5}\right)^7$

LESSON 6.2 – POLYNOMIAL OPERATIONS I

OVERVIEW

Here's what you'll learn in this lesson:

Adding and Subtracting

a. *Definition of polynomial, term, and coefficient*

b. *Evaluating a polynomial*

c. *The degree of a term and a polynomial*

d. *Writing the terms of a polynomial in descending order*

e. *Definition of a monomial, binomial, and trinomial*

f. *Recognizing like or similar terms*

g. *Combining like or similar terms*

h. *Polynomial addition*

i. *Polynomial subtraction*

Multiplying and Dividing

a. *Multiplying a monomial by a monomial*

b. *Multiplying a monomial by a polynomial*

c. *Dividing a monomial by a monomial*

d. *Dividing a polynomial by a monomial*

Every day, people use algebra to find unknown quantities. For example, you may be interested in figuring out how long it would take you to drive across the country. Or, you may want to know why your checkbook doesn't balance.

To find these unknown quantities, you need to be able to add, subtract, multiply, and divide polynomials. That's what you will learn in this lesson.

 **EXPLAIN**

ADDING AND SUBTRACTING

Summary

Identifying Polynomials

A polynomial is a special kind of algebraic expression which may have one or more variables and one or more terms.

For a polynomial in one variable, x, each term has the form ax^r, where the coefficient, a, is any real number, and the exponent, r, is a nonnegative integer.

For example:

$$x^3 + 7x^2 - 4x + 2$$

$$
\begin{array}{cccc}
\downarrow & \downarrow & \downarrow & \downarrow \\
a = 1 & a = 7 & a = -4 & a = 2 \\
r = 3 & r = 2 & r = 1 & r = 0
\end{array}
$$

Remember, when $x \neq 0$:

$$x^0 = 1$$

$$x^1 = x$$

For a polynomial in two variables, x and y, each term has the form $ax^r y^s$, where a is any real number, and r and s are nonnegative integers.

For example:

$$8x^3y^4 + 3xy^2 - 4$$

$$
\begin{array}{ccc}
\downarrow & \downarrow & \downarrow \\
a = 8 & a = 3 & a = -4 \\
r = 3 & r = 1 & r = 0 \\
s = 4 & s = 2 & s = 0
\end{array}
$$

Polynomials with one, two, or three terms have special names:

A monomial has one term: $\qquad\qquad\qquad \dfrac{1}{4}t^5u^3v^2$

A binomial has two terms: $\qquad\qquad\quad 4mpd^2 + 2m^3d$

A trinomial has three terms: $\qquad\qquad -87k - 13k^2 + \sqrt{13}k^3$

An algebraic expression is not a polynomial if any of its terms cannot be written in the form ax^r.

For example, these algebraic expressions are **not** polynomials:

$$\dfrac{2}{3w} + 7w^2 \qquad\qquad \sqrt{5t^2r} + 2t^3 \qquad\qquad 6x^2 - 2\sqrt{y}$$

The Degree of a Polynomial

The degree of a term of a polynomial is the sum of the exponents of the variables in that term. The degree of a polynomial is the degree of the term with the highest degree.

For example, to find the degree of the polynomial $8x^3y^4 + 3xy^2 - 3$; find the degree of each term:

$$8x^3y^4 + 3xy^2 - 3 \ = \ \underset{7}{\underset{\vee}{8x^3y^4}} \ + \ \underset{3}{\underset{\vee}{3x^1y^2}} \ - \ \underset{0}{\underset{\vee}{3x^0y^0}}$$

The degree of the polynomial $8x^3y^4 + 3xy^2 - 3$ is the degree of the term with the highest degree, 7.

Evaluating Polynomials

Sometimes the variables in a polynomial are assigned specific numerical values. In these cases you can evaluate the polynomial by replacing the variables with the numbers.

To evaluate a polynomial:

1. Replace each variable with its assigned value.

2. Calculate the value of the polynomial.

For example, to evaluate the polynomial $6b^2 - 4bc + c^2$ when $b = 2$ and $c = 3$:

$$6b^2 \ - \ 4bc \ + \ c^2$$

1. Replace b with 2 and c with 3. $= 6(2)^2 - 4(2)(3) + (3)^2$

2. Calculate. $= 24 \ - \ 24 \ + \ 9$

$$= 9$$

So, when $b = 2$ and $c = 3$, $6b^2 - 4bc + c^2 = 9$.

Adding Polynomials

To add polynomials, combine like terms - terms that have the same variables with the same exponents.

Here is an example of two like terms:

Here is an example of two terms that are not like terms:

$$3x^2y \text{ and } -2x^2y$$

$$3x^2y \text{ and } 3xy^2$$

To add polynomials:

1. Remove the parentheses.

2. Write like terms next to each other.

3. Combine like terms.

For example, to find:

$$(3z^3 + 2zy^2 - 6y^3) + (15z^2 - 5zy^2 + 4z^2)$$

1. Remove the parentheses. $= 3z^3 + 2zy^2 - 6y^3 + 15z^2 - 5zy^2 + 4z^2$

2. Write like terms next to each other. $= 3z^3 + 2zy^2 - 5zy^2 - 6y^3 + 15z^2 + 4z^2$

3. Combine like terms. $= 3z^3 - 3zy^2 - 6y^3 + 19z^2$

Subtracting Polynomials

To subtract one polynomial from another, add the opposite of the polynomial being subtracted.

To subtract polynomials:

1. Multiply the polynomial being subtracted by −1.

2. Distribute the −1.

3. Simplify.

4. Write like terms next to each other.

5. Combine like terms.

To find the opposite of a polynomial, multiply each term by −1.

When you add the opposite, the result is the same as changing the sign of each term in the polynomial being subtracted.

For example, to find: $(6y^3 - 3z^3 + 2zy^2) - (15z^2 - 5zy^2 + 4y^3)$

1. Multiply the second polynomial by −1. $= (6y^3 - 3z^3 + 2zy^2) + (-1)(15z^2 - 5zy^2 + 4y^3)$

2. Distribute the −1. $= (6y^3 - 3z^3 + 2zy^2) + (-1)15z^2 - (-1)5zy^2 + (-1)4y^3$

3. Simplify. $= 6y^3 - 3z^3 + 2zy^2 - 15z^2 + 5zy^2 - 4y^3$

4. Write like terms next to each other. $= 6y^3 - 4y^3 - 3z^3 + 2zy^2 + 5zy^2 - 15z^2$

5. Combine like terms. $= 2y^3 - 3z^3 + 7zy^2 - 15z^2$

Answers to Sample Problems

Sample Problems

1. Evaluate the polynomial $2r^3 + 3s^2r - 3s + 5$ when $r = 5$ and $s = -2$.

   Evaluate: $\qquad\qquad\qquad\qquad\qquad\qquad 2r^3 + 3s^2r - 3s + 5$

   ✓ a. Replace r with 5. $\qquad\qquad\qquad = 2(5)^3 + 3s^2(5) - 3s + 5$

   b. −2, −2

   ☐ b. Replace s with −2. $\qquad\qquad = 2(5)^3 + 3(\underline{\quad})^2(5) - 3(\underline{\quad}) + 5$

   c. 250, 60, −6

   321

   ☐ c. Calculate. $\qquad\qquad\qquad\qquad = \underline{\quad} + \underline{\quad} - \underline{\quad} + 5$

   $\qquad\qquad\qquad\qquad\qquad\qquad\qquad = \underline{\quad}$

2. Find: $(x^2 - 2xy^2 + 2) + (3x^2 - 5xy^2 + 3)$

   ✓ a. Remove the parentheses. $\qquad = x^2 - 2xy^2 + 2 + 3x^2 - 5xy^2 + 3$

   b. 3x², 5xy², 3

   ☐ b. Write like terms next to $\qquad = x^2 + \underline{\quad} - 2xy^2 - \underline{\quad} + 2 + \underline{\quad}$
   each other.

   c. 4x², 7xy², 5

   ☐ c. Combine like terms. $\qquad\qquad = \underline{\quad\quad} - \underline{\quad\quad} + \underline{\quad\quad}$

3. Find: $(x^2 - 2xy^2 + 2) - (3x^2 - 5xy^2 + 3)$

   ✓ a. Multiply each term in the second
   polynomial by −1. $\qquad = (x^2 - 2xy^2 + 2) + (-1)(3x^2 - 5xy^2 + 3)$

   ☐ b. Distribute the −1.
   $\qquad\qquad = (x^2 - 2xy^2 + 2) + (-1)(\underline{\quad}) + (-1)(\underline{\quad}) + (-1)(\underline{\quad})$

   b. 3x², − 5xy², 3

   c. x² − 2xy² + 2 − 3x² + 5xy² − 3

   ☐ c. Simplify. $= \underline{\qquad\qquad\qquad\qquad}$

   ☐ d. Write like terms next to each other.
   $\qquad = \underline{\qquad\qquad\qquad\qquad}$

   d. x² − 3x² − 2xy² + 5xy² + 2 − 3
   (in any order that like terms are next
   to each other)

   ☐ e. Combine like terms.
   $\qquad = \underline{\qquad\qquad\qquad\qquad}$

   e. −2x² + 3xy² − 1 (in any order)

MULTIPLYING AND DIVIDING

Summary

Multiplying Monomials

Monomials are easy to multiply because they each have only one term.

To multiply monomials:

1. Rearrange the factors so that the constants are next to each other and the factors with the same base are next to each other.

2. Multiply.

For example, to find: $-3p^3q^6r^2s \cdot 2p^4q^4s^5$

   1. Rearrange the factors. $= -3 \cdot 2 \cdot p^3 \cdot p^4 \cdot q^6 \cdot q^4 \cdot r^2 \cdot s \cdot s^5$

   2. Multiply. $= -6 \cdot p^{3+4} \cdot q^{6+4} \cdot r^2 \ s^{1+5}$

                             $= -6 \cdot p^7 \cdot q^{10} \cdot r^2 \cdot s^6$

                             $= -6p^7q^{10}r^2s^6$

Multiplying a Monomial by a Polynomial with More Than One Term

When multiplying a monomial by a polynomial with more than one term, you need to multiply every term in the polynomial by the monomial.

To multiply a monomial by a polynomial with more than one term:

1. Distribute the monomial to each term in the other polynomial.

2. Simplify.

For example, to find: $5x^4(3x^2y^2 + 2xy^2 - x^2y)$

In general, to multiply a monomial by a polynomial with more than one term:

$$a(b + c + d) = a \cdot b + a \cdot c + a \cdot d$$

  1. Distribute the $= 5x^4(3x^2y^2) + 5x^4(2xy^2) - 5x^4(x^2y)$
     monomial.

                       $= 5 \cdot 3 \cdot x^4 \cdot x^2 \cdot y^2 + 5 \cdot 2 \cdot x^4 \cdot x \cdot y^2 - 5 \cdot x^4 \cdot x^2 \cdot y$

  2. Simplify. $= 15 \cdot x^{4+2} \cdot y^2 + 10 \cdot x^{4+1} \cdot y^2 - 5 \cdot x^{4+2} \cdot y$

                       $= 15x^6y^2 + 10x^5y^2 - 5x^6y$

When $r > s$, $x^r \div x^s = \dfrac{x^r}{x^s} = x^{r-s}$.

When $r < s$, $x^r \div x^s = \dfrac{x^r}{x^s} = \dfrac{1}{x^{s-r}}$.

The y is on the bottom since the exponent of y in the denominator is bigger than the exponent of y in the numerator.

Dividing Monomials

To divide monomials:

1. Write the division as a fraction.

2. Cancel common numerical factors.

3. Divide factors with the same base by subtracting exponents.

For example, to find $4x^3y^4z^5 \div 10x^2y^6z^2$:

1. Write the division as a fraction. $\qquad = \dfrac{4x^3y^4z^5}{10x^2y^6z^2}$

2. Cancel common numerical factors. $\qquad = \dfrac{\overset{1}{\cancel{2}} \cdot 2x^3y^4z^5}{\underset{1}{\cancel{2}} \cdot 5x^2y^6z^2}$

3. Divide factors with the same base. $\qquad = \dfrac{2x^{3-2}z^{5-2}}{5y^{6-4}}$

$\qquad\qquad\qquad\qquad\qquad\qquad\qquad = \dfrac{2x^1z^3}{5y^2}$

Dividing a Polynomial with More Than One Term by a Monomial

When you divide a polynomial with more than one term by a monomial, you must divide each term of the polynomial by the monomial.

Use this rule: $\dfrac{a+b}{c} = \dfrac{a}{c} + \dfrac{b}{c}$.

To divide a polynomial with more than one term by a monomial:

1. Write the division as a fraction.

2. Rewrite the fraction using the rule $\dfrac{a+b}{c} = \dfrac{a}{c} + \dfrac{b}{c}$.

3. Perform the division on each of the resulting terms.

For example, to find $(15x^6y^2 + 10x^5y^3) \div 5x^4y$:

1. Write the division as a fraction. $\qquad = \dfrac{15x^6y^2 + 10x^5y^3}{5x^4y}$

2. Rewrite the fraction using the rule $\dfrac{a+b}{c} = \dfrac{a}{c} + \dfrac{b}{c}$. $\qquad = \dfrac{15x^6y^2}{5x^4y} + \dfrac{10x^5y^3}{5x^4y}$

3. Divide each of the resulting terms. $\; = \dfrac{3 \cdot \overset{1}{\cancel{5}}x^{6-4}y^{2-1}}{\underset{1}{\cancel{5}}} + \dfrac{2 \cdot \overset{1}{\cancel{5}}x^{5-4}y^{3-1}}{\underset{1}{\cancel{5}}}$

$\qquad\qquad\qquad\qquad\qquad\qquad = 3x^2y + 2xy^2$

Sample Problems

1. Find: $3wx^3y^6 \cdot 7wx^2y^5z^2$

   ☑ a. Rearrange the factors so the constants
   are next to each other and factors with
   the same base are next to each other.
   $$= 3 \cdot 7 \cdot w \cdot w \cdot x^3 \cdot x^2 \cdot y^6 \cdot y^5 \cdot z^2$$

   ☐ b. Multiply factors with the same
   base by adding the exponents.
   $$= 21 \cdot w^{\underline{}} \cdot x^{\underline{}} \cdot y^{\underline{}} \cdot z^{\underline{}}$$

   b. 2, 5, 11, 2

2. Find: $pr^2s(p^2r + pr^3s^6 - 2)$

   ☐ a. Distribute the monomial.
   $$= \underline{} \cdot p^2r + \underline{} \cdot pr^3s^6 - \underline{} \cdot 2$$

   a. pr^2s, pr^2s, pr^2s

   ☐ b. Multiply each of the
   resulting terms.
   $$= p^{1+2}r^{2+1}s + p^{1+1}r^{2+3}s^{1+6} - 2pr^2s$$
   $$= \underline{} + \underline{} - 2pr^2s$$

   b. $p^3r^3s, p^2r^5s^7$

3. Find: $18m^6n^5p^3r \div 10m^3n^7pr$

   ☑ a. Write the division as a fraction.
   $$= \frac{18m^6n^5p^3r}{10m^3n^7pr}$$

   ☑ b. Cancel common numerical factors.
   $$= \frac{\overset{1}{2} \cdot 9m^6n^5p^3r}{\underset{1}{2} \cdot 5m^3n^7pr}$$

   ☐ c. Divide factors with the same base by
   subtracting exponents.
   $$= \underline{}$$

   c. $\dfrac{9m^3p^2}{5n^2}$

4. Find: $(6w^5x^3 + 4w^3x^2y) \div 2w^2xy$

   ☑ a. Write the division as a fraction.
   $$= \frac{6w^5x^3 + 4w^3x^2y}{2w^2xy}$$

   ☐ b. Rewrite the fraction using the rule
   $\dfrac{a+b}{c} = \dfrac{a}{c} + \dfrac{b}{c}.$
   $$= \frac{6w^5x^3}{2w^2xy} + \underline{}$$

   b. $\dfrac{4w^3x^2y}{2w^2xy}$

   ☐ c. Divide each term.
   $$= \frac{3w^3x^2}{y} + \underline{}$$

   c. $2wx$

HOMEWORK

Homework Problems

Circle the homework problems assigned to you by the computer, then complete them below.

☀ Explain
Adding and Subtracting

1. Circle the algebraic expression that is a polynomial.

    $3\frac{1}{4}y^3 + \sqrt{3y^2 - 5}$

    $3\frac{1}{4}y^3 + 3y^2 - \sqrt{5}$

    $\frac{1}{4y^3} + 3y^2 - 5$

2. Write m beside the monomial, b beside the binomial, and t beside the trinomial.

    _____ $34x + x^2 + z$

    _____ wxy^3z^2

    _____ $pn^2 - 13n^3$

3. Given the polynomial $3y - 2y^3 - 4y^5 + 2$:

    a. write the terms in descending order.

    b. find the degree of each term.

    c. find the degree of the polynomial.

4. Find: $(-3w - 12w^3 + 2) + (15w - 2w^3 + 4w^5 - 3)$

5. Find: $(2v^3 + 6v^2 + 2) - (5v + v^3 + 4v^7 - 3)$

6. Evaluate $\frac{1}{4}xy + 3y^2 - 5x^3$ when $x = 2$ and $y = 4$.

7. Find:
    $(-s^2t + s^3t^3 + 4st^2 - 27) + (3st^2 + 2st - 8s^3t^3 - 13t + 36)$

8. Find: $(12x^3y + 9x^2y^2 + 6xy - y + 7) -$
    $(7xy - x + y - 11x^3y + 3x^2y^2 - 4)$

9. Angelina works at a pet store. Today, she is cleaning three fish tanks. These polynomials describe the volumes of the tanks:

    Tank 1: xy^2

    Tank 2: $x^2y - 2y^3 + 4xy^2 + 3$

    Tank 3: $x^2y + 5xy^2 + 6y^3$

    Write a polynomial that describes the total volume of the three tanks.

    Hint: Add the polynomials.

    volume = _____

10. Angelina has three fish tanks to clean. These polynomials describe their volumes.

    Tank 1: xy^2

    Tank 2: $x^2y - 2y^3 + 4xy^2 + 3$

    Tank 3: $x^2y + 5xy^2 + 6y^3$

    What is the total volume of the fish tanks if $x = 3$ feet and $y = 1.5$ feet?

    volume = _____ cubic feet

11. Find:

$(w^2yz + 3w^3 - 2wyz^2 + 4wyz) -$
$\qquad\qquad (4wy^2z - 3w^2yz + 2wyz^2) + (2wyz + 3)$

12. Find:

$(tu^2v - 4t^2u^2v + 9t^3uv + 3tv) + (3t^2u^2 + 2tv - t^3) -$
$\qquad\qquad (4t^2u^2v + 3tv + 2tu^2v) - (6t^3uv + 2tv)$

Multiplying and Dividing

13. Find: $xyz \cdot x^2y^2z^2$

14. Find: $3p^2r \cdot 2p^3qr$

15. Find: $-6t^3u^2v^{11} \cdot \dfrac{1}{2}tu^2v^4$

16. Find: $3y(2x^3 + 3x^2y)$

17. Find: $5p^2r^3(2pr + p^2r^2)$

18. Find: $t^3uv^4(2tu - 3uv + 4tv + 5)$

19. Write $12w^7x^3y^2z^6 \div 4w^2x^2y^3z^6$ as a fraction and simplify.

20. Write $(36x^3y^3 + 15x^2y^5) \div 9x^2y$ as a fraction and simplify.

21. Find: $15a^7b^4d^2 \div 10a^4b^9c^3d$

22. Tony is an algebra student. This is how he answered a question on a test:

$(2t^8u^3 - 4t^4u^9 + 6t^{12}u^6) \div 2t^4u^3 = t^2u - 2tu^3 + 3t^3u^2$

Is his answer right or wrong? Why? Circle the most appropriate response.

The answer is right.

The answer is wrong. Tony divided the exponents rather than adding them. The correct answer is $t^{12}u^6 - t^8u^{12} + t^{16}u^9$.

The answer is wrong. The terms need to be ordered by degree. The correct answer is $3t^3u^2 + t^2u - 2tu^3$.

The answer is wrong. Tony divided the exponents rather than subtracting them. The correct answer is $t^4 - 2u^6 + 3t^8u^3$.

The answer is wrong. Tony shouldn't have canceled the numerical coefficients. The correct answer is $2t^2u - 4tu^3 + 6t^3u^2$.

23. Find: $(16x^2y^4 + 20x^3y^5) \div 12xy^2$

24. Find: $(20t^5u^{11} + 5t^3u^5 + 30tu^6v^5) \div 10t^4u^5$

Practice Problems

Here are some additional practice problems for you to try.

Adding and Subtracting

1. Circle the algebraic expressions below that are polynomials.

    $2xy + 5xz$

    $\dfrac{2}{3x} + 6x$

    $9y^2 + 13yz - 8z^2$

    $\sqrt{24x^5}$

    $\dfrac{15a^3}{5a^8}$

2. Circle the algebraic expressions below that are polynomials.

    $8xy + \dfrac{3}{y}$

    $\sqrt{17x^3}$

    $3w - 7wz - 1$

    $7x^2 - 13x + 8y^2$

    $\dfrac{12x^2}{3x^3}$

3. Identify each polynomial below as a monomial, a binomial, or a trinomial.

    a. $17x + 24z$

    b. $13ab^2 - 5$

    c . $m - n + 10$

    d. $42a^2b^4c$

    e. $73 + 65x - 21y$

4. Identify each polynomial below as a monomial, a binomial, or a trinomial.

    a. $25 - 6xyz - 4x$

    b. $2xyz^3$

    c. $x + y - 1$

    d. $36 - 3xyz$

    e. $32x^2y$

5. Find the degree of the polynomial $8a^3b^5 - 11a^2b^3 + 7b^6$.

6. Find the degree of the polynomial $12m^4n^7 - 16m^{12}$.

7. Find the degree of the polynomial $7x^3y^2z + 3x^2y^3z^4 + 6z^7$.

8. Evaluate $2x^2 - 8x + 11$ when $x = -1$.

9. Evaluate $x^3 + 3x^2 - x + 1$ when $x = -2$.

10. Evaluate $2x^2 - 5x + 8$ when $x = 3$.

11. Evaluate $x^2y + xy^2$ when $x = 2$ and $y = -3$.

12. Evaluate $5mn + 4mn^2 + 8m - n$ when $m = 4$ and $n = -2$.

13. Evaluate $3uv - 6u^2v + 2u - v + 4$ when $u = 2$ and $v = -4$.

14. Find: $(3x^2 + 7x) + (x^2 - 5)$

15. Find: $(5x^2 + 4x - 8) + (x^2 + 7x)$

16. Find: $(6a^2 + 8a - 10) + (-3a^2 - 2a + 7)$

17. Find:

$(12m^2n^3 + 7m^2n^2 - 14mn) + (3m^2n^3 - 5m^2n^2 + 7mn)$

18. Find: $(10x^4y^3 - 9x^2y^3 + 6xy^2 - x) +$
$(-8x^4y^3 + 14x^2y^3 + 3xy^2 + x)$

19. Find: $(13a^3b^2 + 6a^2b - 5ab^3 + b) +$
$(2a^3b^2 - 2a^2b + 4ab^3 - b)$

20. Find: $(11u^5v^4w^3 + 6u^3v^2w) + (6u^5v^4w^3 - 11u^3v^2w)$

21. Find: $(7xy^2z^3 - 19x^2yz^2 + 26x^3y^3z) +$
$(13xy^2z^3 - 11x^2yz^2 - 16x^3y^3z)$

22. Find: $(9a^4b^2c - 3a^2b^3c + 5abc) +$
$(2abc - 6a^4b^2c - 2) + (3a^2b^3c + 5)$

23. Find: $(5x^3 + 7x) - (x^3 + 8)$

24. Find: $(9a^2 + 7ab + 14b) - (3a^2 - 7b)$

25. Find: $(2y^2 + 6xy + 3y) - (y^2 - y)$

26. Find: $(8x^3 + 9x^2 + 17) - (5x^3 - 3x^2 + 15)$

27. Find: $(9a^5b^3 + 8a^4b - 6b) - (-2a^5b^3 + 12a^4b + 3b)$

28. Find: $(7x^4y^2 - 3x^2y + 5x) - (9x^4y^2 + 3x^2y - 2x)$

Multiplying and Dividing

29. Find: $3y^4 \cdot 5y$

30. Find: $5x^3 \cdot 2x$

31. Find: $-5a^5 \cdot 9a^4$

32. Find: $-3x^3 \cdot 12x^4$

33. Find: $4x^3y^5 \cdot 7xy^3$

34. Find: $-7a^5b^6c^3 \cdot 8ab^3c$

35. Find: $-3w^2x^3y^2z \cdot 2x^2yz^2$

36. Find: $4y^3(3y^2 + 5y - 10)$

37. Find: $-2a^3b^2(3a^4b^5 - 5ab^3 + 6a)$

38. Find: $2xy^3(2x^6 - 5x^4 + y^2)$

39. Find: $5a^2b^2(4a^2 + 2a^2b - 7ab^2 - 3b)$

40. Find: $-4mn^3(-3m^2n + 12mn^2 - 6m + 7n^2)$

41. Find: $4x^3y^3(3x^3 - 7xy^2 + 2xy - y)$

42. Find: $\frac{9x^3y}{3x^2}$

43. Find: $\frac{20a^5b^6}{4a^3b}$

44. Find: $\frac{12x^4y^6}{3x^2y}$

45. Find: $\frac{32a^7b^9c}{12a^5b^6c^2}$

46. Find: $\frac{15m^6n^{10}}{10n^4p^3}$

47. Find: $\frac{24x^6y^2z^7}{16wx^3z^2}$

48. Find: $\frac{27a^4b^3c^{12}d}{15ac^7d^3}$

49. Find: $\frac{42mn^6p^3q^4}{28m^2nq^5}$

50. Find: $\frac{36w^2x^3y^7z}{21w^5y^2z^2}$

51. Find: $\frac{32a^3 + 24a^5}{8a^2}$

52. Find: $\frac{21m^2 + 18mn^3}{3mn}$

53. Find: $\frac{14x + 8x^4y^2}{2xy}$

54. Find: $\frac{24a^2b^2c^3 - 4ab^4c^5}{16abc^3}$

55. Find: $\frac{32x^2y^3z^4 - 8x^5yz^7}{16x^3y^3z^4}$

56. Find: $\frac{32r^4st^2 - 3r^2st^5}{12r^3s^2t}$

Practice Test

Take this practice test to be sure that you are prepared for the final quiz in Evaluate.

1. Circle the expressions that are polynomials.

   $-\sqrt{325}$ $\frac{2}{5}p^3r - 3p^2q + \sqrt{2r}$

   $t^2 - s + 5$ $\frac{5}{7}c^{15} + \frac{3}{14}c^{11} - 3\pi$

   $m^5n^4o^3p^2r$ $x^2 + 3xy - \frac{2}{3x} + y^2$

2. Write m beside the monomial(s), b beside the binomial(s), and t beside the trinomial(s).

   a. ____ w^5x^4

   b. ____ $2x^2 - 36$

   c. ____ $\frac{1}{3}x^{17} + \frac{2}{3}x^{12} - \frac{1}{3}$

   d. ____ 27

   e. ____ $27x^3 - 2x^2y^3$

   f. ____ $x^2 + 3xy - \frac{2}{3}y^2$

3. Given the polynomial $3w^3 - 13w^2 + 7w^5 + 8w^8 - 2$, write the terms in descending order by degree.

4. Find:

   a. $(5x^3y - 8x^2y^2 + 3xy - y^3 + 13) +$
   $(-2xy + 6 + y^2 - 4y^3 - 2x^3y)$

   b. $(5x^3y - 8x^2y^2 + 3xy - y^3 + 13) -$
   $(-2xy + 6 + y^2 - 4y^3 - 2x^3y)$

5. Find: $x^3y^2w \cdot x^5yw^4$

6. Find: $n^2p^3(3n + 2n^3p^2 - 35p^4)$

7. Find: $21x^5y^2z^7 \div 14xyz$

8. Find: $(15t^3u^2v - 5t^5uv^2) \div 10tuv^2$

LESSON 6.3 – POLYNOMIAL OPERATIONS II

OVERVIEW

Here's what you'll learn in this lesson:

Multiplying Binomials

a. Multiplying binomials by the "FOIL" method

b. Perfect squares, product of the sum and difference of two terms

Multiplying and Dividing

a. Multiplying a polynomial by a polynomial

b. Dividing a polynomial by a polynomial

Polynomials can be used to solve many types of problems. Some people might use a polynomial to create a household budget. A structural engineer might use a polynomial to find the wind force on a large building. Or, an automobile company might use a polynomial to find the average cost of manufacturing an airbag.

In this lesson, you will learn more about Polynomial Operations. You will multiply and divide polynomials which have more than one term.

 **EXPLAIN**

MULTIPLYING BINOMIALS

Summary

The multiplication of a binomial by a binomial can be simplified by using the "FOIL" method or by using patterns.

Using the FOIL Method to Multiply Two Binomials

The FOIL method can be used to multiply any two binomials. The letters in the word "FOIL" show you the order in which to multiply.

The general format is:

$$(a + b)(c + d) = a \cdot c + a \cdot d + b \cdot c + b \cdot d$$

$$\text{F} + \text{O} + \text{I} + \text{L}$$

First + Outer + Inner + Last

To multiply two binomials using the FOIL method:

1. Multiply the First terms of the binomials.

2. Multiply the Outer terms (the terms next to the outer parentheses).

3. Multiply the Inner terms (the terms next to the inner parentheses).

4. Multiply the Last terms.

5. Add the terms. Be sure to combine like terms.

For example, to find: $(x - 2)(x + 5)$

1. Multiply the First terms:	$x \cdot x$
2. Multiply the Outer terms:	$x \cdot 5$
3. Multiply the Inner terms:	$-2 \cdot x$
4. Multiply the Last terms:	$-2 \cdot 5$
5. Add the terms.	$x^2 + 5x - 2x - 10$
	$= x^2 + 3x - 10$

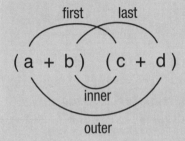

This picture may help you remember how to use the FOIL method. Notice how the connecting lines form a face: F and L make the eyebrows, O makes the smile and I the nose.

Using Patterns to Multiply Two Binomials

Patterns can be used to find certain binomial products.

In general, when you square a binomial you can use one of these patterns:

$$(a + b)^2 = (a + b)(a + b) = a^2 + 2ba + b^2$$

$$(a - b)^2 = (a - b)(a - b) = a^2 - 2ba + b^2$$

These products are called perfect square trinomials.

Another binomial product that follows a pattern has the form:

$$(a + b)(a - b) = a^2 - b^2$$

This product is called a difference of two squares.

To use a pattern to find the product of two binomials:

1. Determine which pattern to use.

2. Determine which values to substitute for a and b in the pattern.

3. Substitute the values into the pattern.

4. Simplify.

For example, to find $(3x^2 + 4)(3x^2 + 4)$:

1. Determine which pattern to use.

   $\checkmark$ $(a + b)(a + b) = a^2 + 2ba + b^2$
   ___ $(a - b)(a - b) = a^2 - 2ba + b^2$
   ___ $(a + b)(a - b) = a^2 - b^2$

2. Determine the values to substitute for a and b.

   $a = 3x^2, b = 4$

3. Substitute $3x^2$ for a and 4 for b.

   $(a + b)^2 = a^2 + 2ba + b^2$

   $(3x^2 + 4)^2 = (3x^2)^2 + 2 \cdot 4 \cdot 3x^2 + 4^2$

4. Simplify.

   $= 9x^4 + 24x^2 + 16$

If you forget the patterns, you can always use the FOIL method to figure them out:

$$F \quad + \quad O \quad + \quad I \quad + \quad L$$

$(a + b)(a + b)$

$\quad = a \cdot a + b \cdot a + b \cdot a + b \cdot b$

$\quad = a^2 + \quad 2ba \quad + b^2$

$(a - b)(a - b)$

$\quad = a \cdot a - b \cdot a - b \cdot a + (-b) \cdot (-b)$

$\quad = a^2 - \quad 2ba \quad + \quad b^2$

$(a + b)(a - b)$

$\quad = a \cdot a - b \cdot a + b \cdot a - b \cdot b$

$\quad = a^2 - b^2$

Sample Problems

1. Use the FOIL method to find: $(x - 6y)(3x + 2y)$

   ☑ a. Multiply the First terms. $x \cdot 3x$

   ☐ b. Multiply the Outer terms. $x \cdot$ ____   b. $2y$

   ☐ c. Multiply the Inner terms.   ____ $\cdot 3x$ c. $-6y$

   ☐ d. Multiply the Last terms.   ____ $\cdot$ ____   d. $-6y, 2y$

   ☐ e. Add the terms, combining like terms. $=$ _____   e. $3x^2 - 16xy - 12y^2$ *(in any order)*

2. Use a pattern to find: $(t - 7)^2$

   ☑ a. Determine which pattern   ____ $(a + b)^2 = a^2 + 2ba + b^2$
      to use.   ✓ $(a - b)^2 = a^2 - 2ba + b^2$
                                          ____ $(a + b)(a - b) = a^2 - b^2$

   ☐ b. Determine the values to
      substitute for a and b. $a = t, b =$ ____   b. 7

   ☐ c. Substitute these values. $(t - 7)^2 = t^2 - 2 \cdot$ ___ $\cdot t +$ ___   c. $7; 7^2$ or 49

   ☐ d. Simplify. $=$ _____   d. $t^2 - 14t + 49$ *(in any order)*

3. Use a pattern to find: $(x + 5y)(x - 5y)$

   ☐ a. Determine which pattern   ____ $(a + b)^2 = a^2 + 2ba + b^2$ a. $(a + b)(a - b) = a^2 - b^2$
      to use.   ____ $(a - b)^2 = a^2 - 2ba + b^2$
                                          ____ $(a + b)(a - b) = a^2 - b^2$

   ☑ b. Determine which values to
      substitute for a and b. $a = x, b = 5y$

   ☐ c. Substitute these values. $(x + 5y)(x - 5y) =$ _____   c. $x^2 - (5y)^2$

   ☐ d. Simplify. $=$ _____   d. $x^2 - 25y^2$

MULTIPYING AND DIVIDING

Summary

In general, to multiply a polynomial by a polynomial when each has more than one term:

$(a + b)(c + d + e)$

$\quad = a(c + d + e) + b(c + d + e)$

$\quad = ac + ad + ae + bc + bd + be$

Multiplying Two Polynomials When Each Has More Than One Term

You can multiply two polynomial each of which has more than one term.

To multiply a polynomial by a polynomial:

1. Distribute each term in the first polynomial to the second polynomial.

2. Distribute again to remove the parentheses.

3. Multiply each of the resulting terms.

4. Combine like terms.

For example, to find: $(x^2 + y)(3x^2 - 2y + xy)$

1. Distribute each term in the first polynomial to the second polynomial.
$$= x^2(3x^2 - 2y + xy) + y(3x^2 - 2y + xy)$$

2. Distribute again.
$$= (x^2)(3x^2) - x^2(2y) + (x^2)(xy) + y(3x^2) - y(2y) + y(xy)$$

3. Multiply.
$$= 3x^4 - 2x^2y + x^3y + 3x^2y - 2y^2 + xy^2$$
$$= 3x^4 - 2x^2y + 3x^2y + x^3y - 2y^2 + xy^2$$

4. Combine like terms.
$$= 3x^4 + x^2y + x^3y - 2y^2 + xy^2$$

Dividing a Polynomial with More Than One Term by Another Polynomial with More Than One Term

To divide a polynomial (dividend) by another polynomial (divisor) where each has more than one term, use long division.

Before you can divide, both polynomials should be arranged so their terms are in descending order by degree. To arrange the terms of a polynomial in descending order:

1. Determine the degree of each term by looking at the exponent of the variable(s).

2. Arrange the terms so they are in descending order by degree.

For example, to rearrange the terms of $x^3 - x + x^4$ in descending order:

1. Determine the degree of each term.

degree 3   degree 1   degree 4

$$x^3 - x^1 + x^4$$

2. Arrange the terms in descending order by degree.
$$x^4 + x^3 - x^1$$

Once the terms of the polynomials are correctly arranged, you are ready to divide.

To divide a polynomial by a polynomial where each has more than one term:

1. Arrange the terms of each polynomial in descending order. In the dividend, write missing terms as $0x^r$ where r is the exponent of the missing term.

2. Write the problem in long division form.

3. Divide the first term of the dividend by the divisor.

4. Multiply the divisor by the term you found in step (3).

5. Subtract the expression you found in step (4) from the dividend.

6. Continue dividing until the degree of the remainder is less than the degree of the divisor.

7. The answer is the expression that appears above the division sign plus the fraction $\frac{remainder}{divisor}$.

8. Check your division by multiplying the expression that appears above the division sign. Then add the remainder (not as a fraction).

For example, to find: $(x^2 + 3x^3 - 2) \div (x^2 + 2)$

1. Arrange the terms of the dividend in descending order. Include missing terms.

$$3x^3 + x^2 + 0x^1 - 2$$

2. Write the problem in long division form.

$$x^2 + 2 \,\overline{)3x^3 + x^2 + 0x^1 - 2}$$

3. Divide the first term of the dividend by the divisor.

4. Multiply the divisor by the term in (3).

5. Subtract the expression you found in step (4) from the dividend.

$$
\begin{array}{r}
3x \\
x^2 + 2 \,\overline{)3x^3 + x^2 + 0x^1 - 2} \\
-(3x^3 + 6x) \\
\hline
x^2 - 6x - 2
\end{array}
$$

6. Continue dividing until the degree of the remainder is less than the degree of the divisor.

$$
\begin{array}{r}
3x + 1 \\
x^2 + 2 \,\overline{)3x^3 + x^2 + 0x^1 - 2} \\
-(3x^3 + 6x) \\
\hline
x^2 - 6x - 2 \\
-(x^2 + 2) \\
\hline
-6x - 4
\end{array}
$$

7. Write your answer.

$$3x + 1 + \frac{-6x - 4}{x^2 + 2}$$

8. Check your answer by multiplying.

$$(3x + 1)(x^2 + 2) + (-6x - 4)$$
$$= 3x^3 + 6x + x^2 + 2 - 6x - 4$$
$$= 3x^3 + x^2 - 2$$

Sample Problems

1. Find: $(t + 2u)(5tu - t^2 - 4u^2)$

   ☑ a. Distribute each term in the
      first polynomial to each term
      in the second polynomial. $= t(5tu - t^2 - 4u^2) + 2u(5tu - t^2 - 4u^2)$

   ☑ b. Distribute again to remove
      parentheses. $= t \cdot 5tu - t \cdot t^2 - t \cdot 4u^2$
      $+ 2u \cdot 5tu - 2u \cdot t^2 - 2u \cdot 4u^2$

   ☐ c. Multiply each of the
      resulting terms. $= 5t^2u - t^3 - 4tu^2 + \underline{\quad} + \underline{\quad} + \underline{\quad}$

   ☐ d. Combine like terms. $= \underline{\hspace{4cm}}$

2. Find: $(6x + 15x^3 - 5) \div (3x - 3)$

   ☑ a. Arrange the terms of the
      dividend in descending order.
      Include "missing" terms. $15x^3 + 0x^2 + 6x - 5$

   ☑ b. Write the division in long division
      form.

$$5x^2 + 5x + \underline{\quad}$$
$$3x - 3 \overline{)15x^3 + 0x^2 + 6x - 5}$$
$$\underline{-(15x^3 - 15x^2)}$$

   ☐ c. Divide.

$$15x^2 + 6x - 5$$
$$\underline{- (15x^2 - 15x)}$$
$$21x - \underline{\quad}$$
$$\underline{- (\underline{\hspace{2cm}})}$$
$$\underline{\quad}$$

   ☐ d. Write the quotient. $\underline{\quad} + \underline{\quad} + \underline{\quad} + \dfrac{16}{3x - 3}$

   ☐ e. Check your division by
      multiplying the quotient by
      the divisor. $(5x^2 + \underline{\quad} + \underline{\quad})(3x - 3) + \underline{\quad}$

e. 5x, 7, 16

Here's one way to do the check:

$(5x^2 + 5x + 7)(3x - 3) + 16$
$= 15x^3 - 15x^2 + 15x^2 - 15x$
$\qquad\qquad + 21x - 21 + 16$
$= 15x^3 + 6x - 5$

 **EXPLORE**

Sample Problems

On the computer, you found the products of two binomial factors. In particular, you found patterns such as a perfect square trinomial and the difference of two squares to help you multiply binomials without using the FOIL method. Below are some additional problems using these patterns.

1. Find $(x + y)^2 + (x - y)^2$.

    ☐ a. First find $(x + y)^2$. This follows $(x + y)^2 = $ _____   *a.* $x^2 + 2xy + y^2$
       the pattern $(a + b)^2 = a^2 + 2ab + b^2$.

    ☐ b. Then find $(x - y)^2$. This follows $(x - y)^2 = $ _____   *b.* $x^2 - 2xy + y^2$
       the pattern $(a - b)^2 = a^2 - 2ab + b^2$.

    ☐ c. Now combine terms.

       $(x + y)^2 + (x - y)^2 = (x^2 + 2xy + y^2) + (x^2 - 2xy + y^2)$

                                       $= $ _____   *c.* $2x^2 + 2y^2$ or $2(x^2 + y^2)$

2. Find $(x + y)^2 - (x - y)^2$ first by using the perfect square trinomial patterns and then by using the pattern of a difference of two squares.

    ☐ a. First find $(x + y)^2$. This follows $(x + y)^2 = $ _____   *a.* $x^2 + 2xy + y^2$
       the pattern $(a + b)^2 = a^2 + 2ab + b^2$.

    ☐ b. Then find $(x - y)^2$. This follows $(x - y)^2 = $ _____   *b.* $x^2 - 2xy + y^2$
       the pattern $(a - b)^2 = a^2 - 2ab + b^2$.

    ☐ c. Now simplify.

       $(x + y)^2 - (x - y)^2 = (x^2 + 2xy + y^2) - (x^2 - 2xy + y^2)$

                                       $= $ _____   *c.* $4xy$

    ☐ d. Now solve the same problem using the pattern of a difference of two squares.

       Hint: Use the pattern $a^2 - b^2 = (a + b)(a - b)$. For this example, $a = (x + y)$ and $b = (x - y)$.

       $(x + y)^2 - (x - y)^2 = [(x + y) + (x - y)][(x + y) - (x - y)]$

                         $= \quad 2x \quad \cdot \quad$ _____   *d.* $2y$

                         $= \quad$ _____ $4xy$

Refer to this diagram of Pascal's triangle as you complete the additional exploration problems below.

$(a + b)^0$ 1

$(a + b)^1$ $1a + 1b$

$(a + b)^2$ $1a^2 + 2ab + 1b^2$

$(a + b)^3$ $1a^3 + 3a^2b + 3ab^2 + 1b^3$

$(a + b)^4$ $1a^4 + 4a^3b + 6a^2b^2 + 4ab^3 + 1b^4$

3. Expand $(a + b)^5$.

   ☐ a. Find the exponents and
       coefficients for each term.

   $1a^5 + 5a^4b^1 + \_\_\_a\overline{}b\overline{} + \_\_\_a\overline{}b\overline{} + \_\_\_a\overline{}b\overline{} + 1b^5$

   Hint: Use the coefficients for the expansion of $(a + b)^4$ to help you find
   the coefficient for each term in $(a + b)^5$.

   $$1a^4 + 4a^3b + 6a^2b^2 + 4ab^3 + 1b^4$$
   $$1a^5 + 5a^4b + \_\_a^-b^- + \_\_a^-b^- + \_\_a^-b^- + 1b^5$$

4. Use Pascal's triangle to find: $(3x + 2y)^4$

   ☑ a. Determine which row of Pascal's
       triangle will help. $(3x + 2y)^4$
                                    So use the row that expands $(a + b)^4$

   ☑ b. Write down the expansion
       for $(a + b)^4$. $(a + b)^4 = 1a^4 + 4a^3b + 6a^2b^2 + 4ab^3 + 1b^4$

   ☐ c. Replace a with $3x$
       and b with $2y$.

   $1(3x)^4 + 4(\_\_)^3(\_\_) + 6(\_\_)^2(\_\_)^2 + 4(\_\_)(\_\_)^3 + 1(2y)^4$

   ☐ d. Simplify. $= 81x^4 + \_\_\_x^3y + \_\_\_x^2y^2 + \_\_\_xy^3 + 16y^4$

HOMEWORK

Homework Problems

Circle the homework problems assigned to you by the computer, then complete them below.

☀ **Explain**
Multiplying Binomials

1. Given $(2p + 3)(p - p^2)$, find the:

     First terms: _____ and _____

     Outer terms: _____ and _____

     Inner terms: _____ and _____

     Last terms: _____ and _____

2. Which pattern could you use to find each of the products (a) - (f) below? Write the appropriate pattern number next to each polynomial.

  a. ____ $(2x + 5y)^2$ I. $(a + b)^2 = a^2 + 2ba + b^2$

  b. ____ $(2x + 5y)(2x - 5y)$ II. $(a - b)^2 = a^2 - 2ba + b^2$

  c. ____ $(3t - 2)^2$ III. $(a + b)(a - b) = a^2 - b^2$

  d. ____ $(3t + 2)^2$

  e. ____ $(3t - 2)(3t - 2)$

  f. ____ $(2x^2 - 5y^3)^2$

3. Given $(2s^3 + 5)^2$ and the pattern $(a + b)^2 = a^2 + 2ba + b^2$:

  a. What would you replace a with in the pattern? _____

  b. What would you replace b with? _____

4. Use a pattern to find: $(3s + 5)^2$

5. Use the FOIL method to find: $(4x - 2y)(3x + 6)$

6. Use a pattern to find: $(3t + 4u)(3t - 4u)$

7. Use patterns to find these products:

  a. $(3x^2 - 2)(3x^2 + 2)$ c. $(3x^2 + 2)(3x^2 + 2)$

  b. $(3x^2 - 2)(3x^2 - 2)$

8. Find: $(5x^3 + 3y^2)^2$

9. A fish tank broke at the pet store where Angelina works, and part of the store was flooded. Since Angelina lost her measuring tape, she used a stick and her handspan to figure out the approximate size of the flooded area. If s equals the length of the stick and h equals the width of her handspan, these are the measurements:

length of flooded space $= 13s + 2h$

width of flooded space $= 13s - 2h$

area of flooded space $= (13s + 2h)(13s - 2h)$

Simplify the equation for the area by multiplying the binomials.

10. The owner of the pet store where Angelina works wants to replace the tile covering the entire floor, not just the flooded area. If the length of the entire floor is $250s - 3h$ and the width is $98s + h$, what is the area of the floor in terms of s and h?

Hint: area = length · width.

11. Find: $(13x^2y^2 - 10x^3)(7x^2y^2 - 6x^3)$

12. Find:

  a. $\left(\frac{1}{2}x^3 - \frac{2}{3}y^5\right)\left(\frac{1}{2}x^3 - \frac{2}{3}y^5\right)$

  b. $\left(\frac{1}{2}x^3 - \frac{2}{3}y^5\right)\left(\frac{1}{2}x^3 + \frac{2}{3}y^5\right)$

  c. $\left(\frac{1}{2}x^3 + \frac{2}{3}y^5\right)\left(\frac{1}{2}x^3 + \frac{2}{3}y^5\right)$

Multiplying and Dividing

13. Find: $(x + 2)(3x + 4xy + 1)$

14. Find: $(p^2 + 2r + 2)(3r^4 - 2p^4)$

15. Find: $(x + y + 1)(x - y)$

16. Find: $(2t + u)(t + 2u - 1)$

17. Angelina is cleaning the windows of the guinea pig case at the pet store where she works. The surface area of the outside of the windows can be described as follows:

surface area $= 2(x + 3)(x - 2) + 2(x - 2)(x - 3)$

Simplify this equation by multiplying the polynomials.

18. The pet store where Angelina works sells an exercise arena for guinea pigs, consisting of two spheres connected by a tube. The volume of the exercise arena can be described by this equation:

volume $= 4\pi r^3 + 3\pi(r^2 + 2r + 4)(r + 2) + \pi r(r - 5)(r - 5)$

Simplify the equation by multiplying the polynomials.

19. Find: $(12x^3 - 2x^2 - 7x)\left(4x^2 - \dfrac{10}{3}x - \dfrac{1}{3} + \dfrac{7}{12x^3 - 2x^2 - 7x}\right)$

20. Find: $\left(\dfrac{1}{3}t^2 + \dfrac{2}{3}v^3\right)\left(\dfrac{1}{3}t^2 + \dfrac{2}{3}v^3\right)\left(\dfrac{1}{3}t^2 + \dfrac{2}{3}v^3\right)$

21. Find: $(3x^2 + 2x - 1) \div (x + 3)$

22. Here is how Tony answered a question on his algebra test.
$(12x^3 - 17x^2 + 3) \div (3x - 2) = 4x^2 - 3x - 2$ remainder -1.

Is his answer right or wrong? Why? Circle the most appropriate response.

His answer is right.

His answer is wrong. When doing the long division, he sometimes added negative terms rather than subtracting them. The right answer is $4x^2 + 2x - 1$.

His answer is wrong. He did not include missing terms in the quotient. The right answer is $0x^3 + 4x^2 - 3x - 2$ remainder -1.

His answer is wrong. He did not put the remainder over the dividend. The right answer is $4x^2 - 3x - 2 + \dfrac{-1}{3x - 2}$.

23. Find: $(15x^5 + x^4 + 5x^2 - 2) \div (3x^2 + 2x)$

24. Find: $(8y^6 + 4y^4 - 10y^3 - 12) \div (2y^3 - 2y + 4)$

 Explore

25. Find: $(3a - 1)(3a + 1)$

26. Use the table below to find a general form for multiplying two polynomials: $(ax^2 + bx + c)(dx - e)$

terms	dx	$-e$
ax^2		
bx		
c		

$(ax^2 + bx + c)(dx - e) =$

27. Use the table below to find the general form for a difference of two squares: $(a + b)(a - b)$. Then use this pattern to find $(2x + 3y)(2x - 3y)$.

terms	a	$-b$
a		
b		

$(a + b)(a - b) =$

$(2x + 3y)(2x - 3y) =$

28. Find: $(x^2 + 3y)^2$

29. Use the table below to find the general form for a perfect square trinomial: $(a - b)(a - b)$. Then use the pattern to find $(2t^3 - 4u^2)(2t^3 - 4u^2)$.

terms	a	$-b$
a		
$-b$		

$(a - b)(a - b) =$

$(2t^3 - 4u^2)(2t^3 - 4u^2) =$

30. Use the table below to find the general form for a perfect square trinomial: $(a + b)(a + b)$. Then use this general form to find $(x^2 + 3y)(x^2 + 3y)$.

terms	a	b
a		
b		

$(a + b)(a + b) =$

$(x^2 + 3y)(x^2 + 3y) =$

APPLY

Practice Problems

Here are some additional practice problems for you to try.

Multiplying Binomials

1. Find: $(a + 2)(a + 5)$

2. Find: $(m - 3)(m - 7)$

3. Find: $(x - 4)(x - 11)$

4. Find: $(3b + 2)(b - 6)$

5. Find: $(5y - 8)(y + 3)$

6. Find: $(6t + 1)(t - 7)$

7. Find: $(4a + 3b)(2a + 5b)$

8. Find: $(3m - 4n)(7m + 2n)$

9. Find: $(6y + 5x)(3y - x)$

10. Find: $(p + 9)(p + 9)$

11. Find: $(x + 3)(x + 3)$

12. Find: $(3z + 2)(3z + 2)$

13. Find: $(5q + 3)(5q + 3)$

14. Find: $(4x + 1)(4x + 1)$

15. Find: $(z - 5)(z - 5)$

16. Find: $(m - 11)(m - 11)$

17. Find: $(t - 6)(t - 6)$

18. Find: $(3x - 2y)(3x - 2y)$

19. Find: $(4a - 7c)(4a - 7c)$

20. Find: $(5r - 8s)(5r - 8s)$

21. Find: $(5m + n)(5m - n)$

22. Find: $(a + 7b)(a - 7b)$

23. Find: $(2x + y)(2x - y)$

24. Find: $(3y + 8)(3y - 8)$

25. Find: $(5x + 3)(5x - 3)$

26. Find: $(m + 12n)(m - 12n)$

27. Find: $(2a + 7b)(2a - 7b)$

28. Find: $(x + 7y)(x - 7y)$

Multiplying and Dividing

29. Find: $(4a - 3b)(2a - 7b)$

30. Find: $(3x + 5)(y + 8)$

31. Find: $(6m - 5n)(3m + 4n)$

32. Find: $(8y + 3z)(2y - 9z)$

33. Find: $(7x - 4)(2y + 3)$

34. Find: $(a + 2b)(a^2 + 6a - 3b)$

35. Find: $(3mn - n)(m^2 - 3n + 4m)$

36. Find: $(2xy - y)(x^2 + 5y - 6x)$

37. Find: $(3ab + 4b)(7a^2 + 3b - 4a)$

38. Find: $(7uv - 3v)(2u^2 - 5v + 8u)$

39. Find: $(5xy + 2y)(2x^2 - 6y + 3x)$

40. Find: $(3a^2 - 4b^2)(2a^3 + 5a^2b - 11ab - b)$

41. Find: $(5m^2n + 3n)(4m^3 - 3m^2n + 8mn^2 - 3n^2)$

42. Find: $(7x^2y + 2y)(3x^3 - 6x^2y + 8xy + y)$

43. Find: $(x^3 + x^2 - 13x + 14) \div (x - 2)$

44. Find: $(x^3 + 11x^2 + 22x - 24) \div (x + 4)$

45. Find: $(x^3 + 10x^2 + 23x + 6) \div (x + 3)$

46. Find: $(x^3 + 7x^2 - 36) \div (x + 6)$

47. Find: $(x^3 - 26x + 5) \div (x - 5)$

48. Find: $(3x^3 + 17x^2 - 58x + 40) \div (3x - 4)$

49. Find: $(4x^3 + 4x^2 - 13x + 5) \div (2x + 5)$

50. Find: $(2x^3 + 7x^2 - x - 2) \div (2x + 1)$

51. Find: $(4x^3 + 7x^2 - 14x + 6) \div (4x - 1)$

52. Find: $(2x^3 - 9x^2 + 12x - 8) \div (2x + 3)$

53. Find: $(3x^3 + 14x^2 + 11x - 8) \div (3x + 2)$

54. Find: $(6x^3 - 7x^2 - 34x + 35) \div (2x - 5)$

55. Find: $(10x^3 - 26x^2 - 7x + 2) \div (5x + 2)$

56. Find: $(8x^3 - 18x^2 + 25x - 12) \div (4x - 3)$

EVALUATE

Practice Test

Take this practice test to be sure that you are prepared for the final quiz in Evaluate.

1. Use the FOIL method to find: $(2x^2 + 3xy)(3x^3y - 2)$

2. Use a pattern to find: $(2x - 3y)^2$

3. Find: $(2x + 3y)^2$

4. Use a pattern to find: $(2x - 3y)(2x + 3y)$

5. Find: $(3x - 2)(5x^2 + 8x - 2)$

6. Find: $(3p^2 + 4r^4 - 5)(3r^4 - 6p^2 + 2)$

7. Find: $(6t^3 + 5t^2 - 3t + 1) \div (2t + 1)$

8. Find: $(8x^3 + 6x - 2) \div (4x + 2)$

9a. Find: $(a^3 - a^5)(a + a^2)$

  b. What is the degree of the resulting polynomial?

10. Find: $(5y^4 - 2y^2 + y)(3y^2 - y + 2)$

11. Use the table in Figure 6.3.1 to find:
    $(2x^3 - 3x + 7)(5x^4 + 8)$

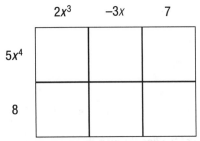

Figure 6.3.1

12. Use the table in Figure 6.3.2 to find:
    $(5x^4 - 7x^3 + 7x^2 - 8x)(x^2 + 1)$

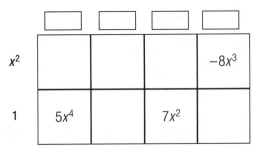

Figure 6.3.2

TOPIC 6 CUMULATIVE ACTIVITIES

CUMULATIVE REVIEW PROBLEMS

These problems combine all of the material you have covered so far in this course. You may want to test your understanding of this material before you move on to the next topic. Or you may wish to do these problems to review for a test.

1. Find:

   a. $2^7 \cdot 2^9$ b. $\dfrac{x^{12}}{x^5}$ c. $(a^5 b^2)^4$

2. Solve $-3 \le 6 + 2y < 4$ for y, then graph its solution on the number line below.

3. Find the equation of the line through the point (3, 7) with slope $-\dfrac{2}{7}$:

   a. in point-slope form.   c. in standard form.

   b. in slope-intercept form.

4. Solve this system:

   $$x + 2y = 5$$
   $$x - 2y = -13$$

5. The difference of two numbers is –32. The sum of three times the smaller number and twice the larger number is 134. What are the two numbers?

6. Circle the true statements.

   The GCF of two numbers that have no factors in common is 1.

   $\dfrac{2}{9} - \dfrac{1}{5} = \dfrac{1}{4}$

   The LCM of 4 and 8 is 4.

   $3^2(4 + 2) = 9(4 + 2)$

   $\dfrac{1}{2} + \dfrac{1}{3} = \dfrac{5}{6}$

7. Write the equation of the line through the point (20, –9) with slope $-\dfrac{8}{5}$:

   a. in point-slope form.   c. in standard form.

   b. in slope-intercept form.

8. Graph the system of inequalities below to find its solution.

   $$2x + y \ge 3$$
   $$x - y < 4$$

9. Find:

   a. 3^0 b. -3^0 c. $(-3)^0$

10. Graph the inequality $4x + y \le 6$.

11. Find: $15x^3 y^8 z^5 \div 10xy^4 z^{11}$

12. Lisa emptied a vending machine and got a total of 279 quarters and dimes worth $57.30. How many quarters did she get?

13. Find the slope and y-intercept of the line $4x - y = 7$.

14. Evaluate the expression $3x^2 - 4xy + 2y$ when $x = 3$ and $y = -5$.

15. Solve $-6 < 4 + 2x < -2$ for x.

16. Find:

   a. $-5x^0 + y^2$ c. $b^4 \cdot b^2 \cdot b \cdot b^6$

   b. $\left(\dfrac{a^3 \cdot b^7 \cdot c}{b^4 \cdot c^2}\right)^3$

17. Circle the true statements.

   $4(3 - 5) = 4 \cdot 3 - 5$

   $\dfrac{26}{117} = \dfrac{2}{9}$

   The LCM of 72 and 108 is 36.

   The GCF of 72 and 108 is 36.

   $\dfrac{4}{7} - \dfrac{2}{3} = -\dfrac{2}{4}$

18. Write the equation of the line through the point (5, 2) with slope $-\frac{7}{3}$:

    a. in point-slope form.   c. in standard form.

    b. in slope-intercept form.

19. Graph the line $y = 6$.

20. Find: $(11p^2 - 3pr - 6r)(3p - 9r)$

21. Solve this system:

$$4x - y = 9$$
$$6x + 5y = -6$$

22. Find the equation of the line that is **parallel** to the line $x + 3y = 4$ and passes through the point (2, 2).

23. Find the equation of the line that is **perpendicular** to the line $x + 3y = 4$ and passes through the point (2, 2).

24. Graph the system of inequalities below to find its solution.

$$y < 2x + 3$$
$$4x - y \geq 1$$

25. Find the slope of the line perpendicular to the line through the points (8, 9) and (6, −4).

26. Find: $(x^3 + 5x^2 + x - 10) \div (x + 2)$

27. Solve $2y + 5 = 4(\frac{1}{2}y + 2)$ for y.

28. Find: $3xy(x^2y - 4)$

Use Figure 6.1 to answer questions 29, 30, and 31.

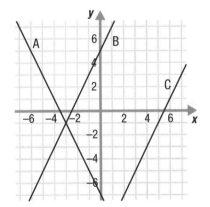

Figure 6.1

29. Which two lines form a system that has a solution of (−3, −1)?

30. Which two lines form a system that has no solution?

31. Which two lines form a system that has a solution that is not shown on the grid?

32. Find:

    a. $(x^2yz^3)^4$ b. $\frac{x^5y^3}{xy^6}$ c. $(x^5)^9$

33. Evaluate the expression $-7a^4 + 3ab^2 + b - 4$ when $a = -2$ and $b = 5$.

34. Find the slope of the line through the points (9, −4) and (2, 7).

35. Graph the system of inequalities below to find its solution.

$$9x - 4y < 20$$
$$9x - 4y \leq 8$$

36. Graph the inequality $\frac{5}{2}x - y \geq 2$.

37. Last year Manuel split $2565 between his savings account, which paid 5% in interest, and his checking account, which paid 3.5% in interest. If he earned a total of $113.49 in interest, how did he split his money between the two accounts?

38. Graph the inequality $\frac{2}{3}x - \frac{1}{3}y \geq 2$.

39. Next to each polynomial below, write whether it is a monomial, a binomial, or a trinomial.

    a. $2 + x$

    b. $8ab^3 - 9abc + 1$

    c. $3x^7yz^5$

    d. $a^5b^2c^3d - 103a^7cd^4$

    e. 10

    f. $a + 7b - 4c$

40. Solve $-8(1 - \frac{1}{2}x) = 4(x - 2)$ for x.

41. Find: $(4a^2b + 3a - 9b) + (7a + 2b - 8a^2b)$

42. The perimeter of a square is the same as the perimeter of a regular hexagon. If each side of the square is 7 feet longer than each side of the hexagon, what is the perimeter of each figure?

LESSON 7.1 – FACTORING POLYNOMIALS I

OVERVIEW

Here's what you'll learn in this lesson:

Greatest Common Factor

a. Finding the greatest common factor (GCF) of a set of monomials

b. Factoring a polynomial by finding the GCF when the GCF is a monomial

Grouping

a. Factoring a polynomial by finding the GCF when the GCF is a binomial

b. Factoring a polynomial with four terms by grouping

You have learned how to multiply polynomials. Now you will learn how to factor them. When you factor a polynomial, you write it as the product of other polynomials.

In this lesson you will learn several different techniques for factoring polynomials.

 **EXPLAIN**

GREATEST COMMON FACTOR

Summary

Factoring Polynomials

You already know how to factor numbers by writing them as the product of other numbers. Now you will learn how to factor polynomials by writing them as the product of other polynomials.

Finding the GCF of a Collection of Monomials

To find the GCF of a collection of monomials:

1. Factor each monomial into its prime factors.

2. List each common prime factor the **smallest** number of times it appears in any factorization.

3. Multiply all the prime factors in the list.

For example, to find the GCF of the monomials $16x^2y^2$, $4x^3y^2$, and $12xy^4$:

1. Factor each monomial into its prime factors.

$$16x^2y^2 = 2 \cdot 2 \cdot 2 \cdot 2 \cdot x \cdot x \cdot y \cdot y$$

$$4x^3y^2 = 2 \cdot 2 \cdot x \cdot x \cdot x \cdot y \cdot y$$

$$12xy^4 = 2 \cdot 2 \cdot 3 \cdot x \cdot y \cdot y \cdot y \cdot y$$

2. List each common prime factor the **smallest** number of times it appears in any factorization.   2, 2, x, y, y

3. Multiply all the prime factors in the list. GCF $= 2 \cdot 2 \cdot x \cdot y \cdot y = 4xy^2$

Factoring a Polynomial By Finding The Greatest Common Factor

One way to factor a polynomial is to find the greatest common factor of its monomial terms. Here are the steps:

1. Identify the monomial terms of the polynomial.

2. Factor each monomial term.

3. Find the GCF of the monomial terms.

4. Rewrite each term of the polynomial using the GCF.

5. Factor out the GCF.

6. Use the distributive property to check your factoring.

Remember that a monomial is a polynomial with only one term. For example: $14x^5y^3$, 32, 6x, and 9xyz are monomials; but $12x^5y + 1$ and $14y + 3x$ are not monomials.

Before deciding if a polynomial is a monomial, binomial, etc., be sure you first combine any like terms and apply the distributive property, if possible.

The GCF of a collection of monomials is the GCF of the coefficients of all the monomials multiplied by the smallest power of each variable in all the monomials.

The GCF of a collection of monomials evenly divides each monomial in the collection.

$$\frac{16x^2y^2}{4xy^2} = 4x$$

$$\frac{4x^3y^2}{4xy^2} = x^2$$

$$\frac{12xy^4}{4xy^2} = 3y^2$$

For example, to factor the polynomial $6x^2y^2 + 8y^2$:

1. Identify the terms of the polynomial. $\qquad$ $6x^2y^2, 8y^2$

2. Factor each monomial term. $\qquad$ $6x^2y^2 = 2 \cdot 3 \cdot x \cdot x \cdot y \cdot y$
   $8y^2 = 2 \cdot 2 \cdot 2 \cdot y \cdot y$

3. Find the GCF of the monomial terms. $\qquad$ $GCF = 2 \cdot y \cdot y = 2y^2$

4. Rewrite each term of the
   polynomial using the GCF $\qquad$ $6x^2y^2 = 2y^2 \cdot 3x^2$
   $8y^2 = 2y^2 \cdot 4$

5. Factor out the GCF. $\qquad$ $6x^2y^2 + 8y^2 = (2y^2)(3x^2 + 4)$

6. Use the distributive property
   to check your factoring. $\qquad$ Is $6x^2y^2 + 8y^2 = (2y^2)(3x^2 + 4)$ $\qquad$?

   Is $6x^2y^2 + 8y^2 = (2y^2)(3x^2) + (2y^2)(4)$?
   Is $6x^2y^2 + 8y^2 = 6x^2y^2 + 8y^2$ $\qquad$? Yes.

Sample Problems

1. Find the GCF of $9x^3y$, $3xy^4$, and $6y^2$.

   ☐ a. Factor each monomial $\qquad$ $9x^3y = 3 \cdot 3 \cdot x \cdot x \cdot x \cdot y$
   into its prime factors. $\qquad$ $3xy^4 = $ _____
   $6y^2 = $ _____

   ☐ b. List each common factor the smallest number
   of times it appears in any factorization. $\qquad$ _____

   ☐ c. Multiply all the prime factors in the list. $\qquad$ $GCF = $ _____

2. Factor: $6x^4y^4 + 30x^2y^3 + 10x^5y^2$

   ☑ a. Find the terms of the polynomial. $\qquad$ $6x^4y^4, 30x^2y^3$, and $10x^5y^2$

   ☐ b. Factor each monomial. $\qquad$ $6x^4y^4 = $ _____
   $30x^2y^3 = $ _____
   $10x^5y^2 = $ _____

   ☐ c. Find the GCF of the monomial terms. $\qquad$ $GCF = $ _____

   ☐ d. Rewrite each term of the $\qquad$ $6x^4y^4 = $ _____
   polynomial using the GCF. $\qquad$ $30x^2y^3 = $ _____
   $10x^5y^2 = $ _____

   ☐ e. Factor out the GCF. $\qquad$ $= ($_____$)($_____$)$

   ☐ f. Use the distributive property to
   check your factoring.

Answers to Sample Problems

a. $3 \cdot x \cdot y \cdot y \cdot y \cdot y$
   $2 \cdot 3 \cdot y \cdot y$

b. $3, y$

c. $3y$

b. $2 \cdot 3 \cdot x \cdot x \cdot x \cdot x \cdot y \cdot y \cdot y \cdot y$
   $2 \cdot 3 \cdot 5 \cdot x \cdot x \cdot y \cdot y \cdot y$
   $2 \cdot 5 \cdot x \cdot x \cdot x \cdot x \cdot x \cdot y \cdot y$

c. $2x^2y^2$

d. $(2x^2y^2)(3x^2y^2)$
   $(2x^2y^2)(15y)$
   $(2x^2y^2)(5x^3)$

e. $(2x^2y^2)(3x^2y^2 + 15y + 5x^3)$

f. $(2x^2y^2)(3x^2y^2 + 15y + 5x^3)$
   $= (2x^2y^2)(3x^2y^2) +$
   $(2x^2y^2)(15y) + (2x^2y^2)(5x^3)$
   $= 6x^4y^4 + 30x^2y^3 + 10x^5y^2$

GROUPING

Summary

Factoring a Polynomial by Finding the Binomial GCF

You have already learned how to factor a polynomial when the GCF of the terms of the polynomial is a monomial. You can use the same steps to factor a polynomial when the GCF of the terms is a binomial.

There are six steps in this procedure:

1. Identify the terms of the polynomial.

2. Factor each term.

3. Find the GCF of the terms.

4. Rewrite each term of the polynomial using the GCF.

5. Factor out the GCF.

6. Check your answer.

For example, to factor the polynomial $5(3x + 2) + x^2(3x + 2)$:

1. Identify the terms of the polynomial. $5(3x + 2)$ and $x^2(3x + 2)$

2. Factor each term. (Here each $5(3x + 2) = 5 \cdot (3x + 2)$
   term is already factored.) $x^2(3x + 2) = x^2 \cdot (3x + 2)$

3. Find the GCF of the terms. $\text{GCF} = 3x + 2$

4. Rewrite each term of the $5(3x + 2) = 5 \cdot (3x + 2)$
   polynomial using the GCF. $x^2(3x + 2) = x^2 \cdot (3x + 2)$

5. Factor out the GCF. $5(3x + 2) + x^2(3x + 2) = 5 \cdot (3x + 2) + x^2 \cdot (3x + 2)$
   $$= (3x + 2)(5 + x^2)$$

6. Check your answer.

   Is $(3x + 2)(5 + x^2) = 5(3x + 2) + x^2(3x + 2)$?

   Is $(3x + 2)(5) + (3x + 2)(x^2) = 5(3x + 2) + x^2(3x + 2)$? Yes.

Factoring By Grouping

Sometimes the GCF of the terms of a polynomial is 1.

For example find the GCF of the terms of $3x^2 + 9 + bx^2 + 3b$:

$$3x^2 = 1 \cdot 3 \cdot x \cdot x$$

$$9 = 1 \cdot 3 \cdot 3$$

$$bx^2 = 1 \cdot b \cdot x \cdot x$$

$$3b = 1 \cdot 3 \cdot b$$

You see that the GCF of the 4 terms $3x^2$, 9, bx^2, and $3b$ is 1. If you try to use the GCF to factor $3x^2 + 9 + bx^2 + 3b$ you get the following factorization:

$$3x^2 + 9 + bx^2 + 3b = 1 \cdot (3x^2 + 9 + bx^2 + 3b)$$

This isn't very interesting!

To factor the polynomial $3x^2 + 9 + bx^2 + 3b$ you need a technique other than finding the GCF of the terms. One such technique is called factoring by grouping. This procedure has 5 steps:

1. Factor each term.

2. Group terms with common factors.

3. Factor out the GCF in each grouping.

4. Factor out the binomial GCF of the polynomial.

5. Check your answer.

For example, use this technique to factor the polynomial $3x^2 + 9 + bx^2 + 3b$:

1. Factor each term.

$$3x^2 = 3 \cdot x \cdot x$$
$$9 = 3 \cdot 3$$
$$bx^2 = b \cdot x \cdot x$$
$$3b = 3 \cdot b$$

This isn't the only way to group the terms. For example, you could also have grouped the terms like this:

$$(3x^2 + bx^2) + (9 + 3b)$$

Try it; you'll get the same answer.

2. Group terms with common factors.

$$= \quad 3x^2 + 9 + bx^2 + 3b$$
$$= (\mathbf{3} \cdot x \cdot x + \mathbf{3} \cdot 3) + (\mathbf{b} \cdot x \cdot x + 3 \cdot \mathbf{b})$$

3. Factor out the GCF in each grouping.

$$= \mathbf{3}(x \cdot x + 3) + \mathbf{b}(x \cdot x + 3)$$

Notice that in steps (1) – (3) you have written the polynomial so that we can see its binomial GCF. In step (4) we are really doing all of steps (1) – (5) from before.

4. Factor out the binomial GCF of the polynomial.

$$= \mathbf{3}(x^2 + 3) + \mathbf{b}(x^2 + 3)$$
$$= (\mathbf{3} + \mathbf{b})(x^2 + 3)$$

5. Check your answer.

Is $\quad (3 + b)(x^2 + 3) \quad = \quad 3x^2 + 9 + bx^2 + 3b$?

Is $3(x^2 + 3) + b(x^2 + 3) = 3x^2 + 9 + bx^2 + 3b$?

Is $\quad 3x^2 + 9 + bx^2 + 3b \quad = \quad 3x^2 + 9 + bx^2 + 3b$? Yes.

Sample Problems

1. Factor: $x(x^2 + y) + (-3)(x^2 + y)$

   ☐ a. Identify the terms
   of the polynomial.

   $x(x^2 + y)$ and _____

   ☐ b. Factor each term.

   $x(x^2 + y) = x \cdot (x^2 + y)$
   $(-3)(x^2 + y) =$ _____ $\cdot$ _____

   ☐ c. Find the GCF of the terms.

   GCF = _____

   ☐ d. Rewrite each term of the
   polynomial using
   the GCF.

   $x(x^2 + y) =$ _____ $\cdot (x^2 + y)$
   $(-3)(x^2 + y) =$ _____ $\cdot (x^2 + y)$

   ☐ e. Factor out the GCF.

   = (_____)(_____)

   ☐ f. Check your answer.

2. Factor: $x^2 + xy + 3x + 3y$

   ☐ a. Factor each term.

   $x^2 = x \cdot x$
   $xy = x \cdot y$
   $3x =$ ___ $\cdot$ ___
   $3y =$ ___ $\cdot$ ___

   ☐ b. Group terms with
   common factors.

   $x^2 \quad + \quad xy \quad + \quad 3x \quad + \quad 3y$
   $= (x \cdot x + x \cdot \underline{\quad}) + (3 \cdot \underline{\quad} + 3 \cdot y)$

   ☐ c. Factor out the GCF
   in each grouping.

   $=$ ___$(x + y) +$ ___$(x + y)$

   ☐ d. Factor out the binomial
   GCF of the polynomial.

   $= (x + y)($_____$)$

   ☐ e. Check your answer.

Answers to Sample Problems

a. $(-3)(x^2 + y)$

b. -3, $(x^2 + y)$ *(in either order)*

c. $x^2 + y$

d. x
   -3

e. $x^2 + y$; $x + (-3)$ or $x - 3$
   (in either order)

f. $(x^2 + y)[x + (-3)] =$
   $(x^2 + y)(x) + (x^2 + y)(-3)$

a. 3, x *(in either order)*
   3, y *(in either order)*

b. y, x

c. x, 3

d. $x + 3$

e. $(x + y)(x + 3)$
   $= x(x + 3) + y(x + 3)$
   $= x^2 + 3x + xy + 3y$
   $= x^2 + xy + 3x + 3y$

HOMEWORK

Homework Problems

Circle the homework problems assigned to you by the computer, then complete them below.

☼ Explain

Greatest Common Factor

1. Circle the expressions below that are monomials.

$$x^2 + 2 \qquad xy^2 + y^2x$$

$$x^3yz^2 \qquad x$$

2. Circle the expressions below that are **not** monomials.

$$xzy^8 \qquad \frac{4}{x}$$

$$\frac{13x}{12} \qquad x^2z + zy^2$$

3. Find the GCF of $12x^3y$ and $6xy^2$.

4. Find the GCF of $3xyz^3$, z, and $16yz$.

5. Factor: $x^2y + 6y^2$

6. Factor: $3x^2 + 9xy^3 - 12xy$

7. Factor: $4a^2 - 4b^2$

8. Factor: $3x^4yz + 3xyz + 9yz$

9. Factor: $6xy^3 - 4x^2y^2 + 2xy$

10. Factor: $16a^3b^2 + 20a^2b^4 - 8a^3b^3$

11. Factor: $17x^2y^2z^2 + 68x^{10}y^{32}z + 153x^9y^4z^{12}$

12. Factor: $x^2 + xy + xz$

Factoring by Grouping

13. Find the binomial GCF: $(x^5 + y) + 6x^2(x^5 + y)$

14. Factor: $(x^5 + y) + 6x^2(x^5 + y)$

15. Find the binomial GCF:
$(3x + y)(xy + yz) + x^2y(xy + yz) + z^3(xy + yz)$

16. Factor: $(3x + y)(xy + yz) + x^2y(xy + yz) + z^3(xy + yz)$

17. Factor: $a^3 - a^2b + ab^2 - b^3$

18. Factor: $3x^2 - 3xy + 3xy^3z^4 - 3y^4z^4$

19. Factor: $x^5y + zx + x^4y^2 + yz + x^4yz + z^2$

20. Factor: $15m^3 + 21m^2n + 10mn + 14n^2$

21. Factor: $x^2z + 3x^2 + y^2z + 3y^2$

22. Factor: $x^3 + x^2y + x^2z + 3x + 3z + 3y$

23. Factor: $3x + yz + xz + 3y$

24. Factor: $x^2 - 3x + 2$
(Hint: rewrite the polynomial as $x^2 - x - 2x + 2$)

APPLY

Practice Problems

Here are some additional practice problems for you to try.

Greatest Common Factor

1. Circle the expressions below that are monomials.

    $8m^3n$ $7y - 2y^2 + 14$ $x - y$

    23 $\dfrac{3}{z}$

2. Circle the expressions below that are monomials.

    $3x + 4x^2 - 7$ 17 $5xyz^3$

    $y + z$ $\dfrac{1}{x}$

3. Find the GCF of $12a^3b$ and $16ab^4$.

4. Find the GCF of $18m^3n^5$ and $24m^4n^3$.

5. Find the GCF of $10xy^4$, and $15x^3y^2$.

6. Find the GCF of $9xy^2z^3$, $24x^5y^3z^6$, and $18x^3yz^4$.

7. Find the GCF of $6abc^4$, $12ac^3$, and $9a^5b^4c^2$.

8. Factor: $5a^3b + 10b$

9. Factor: $16mn^4 + 8m$

10. Factor: $6xy^2 + 12x$

11. Factor: $6x^4y^3 + 14xy$

12. Factor: $24mn - 16m^6n^2$

13. Factor: $8a^3b^2 - 10ab$

14. Factor: $24a^3b^4 + 42a^6b^5$

15. Factor: $36y^7z^8 - 45y^3z^5$

16. Factor: $25x^5y^7 + 35x^2y^4$

17. Factor: $4mn + 10mn^3 - 18m^4n$

18. Factor: $6xy + 9x^3y - 15xy^2$

19. Factor: $8a^3b^4 - 12ab + 20a^3b$

20. Factor: $15a^3b^4c^7 + 25a^5b^3c^2$

21. Factor: $32p^7q^3r^4 - 40p^5q^5r$

22. Factor: $24x^2y^5z^8 - 32x^4y^6z^4$

23. Factor: $9xy^2z^3 - 15x^3y^5z^4 + 21x^4y^2z^5$

24. Factor: $10h^4j^3k^6 + 25h^3j^2k - 40hj^5k^2$

25. Factor: $20a^3b^5c^2 + 12a^4b^2c^3 - 8a^2bc^3$

26. Factor: $20x^2y^4 + 10x^5y^3 - 18x^3y^4 + 12xy^3$

27. Factor: $6a^3b^5c^2 - 9a^4b^4c^3 + 18a^2b^3c^2 - 21a^6b^2c^3$

28. Factor: $18x^2y^4z^3 - 16x^5y^3z + 6x^4y^2z^3 - 10x^3y^4z^2$

Factoring by Grouping

29. Factor: $x(z + 3) + y(z + 3)$

30. Factor: $a(b - 2) + c(b - 2)$

31. Factor: $a(3b - 4) + 9(3b - 4)$

32. Factor: $z(2w + 3) - 12(2w + 3)$

33. Factor: $8m(3n^3 - 4) + 17(3n^3 - 4)$

34. Factor: $12b(2c^4 + 5) - 23(2c^4 + 5)$

35. Factor: $7x(2x^2 + 3) - 11(2x^2 + 3)$

36. Factor: $a(3a - b) - b(3a - b)$

37. Factor: $m(5m + 2n) - 3n(5m + 2n)$

38. Factor: $y(2x + y) + x(2x + y)$

39. Factor: $xw + xz + yw + yz$

40. Factor: $mp - mq + np - nq$

41. Factor: $ac + ad - bc - bd$

42. Factor: $8a^2 + 4a + 10a + 5$

43. Factor: $4a^2 + 2a - 14a - 7$

44. Factor: $6x^2 - 2x + 12x - 4$

45. Factor: $12a^2 + 18a + 10ab + 15b$

46. Factor: $21m^2 - 14m + 24mn - 16n$

47. Factor: $15x^2 + 35x + 6xy + 14y$

48. Factor: $3u^2 + 6u + uv + 2v$

49. Factor: $8z^2 - 2z + 4zw - w$

50. Factor: $2x^2 + 4x - xy - 2y$

51. Factor: $12a^2 - 10b - 15ab + 8a$

52. Factor: $8m^2 + 21n + 12m + 14mn$

53. Factor: $18x^2 - 10y - 15xy + 12x$

54. Factor: $16uv^2 + 10vw + 25w + 40uv$

55. Factor: $12pr^2 - 16rs - 20s + 15pr$

56. Factor: $20ab^2 + 15bc - 6c - 8ab$

 **EVALUATE**

Practice Test

Take this practice test to be sure that you are prepared for the final quiz in Evaluate.

1. Find the GCF of $6xz$, $3xy$, and $2x$.

2. Find the GCF of $16xyz$, $x^2y^2z^2$, and $4x^3y^2z$.

3. Factor: $3x^2y - 3xy^2$

4. Factor: $3xy^3 - 6xy^2 + 3x^3y^4$

5. Factor: $13(x^2 + 4) + 6y(x^2 + 4)$

6. Factor: $17x^2(3xyz + 4z) - 3yz(3xyz + 4z)$

7. Factor: $39rs - 13s + 9r - 3$

8. Factor: $12wz - 44z + 18w - 66$

LESSON 7.2 – FACTORING POLYNOMIALS II

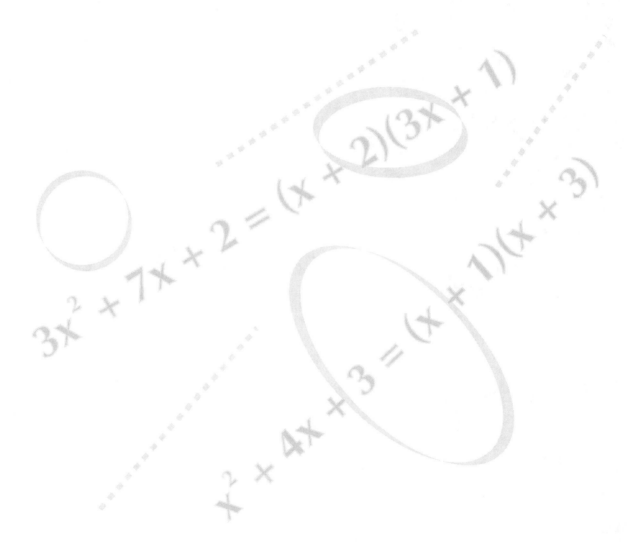

$$3x^2 + 7x + 2 = (x + 2)(3x + 1)$$

$$x^2 + 4x + 3 = (x + 1)(x + 3)$$

OVERVIEW

Here's what you'll learn in this lesson:

Trinomials I

a. Factoring trinomials of the form $x^2 + bx + c$; $x^2 + bxy + cy^2$

Trinomials II

a. Factoring trinomials of the form $ax^2 + bx + c$, $a \neq 1$, by trial-and-error

b. Factoring trinomials of the form $ax^2 + bx + c$, $a \neq 1$, by grouping

c. Solving quadratic equations by factoring

You have already learned how to factor certain polynomials by finding the greatest common factor (GCF) and by grouping.

In this lesson, you will learn techniques for factoring trinomials. Then you will see how to use factoring to solve certain equations.

 **EXPLAIN**

TRINOMIALS I

Summary

Factoring Polynomials of the Form $x^2 + bx + c$

One way to factor a polynomial of the form $x^2 + bx + c$ as a product of binomials is to use the FOIL method, but work backwards. Here's an example.

The product of the first terms is x^2

$$x^2 - 3x - 4 = (x \quad)(x \quad)$$

The product of the last terms is -4

$$x^2 - 3x - 4 = (x \quad)(x \quad)$$

Try all the possible factorizations for which the product of the first terms is x^2 and the product of the last terms is -4. Since the product of the last terms is negative, one of the last terms is positive and the other is negative. Use the FOIL method to find factors whose "inner" and "outer" products add together to make $-3x$.

1. Make a chart of the possibilities for the binomial factors. These are shown in the table.

possible factorizations
$(x + 4)(x - 1)$
$(x - 4)(x + 1)$
$(x + 2)(x - 2)$

2. Use the FOIL method to multiply the possible factorizations you listed in step (1). These are shown in the table.

possible factorizations
$(x + 4)(x - 1) \ = \ x^2 + 3x - 4$
$(x - 4)(x + 1) \ = \ x^2 - 3x - 4$
$(x + 2)(x - 2) \ = \ x^2 - 4$

3. Find the factorization that gives the original polynomial, $x^2 - 3x - 4$. In the second row you see that $x^2 - 3x - 4 = (x - 4)(x + 1)$.

    So the factorization is: $x^2 - 3x - 4 = (x - 4)(x + 1)$.

Sample Problems

1. Factor: $x^2 + 3x + 2$

   ☑ a. List all the possible factorizations where:

   - the product of the first terms is x^2
   - the product of the last terms is $+2$

   Since the product of the last terms is positive, both of the last terms are positive or both are negative.

possible factorizations
$(x + 1)(x + 2)$
$(x - 1)(x - 2)$

   ☐ b. Multiply the possible factorizations. Identify the factorization that gives the middle term $+3x$.

possible factorizations
$(x + 1)(x + 2)$ = _____
$(x - 1)(x - 2)$ = $x^2 - 3x + 2$

b. $x^2 + 3x + 2$

   ☐ c. Write the correct factorization. $x^2 + 3x + 2 =$ _____

c. $(x + 1)(x + 2)$ (in either order)

2. Factor: $x^2 - 7x + 12$

   ☐ a. List all the possible factorizations where:

   - the product of the first terms is x^2
   - the product of the last terms is $+12$

   Since the product of the last terms is positive, both of the last terms are positive or both are negative.

possible factorizations
$(x + 1)(x + 12)$
$(x - 1)(x - 12)$
$(x + 2)(x + 6)$
$(x - 2)(x - \_\_)$
$(x + \_\_)(x + 4)$
$(x - 3)(x - 4)$

a. 6

3

b. Multiply the possible factorizations. Identify the factorization that gives the middle term $-7x$.

possible factorizations
$(x + 1)(x + 12)$ = $x^2 + 13x + 12$
$(x - 1)(x - 12)$ = _____
$(x + 2)(x + 6)$ = _____
$(x - 2)(x - \underline{})$ = $x^2 - 8x + 12$
$(x + \underline{})(x + 4)$ = $x^2 + 7x + 12$
$(x - 3)(x - 4)$ = _____

c. Write the correct factorization. $x^2 - 7x + 12 =$ _____

3. Factor: $x^2 + x - 2$

a. List all the possible factorizations where:
   • the product of the first terms is x^2
   • the product of the last terms is -2

   Since the product of the last terms is negative, one of the last terms is positive and the other is negative.

possible factorizations
$(x + 1)(x - 2)$
$(x - 1)(\underline{})$

b. Multiply the possible factorizations.
   Identify the factorization that gives the middle term $+1x$.

possible factorizations
$(x + 1)(x - 2)$ = _____
$(x - 1)()$ = _____

c. Write the correct factorization. $x^2 + x - 2 =$ _____ .

4. Factor: $x^2 + 2x - 2$

☐ a. List all the possible factorizations where:
- the product of the first terms is x^2
- the product of the last terms is -2

Since the product of the last terms is negative, one of the last terms is positive and the other is negative.

possible factorizations
$(x + 1)(x - 2)$
$(x - 1)(\underline{\quad\quad})$

☐ b. Multiply the possible factorizations.
Identify the factorization that gives the middle term $+2x$.

possible factorizations
$(x + 1)(x - 2)$ = _____
$(x - 1)(\underline{\quad\quad})$ = _____

☑ c. Write the correct factorization. Neither of the possible factorizations gives the original polynomial, $x^2 + 2x - 2$. So, $x^2 + 2x - 2$ cannot be factored using integers.

TRINOMIALS II

Summary

Factoring Polynomials of the Form $ax^2 + bx + c$ by Trial and Error

You have learned how to factor trinomials of the form $x^2 + bx + c$, where b and c are integers. Notice that the coefficient of x^2 is 1.

Now you will see how to factor trinomials of the form $ax^2 + bx + c$, where a, b, and c are integers. Notice that the coefficient of x^2 can be an integer other than 1.

One way to factor a trinomial of the form $ax^2 + bx + c$ as a product of binomials is by trial and error. Here's an example.

Factor the trinomial $3x^2 - 14x - 5$ using trial and error. Notice that any factorization of this trinomial must look like this:

$$3x^2 - 14x - 5 = (?x \quad ?)(?x \quad ?)$$

The product of the x-terms must be $3x^2$ and the product of the constants must be -5. Since the product of the constants is negative, one of the constants is positive and the other is negative.

1. Make a chart of the possibilities for the x-terms in the binomial factors and possibilities for the constant terms in the binomial factors. These are shown in the table below.

x-terms	constants
$3x, x$	$1, -5$
$3x, x$	$5, -1$
$3x, x$	$-1, 5$
$3x, x$	$-5, 1$

2. Use the values from step (1) to list possible factorizations. These are shown in the table below.

x-terms	constants	possible factorizations
$3x, x$	$1, -5$	$(3x + 1)(x - 5)$
$3x, x$	$5, -1$	$(3x + 5)(x - 1)$
$3x, x$	$-1, 5$	$(3x - 1)(x + 5)$
$3x, x$	$-5, 1$	$(3x - 5)(x + 1)$

3. Use the FOIL method to do the multiplication of the possible factorizations you listed in step (2). These are shown in the table below.

x-terms	constants	possible factorizations
3x, x	1, −5	$(3x + 1)(x - 5) = 3x^2 - 14x - 5$
3x, x	5, −1	$(3x + 5)(x - 1) = 3x^2 + 2x - 5$
3x, x	−1, 5	$(3x - 1)(x + 5) = 3x^2 + 14x - 5$
3x, x	−5, 1	$(3x - 5)(x + 1) = 3x^2 - 2x - 5$

4. Find the factorization that equals the original polynomial, $3x^2 - 14x - 5$. You can see that the shaded row is $3x^2 - 14x - 5$. So the factorization is:
$3x^2 - 14x - 5 = (3x + 1)(x - 5)$

Here's another example. Factor the trinomial $15x^2 - 16x + 4$ using trial and error. Notice that any factorization of this trinomial must look like this:

$15x^2 - 16x + 4 = (?x \quad ?)(?x \quad ?)$

The product of the x-terms must be $15x^2$ and the product of the constant terms must be $+4$. Since the product of the last terms is positive, both of the last terms are positive or both are negative.

1. Make a chart of the possibilities for the x-terms in the binomial factors and possibilities for the constant terms in the binomial factors. These are shown in the table below.

x-terms	constants
x, 15x	1, 4
x, 15x	2, 2
x, 15x	4, 1
x, 15x	−1, −4
x, 15x	−2, −2
x, 15x	−4, −1
3x, 5x	1, 4
3x, 5x	2, 2
3x, 5x	4, 1
3x, 5x	−1, −4
3x, 5x	−2, −2
3x, 5x	−4, −1

2. Use the values from step (1) to list possible factorizations. These are shown in the table that follows.

x-terms	constants	possible factorizations
$x, 15x$	1, 4	$(x + 1)(15x + 4)$
$x, 15x$	2, 2	$(x + 2)(15x + 2)$
$x, 15x$	4, 1	$(x + 4)(15x + 1)$
$x, 15x$	$-1, -4$	$(x - 1)(15x - 4)$
$x, 15x$	$-2, -2$	$(x - 2)(15x - 2)$
$x, 15x$	$-4, -1$	$(x - 4)(15x - 1)$
$3x, 5x$	1, 4	$(3x + 1)(5x + 4)$
$3x, 5x$	2, 2	$(3x + 2)(5x + 2)$
$3x, 5x$	4, 1	$(3x + 4)(5x + 1)$
$3x, 5x$	$-1, -4$	$(3x - 1)(5x - 4)$
$3x, 5x$	$-2, -2$	$(3x - 2)(5x - 2)$
$3x, 5x$	$-4, -1$	$(3x - 4)(5x - 1)$

3. Use the FOIL method to do the multiplication of the possible factorizations you listed in step (2). These are shown in the table.

x-terms	constants	possible factorizations
$x, 15x$	1, 4	$(x + 1)(15x + 4) = 15x^2 + 19x + 4$
$x, 15x$	2, 2	$(x + 2)(15x + 2) = 15x^2 + 32x + 4$
$x, 15x$	4, 1	$(x + 4)(15x + 1) = 15x^2 + 61x + 4$
$x, 15x$	$-1, -4$	$(x - 1)(15x - 4) = 15x^2 - 19x + 4$
$x, 15x$	$-2, -2$	$(x - 2)(15x - 2) = 15x^2 - 32x + 4$
$x, 15x$	$-4, -1$	$(x - 4)(15x - 1) = 15x^2 - 61x + 4$
$3x, 5x$	1, 4	$(3x + 1)(5x + 4) = 15x^2 + 17x + 4$
$3x, 5x$	2, 2	$(3x + 2)(5x + 2) = 15x^2 + 16x + 4$
$3x, 5x$	4, 1	$(3x + 4)(5x + 1) = 15x^2 + 23x + 4$
$3x, 5x$	$-1, -4$	$(3x - 1)(5x - 4) = 15x^2 - 17x + 4$
$3x, 5x$	$-2, -2$	$(3x - 2)(5x - 2) = 15x^2 - 16x + 4$
$3x, 5x$	$-4, -1$	$(3x - 4)(5x - 1) = 15x^2 - 23x + 4$

4. Find the factorization that equals the original polynomial, $15x^2 - 16x + 4$. You can see that the shaded row is $15x^2 - 16x + 4$. So the factorization is:
$$15x^2 - 16x + 4 = (3x - 2)(5x - 2)$$

Here's another example. Factor the trinomial $3x^2 - 8x - 5$ using trial and error. Notice that any factorization of this trinomial must look like this:
$$3x^2 - 8x - 5 = (?x \quad ?)(?x \quad ?)$$

The product of the x-terms must be $3x^2$ and the product of the constants must be -5. Since the product of the constants is negative, one of the constants is positive and the other is negative.

1. Make a chart of the possibilities for the x-terms in the binomial factors and possibilities for the constant terms in the binomial factors. These are shown in the table below.

x-terms	constants
$3x$, x	1, −5
$3x$, x	5, −1
$3x$, x	−1, 5
$3x$, x	−5, 1

2. Use the values from step (1) to list possible factorizations. These are shown in the table below.

x-terms	constants	possible factorizations
$3x$, x	1, −5	$(3x + 1)(x − 5)$
$3x$, x	5, −1	$(3x + 5)(x − 1)$
$3x$, x	−1, 5	$(3x − 1)(x + 5)$
$3x$, x	−5, 1	$(3x − 5)(x + 1)$

3. Use the FOIL method to do the multiplication of the possible factorizations you listed in step (2). These are shown in the table below.

x-terms	constants	possible factorizations
$3x$, x	1, −5	$(3x + 1)(x − 5) = 3x^2 − 14x − 5$
$3x$, x	5, −1	$(3x + 5)(x − 1) = 3x^2 + 2x − 5$
$3x$, x	−1, 5	$(3x − 1)(x + 5) = 3x^2 + 14x − 5$
$3x$, x	−5, 1	$(3x − 5)(x + 1) = 3x^2 − 2x − 5$

4. Find the factorization that equals the original polynomial, $3x^2 − 8x − 5$. You can see that no row is $3x^2 − 8x − 5$. So, $3x^2 − 8x − 5$ cannot be factored using integers.

Factoring Polynomials of the Form $ax^2 + bx + c$ by Grouping

Another way to factor a trinomial of the form $ax^2 + bx + c$ is by grouping.

Remember how to multiply binomials using the FOIL method.

$$(x + 2)(3x + 1) = 3x^2 + x + 6x + 2$$
$$= 3x^2 + 7x + 2$$

To factor $3x^2 + 7x + 2$, we go the other way. We first write $3x^2 + 7x + 2$ using two x-terms, like this:

$$3x^2 + x + 6x + 2$$

Now, factor $3x^2 + x + 6x + 2$ by grouping:

1. Factor each term.

    $$3x^2 = 3 \cdot x \cdot x$$
    $$x = x$$
    $$6x = 2 \cdot 3 \cdot x$$
    $$2 = 2$$

2. Group terms with common factors.

    $= (3x^2 + x) + (6x + 2)$

3. Factor out the GCF in each grouping.

    $= x(3x + 1) + 2(3x + 1)$

4. Factor out the binomial GCF of the polynomial.

    $= (3x + 1)(x + 2)$

5. Check your answer.

    Is $(3x + 1)(x + 2) = 3x^2 + 7x + 2$?

    Is $3x^2 + 7x + 2 = 3x^2 + 7x + 2$? Yes.

In order to use grouping to factor this trinomial, you had to find two integers whose sum was 7 and whose product was 6.

To factor a trinomial of the form $ax^2 + bx + c$, you need to find two integers whose sum is b and whose product is ac. Then you can split the x-term into two terms and factor by grouping.

For example, to factor $6x^2 + 7x + 2$ by grouping:

1. Make a chart of possible pairs of integers product is $6 \cdot 2 = 12$.

possibilities	product	sum
1, 12	12	13
2, 6	12	8
3, 4	12	7

2. Identify the numbers that work. Here, the last choice works since $3 + 4 = 7$ and $3 \cdot 4 = 12$.

3. Rewrite the trinomial.

    $6x^2 + 7x + 2 = 6x^2 + 3x + 4x + 2$

4. Group the terms.

    $= (6x^2 + 3x) + (4x + 2)$

5. Factor out the GCF in each grouping.

    $= 3x(2x + 1) + 2(2x + 1)$

6. Factor out the binomial GCF of the polynomial.

    $= (2x + 1)(3x + 2)$

7. Check your answer. Is $(2x + 1)(3x + 2) = 6x^2 + 7x + 2$?

    Is $6x^2 + 4x + 3x + 2 = 6x^2 + 7x + 2$? Yes.

Notice that the chart doesn't include negative factors of 12. Can you see why not? Since the product of the two numbers has to be +12, if one factor is negative, both would have to be negative. But since the sum of the integers needs to be +7, a positive number, you know both factors can't be negative.

Solving Quadratic Equations of the Form $ax^2 + bx + c = 0$ by Factoring

You can use what you have learned about factoring to solve some quadratic equations.

A quadratic (or second-degree) equation in one variable is an equation that can be written in this form:

$$ax^2 + bx + c = 0$$

This is called standard form. Here, a, b, and c are real numbers, and $a \neq 0$. The terms on the left side of the equation are in descending order by degree. The right side of the equation is zero.

If the left side of a quadratic equation in standard form can be factored, then you can solve the quadratic equation by factoring. To solve such an equation, you'll use a property called the Zero Product Property, which states the following: if P and Q are polynomials and if $P \cdot Q = 0$, then $P = 0$ or $Q = 0$ or both P and Q are 0.

Here's how to solve a quadratic equation in standard form when the left side can be factored:

1. Make sure the equation is in standard form.

2. Factor the left side.

3. Use the Zero Product Property. Set each factor equal to zero.

4. Finish solving for x.

5. Check your answer.

For example, to solve the equation $x^2 = 4x$:

1. Write the equation in standard form.		$x^2 - 4x = 0$
2. Factor the left side.		$x(x - 4) = 0$
3. Use the Zero Product Property to set each factor equal to zero.		$x = 0$ or $x - 4 = 0$
4. Finish solving for x.		$x = 0$ or $x = 4$
5. Check your answer.		

Check $x = 0$:   Check $x = 4$:

Is $0^2 = 4(0)$?   Is $4^2 = 4(4)$?

Is $0 = 0$? Yes.   Is $16 = 16$? Yes.

So, both 0 and 4 are valid solutions of the equation $x^2 = 4x$.

Sample Problems

1. Use trial and error to factor the polynomial $35x^2 + 73x + 6$.

   ☐ a. Write possible x-terms whose product is $35x^2$ and write possible constant terms whose product is 6.

   ☐ b. List the possible factorizations.

   ☐ c. Multiply the possible factorizations.

x-terms	constants	possible factorizations
$x, 35x$	1, 6	$(x + 1)(35x + 6) = 35x^2 + 41x + 6$
$x, 35x$	2, __	$(\_\_\_)(\_\_\_) = _____$
$x, 35x$	3, 2	$(x + 3)(35x + 2) = 35x^2 + 107x + 6$
$x, 35x$	6, __	$(\_\_\_)(\_\_\_) = _____$
$x, 35x$	__, −6	$(x − 1)(35x − 6) = 35x^2 − 41x + 6$
$x, 35x$	−2, __	$(\_\_\_)(\_\_\_) = _____$
$x, 35x$	−3, −2	$(x − 3)(35x − 2) = 35x^2 − 107x + 6$
$x, 35x$	−6, __	$(\_\_\_)(\_\_\_) = _____$
$5x, 7x$	1, 6	$(5x + 1)(7x + 6) = _____$
$5x, 7x$	2, __	$(\_\_\_)(\_\_\_) = 35x^2 + 29x + 6$
$5x, 7x$	3, 2	$(5x + 3)(7x + 2) = 35x^2 + 31x + 6$
$5x, 7x$	6, __	$(5x + 6)(7x + 1) = _____$
$5x, 7x$	−1, __	$(\_\_\_)(\_\_\_) = _____$
$5x, 7x$	__, −3	$(\_\_\_)(\_\_\_) = _____$
$5x, 7x$	−3, −2	$(5x − 3)(7x − 2) = 35x^2 − 31x + 6$
$5x, 7x$	__, −1	$(\_\_\_)(\_\_\_) = _____$

a., b., c.

3, $(x + 2)(35x + 3) = 35x^2 + 73x + 6$

1, $(x + 6)(35x + 1) = 35x^2 + 211x + 6$

−1

−3, $(x − 2)(35x − 3) = 35x^2 − 73x + 6$

−1, $(x − 6)(35x − 1) = 35x^2 − 211x + 6$

$35x^2 + 37x + 6$

3, $(5x + 2)(7x + 3)$

1, $35x^2 + 47x + 6$

−6, $(5x − 1)(7x − 6) = 35x^2 − 37x + 6$

−2, $(5x − 2)(7x − 3) = 35x^2 − 29x + 6$

−6, $(5x − 6)(7x − 1) = 35x^2 − 47x + 6$

   ☐ d. Write the correct factorization. $35x^2 + 73x + 6 = _____$ ("in either order")

2. Use trial and error to factor $4x^2 − 4x − 15$.

   ☐ a. Write possible x-terms whose product is $4x^2$ and the possible constant terms whose product is −15.

   ☐ b. List the possible factorizations.

   ☐ c. Multiply the possible factorizations.

x-terms	constants	possible factorizations
x, 4x	1, −15	$(x + 1)(4x − 15) = 4x^2 − 11x − 15$
2x, 2x	1, −15	$(2x + 1)(2x − 15) = 4x^2 − 28x − 15$
x, 4x	3, −5	$(x + 3)(4x − 5) = 4x^2 + 7x − 15$
2x, 2x	3, −5	$(2x + 3)(2x − 5) = 4x^2 − 4x − 15$
x, 4x	5, −3	$(x + 5)(4x − 3) = 4x^2 + 17x − 15$
2x, 2x	5, −3	$(2x + 5)(2x − 3) = 4x^2 + 4x − 15$
x, 4x	15, ___	(____)(____) = _____
2x, 2x	15, ___	(____)(____) = _____
x, 4x	−1, ___	(____)(____) = _____
2x, 2x	−1, ___	(____)(____) = _____
x, 4x	___, 5	(____)(____) = _____
2x, 2x	___, 5	(____)(____) = _____
x, 4x	___, 3	(____)(____) = _____
2x, 2x	___, 3	(____)(____) = _____
x, 4x	−15, ___	(____)(____) = _____
2x, 2x	−15, ___	(____)(____) = _____

a., b., c.

$-1, x +15, 4x − 1, 4x^2 + 59x −15$

$-1, 2x +15, 2x − 1, 4x^2 + 28x −15$

$15, x − 1, 4x + 15, 4x^2 + 11x −15$

$15, 2x − 1, 2x + 15, 4x^2 + 28x −15$

$-3, x − 3, 4x + 5, 4x^2 − 7x −15$

$-3, 2x − 3, 2x + 5, 4x^2 + 4x −15$

$-5, x − 5, 4x + 3, 4x^2 − 17x −15$

$-5, 2x − 5, 2x + 3, 4x^2 − 4x −15$

$1, x − 15, 4x + 1, 4x^2 − 59x −15$

$1, 2x − 15, 2x + 1, 4x^2 − 28x −15$

☐ d. Write the correct factorization. $4x^2 − 4x − 15 = ($ _____ $)($ _____ $)$

d. 2x − 5, 2x + 3 (in either order)

3. Use grouping to factor $6x^2 + 11x + 4$.

☑ a. Make a chart of pairs of integers whose product is $6 \cdot 4 = 24$.

possibilities	product	sum
1, 24	24	25
2, 12	24	14
3, 8	24	11
4, 6	24	10

☑ b. Identify the two integers whose product is 24 and whose sum is 11.   The two integers are 3 and 8.

☑ c. Rewrite the trinomial by $6x^2 + 11x + 4 = 6x^2 + 3x + 8x + 4$
splitting the x-term.

☐ d. Group the terms. $= ($ _____ $) + ($ _____ $)$

d. $6x^2 + 3x, 8x + 4$

☐ e. Factor out the GCF in each grouping. $=$ ___ $(2x + 1) +$ ___ $(2x + 1)$

e. 3x, 4

☐ f. Factor out the binomial $= (2x + 1)($ _____ $)$
GCF of the polynomial.

f. $(3x + 4)$

☐ g. Check your answer.

4. Use grouping to factor $3x^2 - 4x - 15$.

   ☐ a. Make a chart of pairs of integers whose product is $3 \cdot (-15) = -45$.

possibilities	product	sum
−1, 45	−45	44
−3, 15	−45	___
−5, ___	−45	___
1, ___	−45	___
___, −15	−45	___
___, ___	−45	___

   ☐ b. Identify the two integers whose product is −45 and whose sum is −4.   The two integers are _____ and _____.

   ☐ c. Rewrite the trinomial by splitting the x-term. $3x^2 - 4x - 15 = 3x^2 + $ ___$x + $ ___$x - 15$

   ☐ d. Group the terms. $= ($ _____ $) + ($ _____ $)$

   ☐ e. Factor out the GCF in each grouping. $= $ __$($ _____ $) + $ __$($ _____ $)$

   ☐ f. Factor out the binomial GCF of the polynomial. $= ($ _____ $)($ _____ $)$

   ☐ g. Check your answer.

5. Solve this quadratic equation for x by factoring: $8x^2 = 26x + 45$

   ☑ a. Write the equation in standard form. $8x^2 - 26x - 45 = 0$

   ☐ b. Factor the left side. $($ _____ $)($ _____ $) = 0$

   ☐ c. Use the Zero Product Property. $4x + 5 = $ ___ or $2x - 9 = $ ___

   ☐ d. Finish solving for x. $4x = -5$ or $2x = 9$

                         $x = $ ___ or $x = $ ___

   ☐ e. Check your answer.

EXPLORE

Sample Problems

On the computer you used overlapping circles to help find the GCF of a collection of monomials. You used a table to help factor polynomials. Below are some additional problems.

1. Use overlapping circles to find the GCF of $3x$ and $-9xy^3$.

☑ a. Factor each monomial.
$$3x = 3 \cdot x$$
$$-9xy^3 = -1 \cdot 3 \cdot 3 \cdot x \cdot y \cdot y \cdot y$$

☑ b. Write the factorizations in the overlapping circles.

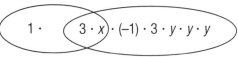

c. 3x

☐ c. Find the GCF from the overlapping circles.

$$GCF = \underline{\hspace{2cm}}$$

2. Factor: $\frac{1}{5}x^2y - \frac{3}{5}xy$

a. $\frac{1}{5} \cdot x \cdot x \cdot y$

$\frac{1}{5} \cdot 3 \cdot x \cdot y$

☐ a. Factor each monomial.
$$\frac{1}{5}x^2y = \underline{\hspace{2cm}}$$
$$\frac{3}{5}xy = \underline{\hspace{2cm}}$$

b. $\frac{1}{5}xy$

☐ b. Find the GCF of $\frac{1}{5}x^2y$ and $\frac{3}{5}xy$.
$$GCF = \underline{\hspace{2cm}}$$

c. $\frac{1}{5}xy(x-3)$

☐ c. Factor the polynomial $\frac{1}{5}x^2y - \frac{3}{5}xy$.
$$(\underline{\hspace{1.5cm}})(\underline{\hspace{1.5cm}})$$

3. Find the GCF of the polynomials below.

A: $22x^2z + 22yz$

B: $11x^3 + 11xy$

C: $2x^2 + 2y$

☐ a. Factor each polynomial. $22x^2z + 22yz = 2 \cdot 11 \cdot z(x^2 + y)$

$11x^3 + 11xy = $ _____

$2x^2 + 2y = $ _____

☐ b. Finish writing the
factorizations in the
overlapping circles.

4. A trinomial with a missing constant term has been partially factored in the table below. Complete the table and write the polynomial and its factorization.

☐ a. What times x gives $7x$?
Use this to fill in box a.

☐ b. What times $3x$ gives $-9x$?
Use this to fill in box b.

☐ c. Multiply boxes a and b.
Use this to fill in box c.

☐ d. Write the polynomial
and its factorization.

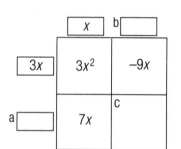

$($ _____ $) = ($ _____ $) + ($ _____ $)$

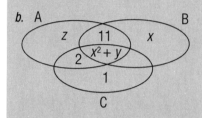
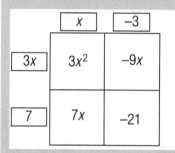

 **HOMEWORK**

Homework Problems

Circle the homework problems assigned to you by the computer, then complete them below.

 Explain

Trinomials I

1. Factor: $x^2 + 7x + 12$

2. Factor: $y^2 + 9y + 18$

3. Factor: $x^2 + 12x + 35$

4. Factor: $z^2 + 10z + 16$

5. Factor: $x^2 - 5x - 24$

6. Factor: $a^2 - 15a - 16$

7. Factor: $x^2 - x - 6$

8. Factor: $x^2 + 10x - 11$

9. Factor: $x^2 - 4x - 21$

10. Factor: $y^2 + 3y - 40$

11. Factor: $x^2 + 35x - 36$

12. Factor: $a^2 - 9a + 14$

Trinomials II

13. Factor: $2x^2 + 11x + 5$

14. Factor: $3x^2 + 13x + 4$

15. Factor: $4y^2 - 8y - 21$

16. Factor: $3z^2 - 17z + 20$

17. Factor: $15a^2 - 30a + 15$

18. Solve for x by factoring: $6x^2 = 63 - 13x$

19. Solve for x by factoring: $25x^2 + 5x = 2$

20. Factor: $4x^2 - 12x + 9$

21. Factor: $13x^2 + 37x + 22$

22. Factor: $x^2 - a^2$

23. Factor: $x^2 + 2xy + y^2$

24. Factor: $x^4 - 2ax^2 + a^2$

Explore

25. Circle the monomial(s) below that might appear in the factorization of

    $$3x^3y^2 + 2x^2y - 3xy$$

    $3x^2y$ $2x^2y$ xy $3x$

26. If the GCF of the terms of a polynomial is $4x^2y^3$, which of the monomials below could be terms in the polynomial?

    $4xy^3$ $8x^3y^4$ $4x^2y^3$ $4x^2$

27. Factor this polynomial using overlapping circles: $\frac{x^2y}{2} - \frac{2y}{4}$

28. A trinomial with a missing constant term has been partially factored in the table below. Complete the table and write the polynomial and its factorization.

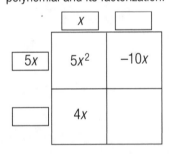

29. Complete the diagram below to find the GCF of the polynomials A, B, and C.

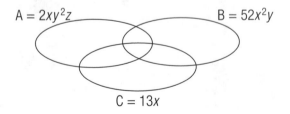

30. Factor this polynomial using overlapping circles:

    $$\frac{1}{2}x^2y^2 + \frac{3}{2}x^3y^3 - 3x^2y$$

 APPLY

Practice Problems

Here are some additional practice problems for you to try.

Trinomials I

1. Factor: $x^2 + 5x + 4$
2. Factor: $x^2 - 4x + 5$
3. Factor: $x^2 + 15x + 14$
4. Factor: $x^2 + 11x + 10$
5. Factor: $x^2 + 8x + 15$
6. Factor: $x^2 + 9x + 18$
7. Factor: $x^2 + 7x + 12$
8. Factor: $x^2 - 13x + 30$
9. Factor: $x^2 - 8x + 12$
10. Factor: $x^2 - 7x + 10$
11. Factor: $x^2 - 15x + 44$
12. Factor: $x^2 - 11x + 30$
13. Factor: $x^2 - 10x + 21$
14. Factor: $x^2 - 6x - 27$
15. Factor: $x^2 - 7x - 30$
16. Factor: $x^2 - 5x - 14$
17. Factor: $x^2 + 4x - 21$
18. Factor: $x^2 + 10x - 24$
19. Factor: $x^2 + 5x - 36$
20. Factor: $x^2 + 2x - 15$
21. Factor: $x^2 - 7x - 18$
22. Factor: $x^2 + 9x - 36$
23. Factor: $x^2 - 4x - 21$
24. Factor: $x^2 + 10x + 24$
25. Factor: $x^2 - 2x - 63$
26. Factor: $x^2 + 9x - 22$
27. Factor: $x^2 - 7x - 60$
28. Factor: $x^2 - 6x - 91$

Trinomials II

29. Factor: $2x^2 + 7x + 5$
30. Factor: $2x^2 + 9x + 9$
31. Factor: $3x^2 - 19x - 14$
32. Factor: $2x^2 - 3x - 20$
33. Factor: $2x^2 - x - 28$
34. Factor: $3x^2 + 16x - 35$
35. Factor: $2x^2 + 5x - 12$
36. Factor: $2x^2 + 9x - 5$
37. Factor: $2x^2 + 13x + 15$
38. Factor: $2x^2 + 15x + 28$
39. Factor: $3x^2 + 11x + 6$
40. Factor: $12x^2 - 7x + 1$
41. Factor: $10x^2 - 9x + 2$
42. Factor: $6x^2 - 5x + 1$
43. Factor: $6x^2 - 11x - 10$
44. Factor: $9x^2 - 18x - 7$
45. Factor: $8x^2 - 2x - 3$
46. Factor: $6x^2 + 21x - 28$
47. Factor: $9x^2 - 3x - 20$
48. Factor: $4x^2 - 4x - 15$
49. Factor: $36x^2 + 13x + 1$
50. Factor: $30x^2 + 11x + 1$
51. Factor: $5x^2 + 14xy - 3y^2$
52. Factor: $4x^2 - 7xy - 2y^2$
53. Factor: $3x^2 - 5xy - 2y^2$
54. Factor: $6x^2 + xy - 12y^2$
55. Factor: $9x^2 - 3xy - 2y^2$
56. Factor: $4x^2 - 4xy - 3y^2$

EVALUATE

Practice Test

Take this practice test to be sure that you are prepared for the final quiz in Evaluate.

1. Factor: $x^2 - 10x + 24$

2. Circle the statement(s) below that are true.

$x^2 + 2x - 1 = (x - 1)(x - 1)$

$x^2 + 2x - 1 = (x + 2)(x - 1)$

$x^2 + 2x - 1 = (x - 1)(x + 1)$

$x^2 + 2x - 1 = (x + 1)(x + 1)$

$x^2 + 2x - 1$ cannot be factored using integers

3. Factor: $t^2 - 16t - 17$

4. Factor: $r^2 + 10rt + 25t^2$

5. Factor: $5x^2 + 8x - 4$

6. Factor: $27v^2 - 57v + 28$

7. Factor: $4x^2 + 57x + 108$

8. Solve for x by factoring: $7x^2 - 5x - 12 = 0$

9. The overlapping circles contain the factors of three monomials, A, B, and C.
Circle the true statements below.

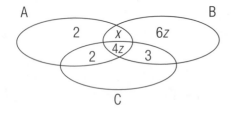

Two factors of C are z and 2.

$B = 72xz$

The GCF of A and B is x.

The GCF of A, B, and C is $4z$.

10. The overlapping circles contain the factors of two binomials, A and B. Their GCF is $(3u + 4v)$. What are A and B?

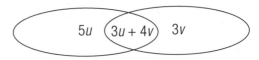

11. The polynomial $14xy + 21y - 6x^2 - 9x$ can be grouped as two binomials: $(14xy - 6x^2) + (21y - 9x)$. Find the GCF of the two binomials by factoring the polynomial using the overlapping circles below.

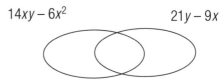

12. Finish factoring the trinomial $6x^2 - 7xy - 3y^2$ using the table below.

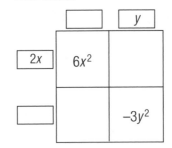

OVERVIEW

Here's what you'll learn in this lesson:

Recognizing Patterns

a. Factoring a perfect square trinomial

b. Factoring a difference of two squares

c. Factoring a sum and difference of two cubes

d. Factoring using a combination of methods

A shortcut can make you more efficient by reducing the amount of time it takes to accomplish a task. It's always nice when you discover a shortcut: for taking notes in class, for programming a VCR, or for getting to a friend's house.

There are shortcuts that you can use in algebra to help you solve problems. For instance, when factoring polynomials, there are patterns you can look for that will help you factor the polynomials more quickly and accurately.

In this lesson you will learn how to recognize patterns for factoring polynomials.

 EXPLAIN

RECOGNIZING PATTERNS

Summary

Factoring by patterns means recognizing that polynomials having a certain form will always factor in a specific way. Perfect square trinomials, differences of two squares, and differences and sums of two cubes can all be factored using patterns. These patterns are described below.

The patterns are just a shortcut. You can always factor using algebra tiles or trial and error.

Perfect Square Trinomials

One type of polynomial that's easy to factor using a pattern is a perfect square trinomial.

A perfect square trinomial is a polynomial that can be written so that it:

- has three terms
- has a first term that is a perfect square: a^2
- has a third term that is a perfect square: b^2
- has a second term that is twice the product of a and b: $2ba$

For example, the polynomials below are perfect square trinomials:

$x^2 + 6x + 9$ $\qquad\qquad$ $w^2 - 12w + 36$

$x^2 - 2xy + y^2$ $\qquad\qquad$ $4y^4 + 24xy^2 + 36x^2$

The patterns for factoring perfect square trinomials are:

$$\triangle^2 + 2\,\square\,\triangle + \square^2 = (\triangle + \square)^2$$

$$a^2 + \quad 2ba \quad + b^2 = (a + b)^2$$

$$\triangle^2 - 2\,\square\,\triangle + \square^2 = (\triangle - \square)^2$$

$$a^2 - \quad 2ba \quad + b^2 = (a - b)^2$$

Notice that the only difference between these two patterns is the sign of the middle term.

For example, to factor $x^2 + 2x + 1$:

1. Decide which pattern to use. $\qquad a^2 + \quad 2ba \quad + b^2 = (a + b)^2$

2. Substitute x for a and 1 for b. $\qquad x^2 + 2(1)(x) + 1^2 = (x + 1)^2$

So, $x^2 + 2x + 1 = (x + 1)^2$.

As another example, to factor $x^2 - 4x + 4$:

1. Decide which pattern to use. $a^2 - 2ba + b^2 = (a - b)^2$

2. Substitute x for a and 2 for b. $x^2 - 2(2)(x) + 2^2 = (x - 2)^2$

So, $x^2 - 4x + 4 = (x - 2)^2$.

Difference of Two Squares

Another type of polynomial that can be factored using a pattern is a difference of two squares.

A difference of two squares is a polynomial that can be written so that it:

- has two terms
- has a first term that is a perfect square: a^2
- has a second term that is a perfect square: b^2
- has a minus sign between the terms

For example, the polynomials below are differences of two squares.

$x^2 - 25$ $\qquad\qquad\qquad$ $y^2 - 100$

$w^4 - x^2$ $\qquad\qquad\qquad$ $9x^2 - 81z^6$

The pattern for factoring a difference of two squares is:

$a^2 - b^2 = (a + b)(a - b)$

For example, to factor $9y^2 - 25$:

1. Use this pattern. $a^2 - b^2 = (a + b)(a - b)$

2. Substitute $3y$ for a and 5 for b. $(3y)^2 - 5^2 = (3y + 5)(3y - 5)$

So, $9y^2 - 25 = (3y + 5)(3y - 5)$.

Differences or Sums of Two Cubes

You can also use patterns to factor a difference of two cubes or a sum of two cubes. These are polynomials that:

- have two terms
- have a first term that is a perfect cube: a^3
- have a second term that is a perfect cube: b^3

For example, the polynomials below are differences or sums of two cubes.

$x^3 - 27$ $\qquad\qquad\qquad$ $64y^3 + 1$

$x^3 + y^3$ $\qquad\qquad\qquad$ $8w^3 - 125x^6$

Why don't these polynomials have a middle term? Try multiplying $(a + b)(a - b)$ using the FOIL method and see what happens.

*You can only use this pattern for factoring a **difference** of two squares. A **sum** of two squares can't be factored using integers.*

The patterns that can be used to factor a difference or sum of two cubes are:

$$\triangle^3 - \square^3 = (\triangle - \square)(\triangle^2 + \triangle\square + \square^2)$$
$$a^3 - b^3 = (a - b)\ (a^2 + ab + b^2)$$

$$\triangle^3 + \square^3 = (\triangle + \square)(\triangle^2 - \triangle\square + \square^2)$$
$$a^3 + b^3 = (a + b)\ (a^2 - ab + b^2)$$

For example, to factor $x^3 - 8$:

1. Decide which pattern to use. $\quad a^3 - b^3 = (a - b)(a^2 + ab + b^2)$

2. Substitute x for a and 2 for b. $\quad x^3 - 2^3 = (x - 2)(x^2 + 2x + 2^2)$

$$= (x - 2)(x^2 + 2x + 4)$$

So, $x^3 - 8 = (x - 2)(x^2 + 2x + 4)$.

As another example, to factor $27y^3 + z^6$:

1. Decide which pattern to use. $\quad a^3 + b^3 = (a + b)(a^2 - ab + b^2)$

2. Substitute $3y$ for a and z^2 for b. $\quad (3y)^3 + (z^2)^3 = (3y + z^2)[(3y)^2 - 3yz^2 + (z^2)^2]$

$$= (3y + z^2)(9y^2 - 3yz^2 + z^4)$$

So, $27y^3 + z^6 = (3y + z^2)(9y^2 - 3yz^2 + z^4)$.

Combining Patterns

When a polynomial doesn't look like it fits into one of the patterns you've seen, don't give up. Try to factor using another method first, then see if one of the polynomials in the factorization fits a pattern you have learned. For example, sometimes you need to factor out the greatest common factor of the terms before the polynomial will fit one of the patterns.

For example, to factor $2x^2y - 28xy + 98y$:

1. Factor out the GCF of the terms. $\quad 2y(x^2 - 14x + 49)$

2. Decide which pattern to use. $\quad a^2 - 2ba + b^2 = (a - b)^2$

3. Substitute x for a and 7 for b. $\quad x^2 - 2(7)(x) + 7^2 = (x - 7)^2$

So, $2x^2y - 28xy + 98y = 2y(x - 7)^2$.

Whatever factoring technique you use, look at the final product to be sure it cannot be factored any further.

For example, to factor $x^4 - y^4$:

1. Decide which pattern to use. $\quad a^2 - b^2 = (a + b)\ (a - b)$

2. Substitute x^2 for a and y^2 for b. $\quad (x^2)^2 - (y^2)^2 = (x^2 + y^2)(x^2 - y^2)$

3. Determine if any factor can be factored further. $x^2 - y^2$ can be factored further

4. Decide which pattern to use. $a^2 - b^2 = (a+b)(a-b)$

5. Substitute x for a and y for b. $x^2 - y^2 = (x+y)(x-y)$

So, $x^4 - y^4 = (x^2 + y^2)(x^2 - y^2) = (x^2 + y^2)(x+y)(x-y)$.

Sample Problems

1. Factor: $w^2 - 10w + 25$

   ☑ a. Decide which pattern to use.   __ $a^2 + 2ba + b^2 = (a+b)^2$

                                    ✓ $a^2 - 2ba + b^2 = (a-b)^2$

                                      __ $a^2 - b^2 = (a+b)(a-b)$

                                      __ $a^3 - b^3 = (a-b)(a^2 + ab + b^2)$

                                      __ $a^3 + b^3 = (a+b)(a^2 - ab + b^2)$

   ☑ b. Substitute w for a and 5 for b. $a^2 - 2ba + b^2 = (a-b)^2$

                                      $w^2 - 2(5)(w) + 5^2 = (w-5)^2$

2. Factor: $y^2 - 9$

   ☑ a. Decide which pattern to use. $a^2 - b^2 = (a+b)(a-b)$

   ☐ b. Substitute y for a and 3 for b. $y^2 - 9 \ = (\underline{\hspace{1cm}})(\underline{\hspace{1cm}})$

3. Factor: $27x^3 - 8$

   ☐ a. Decide which pattern to use. $\underline{\hspace{1cm}} = (\underline{\hspace{1cm}})(\underline{\hspace{1cm}})$

   ☐ b. Substitute $3x$ for a and 2 for b. $27x^3 - 8 = (\underline{\hspace{1cm}})(\underline{\hspace{1cm}})$

4. Factor: $12wx^2 - 27wy^2$

   ☐ a. Factor out the GCF of the terms. $12wx^2 - 27wy^2 = 3w(\underline{\hspace{1.5cm}})$

   ☐ b. Decide which pattern to use. $\underline{\hspace{1.5cm}} = (\underline{\hspace{1cm}})(\underline{\hspace{1cm}})$

   ☐ c. Substitute for a and b. $3w(\underline{\hspace{1cm}}) = 3w(\underline{\hspace{1cm}})(\underline{\hspace{1cm}})$

HOMEWORK

Homework Problems

Circle the homework problems assigned to you by the computer, then complete them below.

 Explain
Recognizing Patterns

Factor the polynomials in problems 1 through 12.

1. $x^2 + 14x + 49$

2. $w^2 - 16$

3. $x^3 + 125$

4. $25y^2 - 30y + 9$

5. $9xy^2 - x$

6. $64y^3 - 27w^9$

7. $x^2 + 8w^2x + 16w^4$

8. $2x^6 - 72y^2$

9. $49y^2 - 28xy + 4x^2$

10. $x^3y^2 + 8y^2$

11. $2x^3 + 12x^2 + 18x$

12. $y^6 - 16y^2$

 APPLY

Practice Problems

Here are some additional practice problems for you to try.

Recognizing Patterns

1. Factor: $a^2 + 18a + 81$

2. Factor: $y^2 + 14y + 49$

3. Factor: $9x^2 + 42x + 49$

4. Factor: $25m^2 + 30m + 9$

5. Factor: $4a^2 + 20a + 25$

6. Factor: $b^2 - 16b + 64$

7. Factor: $z^2 - 22z + 121$

8. Factor: $y^2 - 18y + 81$

9. Factor: $16a^2 - 40a + 25$

10. Factor: $4c^2 + 28c + 49$

11. Factor: $9x^2 - 12x + 4$

12. Factor: $m^2 - 144$

13. Factor: $x^2 - 36$

14. Factor: $9m^2 - 81n^2$

15. Factor: $25a^2 - 625b^2$

16. Factor: $16x^2 - 64y^2$

17. Factor: $a^3 - 216$

18. Factor: $m^3 - 1000$

19. Factor: $x^3 - 125$

20. Factor: $8b^3 - 125$

21. Factor: $27z^3 - 343$

22. Factor: $64a^3 - 216$

23. Factor: $c^3 + 64$

24. Factor: $p^3 + 512$

25. Factor: $y^3 + 27$

26. Factor: $3a^3 + 42a^2b + 147b^2$

27. Factor: $50m^3n - 128mn^3$

28. Factor: $5x^3 - 20xy^2$

 EVALUATE

Practice Test

1. Circle the expressions below that are perfect square trinomials.

   $9x^2 + 12x + 4$

   $0.25x^2 + 8x + 64$

   $25x^2 - 9$

   $9x^2 + 20x + 4$

   $x^2 - 2x + 1$

   $x^2 - 7x + 6$

2. Factor the polynomials below.

   a. $x^2 - 10x + 25$

   b. $49y^2 + 28y + 4$

   c. $16x^2 - 1$

   d. $9y^2 - 36$

3. Circle the polynomials below that **cannot** be factored any further using integers.

   $x^2 - 1000$

   $4y^2 - 4y + 1$

   $3x^2 - 27x + 9$

   $9m^2 - 24mn - 16n^2$

   $12x^3 - 8xy + 2y$

4. Factor: $12x^3 - 60x^2 + 75x$

5. Circle the expressions below that are perfect square trinomials.

   $36x^2 - 1$

   $4x^2 - 2x - 56$

   $x^2 + 8x + 16$

   $4x^2 - 12x + 9$

   $x^2 - 16x + 4$

6. Factor the polynomials below.

   a. $4x^2 - 24x + 36$

   b. $64z^2 + 16z + 1$

   c. $4w^2 - 49$

   d. $9m^2 - n^2$

7. Factor the polynomials below.

   a. $x^3 + 1000$

   b. $216y^3 - 1$

   c. $343x^3 + 8y^3$

8. Factor: $27w^3 + 90w^2 + 75w$

TOPIC 7 CUMULATIVE ACTIVITIES

CUMULATIVE REVIEW PROBLEMS

These problems combine all of the material you have covered so far in this course. You may want to test your understanding of this material before you move on to the next topic. Or you may wish to do these problems to review for a test.

1. Find: $(a^5 - 9a^3 + 5a^2 + 14a - 35) \div (a^2 - 7)$

2. Find the slope of the line perpendicular to the line through the points $(-3, 7)$ and $(9, -5)$.

3. Simplify this expression: $11x^2 + 6y + 2 - 4x^2 - y$

4. Graph the inequality $2x + 3y \leq 6$.

5. Alfredo needs to make 250 ml of a 27% alcohol solution using a 15% solution and a 40% solution. How much of each should he use?

6. The point $(-2, -3)$ lies on a line with slope 2. Graph this line by finding another point that lies on the line.

7. Factor: $x^2 - 6x + 9$

8. Find the GCF of $6x^2y^2$ and $8xy^4$.

9. Circle the true statements below.

    $5(7 + 3) = 5(10)$

    $|9 - 20| = -11$

    The fraction $\frac{4}{6}$ is in lowest terms.

    The LCM of 45 and 75 is 225.

    $\frac{9}{11} \cdot \frac{22}{3} = 6$

10. Solve for x: $3(x + 1) = x + 2\left(x + \frac{3}{2}\right)$

11. Graph the system of inequalities below to find its solution.

    $3x - 2y \leq 7$
    $4x + y > 3$

12. Find the slope of the line parallel to the line through the points $(7, -1)$ and $(2, 8)$.

13. Factor: $a^4 + 4a^2 + 4$

14. Solve this system:

    $3x - y = 23$
    $2x + y = 22$

15. Solve $5 \leq 3x - 13 < 17$ for x, then graph the solution on the number line below.

16. Factor: $6x + 3ax + 2b + ab$

17. Find the equation of the line through the point $(-7, 12)$ with slope $m = -2$:

    a. in point-slope form.

    b. in slope-intercept form.

    c. in standard form.

18. Find:

    a. $-2x^0 - \frac{4}{y^0}$ b. $\left(\frac{x^7yz^4}{x^2z}\right)^2$

    c. $a^0 \cdot a^0 \cdot a^0$

19. Solve for x by factoring: $x^2 - 5x - 14 = 0$

20. Find:

    a. $11^3 \cdot 11^5$

    b. $\frac{x}{x^8}$

    c. $(ab^6)^3$

21. Find the equation of the line through the point $(4, -3)$ with slope $-\frac{8}{5}$:

    a. in point-slope form.   b. in slope-intercept form.

    c. in standard form.

22. Factor: $4y^2 - 28y + 49$

23. Find the slope of the line through the points $(31, 16)$ and $(-2, 8)$.

24. Find: $(5y - 3)^2$

25. Solve $5y + 5 = 5(2 + y)$ for y.

26. Find the slope and y-intercept of the line $\frac{9}{4}x - \frac{2}{3}y = 2$.

27. Factor: $5x^2 + 2x - 7$

28. Find: $(xy^3 - 5x^2y + 11xy - 1) - (4xy^3 - 7 - x^2y + 3xy)$

29. Evaluate the expression $2a^3 - 8ab + 5b^2$ when $a = -2$ and $b = 4$.

30. Factor: $2x^2 + 9x - 18$

31. Use the FOIL method to find: $(a + 4)(a - 2)$

32. Graph the inequality $3x + 5y > 8$.

33. Jerome owed a total of $1820 on his two credit cards last year for which he paid $278.60 in interest. If one card charged 14% in interest and the other card charged 16% in interest, how much did he owe on each card?

34. Solve $-6 \le 4y - 3 \le 5$ for y.

35. Find the equation of the line through the point $(5, 3)$ with slope $m = \frac{5}{6}$:

    a. in point-slope form.

    b. in slope-intercept form.

    c. in standard form.

36. Factor: $x^2 - 7x + 12$

37. Graph the system of inequalities below to find its solution.

$$4x - 3y \le 6$$
$$y \le \frac{1}{2}x + 5$$

38. Circle the expressions below that are monomials.

        15 $2x^4 + x$

        $9y$ $a^3b^4c^2$

39. Graph the inequality $2x - 3y > 12$.

40. Hye was cleaning out her car and found a total of 44 nickels and quarters worth $4.80. How many of each did she find?

41. Factor: $36b^2 + 60b + 25$

42. Circle the true statements below.

$$\frac{17}{21} - \frac{7}{18} = \frac{10}{3}$$
$$\frac{15}{33} = \frac{5}{11}$$
$$|8| - |13| = |8 - 13|$$

    The GCF of 120 and 252 is 12.

$$11^2(15 - 2) = 121(15 - 2)$$

43. Simplify this expression: $8x^3 + 5xy^2 + 7 - 4x^3 + 2xy^2$

44. Graph the system of inequalities below to find its solution.

$$x + 2y < 6$$
$$x + 2y \ge -5$$

45. Solve this system:

$$5x + 7y = 25$$
$$x - 3y = -17$$

46. Find the slope of the line through the points $(2, 11)$ and $(6, -8)$.

47. Find: $(x + 5)(5x + xy - 3y)$

48. Factor: $3x^2 - 5x - 2$

49. The length of a rectangle is 3 times its width. If the perimeter of the rectangle is 136 feet, what are its dimensions?

50. Solve this system:

$$6x + y = 1$$
$$2x + 5y = -9$$

LESSON 8.1 – RATIONAL EXPRESSIONS I

Here is what you'll learn in this lesson:

Multiplying and Dividing

a. Determining when a rational expression is undefined

b. Writing a rational expression in lowest terms

c. Multiplying rational expressions

d. Dividing rational expressions

e. Simplifying a complex fraction

Adding and Subtracting

a. Adding rational expressions with the same denominator

b. Subtracting rational expressions with the same denominator

Almost 2000 years ago an Alexandrian mathematician named Eratosthenes figured out the circumference of the earth. He did this by setting up an equation that involved rational expressions, fractions in which the numerator and denominator are polynomials.

In this lesson, you will learn how to multiply, divide, add, and subtract rational expressions.

 EXPLAIN

MULTIPLYING AND DIVIDING

Summary

Determining When a Rational Expression is Undefined

You have divided integers to form fractions, or rational numbers. Similarly, you can divide one polynomial by another to form an algebraic fraction. Algebraic fractions are also called rational expressions.

For example, here are some rational expressions:

$$\frac{5x^2 - 7x + 2}{x + 5} \qquad \frac{1}{7x^2 + 3} \qquad \frac{8x^2 - 9}{3x^5 + 4x - 1}$$

In general, a rational expression is written in the form $\frac{P}{Q}$ where P and Q are polynomials and $Q \neq 0$.

A rational expression is undefined when the value of its denominator is zero. To find the values of the variable for which a rational expression is undefined:

1. Set the polynomial in the denominator equal to 0.

2. Solve the equation in step 1.

For example, to determine when $\frac{2x^2 + 5x - 6}{x - 4}$ is undefined:

   1. Set the denominator equal to zero. $\qquad x - 4 = 0$

   2. Solve the equation $x - 4 = 0$ for x. $\qquad x = 4$

So, $\frac{2x^2 + 5x - 6}{x - 4}$ is undefined when $x = 4$.

Writing a Rational Expression in Lowest Terms

You have learned how to reduce fractions to lowest terms. In a similar way, you can reduce rational expressions to lowest terms.

To reduce a rational expression to lowest terms:

1. Factor the numerator.

2. Factor the denominator.

3. Cancel pairs of factors that are common to both the numerator and the denominator.

A rational number can be written as a ratio of two integers, $\frac{a}{b}$, where $b \neq 0$.

A rational number such as $\frac{5}{19}$ is a special case of a rational expression. Here, the polynomials in the numerator and the denominator each consist of one term, an integer.

It's okay if the numerator of a rational expression is zero. But the denominator can't be zero since division by 0 is undefined.

Remember, to reduce a fraction to lowest terms:

1. *Find the prime factorization of the numerator.* $\quad \frac{6}{9} = \frac{2 \cdot 3}{9}$

2. *Find the prime factorization of the denominator.* $\quad = \frac{2 \cdot 3}{3 \cdot 3}$

3. *Cancel pairs of prime factors common to both the numerator and the denominator.* $\quad = \dfrac{2 \cdot \overset{1}{\cancel{3}}}{\underset{1}{\cancel{3}} \cdot 3}$

So, $\frac{6}{9} = \frac{2}{3}$.

For example, to reduce $\frac{21a^4b^3c^2}{3a^3cd^2}$ to lowest terms:

1. Factor the numerator.

$$= \frac{3 \cdot 7 \cdot a \cdot a \cdot a \cdot a \cdot b \; b \cdot b \cdot c \cdot c}{3a^3cd^2}$$

2. Factor the denominator.

$$= \frac{3 \cdot 7 \cdot a \cdot a \cdot a \cdot a \cdot b \; b \cdot b \cdot c \cdot c}{3 \cdot a \cdot a \cdot a \cdot c \cdot d \cdot d}$$

3. Cancel common factors.

$$= \frac{\overset{1}{\cancel{3}} \cdot 7 \cdot \overset{1}{\cancel{a}} \cdot \overset{1}{\cancel{a}} \cdot \overset{1}{\cancel{a}} \cdot a \cdot b \; b \cdot b \cdot \overset{1}{\cancel{c}} \cdot c}{\underset{1}{\cancel{3}} \cdot \underset{1}{\cancel{a}} \cdot \underset{1}{\cancel{a}} \cdot \underset{1}{\cancel{a}} \cdot \underset{1}{\cancel{c}} \cdot d \cdot d}$$

$$= \frac{7ab^3c}{d^2}$$

As another example, to reduce $\frac{x^2 - 3x - 10}{x^2 - 2x - 15}$ to lowest terms:

1. Factor the numerator.

$$= \frac{(x+2)(x-5)}{x^2 - 2x - 15}$$

2. Factor the denominator.

$$= \frac{(x+2)(x-5)}{(x+3)(x-5)}$$

3. Cancel common factors.

$$= \frac{(x+2)\overset{1}{\cancel{(x-5)}}}{(x+3)\underset{1}{\cancel{(x-5)}}}$$

$$= \frac{x+2}{x+3}$$

So, $\frac{x^2 - 3x - 10}{x^2 - 2x - 15} = \frac{x+2}{x+3}$.

Multiplying Rational Expressions

You have learned how to multiply fractions. You can multiply rational expressions in the same way.

To multiply rational expressions:

1. Factor the numerators and denominators.

2. Cancel all pairs of factors common to the numerators and denominators.

3. Multiply the numerators. Multiply the denominators.

For example, to find $\frac{9x^2y^2}{4wz^2} \cdot \frac{8wz}{3x^2y}$:

1. Factor the numerators and denominators.

$$= \frac{3 \cdot 3 \cdot x \cdot x \cdot y \cdot y}{2 \cdot 2 \cdot w \cdot z \cdot z} \cdot \frac{2 \cdot 2 \cdot 2 \cdot w \cdot z}{3 \cdot x \cdot x \cdot y}$$

2. Cancel pairs of factors common to the numerators and denominators.

$$= \frac{\overset{1}{\cancel{3}} \cdot 3 \cdot \overset{1}{\cancel{x}} \cdot \overset{1}{\cancel{x}} \cdot \overset{1}{\cancel{y}} \cdot y}{\underset{1}{\cancel{2}} \cdot \underset{1}{\cancel{2}} \cdot w \cdot \underset{1}{\cancel{z}} \cdot z} \cdot \frac{\overset{1}{\cancel{2}} \cdot \overset{1}{\cancel{2}} \cdot 2 \cdot \overset{1}{\cancel{w}} \cdot \overset{1}{\cancel{z}}}{3 \cdot \underset{1}{\cancel{x}} \cdot \underset{1}{\cancel{x}} \cdot \underset{1}{\cancel{y}}}$$

3. Multiply the numerators. Multiply the denominators.

$$= \frac{3 \cdot y}{z} \cdot \frac{2}{1}$$

$$= \frac{6y}{z}$$

So, $\frac{9x^2y^2}{4wz^2} \cdot \frac{8wz}{3x^2y} = \frac{6y}{z}$.

Remember, to multiply fractions:

1. *Factor the numerators and denominators into prime factors.*

$$\frac{7}{18} \cdot \frac{9}{14} = \frac{7}{2 \cdot 3 \cdot 3} \cdot \frac{3 \cdot 3}{2 \cdot 7}$$

2. *Cancel pairs of factors common to the numerators and denominators.*

$$= \frac{\overset{1}{\cancel{7}}}{2 \cdot \underset{1}{\cancel{3}} \cdot \underset{1}{\cancel{3}}} \cdot \frac{\overset{1}{\cancel{3}} \cdot \overset{1}{\cancel{3}}}{2 \cdot \underset{1}{\cancel{7}}}$$

3. *Multiply the numerators and the denominators.*

$$= \frac{1}{4}$$

Dividing Rational Expressions

You have learned how to divide fractions. You can divide rational expressions in the same way.

To divide rational expressions:

1. Invert the second fraction and change "÷" to "·". Then multiply.

2. Factor the numerators and denominators.

3. Cancel pairs of factors common to the numerators and denominators.

4. Multiply the numerators. Multiply the denominators.

For example, to find $\frac{45xy^2}{2w^2} \div \frac{30x^2y^3}{w^3}$:

1. Invert the second fraction and change ÷ to ·. Then multiply.
$$= \frac{45xy^2}{2w^2} \cdot \frac{w^3}{30x^2y^3}$$

2. Factor the numerators and denominators.
$$= \frac{3 \cdot 3 \cdot 5 \cdot x \cdot y \cdot y}{2 \cdot w \cdot w} \cdot \frac{w \cdot w \cdot w}{2 \cdot 3 \cdot 5 \cdot x \cdot x \cdot y \cdot y \cdot y}$$

3. Cancel pairs of factors common to the numerator and denominator.
$$= \frac{\overset{1}{\cancel{3}} \cdot 3 \cdot \overset{1}{\cancel{5}} \cdot \overset{1}{\cancel{x}} \cdot \overset{1}{\cancel{y}} \cdot \overset{1}{\cancel{y}}}{2 \cdot \underset{1}{\cancel{w}} \cdot \underset{1}{\cancel{w}}} \cdot \frac{\overset{1}{\cancel{w}} \cdot \overset{1}{\cancel{w}} \cdot w}{2 \cdot \underset{1}{\cancel{3}} \cdot \underset{1}{\cancel{5}} \cdot \underset{1}{\cancel{x}} \cdot x \cdot \underset{1}{\cancel{y}} \cdot \underset{1}{\cancel{y}} \cdot y}$$

4. Multiply the numerators. Multiply the denominators.
$$= \frac{3}{2} \cdot \frac{w}{2xy}$$
$$= \frac{3w}{4xy}$$

So, $\frac{45xy^2}{2w^2} \div \frac{30x^2y^3}{w^3} = \frac{3w}{4xy}$.

Simplifying a Complex Fraction

When a fraction contains other fractions or rational expressions it is called a complex fraction. One way to simplify a complex fraction is to use division.

To simplify a complex fraction using division:

1. Rewrite the complex fraction as a division problem.

2. To divide, invert the second fraction and multiply.

3. Factor the numerators and denominators.

4. Cancel pairs of factors common to the numerator and denominator.

5. Multiply the numerators. Multiply the denominators.

For example, to simplify this complex fraction $\dfrac{\frac{3m^3}{5n^5}}{\frac{6m^2}{10n^2}}$:

1. Rewrite the complex fraction as a division problem.

$$= \frac{3m^3}{5n^5} \div \frac{6m^2}{10n^2}$$

2. Divide by inverting the second fraction and multiplying.

$$= \frac{3m^3}{5n^5} \cdot \frac{10n^2}{6m^2}$$

3. Factor the numerators and denominators.

$$= \frac{3 \cdot m \cdot m \cdot m}{5 \cdot n \cdot n \cdot n \cdot n \cdot n} \cdot \frac{2 \cdot 5 \cdot n \cdot n}{2 \cdot 3 \cdot m \cdot m}$$

4. Cancel pairs of factors common to the numerator and denominator.

$$= \frac{\overset{1}{3} \cdot \overset{1}{\cancel{m}} \cdot \overset{1}{\cancel{m}} \cdot m}{\underset{1}{\cancel{5}} \cdot \underset{1}{\cancel{n}} \cdot \underset{1}{\cancel{n}} \cdot n \cdot n \cdot n} \cdot \frac{\overset{1}{2} \cdot \overset{1}{\cancel{5}} \cdot \overset{1}{\cancel{n}} \cdot \overset{1}{\cancel{n}}}{\underset{1}{2} \cdot \underset{1}{3} \cdot \underset{1}{\cancel{m}} \cdot \underset{1}{\cancel{m}}}$$

5. Multiply the numerators. Multiply the denominators.

$$= \frac{m}{n^3}$$

So, $\dfrac{\frac{3m^3}{5n^5}}{\frac{6m^2}{10n^2}} = \dfrac{m}{n^3}$.

Sample Problems

1. Find the values of x for which this rational expression is undefined: $\dfrac{x^2 - 4}{(x + 8)(x - 5)}$

   ☑ a. Set the polynomial in the denominator equal to 0.

   $$(x + 8)(x - 5) = 0$$

   ☐ b. Solve the equation $(x + 8)(x - 5) = 0$.

   $x = $ ___ or $x = $ ___

2. Reduce to lowest terms: $\dfrac{24x^3yz^2}{6x^2z^3w}$

   ☑ a. Factor the numerator.

   $$= \frac{2 \cdot 2 \cdot 2 \cdot 3 \cdot x \cdot x \cdot x \cdot y \cdot z \cdot z}{6x^2z^3w}$$

   ☐ b. Factor the denominator.

   $$= \frac{2 \cdot 2 \cdot 2 \cdot 3 \cdot x \cdot x \cdot x \cdot y \cdot z \cdot z}{\underline{\ }\cdot\underline{\ }\cdot\underline{\ }\cdot\underline{\ }\cdot\underline{\ }\cdot\underline{\ }\cdot\underline{\ }\cdot\underline{\ }}$$

   ☐ c. Cancel common factors.

   $$= \underline{\qquad}$$

3. Reduce to lowest terms: $\dfrac{x^2 + 3x - 4}{x^2 - 3x + 2}$

   ☑ a. Factor the numerator.

   $$= \frac{(x - 1)(x + 4)}{x^2 - 3x + 2}$$

   ☐ b. Factor the denominator.

   $$= \frac{(x - 1)(x + 4)}{(\underline{\quad})(\underline{\quad})}$$

   ☐ c. Cancel common factors.

   $$= \frac{x + 4}{\underline{\qquad}}$$

4. Find: $\frac{3a^2b}{2cd^3} \cdot \frac{cd^2}{6a^3b^2}$

☑ a. Factor the numerators and denominators.

$$= \frac{3 \cdot a \cdot a \cdot b}{2 \cdot c \cdot d \cdot d \cdot d} \cdot \frac{c \cdot d \cdot d}{2 \cdot 3 \cdot a \cdot a \cdot a \cdot b \cdot b}$$

☐ b. Cancel pairs of factors common to the numerators and denominators.

$$= \frac{1}{\_ \cdot \_} \cdot \frac{1}{\_ \cdot \_ \cdot \_}$$

☐ c. Multiply the numerators. Multiply the denominators.

$$= \underline{}$$

5. Find: $\frac{6a^2b^2}{5c^3} \div \frac{3ab}{10c^2d}$

☑ a. Invert the second fraction and change ÷ to ·. Then multiply.

$$= \frac{6a^2b^2}{5c^3} \cdot \frac{10c^2d}{3ab}$$

☐ b. Factor the numerators and denominators.

$$= \frac{2 \cdot 3 \cdot a \cdot a \cdot b \cdot b}{\_ \cdot \_ \cdot \_ \cdot \_} \cdot \frac{2 \cdot 5 \cdot c \cdot c \cdot d}{\_ \cdot \_ \cdot \_}$$

☐ c. Cancel pairs of factors common to the numerator and denominator.

$$= \underline{} \cdot \underline{}$$

☐ d. Multiply the numerators. Multiply the denominators.

$$= \underline{}$$

6. Simplify this complex fraction: $\dfrac{\frac{a^3}{3b^4}}{\frac{7a^2}{6b^2}}$

☑ a. Rewrite as a division problem.

$$= \frac{a^3}{3b^4} \div \frac{7a^2}{6b^2}$$

☐ b. Divide. (Invert the second fraction and multiply.)

$$= \frac{a^3}{3b^4} \cdot \underline{}$$

☐ c. Factor the numerators and denominators.

$$= \underline{}$$

☐ d. Cancel pairs of factors common to the numerator and denominator.

$$= \underline{}$$

☐ e. Multiply the numerators. Multiply the denominators.

$$= \underline{}$$

Answers to Sample Problems

Answers to Sample Problems

b. 2, d; 2, a, b

c. $\frac{1}{4abd}$

b. 5, c, c, c; 3, a, b

c. $\frac{2ab}{c}$, $\frac{2d}{1}$ or 2d

d. $\frac{4abd}{c}$

b. $\frac{6b^2}{7a^2}$

c. $\frac{a \cdot a \cdot a}{3 \cdot b \cdot b \cdot b \cdot b} \cdot \frac{2 \cdot 3 \cdot b \cdot b}{7 \cdot a \cdot a}$

d. $\frac{a}{b \cdot b} \cdot \frac{2}{7}$

e. $\frac{2a}{7b^2}$

ADDING AND SUBTRACTING

Summary

Adding Rational Expressions with the Same Denominator

Remember, to add fractions with the same denominator:

1. Add the numerators. The denominator stays the same.

$$\frac{3}{4} + \frac{7}{4} = \frac{3+7}{4}$$
$$= \frac{10}{4}$$

2. Factor the numerator and denominator.

$$= \frac{2 \cdot 5}{\overset{1}{\underset{1}{2}} \cdot 2}$$

3. Reduce to lowest terms. $= \frac{5}{2}$

You have learned how to add fractions with the same denominator.
You can add rational expressions with the same denominator in a similar way.

To add rational expressions with the same denominator, add the numerators. The denominator stays the same.

For example, to find $\dfrac{3x}{x+5} + \dfrac{11}{x+5}$:

1. Add the numerators.
   The denominator stays the same.

$$= \frac{3x + 11}{x + 5}$$

So, $\dfrac{3x}{x+5} + \dfrac{11}{x+5} = \dfrac{3x+11}{x+5}$.

After you add rational expressions, you often simplify the resulting rational expression by reducing it to lowest terms.

For example, to find $\dfrac{x+1}{x^2 - 3x - 4} + \dfrac{2x+2}{x^2 - 3x - 4}$:

1. Add the numerators. The denominator stays the same.

$$= \frac{x + 1 + 2x + 2}{x^2 - 3x - 4}$$
$$= \frac{3x + 3}{x^2 - 3x - 4}$$

2. Factor the numerator and denominator.

$$= \frac{\overset{1}{3(x+1)}}{\underset{1}{(x+1)}(x-4)}$$

3. Reduce to lowest terms.

$$= \frac{3}{x - 4}$$

So, $\dfrac{x+1}{x^2 - 3x - 4} + \dfrac{2x+2}{x^2 - 3x - 4} = \dfrac{3}{x-4}$.

Subtracting Rational Expressions with the Same Denominator

Remember, to subtract fractions with the same denominator:

*1. Subtract the numerators.
The denominator stays the same.*

$$\frac{5}{6} - \frac{1}{6} = \frac{5-1}{6}$$
$$= \frac{4}{6}$$

2. Factor the numerator and denominator.

$$= \frac{\overset{1}{2} \cdot 2}{\underset{1}{2} \cdot 3}$$

3. Reduce to lowest terms. $= \frac{2}{3}$

You have learned how to subtract fractions with the same denominator.
You can subtract rational expressions with the same denominator in a similar way.

To subtract rational expressions with the same denominator, subtract the numerators. The denominator stays the same.

For example, to find $\dfrac{4y}{17w} - \dfrac{5}{17w}$:

1. Subtract the numerators.
   The denominator stays the same.

$$= \frac{4y - 5}{17w}$$

So, $\dfrac{4y}{17w} - \dfrac{5}{17w} = \dfrac{4y-5}{17w}$.

After you subtract rational expressions, you often simplify the resulting rational expression by reducing it to lowest terms.

For example, to find $\dfrac{5x-6}{x^2+6x+5} - \dfrac{4x-7}{x^2+6x+5}$:

1. Subtract the numerators. The denominator stays the same.

$$= \dfrac{5x-6-(4x-7)}{x^2+6x+5}$$

2. Distribute. Be careful with the signs.

$$= \dfrac{5x-6-4x+7}{x^2+6x+5}$$

$$= \dfrac{x+1}{x^2+6x+5}$$

3. Factor the numerator and denominator.

$$= \dfrac{\overset{1}{\cancel{x+1}}}{\cancel{(x+1)}(x+5)}_{1}$$

4. Reduce to lowest terms.

$$= \dfrac{1}{x+5}$$

So, $\dfrac{5x-6}{x^2+6x+5} - \dfrac{4x-7}{x^2+6x+5} = \dfrac{1}{x+5}$.

Sample Problems

1. Find: $\dfrac{2a}{5b} + \dfrac{6a}{5b}$

☐ a. Add the numerators. The denominator stays the same.

$$= \dfrac{}{5b}$$

$$= \underline{}$$

2. Find: $\dfrac{2x}{x-1} + \dfrac{13}{x-1}$:

☐ a. Add the numerators. The denominator stays the same.

$$= \dfrac{}{x-1}$$

3. Find: $\dfrac{x-12}{x^2-2x-15} + \dfrac{3x-8}{x^2-2x-15}$

☐ a. Add the numerators. The denominator stays the same.

$$= \dfrac{(x-12)+(3x-8)}{x^2-2x-15}$$

$$= \dfrac{}{x^2-2x-15}$$

☐ b. Factor the numerator and denominator.

$$= \underline{}$$

☐ c. Reduce to lowest terms.

$$= \underline{}$$

a. $7 - 2$

$\dfrac{5}{11y}$

a. $13 - (2 + x)$

b. $13 - 2 - x$

$\dfrac{11 - x}{4x}$

a. $\dfrac{x - 5}{x^2 - 25}$

b. $\dfrac{x - 5}{(x + 5)(x - 5)}$

c. $\dfrac{1}{x + 5}$

4. Find: $\dfrac{7}{11y} - \dfrac{2}{11y}$

☐ a. Subtract the numerators. The denominator stays the same.

$= \dfrac{}{11y}$

$= \dfrac{}{\underline{}}$

5. Find: $\dfrac{13}{4x} - \dfrac{2 + x}{4x}$

☐ a. Subtract the numerators. The denominator stays the same.

$= \dfrac{}{4x}$

☐ b. Distribute. Be careful with the signs.

$= \dfrac{}{4x}$

$= \dfrac{}{\underline{}}$

6. Find: $\dfrac{5x + 2}{x^2 - 25} - \dfrac{4x + 7}{x^2 - 25}$

☑ a. Subtract the numerators. The denominator stays the same.

$= \dfrac{5x + 2 - (4x + 7)}{x^2 - 25}$

$= \dfrac{}{\underline{}}$

☐ b. Factor the numerator and denominator.

$= \dfrac{}{\underline{}}$

☐ c. Reduce to lowest terms.

$= \dfrac{}{\underline{}}$

HOMEWORK

Homework Problems

Circle the homework problems assigned to you by the computer, then complete them below.

☀ Explain

Multiplying and Dividing

1. For what values of x is the rational expression below undefined?

    $\dfrac{(x+7)(x-8)}{(x-14)(x+2)}$

2. For what values of x is the rational expression below undefined?

    $\dfrac{x^2-9}{x^2-4}$

3. Reduce to lowest terms: $\dfrac{3a^2b^5}{27ab^7}$

4. Reduce to lowest terms: $\dfrac{x^2-3x-28}{x^2+5x+4}$

5. Find: $\dfrac{3y^3}{z} \cdot \dfrac{yz}{7y^2}$

6. Find: $\dfrac{15a^2b^2}{2c^3d} \cdot \dfrac{2cd^3}{5ab^2}$

7. Find: $\dfrac{12xy}{w^2} \div \dfrac{4xy^3}{w^3}$

8. Find: $\dfrac{3ab^2}{13d^4} \div \dfrac{6a^2b}{11d^2}$

9. Simplify this complex fraction: $\dfrac{\frac{5}{a^3}}{\frac{9}{a^2}}$. Write your answer in lowest terms.

10. The ratio of the area of a circle to its circumference is given by $\dfrac{\pi r^2}{2\pi r}$.

    a. What value of r makes this ratio undefined?

    b. Simplify this expression and then determine what value of r will make the ratio equal to 2. That is, find the radius that will yield an area that is twice the circumference.

11. Simplify this complex fraction: $\dfrac{\frac{x}{x-1}}{\frac{1}{x+1}}$. Write your answer in lowest terms.

12. Simplify this complex fraction: $\dfrac{\frac{3y^4}{y+2}}{\frac{9y^2}{y-2}}$. Write your answer in lowest terms.

Adding and Subtracting

13. Find: $\dfrac{3x}{5y} + \dfrac{18x}{5y}$

14. Find: $\dfrac{3+5a}{8-a} + \dfrac{2a+1}{8-a}$

15. Find: $\dfrac{x}{x^2-4} + \dfrac{4}{x^2-4}$

16. Find: $\dfrac{3z+4}{z-11} + \dfrac{2z}{z-11}$

17. Find: $\dfrac{2x-5}{x^2-5x-14} + \dfrac{3x+15}{x^2-5x-14}$

18. Find: $\dfrac{2z+11}{z^2-3z-18} + \dfrac{3z+4}{z^2-3z-18}$

19. Find: $\dfrac{9}{15x} - \dfrac{2}{15x}$

20. Find: $\dfrac{17}{13y} - \dfrac{5+y}{13y}$

21. The volume, V, of a sphere of radius r is defined by the formula $V = \dfrac{4\pi r^3}{3}$. Find the volume of two identical spheres. That is, find $\dfrac{4\pi r^3}{3} + \dfrac{4\pi r^3}{3}$.

22. Find: $\dfrac{y}{y^2-81} - \dfrac{9}{y^2-81}$

23. Find: $\dfrac{4y+6}{3y+6} - \dfrac{3y+4}{3y+6}$

24. Find: $\dfrac{4x-4}{x^2-2x-15} - \dfrac{3x-7}{x^2-2x-15}$

APPLY

Practice Problems

Here are some additional practice problems for you to try.

Multiplying and Dividing

1. For what value(s) of x is the rational expression below undefined?

   $$\frac{1}{x+5}$$

2. For what value(s) of x is the rational expression below undefined?

   $$\frac{2}{(x-3)(x+5)}$$

3. For what value(s) of x is the rational expression below undefined?

   $$\frac{25}{3x^2-12}$$

4. For what value(s) of x is the rational expression below undefined?

   $$\frac{17}{2x^2-18}$$

5. Reduce to lowest terms: $\dfrac{36m^5n^3}{27mn^6}$

6. Reduce to lowest terms: $\dfrac{75xy^2z^7}{45x^4y^3z^6}$

7. Reduce to lowest terms: $\dfrac{44a^2b^4c}{77a^7bc}$

8. Reduce to lowest terms: $\dfrac{x^2+10x+21}{x^2+5x+6}$

9. Reduce to lowest terms: $\dfrac{x^2+4x-5}{x^2-3x+2}$

10. Reduce to lowest terms: $\dfrac{x^2-x-12}{x^2+5x+6}$

11. Find: $\dfrac{6a}{b^3c^2} \cdot \dfrac{7b}{3a^3}$

12. Find: $\dfrac{8x}{y^2z} \cdot \dfrac{5y}{4x^2}$

13. Find: $\dfrac{10m^3n^5}{9mn^3} \cdot \dfrac{21m^5}{15n^6}$

14. Find: $\dfrac{5ab^4}{3c^2} \cdot \dfrac{6c}{10b^3}$

15. Find: $\dfrac{8m^5n}{7p^2} \cdot \dfrac{14mp}{24m^2n^4}$

16. Find: $\dfrac{3xy^3}{4z} \cdot \dfrac{2z^2}{9xy^2}$

17. Find: $\dfrac{4a^2b}{5c^3} \div \dfrac{8ab^2}{15c}$

18. Find: $\dfrac{12m^3n^4}{7p} \div \dfrac{18mn^5}{21p^4}$

19. Find: $\dfrac{3xy^2}{7z} \div \dfrac{6x^2y}{14z^2}$

20. Find: $\dfrac{5x^2y}{12z^3w} \div \dfrac{xy}{4z}$

21. Find: $\dfrac{9m^3n^4}{11pq^2} \div \dfrac{12mn^3}{22p^2q}$

22. Find: $\dfrac{7a^2b}{9c^2d^2} \div \dfrac{ab}{3cd}$

23. Simplify the complex fraction below.

   $$\frac{\frac{4m^2}{n^3}}{\frac{2m}{n}}$$

24. Simplify the complex fraction below.

   $$\frac{\frac{6x^5}{5y^3}}{\frac{3x^3}{10y}}$$

25. Simplify the complex fraction below.

   $$\frac{\frac{6a^2}{b^3}}{\frac{3a}{2b}}$$

26. Simplify the complex fraction below.

   $$\frac{\frac{6a^3}{a+5}}{\frac{12a}{a-4}}$$

27. Simplify the complex fraction below.

   $$\frac{\frac{5x^2}{x+7}}{\frac{10x}{x-3}}$$

28. Simplify the complex fraction below.

   $$\frac{\frac{15y^5}{y-3}}{\frac{18y^3}{y+3}}$$

Adding and Subtracting

29. Find: $\dfrac{3a}{7b} + \dfrac{2a}{7b}$

30. Find: $\dfrac{2x}{5y} + \dfrac{7x}{5y}$

31. Find: $\dfrac{3b}{2b+1} + \dfrac{5}{2b+1}$

32. Find: $\dfrac{9n}{4n-7} + \dfrac{2}{4n-7}$

33. Find: $\dfrac{7x}{5x-1} + \dfrac{2}{5x-1}$

34. Find: $\dfrac{5y+2}{3y-2} - \dfrac{4y-5}{3y-2}$

35. Find: $\dfrac{7b+1}{2b+9} - \dfrac{5b+5}{2b+9}$

36. Find: $\dfrac{6x+1}{2x+3} - \dfrac{5x-3}{2x+3}$

37. Find: $\dfrac{15}{7n} - \dfrac{4-5n}{7n}$

38. Find: $\dfrac{11}{9x} - \dfrac{7-5x}{9x}$

39. Add and reduce your answer to lowest terms:
$\dfrac{3x+4}{x^2+5x+6} + \dfrac{5}{x^2+5x+6}$

40. Add and reduce your answer to lowest terms:
$\dfrac{4x+7}{x^2+7x+10} + \dfrac{1}{x^2+7x+10}$

41. Add and reduce your answer to lowest terms:
$\dfrac{2x+5}{x^2+x-12} + \dfrac{3}{x^2+x-12}$

42. Add and reduce your answer to lowest terms:
$\dfrac{3x+12}{x^2+2x-3} + \dfrac{2x+3}{x^2+2x-3}$

43. Add and reduce your answer to lowest terms:
$\dfrac{3x+19}{x^2+3x-10} + \dfrac{x+1}{x^2+3x-10}$

44. Add and reduce your answer to lowest terms:
$\dfrac{2x+15}{x^2+5x-14} + \dfrac{x+6}{x^2+5x-14}$

45. Add and reduce your answer to lowest terms:
$\dfrac{x^2-5x+2}{x^2+7x+12} + \dfrac{2(5x+1)}{x^2+7x+12}$

46. Add and reduce your answer to lowest terms:
$\dfrac{x^2-3x+2}{x^2+3x+2} + \dfrac{4(2x+1)}{x^2+3x+2}$

47. Add and reduce your answer to lowest terms:
$\dfrac{x^2-7x+12}{x^2+9x+18} + \dfrac{5(3x+1)-2}{x^2+9x+18}$

48. Subtract and reduce your answer to lowest terms:
$\dfrac{4x+7}{3x-9} - \dfrac{3x+10}{3x-9}$

49. Subtract and reduce your answer to lowest terms:
$\dfrac{5x+6}{2x+10} - \dfrac{2x-9}{2x+10}$

50. Subtract and reduce your answer to lowest terms:
$\dfrac{3x+2}{2x+14} - \dfrac{2x-5}{2x+14}$

51. Subtract and reduce your answer to lowest terms:
$\dfrac{2x+5}{x^2+3x-10} - \dfrac{x+7}{x^2+3x-10}$

52. Subtract and reduce your answer to lowest terms:
$\dfrac{4x+2}{x^2-x-20} - \dfrac{3x-2}{x^2-x-20}$

53. Subtract and reduce your answer to lowest terms:
$\dfrac{3x+5}{x^2-4x+3} - \dfrac{2x+8}{x^2-4x+3}$

54. Subtract and reduce your answer to lowest terms:
$\dfrac{8x+5}{4x-4} - \dfrac{5x+8}{4x-4}$

55. Subtract and reduce your answer to lowest terms:
$\dfrac{5x+9}{5x+20} - \dfrac{2x-3}{5x+20}$

56. Subtract and reduce your answer to lowest terms:
$\dfrac{7x-2}{3x+3} - \dfrac{5x-4}{3x+3}$

 EVALUATE

Practice Test

Take this practice test to be sure that you are prepared for the final quiz in Evaluate.

1. For what values of x is the following expression undefined?

$$\frac{x^2 - 16}{(x+3)(x-2)}$$

2. Reduce to lowest terms: $\dfrac{x^2 - 3x - 28}{x^2 + 10x + 24}$

3. Find:

   a. $\dfrac{5y^2}{9z^2} \cdot \dfrac{z}{y^2}$

   b. $\dfrac{12x^2y^2}{z^3w} \cdot \dfrac{2zw}{3xy^2}$

4. a. Find: $\dfrac{3x^2}{yz} \div \dfrac{7x}{2yz}$

   b. Simplify this complex fraction: $\dfrac{\dfrac{2x^2y}{9w}}{\dfrac{10xy^2}{3w^2}}$. Write your answer in lowest terms.

5. Find:

   a. $\dfrac{5x}{13y} + \dfrac{2x}{13y}$

   b. $\dfrac{3w}{z-8} + \dfrac{14}{z-8}$

6. Find: $\dfrac{15y}{7x} - \dfrac{3+6y}{7x}$

7. Find the following. Reduce your answer to lowest terms.

$$\frac{x+7}{x^2 - 3x - 18} + \frac{3x+5}{x^2 - 3x - 18}$$

8. Find the following. Reduce your answer to lowest terms.

$$\frac{9y}{y-7} - \frac{3y-4}{y-7}$$

LESSON 8.2 – RATIONAL EXPRESSIONS II

OVERVIEW

Rational expressions are fractions in which the numerator and denominator are polynomials.

In this lesson, you will learn more about how to multiply, divide, add, and subtract rational expressions. You will also learn about negative exponents.

Here is what you'll learn in this lesson:

Negative Exponents

a. Notation

b. Scientific notation

Multiplying and Dividing

a. Reducing a rational expression of the form $\frac{a-b}{b-a}$

b. Multiplying rational expressions

c. Dividing rational expressions

d. Simplifying a complex fraction

Adding and Subtracting

a. Finding the least common denominator of rational expressions

b. Adding rational expressions with different denominators

c. Subtracting rational expressions with different denominators

d. Simplifying a complex fraction

 EXPLAIN

NEGATIVE EXPONENTS

Summary

Negative Exponents

You have seen how to work with exponents that are positive integers or 0. Now you will learn about negative integer exponents.

For example, you have previously found $\frac{3^5}{3^7}$ by canceling:

$$\frac{3^5}{3^7} = \frac{\overset{1}{\cancel{3}} \cdot \overset{1}{\cancel{3}} \cdot \overset{1}{\cancel{3}} \cdot \overset{1}{\cancel{3}} \cdot \overset{1}{\cancel{3}}}{\underset{1}{\cancel{3}} \cdot \underset{1}{\cancel{3}} \cdot \underset{1}{\cancel{3}} \cdot \underset{1}{\cancel{3}} \cdot \underset{1}{\cancel{3}} \cdot 3 \cdot 3} = \frac{1}{3^2}$$

You can also find $\frac{3^5}{3^7}$ by subtracting exponents:

$$\frac{3^5}{3^7} = 3^{5-7} = 3^{-2}$$

So, $3^{-2} = \frac{1}{3^2}$.

In general:

$$x^{-n} = \frac{1}{x^n}$$

Here, $x \neq 0$ and n is a nonnegative integer.

You can rewrite an expression with a negative exponent as a fraction by following these steps:

1. Write the numerator as 1.

2. Write the denominator as the original expression, except change the negative exponent to a positive exponent.

3. Do the multiplication in the denominator.

For example, to find 5^{-2}:

1. Write the numerator as 1.

2. Write the denominator as the original expression, except change the negative exponent to a positive exponent.
$$\frac{1}{5^2}$$

3. Do the multiplication in the denominator.
$$= \frac{1}{25}$$

So, $5^{-2} = \frac{1}{25}$.

Even though the 5 is raised to a negative power, the result is a positive number.

Similarly, here's how to simplify an expression raised to a negative exponent that is in the denominator of a fraction:

1. Take the expression out of the denominator and change the negative exponent to a positive exponent.

2. Do the multiplication.

For example, to find $\dfrac{1}{4^{-3}}$:

1. Take the expression out of the denominator and change the negative exponent to a positive exponent. 4^3

2. Do the multiplication. $= 64$

In general:

$$\frac{1}{x^{-n}} = x^n$$

Here, $x \neq 0$ and n is a nonnegative integer.

Properties of Negative Exponents

The basic properties of nonnegative exponents also hold for negative exponents. The table below summarizes some of these properties.

Property	Positive Integer Exponents	Negative Integer Exponents
Multiplication	$3^2 \cdot 3^4 = 3^{2+4} = 3^6$	$3^{-2} \cdot 3^{-4} = 3^{(-2)+(-4)} = 3^{-6}$
Division	$\dfrac{4^7}{4^5} = 4^{7-5} = 4^2$	$\dfrac{4^{-7}}{4^{-5}} = 4^{(-7)-(-5)} = 4^{-7+5} = 4^{-2}$
Power of a Power	$(5^2)^3 = 5^{2\cdot3} = 5^6$	$(5^{-2})^{-3} = 5^{(-2)(-3)} = 5^6$
Power of a Product	$(5 \cdot 7)^3 = 5^3 \cdot 7^3$	$(5 \cdot 7)^{-3} = 5^{-3} \cdot 7^{-3}$
Power of a Quotient	$\left(\dfrac{3}{5}\right)^4 = \dfrac{3^4}{5^4}$	$\left(\dfrac{3}{5}\right)^{-4} = \dfrac{3^{-4}}{5^{-4}}$

Some other properties that hold for negative exponents are given below.

One property involving negative exponents shows how to rewrite a fraction in which the numerator and denominator are each raised to a negative power.

For example, here's one way to rewrite $\dfrac{2^{-4}}{5^{-3}}$ using only positive exponents:

$$\frac{2^{-4}}{5^{-3}} = \frac{\frac{1}{2^4}}{\frac{1}{5^3}}$$

$$= \frac{5^3}{2^4}$$

$$= \frac{125}{16}$$

Why does this work?

Well, since $4^{-3} = \dfrac{1}{4^3}$, then

$$\frac{1}{4^{-3}} = \frac{1}{\frac{1}{4^3}}$$

$$= 1 \div \frac{1}{4^3}$$

$$= 1 \cdot 4^3$$

$$= 64$$

Notice that the 2^{-4} in the numerator became the 2^4 in the denominator and that the 5^{-3} in the denominator became the 5^3 in the numerator.

In general:

$$\frac{x^{-m}}{y^{-n}} = \frac{y^n}{x^m}$$

Here, $x \neq 0$, $y \neq 0$, and m and n are nonnegative integers.

Another property involving negative exponents shows how to rewrite a fraction that is raised to a negative power.

For example, here's one way to simplify $\left(\frac{2}{3}\right)^{-4}$:

$$\left(\frac{2}{3}\right)^{-4} = \frac{2^{-4}}{3^{-4}}$$

$$= \frac{3^4}{2^4}$$

$$= \left(\frac{3}{2}\right)^4$$

In general:

$$\left(\frac{x}{y}\right)^{-n} = \left(\frac{y}{x}\right)^n$$

Here, $x \neq 0$, $y \neq 0$, and n is a nonnegative integer.

When a fraction is raised to a negative power, flip the fraction and change the negative power to positive.

Simplifying Expressions with Negative Exponents

You can combine different properties to simplify more complicated expressions containing negative exponents.

For example, to rewrite $(3r^4s^{-5}t)^{-2}$ using only positive exponents:

1. Use the Power of a Product Property.

   $$(3r^4s^{-5}t)^{-2} = (3)^{-2} \cdot (r^4)^{-2} \cdot (s^{-5})^{-2} \cdot (t)^{-2}$$

2. Use the Power of a Power Property.

   $$= 3^{-2} \cdot r^{-8} \cdot s^{10} \cdot t^{-2}$$

3. Rewrite the negative exponents as positive exponents.

   $$= \frac{1}{3^2} \cdot \frac{1}{r^8} \cdot s^{10} \cdot \frac{1}{t^2}$$

4. Simplify.

   $$= \frac{s^{10}}{9r^8t^2}$$

So, $(3r^4s^{-5}t)^{-2} = \frac{s^{10}}{9r^8t^2}$.

As another example, rewrite $\frac{m^3n^{-6}}{6m^{-2}n^3}$ using only positive exponents:

1. Since $\frac{x^{-m}}{y^{-n}} = \frac{y^n}{x^m}$, rewrite $\frac{n^{-6}}{m^{-2}}$ as $\frac{m^2}{n^6}$.

   $$\frac{m^3n^{-6}}{6m^{-2}n^3} = \frac{m^3 \cdot m^2}{6n^6 \cdot n^3}$$

2. Use the Multiplication Property of Exponents.

   $$= \frac{m^5}{6n^9}$$

So, $\frac{m^3n^{-6}}{6m^{-2}n^3} = \frac{m^5}{6n^9}$.

As a third example, to simplify $\dfrac{73}{3^{-2} + 4^{-3}}$:

1. Rewrite using positive exponents.

$$\frac{73}{3^{-2} + 4^{-3}} = \frac{73}{\dfrac{1}{3^2} + \dfrac{1}{4^3}}$$

2. Multiply to eliminate the exponents.

$$= \frac{73}{\dfrac{1}{9} + \dfrac{1}{64}}$$

3. Rewrite each fraction in the denominator with the LCD, 576.

$$= \frac{73}{\dfrac{64}{576} + \dfrac{9}{576}}$$

4. Add the fractions in the denominator.

$$= \frac{73}{\dfrac{73}{576}}$$

5. Simplify.

$$= 73 \div \frac{73}{576}$$
$$= 73 \cdot \frac{576}{73}$$
$$= 576$$

So, $\dfrac{73}{3^{-2} + 4^{-3}} = 576$.

Scientific Notation

Now that you know how to work with negative exponents as well as positive exponents, you can learn how to use scientific notation. Scientific notation is a shorthand that is often used to write very small or very large numbers.

For example here are some very small and very large numbers:
- the gross national debt at the end of 1994 was $4,643,700,000,000
- a number called Planck's constant that relates the energy of a photon to its frequency is 0.000000000000004135 electron volt seconds
- the speed of light is 299,792,500 meters per second

You can rewrite these numbers in scientific notation.

A number written in scientific notation is the product of a number between 1 and 10 and an integer power of 10. To write a number in scientific notation:

1. Move the decimal point to the left or to the right until you have a number between 1 and 10.

2. Multiply this number by a power of 10. To find the power, count the number of places you moved the decimal point.

- If you moved the decimal point to the left, the power is positive.

- If you moved the decimal point to the right, the power is negative.

Remember the decimal point is at the end of a whole number. For example, the number 239 can be written like this: 239.0

For example, to write $4,643,700,000,000, in scientific notation:

1. Move the decimal point to the left.

$$4.643700000000$$
12 11 10 9 8 7 6 5 4 3 2 1

2. Multiply by the appropriate power of 10.

$$4.6437 \times 10^{12}$$

So, $4,643,700,000,000 = 4.6437 \times 10^{12}$ dollars.

As another example. To write 0.000000000000004135 eVsec (electron volt seconds), in scientific notation:

1. Move the decimal point to the right.

$$0.000000000000004.135$$
1 2 3 4 5 6 7 8 9 10 11 12 13 14 15

2. Multiply by the appropriate power of 10.

$$4.135 \times 10^{-15}$$

So, 0.000000000000004135 eVsec $= 4.135 \times 10^{-15}$ eVsec.

You can also reverse this process to write a number given in scientific notation in expanded form.

To write a number in expanded form you must perform the multiplication indicated by the power of 10 in scientific notation. So:

1. Move the decimal point the number of places indicated by the power of 10.
   - If the power is positive, move the decimal point to the right.
   - If the power is negative, move the decimal point to the left.

2. Fill in additional zeros as needed.

For example, to write $2.997925 \times 10^{8} \frac{m}{s}$, in expanded form:

1. Move the decimal point to the right 8 places.

$$2997925$$
1 2 3 4 5 6 7 8

2. Fill in additional zeros.

$$299792500.$$
1 2 3 4 5 6 7 8

So, $2.997925 \times 10^{8} \frac{m}{s} = 299,792,500 \frac{m}{s}$.

Sample Problems

1. Rewrite $(5a^{-2}b^7c^{-1})^3$ using only positive exponents.

    ☑ a. Use the Power of a Product Property. $= (5)^3 \cdot (a^{-2})^3 \cdot (b^7)^3 \cdot (c^{-1})^3$

    ☑ b. Use the Power of a Power Property. $= 125 \cdot a^{-6} \cdot b^{21} \cdot c^{-3}$

    ☐ c. Rewrite using only positive exponents. $= 125 \cdot \underline{\ \ \ } \cdot b^{21} \cdot \underline{\ \ \ }$

    ☐ d. Simplify. $= \underline{\ \ \ \ \ \ \ }$

2. Rewrite $\dfrac{3x^{-4}y^3}{x^2 y^{-3}}$ using only positive exponents.

    ☐ a. Use $\dfrac{x^{-m}}{y^{-n}} = \dfrac{y^n}{x^m}$ to $\dfrac{3x^{-4}y^3}{x^2 y^{-3}} = \dfrac{3\underline{\ \ \ } \cdot y^3}{x^2 \cdot \underline{\ \ \ }}$

        rewrite $\dfrac{x^{-4}}{y^{-3}}$.

    ☐ b. Use the Multiplication Property $= \underline{\ \ \ \ \ \ \ }$
        of Exponents.

3. Simplify this expression: $\dfrac{17}{2^{-3} + 3^{-2}}$

    ☑ a. Rewrite using only positive exponents. $= \dfrac{17}{\dfrac{1}{2^3} + \dfrac{1}{3^2}}$

    ☐ b. Multiply to eliminate the exponents. $= \underline{\ \ \ \ \ \ \ \ \ \ }$

    ☐ c. Rewrite each fraction in the $= \underline{\ \ \ \ \ \ \ \ \ \ }$
        denominator with the LCD.

    ☐ d. Add the fractions in the denominator. $= \underline{\ \ \ \ \ \ \ \ \ \ }$

    ☐ e. Simplify. $= \underline{\ \ \ \ \ \ \ \ \ \ }$

        $= \underline{\ \ \ \ \ \ \ \ \ \ }$

        $= \underline{\ \ \ \ \ \ \ }$

4. Write 45,318,000 in scientific notation.

    ☑ a. Move the decimal point.   4.5318000

    ☐ b. Multiply by the appropriate power of 10. $= \underline{\ \ \ \ \ \ \ \ \ \ }$

5. Write 9.68×10^{-5} in expanded form.

    ☐ a. Move the decimal point. $\underline{\ \ \ \ \ \ \ \ \ \ }$

    ☐ b. Fill in additional zeros. $= \underline{\ \ \ \ \ \ \ \ \ \ }$

MULTIPLYING AND DIVIDING

Summary

Reducing a Rational Expression of the Form $\frac{a-b}{b-a}$

Look at this rational expression: $\frac{7-x}{x-7}$

Here, the numerator and denominator are already factored, and at first glance the rational expression may appear to be in lowest terms.

But notice that the numerator and denominator are the same except for their signs. To reduce this rational expresson further:

1. Factor -1 out of **either** the numerator or the denominator. Here, the numerator is factored.

$$\frac{7-x}{x-7} = \frac{-1(-7+x)}{x-7}$$

2. Rewrite $(-7+x)$ as $(x-7)$.

$$= \frac{-1(x-7)}{x-7}$$

3. Cancel common factors.

$$= \frac{-1\cancel{(x-7)}^{1}}{\cancel{x-7}_{1}}$$

$$= -1$$

So, $\frac{7-x}{x-7} = -1$.

In general:

$$\frac{a-b}{b-a} = -1$$

Here, $a \neq b$.

Here's another example. To reduce $\frac{x^2+6x-27}{9-x^2}$ to lowest terms:

1. Factor the numerator and denominator.

$$= \frac{(x-3)(x+9)}{(3-x)(3+x)}$$

   In the numerator, notice $(x-3)$.
   In the denominator, notice $(3-x)$.

2. Recall that $\frac{x-3}{3-x} = -1$.

$$= -1 \cdot \frac{(x+9)}{(3+x)}$$

   So, you can rewrite the expression.

$$= -\frac{x+9}{3+x}$$

Multiplying Rational Expressions

You have learned how to multiply rational expressions. Here is an example.

Find $\dfrac{x^2 + 2x - 3}{x^2 + 5x + 6} \cdot \dfrac{x^2 + x - 12}{x^2 - 4x + 3}$:

1. Factor the numerators and denominators.

$$= \dfrac{(x-1)(x+3)}{(x+2)(x+3)} \cdot \dfrac{(x+4)(x-3)}{(x-1)(x-3)}$$

2. Cancel pairs of factors common to the numerators and denominators.

$$= \dfrac{\overset{1}{(x-1)}\overset{1}{(x+3)}}{(x+2)\underset{1}{(x+3)}} \cdot \dfrac{(x+4)\overset{1}{(x-3)}}{\underset{1}{(x-1)}\underset{1}{(x-3)}}$$

3. Multiply the numerators. Multiply the denominators.

$$= \dfrac{x+4}{x+2}$$

So, $\dfrac{x^2 + 2x - 3}{x^2 + 5x + 6} \cdot \dfrac{x^2 + x - 12}{x^2 - 4x + 3} = \dfrac{x+4}{x+2}$.

Dividing Rational Expressions

You have learned how to divide rational expressions by inverting the divisor and then multiplying. Here is an example.

Find $\dfrac{2x^2 + 5x - 3}{x^2 - 6x + 5} \div \dfrac{2x^2 + x - 1}{x^2 - 4x - 5}$:

1. Invert the second fraction and change $\div$ to $\cdot$. Then multiply.

$$= \dfrac{2x^2 + 5x - 3}{x^2 - 6x + 5} \cdot \dfrac{x^2 - 4x - 5}{2x^2 + x - 1}$$

2. Factor the numerators and denominators.

$$= \dfrac{(2x-1)(x+3)}{(x-1)(x-5)} \cdot \dfrac{(x+1)(x-5)}{(2x-1)(x+1)}$$

3. Cancel pairs of factors common to the numerators and denominators.

$$= \dfrac{\overset{1}{(2x-1)}(x+3)}{(x-1)\underset{1}{(x-5)}} \cdot \dfrac{\overset{1}{(x+1)}\overset{1}{(x-5)}}{\underset{1}{(2x-1)}\underset{1}{(x+1)}}$$

4. Multiply the numerators. Multiply the denominators.

$$= \dfrac{x+3}{x-1}$$

So, $\dfrac{2x^2 + 5x - 3}{x^2 - 6x + 5} \div \dfrac{2x^2 + x - 1}{x^2 - 4x - 5} = \dfrac{x+3}{x-1}$.

Simplifying a Complex Fraction

A fraction that contains other fractions is called a complex fraction. To simplify a complex fraction, start by rewriting the complex fraction as a division problem. Then invert the second fraction, and multiply.

For example, to simplify the complex fraction $\dfrac{\dfrac{x^2 - 4}{x^2 - 8x + 15}}{\dfrac{12x + 24}{3x - 15}}$:

1. Rewrite the complex fraction as a division problem.

$$= \dfrac{x^2 - 4}{x^2 - 8x + 15} \div \dfrac{12x + 24}{3x - 15}$$

2. Divide by inverting the second fraction and multiplying.

$$= \frac{x^2 - 4}{x^2 - 8x + 15} \cdot \frac{3x - 15}{12x + 24}$$

3. Factor the numerators and denominators.

$$= \frac{(x + 2)(x - 2)}{(x - 5)(x - 3)} \cdot \frac{3(x - 5)}{12(x + 2)}$$

$$= \frac{(x + 2)(x - 2)}{(x - 5)(x - 3)} \cdot \frac{3(x - 5)}{2 \cdot 2 \cdot 3(x + 2)}$$

4. Cancel pairs of factors common to the numerators and denominators.

$$= \frac{\overset{1}{\cancel{(x + 2)}}(x - 2)}{\underset{1}{\cancel{(x - 5)}}(x - 3)} \cdot \frac{\overset{1}{\cancel{3}}\overset{1}{\cancel{(x - 5)}}}{2 \cdot 2 \cdot \underset{1}{\cancel{3}}\underset{1}{\cancel{(x + 2)}}}$$

5. Multiply the numerators. Multiply the denominators.

$$= \frac{x - 2}{4(x - 3)}$$

So, $\dfrac{\dfrac{x^2 - 4}{x^2 - 8x + 15}}{\dfrac{12x + 24}{3x - 15}} = \dfrac{x - 2}{4(x - 3)}$.

Sample Problems

1. Reduce to lowest terms: $\frac{9 - x}{x - 9}$

   ☑ a. Factor −1 out of the numerator. $= \frac{-1(-9 + x)}{x - 9}$

   ☐ b. Rewrite $(-9 + x)$ as $(x - 9)$ = _____

   ☐ c. Cancel common factors. = _____

2. Reduce to lowest terms: $\frac{x^2 - 9x + 20}{16 - x^2}$

   ☑ a. Factor the numerator and denominator. $= \frac{(x - 4)(x - 5)}{(4 - x)(4 + x)}$

   ☐ b. Recall that $\frac{x - 4}{4 - x} = -1$. So rewrite the expression. = _____ · _____

   ☐ c. Multiply. = _____

a. $x + 5$, $x + 1$

b. $x \cdot \dfrac{x-4}{(x+1)}$

c. $\dfrac{x(x-4)}{x+1}$

b. $x + 2$, $x + 2$

c. $\dfrac{x}{(x+2)} \cdot \dfrac{(x+4)}{(x+2)}$

d. $\dfrac{x(x+4)}{(x+2)^2}$

b. $\dfrac{9}{2x+4}$

c. $\dfrac{x(x+2)}{3 \cdot x} \cdot \dfrac{3 \cdot 3}{2(x+2)}$

d. $\dfrac{1}{1} \cdot \dfrac{3}{2}$

e. $\dfrac{3}{2}$

3. Find: $\dfrac{x^3 + 2x^2 - 15x}{x^2 + 8x + 15} \cdot \dfrac{x^2 - x - 12}{x^2 - 2x - 3}$

☐ a. Factor the numerators and denominators.

$= \dfrac{x(x+5)(x-3)}{(\underline{\quad})(x+3)} \cdot \dfrac{(x+3)(x-4)}{(\underline{\quad})(x-3)}$

☐ b. Cancel common factors.

$= \underline{\hspace{2cm}}$

☐ c. Multiply the numerators. Multiply the denominators.

$= \underline{\hspace{2cm}}$

4. Find: $\dfrac{x^2 - 3x}{x^2 + x - 2} \div \dfrac{x^2 - x - 6}{x^2 + 3x - 4}$

☑ a. Invert the second fraction and change $\div$ to $\cdot$.

$= \dfrac{x^2 - 3x}{x^2 + x - 2} \cdot \dfrac{x^2 + 3x - 4}{x^2 - x - 6}$

☐ b. Factor the numerators and denominators.

$= \dfrac{x(x-3)}{(x-1)(\underline{\quad})} \cdot \dfrac{(x-1)(x+4)}{(x-3)(\underline{\quad})}$

☐ c. Cancel common factors.

$= \underline{\hspace{2cm}}$

☐ d. Multiply the numerators. Multiply the denominators.

$= \underline{\hspace{2cm}}$

5. Simplify this complex fraction: $\dfrac{\dfrac{x^2 + 2x}{3x}}{\dfrac{2x + 4}{9}}$

☑ a. Rewrite as a division problem.

$= \dfrac{x^2 + 2x}{3x} \div \dfrac{2x + 4}{9}$

☐ b. Divide. (Invert the second fraction and multiply.)

$= \dfrac{x^2 + 2x}{3x} \cdot \underline{\hspace{2cm}}$

☐ c. Factor the numerators and denominators.

$= \underline{\hspace{2cm}}$

☐ d. Cancel pairs of factors common to the numerator and denominator.

$= \underline{\hspace{2cm}}$

☐ e. Multiply the numerators. Multiply the denominators.

$= \underline{\hspace{2cm}}$

ADDING AND SUBTRACTING

Summary

Finding the Least Common Denominator (LCD) of Two or More Rational Expressions

In order to add or subtract rational expressions with different denominators, you need to find the least common denominator (LCD) of the rational expressions. This LCD is the least common multiple (LCM) of the polynomials in the denominators. You can find the LCM of a collection of polynomials in the same way you find the LCM of a collection of numbers.

To find the LCM of a collection of polynomials:

1. Factor each polynomial.
2. List each factor the greatest number of times it appears in any factorization.
3. Find the product of these factors. This is the LCM.

For example, to find the LCM of $x^2 - 4$, $x^2 + x - 6$, and $x^2 + 6x + 9$:

1. Factor each polynomial.

$$x^2 - 4 = (x + 2)(x - 2)$$
$$x^2 + x - 6 = (x + 3)(x - 2)$$
$$x^2 + 6x + 9 = (x + 3)(x + 3)$$

2. List each factor the greatest number of times it appears in any factorization.

$(x + 2)$, $(x - 2)$, $(x + 3)$, $(x + 3)$

3. Find the product of these factors.

$(x + 2)(x - 2)(x + 3)(x + 3)$

So, the LCM of $x^2 - 4$, $x^2 + x - 6$, and $x^2 + 6x + 9$ is $(x + 2)(x - 2)(x + 3)(x + 3)$.

Remember, to find the LCM of a collection of numbers:

1. *Find the prime factorization of each number.*

$$
\begin{array}{ccc}
15 & 25 & 18 \\
3 \cdot 5 & 5 \cdot 5 & 2 \cdot 9 \\
& & 3 \cdot 3
\end{array}
$$

2. *List each prime factor the greatest number of times it appears in any factorization.*   2, 3, 3, 5, 5

3. *Find the product of these factors. This is the LCM.*

$$2 \cdot 3 \cdot 3 \cdot 5 \cdot 5 = 450$$

The LCM of a collection of polynomials is usually left in factored form. You don't have to do the multiplication.

Adding Rational Expressions with Different Denominators

You can add rational expressions with different denominators in much the same way you add fractions with different denominators. The key idea is to write the rational expressions with the same denominator.

To add rational expressions with different denominators:

1. Factor each denominator.

2. Find the LCM of the denominators. This is the LCD of the rational expressions.

3. Rewrite each algebraic fraction with this LCD.

4. Add the numerators. The denominator stays the same.

5. Factor and reduce the rational expression to lowest terms as appropriate.

For example, to find $\dfrac{3}{m^3n} + \dfrac{5}{m^2n^2}$:

1. Factor each denominator: $\quad = \dfrac{3}{m \cdot m \cdot m \cdot n} + \dfrac{5}{m \cdot m \cdot n \cdot n}$

2. Find the LCD of the rational expressions. $\quad LCD = m \cdot m \cdot m \cdot n \cdot n$

3. Rewrite each algebraic fraction with the LCD, $m \cdot m \cdot m \cdot n \cdot n$. $\quad = \dfrac{3 \cdot n}{m \cdot m \cdot m \cdot n \cdot n} + \dfrac{5 \cdot m}{m \cdot m \cdot n \cdot n \cdot m}$

4. Add the numerators. The denominator stays the same. $\quad = \dfrac{3n + 5m}{m^3n^2}$

This rational expression is in lowest terms.

So, $\dfrac{3}{m^3n} + \dfrac{5}{m^2n^2} = \dfrac{3n + 5m}{m^3n^2}$.

As another example, to find $\dfrac{x+2}{x^2-4x} + \dfrac{x+2}{x^2-16}$:

1. Factor each denominator: $\quad = \dfrac{x+2}{x(x-4)} + \dfrac{x+2}{(x+4)(x-4)}$

2. Find the LCD of the rational expressions. $\quad LCD = x(x+4)(x-4)$

3. Rewrite each algebraic fraction with the LCD, $x(x-4)(x+4)$. $\quad = \dfrac{(x+2)(x+4)}{x(x-4)(x+4)} + \dfrac{(x+2)x}{(x+4)(x-4)x}$

4. Add the numerators. The denominator stays the same. $\quad = \dfrac{x^2 + 6x + 8 + x^2 + 2x}{x(x+4)(x-4)}$

$\quad = \dfrac{2x^2 + 8x + 8}{x(x+4)(x-4)}$

5. Factor and reduce the rational expression to lowest terms. $\quad = \dfrac{2(x+2)(x+2)}{x(x+4)(x-4)}$

So, $\dfrac{x+2}{x^2-4x} + \dfrac{x+2}{x^2-16} = \dfrac{2(x+2)(x+2)}{x(x+4)(x-4)}$.

Remember, to add fractions with different denominators:

1. Factor each denominator into its prime factors.
$$\frac{5}{6} + \frac{7}{15} = \frac{5}{2 \cdot 3} + \frac{7}{3 \cdot 5}$$

2. Find the LCD of both fractions. This is the LCM of the denominators.
$$LCD = 2 \cdot 3 \cdot 5$$

3. Rewrite each fraction with this LCD.
$$= \frac{5 \cdot 5}{2 \cdot 3 \cdot 5} + \frac{7 \cdot 2}{3 \cdot 5 \cdot 2}$$

4. Add the numerators. Keep the denominator the same.
$$= \frac{25 + 14}{30}$$
$$= \frac{39}{30}$$

5. Factor and reduce the fraction to lowest terms as appropriate.
$$= \frac{\cancel{3} \cdot 13}{\cancel{3} \cdot 2 \cdot 5}$$
$$= \frac{13}{10}$$
So, $\dfrac{5}{6} + \dfrac{7}{15} = \dfrac{13}{10}$.

In the third step you may have been tempted to cancel common factors. But this would get you back to where you started. Remember to add first, then factor and cancel common factors.

Subtracting Rational Expressions with Different Denominators

You can subtract rational expressions with different denominators in much the same way you subtract fractions with different denominators. The key idea is to write the rational expressions with the same denominator.

To subtract rational expressions with different denominators:

1. Factor each denominator.

2. Find the LCM of the denominators. This is the LCD of the rational expressions.

3. Rewrite each algebraic fraction with this LCD.

4. Subtract the numerators. The denominator stays the same.

5. Factor and reduce the rational expression to lowest terms as appropriate.

For example, to find $\dfrac{4}{ab} - \dfrac{7}{a^3b^2}$:

1. Factor each denominator:
$$= \frac{4}{a \cdot b} - \frac{7}{a \cdot a \cdot a \cdot b \cdot b}$$

2. Find the LCD of the rational expressions.
$$LCD = a \cdot a \cdot a \cdot b \cdot b$$

3. Rewrite each algebraic fraction with the LCD, $a \cdot a \cdot a \cdot b \cdot b$.
$$= \frac{4 \cdot a \cdot a \cdot b}{a \cdot b \cdot a \cdot a \cdot b} - \frac{7}{a \cdot a \cdot a \cdot b \cdot b}$$

4. Subtract the numerators. The denominator stays the same.
$$= \frac{4a^2b - 7}{a^3b^2}$$

This rational expression is in lowest terms.

So, $\dfrac{4}{ab} - \dfrac{7}{a^3b^2} = \dfrac{4a^2b - 7}{a^3b^2}$.

As another example, to find $\dfrac{10}{x^2 + x - 6} - \dfrac{3}{x - 2}$:

1. Factor each denominator:
$$= \frac{10}{(x - 2)(x + 3)} - \frac{3}{x - 2}$$

2. Find the LCD of the rational expressions.
$$LCD = (x - 2)(x + 3)$$

3. Rewrite each algebraic fraction with the LCD $(x - 2)(x + 3)$.
$$= \frac{10}{(x - 2)(x + 3)} - \frac{3(x + 3)}{(x - 2)(x + 3)}$$

4. Subtract the numerators. The denominator stays the same.
$$= \frac{10 - 3(x + 3)}{(x - 2)(x + 3)}$$
$$= \frac{10 - 3x - 9}{(x - 2)(x + 3)}$$
$$= \frac{1 - 3x}{(x - 2)(x + 3)}$$

This rational expression is in lowest terms.

So, $\dfrac{10}{x^2 + x - 6} - \dfrac{3}{x - 2} = \dfrac{1 - 3x}{(x - 2)(x + 3)}$.

Remember, to subtract fractions with different denominators:

1. Factor each denominator into its prime factors.
$$\frac{5}{6} - \frac{7}{15} = \frac{5}{2 \cdot 3} - \frac{7}{3 \cdot 5}$$

2. Find the LCD of both fractions. This is the LCM of the denominators.
$$LCD = 2 \cdot 3 \cdot 5$$

3. Rewrite each fraction with this LCD.
$$= \frac{5 \cdot 5}{2 \cdot 3 \cdot 5} - \frac{7 \cdot 2}{3 \cdot 5 \cdot 2}$$

4. Subtract the numerators. Keep the denominator the same.
$$= \frac{25 - 14}{30}$$
$$= \frac{11}{30}$$

So, $\dfrac{5}{6} - \dfrac{7}{15} = \dfrac{11}{30}$.

In the third step you may have been tempted to cancel common factors. But this would get you back to where you started. Remember to subtract first, then factor and cancel common factors.

Simplifying a Complex Fraction

Recall that a fraction that contains other fractions is called a complex fraction. Here's one way to simplify certain complex fractions:

1. Perform any addition or subtraction in the numerator or denominator.

2. Rewrite the complex fraction as a division problem.

3. Divide.

For example, to find $\dfrac{\dfrac{3}{x+1}+5}{\dfrac{3}{x+1}}$:

1. Perform the addition in the numerator. The least common denominator of $\dfrac{3}{x+1}$ and 5 is $(x+1)$.

$$= \dfrac{\dfrac{3}{x+1}+\dfrac{5(x+1)}{x+1}}{\dfrac{3}{x+1}}$$

$$= \dfrac{\dfrac{3+5(x+1)}{x+1}}{\dfrac{3}{x+1}}$$

$$= \dfrac{\dfrac{3+5x+5}{x+1}}{\dfrac{3}{x+1}}$$

$$= \dfrac{\dfrac{8+5x}{x+1}}{\dfrac{3}{x+1}}$$

2. Rewrite the complex fraction as a division problem.

$$= \dfrac{8+5x}{x+1} \div \dfrac{3}{x+1}$$

3. To divide, invert the second fraction and replace $\div$ with $\cdot$.

$$= \dfrac{8+5x}{x+1} \cdot \dfrac{x+1}{3}$$

4. Cancel common factors.

$$= \dfrac{8+5x}{3}$$

So, $\dfrac{\dfrac{3}{x+1}+5}{\dfrac{3}{x+1}} = \dfrac{8+5x}{3}$.

Sample Problems

1. Find the LCM of $x^2 + 7x + 12$, $x^2 - 9$, and $x^2 + 8x + 16$.

   ☐ a. Factor each polynomial.

$$x^2 + 7x + 12 = (\underline{})(\underline{})$$
$$x^2 - 9 \quad\; = (\underline{})(\underline{})$$
$$x^2 + 8x + 16 = (\underline{})(\underline{})$$

a. x + 3, x + 4 (in either order)
x + 3, x – 3 (in either order)
x + 4, x + 4 (in either order)

   ☐ b. List each factor the greatest number of times it appears in any factorization.

$$\underline{}, \; \underline{}, \; \underline{}, \; \underline{}$$

   ☐ c. Find the product of the factors in the list. This is the LCM.

$$(\underline{})(\underline{})(\underline{})(\underline{})$$

2. Find: $\dfrac{2}{x^2 + 2x - 15} + \dfrac{x}{x^2 - 9}$

   ☑ a. Factor each denominator.

$$= \frac{2}{(x+5)(x-3)} + \frac{x}{(x+3)(x-3)}$$

   ☐ b. Find the LCD of the rational expressions.

$$\text{LCD} = (\underline{})(\underline{})(\underline{})$$

   ☐ c. Rewrite each rational expression using this LCD.

$$= \frac{2(\underline{})}{(x+5)(x-3)(\underline{})} + \frac{x(\underline{})}{(x+3)(x-3)(\underline{})}$$

   ☐ d. Add the numerators. The denominator stays the same.

$$= \frac{2(\underline{}) + x(\underline{})}{(x+5)(x+3)(x-3)}$$

$$= \frac{2x + 6 + x^2 + 5x}{(x+5)(x+3)(x-3)}$$

$$= \underline{}$$

3. Find: $\dfrac{x}{x^2 - 6x + 9} - \dfrac{3}{x^2 - 9}$

   ☑ a. Factor each denominator.

$$= \frac{x}{(x-3)(x-3)} - \frac{3}{(x+3)(x-3)}$$

   ☐ b. Find the LCD of the rational expressions.

$$\text{LCD} = (\underline{})(\underline{})(\underline{})$$

   ☐ c. Rewrite each rational expression using this LCD.

$$= \frac{x(\underline{})}{(x-3)(x-3)(\underline{})} - \frac{3(\underline{})}{(x+3)(x-3)(\underline{})}$$

   ☐ d. Subtract the numerators. The denominator stays the same.

$$= \frac{x(\underline{}) - 3(\underline{})}{\underline{}}$$

$$= \underline{}$$

b. $z + 5, z - 5$

c. top: $z + 5, z - 5$
   bottom: $z + 5, z - 5$

d. top: $z + 5, -1, z - 5$
   bottom: $(z + 5)(z - 5)$

$\dfrac{2z + 10 - z + 5 - 10}{(z + 5)(z - 5)}$

$\dfrac{z + 5}{(z + 5)(z - 5)}$

e. $\dfrac{1}{z - 5}$

a. xy^2

b. $\dfrac{2y}{xy^2}, \dfrac{1}{xy^2}$

c. $\dfrac{2y + 1}{xy^2}$

d. $3 \div \dfrac{2y + 1}{xy^2}$

e. $\dfrac{3xy^2}{2y + 1}$

---

4. Find: $\dfrac{2}{z - 5} + \dfrac{-1}{z + 5} - \dfrac{10}{z^2 - 25}$

   ☑ a. Factor each denominator. $\qquad = \dfrac{2}{z - 5} + \dfrac{-1}{z + 5} - \dfrac{10}{(z + 5)(z - 5)}$

   ☐ b. Find the LCD of the rational $\qquad$ LCD = (_____)(_____)
        expressions.

   ☐ c. Rewrite each rational $\qquad = \dfrac{2(\underline{\quad})}{(z - 5)(\underline{\quad})} + \dfrac{-1(\underline{\quad})}{(z + 5)(\underline{\quad})} - \dfrac{10}{(z + 5)(z - 5)}$
        expression using this LCD.

   ☐ d. Add or subtract the $\qquad = \dfrac{2(\underline{\quad}) + (\underline{\ }) (\underline{\quad}) - 10}{\underline{\qquad\qquad}}$
        numerators. The denominator
        stays the same. $\qquad = \underline{\qquad\qquad\qquad}$

        $\qquad = \underline{\qquad\qquad\qquad}$

   ☐ e. Factor and reduce the $\qquad = \underline{\qquad\qquad\qquad}$
        rational expression to lowest
        terms.

5. Simplify this complex fraction: $\dfrac{3}{\dfrac{2}{xy} + \dfrac{1}{xy^2}}$

   ☐ a. To add the fractions in the $\qquad$ LCD = _____
        denominator, find the LCD.

   ☐ b. Rewrite each fraction using $\qquad = \dfrac{3}{\underline{\quad} + \underline{\quad}}$
        this LCD.

   ☐ c. Add the fractions in $\qquad = \dfrac{3}{\underline{\quad}}$
        the denominator.

   ☐ d. Rewrite the complex fraction $\qquad = \underline{\qquad\qquad}$
        as a division problem.

   ☐ e. Divide. $\qquad = \underline{\qquad\qquad}$

HOMEWORK

Homework Problems

Circle the homework problems assigned to you by the computer, then complete them below.

☀ Explain
Negative Exponents

1. Find: 5^{-3}

2. Find: $7^{-2} \cdot 7^3$

3. Find: $\dfrac{50}{10^{-2} + 5^{-2}}$

In problems 4 through 9, simplify each expression. Use only positive exponents in your answers.

4. $\left(\dfrac{2m}{n}\right)^{-3}$

5. $\dfrac{(2m)^{-3}}{n^3}$

6. $(n^4 p^{-3})^5$

7. $(2s^3 t)^{-5} \cdot (2st^3)^6$

8. $\dfrac{(3uv^4)(9u^3 v)^{-2}}{(3u^5 v^{-4})^{-3}}$

9. All matter is made up of tiny particles called atoms. Through experimentation, it has been found that the diameters of atoms range from 1×10^{-8} cm to 5×10^{-8} cm. Rewrite each of these numbers in expanded form.

10. The monthly payment E on a loan of amount P can be computed by using the formula below, where r is the monthly interest rate, and n is the number of months for which the loan is made. Find the monthly payment on a $15,000 loan for 4 years (48 months) if the monthly interest rate is 1%.

$$E = \dfrac{Pr}{1 - (1 + r)^{-n}}$$

11. Write $\left(\dfrac{3x^4 y}{z^5}\right)^{-2}$ using only positive exponents.

12. Suppose $x = 3$ and $y = 5$.

    a. Is $x^{-1} y^{-1} = \dfrac{1}{x} \cdot \dfrac{1}{y}$?

    b. Is $x^{-1} y^{-1} = \dfrac{1}{xy}$?

    c. Is $x^{-1} + y^{-1} = \dfrac{1}{x} + \dfrac{1}{y}$?

    d. Is $x^{-1} + y^{-1} = \dfrac{1}{x + y}$?

Multiplying and Dividing

13. Reduce to lowest terms: $\dfrac{y - 13}{13 - y}$

14. Reduce to lowest terms: $\dfrac{y^2 + y - 20}{16 - y^2}$

15. Find: $\dfrac{6 - x}{x + 7} \cdot \dfrac{x^2 + 7x}{x - 6}$

16. $\dfrac{a^2 - 5a - 14}{a^2 - 2a - 35} \cdot \dfrac{a^2 + 6a + 5}{a^2 - a - 6}$

17. Find: $\dfrac{3x^2 - 7x + 2}{x^2 + 4x - 12} \cdot \dfrac{2x^2 + 12x}{3x^2 + 2x - 1}$

18. Find: $\dfrac{(x + 3)(x - 2)}{x^2 - 4} \div \dfrac{(x + 3)(x - 4)}{x^2 - 16}$

19. Find: $\dfrac{y^2 + 15y + 56}{8y^2 - 10y + 3} \div \dfrac{y + 7}{4y - 3}$

20. Find: $\dfrac{x(x - 9)}{x^2 + 3x + 2} \div \dfrac{x^2 - 81}{x(x + 2)}$

21. Find: $\dfrac{z^2 + 8z + 15}{z^2 - 16} \div \dfrac{z^2 + 10z + 25}{z + 4}$

In Problems 22, 23 and 24, simplify the complex fractions. Reduce your answer to lowest terms.

22. $\dfrac{\dfrac{x^2 - 13x + 42}{x^2 - 4}}{\dfrac{3x - 18}{3x - 6}}$

23. $\dfrac{\dfrac{2z^2 - 4z}{3z^3 + 6z^2}}{\dfrac{z - 2}{z + 2}}$

24. $\dfrac{\dfrac{2w^2 - 4w}{2w^2 - 2}}{\dfrac{w^2 - 5w + 6}{w^2 - 2w - 3}}$

Adding and Subtracting

25. Find the LCM of $x^2 + 10x + 25$ and $x^2 - 3x - 40$.

26. Find the LCM of x, $x^2 - 16$, and $x^2 + 7x + 12$.

27. Find: $\dfrac{5}{xy^3} + \dfrac{14}{x^2y^2}$

28. Find: $\dfrac{-1}{x^2 + 3x + 2} + \dfrac{2}{x^2 + 2x}$

29. Find: $\dfrac{4}{xy^2z} - \dfrac{3}{xyz^2}$

30. Find: $\dfrac{8}{x^2 + 14x + 49} - \dfrac{4}{x^2 - 49}$

31. Find: $\dfrac{-1}{y + 2} + \dfrac{2}{y - 2} - \dfrac{4}{y^2 - 4}$

32. Find: $\dfrac{1}{x^2 + 8x} - \dfrac{3}{x^2 - 64} + \dfrac{2}{x^2 + 6x - 16}$

33. Simplify the left side of the equation below to show it equals $\dfrac{1}{n}$. Then use the equation to find two fractions with 1 in the numerator whose difference is $\dfrac{1}{5}$. (Hint: let $n = 5$).

$$\dfrac{1}{n - 1} - \dfrac{1}{n(n - 1)} = \dfrac{1}{n}$$

34. Optometrists use the formula below to find the strength to be used for the lenses of glasses. Simplify the right side of this formula, then find the value of P that corresponds to $a = 12$ and $b = 0.3$.

$$P = \dfrac{1}{a} + \dfrac{1}{b}$$

35. The total resistance, R, of a circuit that consists of two resistors connected in parallel with resistance R_1 and R_2 is given by the formula below. Simplify this formula, then find the resistance, R, if $R_1 = 3$ ohms and $R_2 = 4$ ohms.

$$R = \dfrac{1}{\dfrac{1}{R_1} + \dfrac{1}{R_2}}$$

36. Simplify this complex fraction: $\dfrac{1}{\dfrac{1}{a} + \dfrac{1}{ab^2}}$

 APPLY

Practice Problems

Here are some additional practice problems for you to try.

Negative Exponents

1. Find: 2^{-3}

2. Find: 4^{-2}

3. Find: 3^{-4}

4. Find: 5^{-2}

5. Find: $\dfrac{1}{3^{-4}}$

6. Find: $\dfrac{1}{5^{-3}}$

7. Find: $4^{-7} \cdot 4^5$

8. Find: $2^8 \cdot 2^{-5}$

9. Find: $5^{-9} \cdot 5^6$

10. Find: $\dfrac{1}{3^{-2} + 2^{-3}}$

11. Find: $\dfrac{3}{4^{-3} + 5^{-2}}$

12. Find: $\dfrac{2}{4^{-2} + 7^{-1}}$

13. Rewrite using only positive exponents: $(a^4 b^6)^{-1}$

14. Rewrite using only positive exponents: $(x^3 y^5)^{-2}$

15. Rewrite using only positive exponents: $(m^6 n^3 p)^{-4}$

16. Rewrite using only positive exponents: $(x^{-5} b^2)^5$

17. Rewrite using only positive exponents: $(a^{-3} b^7)^4$

18. Rewrite using only positive exponents: $(x^{-6} y z^{-3})^5$

19. Rewrite using only positive exponents: $\left(\dfrac{c^4 d^{-5}}{a^2} \right)^{-2}$

20. Rewrite using only positive exponents: $\left(\dfrac{x^3 w^{-2}}{y^4} \right)^{-3}$

21. Rewrite using only positive exponents: $\left(\dfrac{m^{-4} n^5}{p^{-3}} \right)^{-4}$

22. Rewrite using only positive exponents: $(3x^3 y)^{-2} \cdot (3x^2 y^{-1} z^3)^4$

23. Rewrite using only positive exponents: $(4a^4 b^2)^4 \cdot (4a^{-3} bc^2)^{-3}$

24. Rewrite using only positive exponents: $(5x^{-3} y^{-1} z)^{-3} \cdot (5xz^{-5})^4$

25. Write in scientific notation: 0.000057

26. Write in scientific notation: 148,000,000

27. The following number is written in scientific notation. Write it in expanded form: 4.3×10^6

28. The following number is written in scientific notation. Write it in expanded form: 1.785×10^{-4}

Multiplying and Dividing

29. Reduce to lowest terms: $\dfrac{x-5}{5-x}$

30. Reduce to lowest terms: $\dfrac{x-3}{3-x}$

31. Reduce to lowest terms: $\dfrac{x^2 - 2x - 35}{x^2 - 25}$

32. Reduce to lowest terms: $\dfrac{x^2 + 2x - 24}{16 - x^2}$

33. Reduce to lowest terms: $\dfrac{x^2 - 8x + 7}{49 - x^2}$

34. Reduce to lowest terms: $\dfrac{x^2 - 11x + 30}{x^2 - 36}$

35. Reduce to lowest terms: $\dfrac{x^2 - 8x - 9}{81 - x^2}$

36. Find: $\dfrac{x^2 + 9x + 14}{x^2 + 2x - 15} \cdot \dfrac{x^2 - 4x + 3}{x^2 + 6x - 7}$

37. Find: $\dfrac{x^2 - 7x + 12}{x^2 + 2x - 15} \cdot \dfrac{x^2 - x - 30}{x^2 - 3x - 18}$

38. Find: $\dfrac{x^2 + 5x + 6}{x^2 - 5x - 6} \cdot \dfrac{x^2 - 10x + 24}{x^2 - x - 12}$

39. Find: $\dfrac{3x^2 - 6x}{x + 1} \cdot \dfrac{x - 1}{2 - x}$

40. Find: $\dfrac{5x^2 - 25x}{x - 3} \cdot \dfrac{x + 3}{5 - x}$

41. Find: $\dfrac{2x^2 - 6x}{x - 5} \cdot \dfrac{x + 5}{3 - x}$

42. Find: $\dfrac{x^2 + 2x - 15}{x^2 - 25} \cdot \dfrac{x^2 + 3x + 2}{x^2 - 2x - 3}$

43. Find: $\dfrac{x^2 + 4x - 45}{x^2 - 81} \cdot \dfrac{x^2 - 4x - 45}{x^2 - 7x + 10}$

44. Find: $\dfrac{x^2 + 10x + 21}{x^2 - 2x - 15} \cdot \dfrac{x^2 - x - 20}{x^2 - 16}$

45. Find: $\dfrac{x^2 + 2x - 35}{x^2 + x - 90} \div \dfrac{x^2 + 10x + 21}{x^2 + x - 90}$

46. Find: $\dfrac{x^2 + 2x - 35}{x^2 + x - 90} \div \dfrac{x^2 + 10x + 21}{x^2 + x - 90}$

47. Find: $\dfrac{x^2 - 2x - 3}{x^2 + 4x - 5} \div \dfrac{x^2 - 10x + 21}{x^2 + 4x - 5}$

48. Find: $\dfrac{5x - 25}{x^2 - 49} \div \dfrac{x^2 - 9x + 20}{x^2 - 11x + 28}$

49. Find: $\dfrac{x^2 + 7x + 6}{x^2 - 5x - 6} \div \dfrac{3x + 18}{x^2 + 5x - 66}$

50. Find: $\dfrac{4x - 16}{x^2 - 36} \div \dfrac{x^2 - x - 12}{x^2 + 9x + 18}$

51. Simplify the complex fraction below. Write your answer in lowest terms.

$$\dfrac{\dfrac{x^2 + 3x - 70}{x^2 - 49}}{\dfrac{x^2 + 9x - 10}{3x^2 - 3x}}$$

52. Simplify the complex fraction below. Write your answer in lowest terms.

$$\dfrac{\dfrac{x^2 - 4x - 45}{4x^2 + 20x}}{\dfrac{x^2 + 2x - 99}{x^2 - 121}}$$

53. Simplify the complex fraction below. Write your answer in lowest terms.

$$\dfrac{\dfrac{x^2 - 8x - 33}{x^2 - 9}}{\dfrac{x^2 - 9x - 22}{5x^2 + 10x}}$$

54. Simplify the complex fraction below. Write your answer in lowest terms.

$$\dfrac{\dfrac{x^2 - 7x + 6}{x^2 + 8x + 12}}{\dfrac{3x^2 - 3x}{x^2 - 36}}$$

55. Simplify the complex fraction below. Write your answer in lowest terms.

$$\dfrac{\dfrac{x^2 + 2x - 15}{5x^2 + 15x}}{\dfrac{x^2 - 2x - 35}{x^2 - 9}}$$

56. Simplify the complex fraction below. Write your answer in lowest terms.

$$\dfrac{\dfrac{x^2 + x - 20}{x^2 + 5x + 4}}{\dfrac{2x^2 + 10x}{x^2 - 16}}$$

Adding and Subtracting

57. Find the LCM of $x^2 + 7x + 12$ and $x^2 - 3x - 28$.

58. Find the LCM of $x^2 + 11x + 28$ and $x^2 + 2x - 8$.

59. Find the LCM of $x^2 + 4x$, $x^2 + 3x - 4$, and $x^2 - 2x + 1$.

60. Find the LCM of $x^2 - 6x$, $x^2 - 5x - 6$, and $x^2 + 2x + 1$.

61. Find: $\dfrac{4}{9a} + \dfrac{2}{3b}$

62. Find: $\dfrac{3}{8m} + \dfrac{7}{10n}$

63. Find: $\dfrac{2}{3x} + \dfrac{2}{6y}$

64. Find: $\dfrac{4}{x^2 - 4x - 12} + \dfrac{3}{x - 6}$

65. Find: $\dfrac{2}{x + 10} + \dfrac{5}{x^2 + 5x - 50}$

66. Find: $\dfrac{5}{x^2 + 4x - 21} + \dfrac{1}{x + 7}$

67. Find: $\dfrac{x}{x + 1} + \dfrac{x + 4}{x - 4}$

68. Find: $\dfrac{3x}{x + 3} + \dfrac{x + 2}{x - 7}$

69. Find: $\dfrac{5x}{x - 3} + \dfrac{x + 1}{x + 2}$

70. Find: $\dfrac{x + 2}{x^2 + 7x + 12} + \dfrac{x - 1}{x^2 + x - 12}$

71. Find: $\dfrac{x + 4}{x^2 + 5x + 6} + \dfrac{x - 3}{x^2 - 3x - 10}$

72. Find: $\dfrac{x - 5}{x^2 - 6x + 8} + \dfrac{x + 3}{x^2 + 2x - 8}$

73. Find: $\dfrac{4}{m^2 n} - \dfrac{1}{mn^2}$

74. Find: $\dfrac{5}{abc} - \dfrac{7}{b^2}$

75. Find: $\dfrac{3}{xyz} - \dfrac{2}{x^2}$

76. Find: $\dfrac{7x}{x + 6} - \dfrac{5}{x - 1}$

77. Find: $\dfrac{3x}{x-8} - \dfrac{11}{x+3}$

78. Find: $\dfrac{8x}{x-7} - \dfrac{3}{x+1}$

79. Find: $\dfrac{x+2}{x^2+6x+5} - \dfrac{x+3}{x^2+4x-5}$

80. Find: $\dfrac{x+5}{x^2-8x+12} - \dfrac{x-1}{x^2-3x-18}$

81. Find: $\dfrac{x+4}{x^2+x-2} - \dfrac{x+1}{x^2+2x-3}$

82. Simplify this complex fraction: $\dfrac{\dfrac{4}{x}+\dfrac{1}{y}}{\dfrac{3}{x}-\dfrac{5}{y}}$

83. Simplify this complex fraction: $\dfrac{\dfrac{1}{x+1}+\dfrac{2}{x}}{\dfrac{4}{x}-\dfrac{3}{x+1}}$

84. Simplify this complex fraction: $\dfrac{\dfrac{7}{x}-\dfrac{3}{y}}{\dfrac{1}{x}+\dfrac{2}{y}}$

EVALUATE

Practice Test

Take this practice test to be sure that you are prepared for the final quiz in Evaluate.

1. Find: $\left(\dfrac{1}{2}\right)^{-3}$

2. Fill in the blanks below by writing the numbers in either scientific notation or expanded form.

$$73901 = \underline{\hspace{1.5cm}} \times 10^4$$

$$0.00004003 = 4.003 \times 10^{\underline{\hspace{0.5cm}}}$$

$$2.081 \times 10^2 = \underline{\hspace{2.5cm}}$$

$$9.019 \times 10^{-5} = \underline{\hspace{2.5cm}}$$

3. Rewrite $\dfrac{a^{-8}b^{13}}{a^{-2}b^{-2}}$ using only positive exponents.

4. Rewrite $\dfrac{(2xy^5)(8x^2y)^{-3}}{(2x^4y^{-5})^{-2}}$ using only positive exponents.

5. Reduce to lowest terms: $\dfrac{2x^3 - 8x}{4x - 2x^2}$

6. Multiply and reduce to lowest terms:

a. $\dfrac{x+3}{x^2-4} \cdot \dfrac{x^2-10x+16}{x-8}$

b. $\dfrac{x^2+x-2}{x^3-6x^2} \cdot \dfrac{2x^2-14x+12}{x+2}$

7. Divide and reduce your result to lowest terms:

a. $\dfrac{3x^2-75}{x^2-10x+25} \div \dfrac{-x-5}{x-5}$

b. $\dfrac{y^2+5y+4}{y^2+y-30} \div \dfrac{y^2+2y+1}{y-5}$

8. Simplify this complex fraction: $\dfrac{\dfrac{4x+12}{5x-5}}{\dfrac{2x^2-18}{x^2-2x+1}}$

9. Find the LCM of x, $x^2+12x+35$, and x^2-25.

10. Add and reduce your answer to lowest terms.

$$\dfrac{4}{b^2+4b+3} + \dfrac{3}{b^2+3b+2}$$

11. Subtract and reduce your answer to lowest terms.

$$\dfrac{3y}{y^2+7y+10} - \dfrac{2y}{y^2+6y+8}$$

12. Simplify this complex fraction: $\dfrac{\dfrac{3}{5}}{\dfrac{1}{x}+3}$

LESSON 8.3 — EQUATIONS WITH FRACTIONS

Here is what you'll learn in this lesson:

Solving Equations

a. *Solving equations with rational expressions*

b. *Solving for an unknown in a formula involving a rational expression*

Suppose you want to figure out a baseball pitcher's earned run average, or estimate the population of fish in a lake, or figure out how tall a building is based on a scale model. For each of these examples, you can figure out the answer by setting up an equation that involves fractions or ratios.

However, even with all the techniques you have for solving equations, solving an equation with fractions can be tricky. There may not be any solution at all, or the solution you find might not check when you plug it back into the original equation.

In this lesson you will learn how to solve equations that have fractions in them, and you will learn how to identify extraneous, or false solutions.

 **EXPLAIN**

SOLVING EQUATIONS

Summary

Solving Equations with Rational Expressions

When you solve an equation that contains a rational expression, it helps to clear the fraction in the equation. To do this, multiply both sides of the equation by the least common denominator (LCD) of the fractions.

To solve an equation that contains rational expressions:

1. Clear the fractions by multiplying both sides of the equation by the LCD of the fractions.

2. Distribute the LCD and simplify.

3. Finish solving for the variable.

For example, to solve $2x - \frac{1}{3} = \frac{2x}{3} + \frac{2x}{5} - \frac{3}{15}$ for x:

    1. Multiply by the LCD
       of the fractions, 15. $15 \cdot (2x - \frac{1}{3}) = 15 \cdot \left(\frac{2x}{3} + \frac{2x}{5} - \frac{3}{15} \right)$

    2. Distribute the LCD
       and simplify. $15 \cdot 2x - 15 \cdot \frac{1}{3} = 15 \cdot \frac{2x}{3} + 15 \cdot \frac{2x}{5} - 15 \cdot \frac{3}{15}$

$$30x \ - \ \ 5 \ = 10x \ + \ \ 6x \ - \ \ 3$$

$$30x \ - \ \ 5 \ = \ \ \ \ \ \ 16x \ \ \ \ \ - \ 3$$

    3. Finish solving for x.

$$14x = 2$$

$$x = \frac{2}{14}$$

$$x = \frac{1}{7}$$

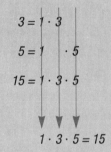

Here's how to find the LCD of
$\frac{1}{3}, \frac{2x}{5},$ *and* $\frac{3}{15}$:

$$3 = 1 \cdot 3$$
$$5 = 1 \quad \cdot 5$$
$$15 = 1 \cdot 3 \cdot 5$$

$$1 \cdot 3 \cdot 5 = 15$$

Checking for Extraneous Solutions

If an equation contains a fraction with a variable in the denominator, the solution of the equation might be extraneous (false).

To check for an extraneous solution:

1. Solve the equation for the variable.

2. Substitute the solution into the original equation and simplify.

3. Look at the denominators of the fractions. If any denominator is zero, the solution is extraneous.

For example, to determine if $\frac{4}{x+3} = \frac{1}{x} - \frac{12}{x(x+3)}$ has an extraneous solution:

1. Solve the equation for x.

$$\frac{4}{x+3} = \frac{1}{x} - \frac{12}{x(x+3)}$$

$$x(x+3) \cdot \frac{4}{x+3} = x(x+3) \cdot \left(\frac{1}{x} - \frac{12}{x(x+3)} \right)$$

$$x(x+3) \cdot \frac{4}{x+3} = x(x+3) \cdot \frac{1}{x} - x(x+3) \cdot \frac{12}{x(x+3)}$$

$$4x = x+3 - 12$$

$$3x = -9$$

$$x = -3$$

The LCD of $\frac{4}{x+3}$, $\frac{1}{x}$, and $\frac{12}{x(x+3)}$ is $x(x+3)$.

2. Substitute -3 for x in the original equation.

$$\frac{4}{-3+3} = \frac{1}{-3} - \frac{12}{-3(-3+3)}$$

3. Check the denominators. When you substitute $x = -3$, two of the fractions have a denominator of zero.

$$\frac{4}{0} = \frac{1}{-3} - \frac{12}{0}$$

So, the solution is extraneous. This equation has no solution.

Using Cross Multiplication to Solve Proportions

An equation that sets one fraction equal to another fraction is called a proportion. An easy way to solve a proportion is to "cross multiply."

To solve a proportion using cross multiplication:

1. Multiply the numerator of the first fraction by the denominator of the second fraction.

2. Multiply the denominator of the first fraction by the numerator of the second fraction.

3. Set the two products equal to each other.

4. Finish solving for the variable.

Proportions have exactly one term on each side of the equation. Here are some examples of proportions:

$$\frac{5}{x} = \frac{10}{20} \qquad \frac{3x-2}{3} = \frac{1}{2} \qquad \frac{-5y}{30} = 5$$

For example, to solve the proportion $\frac{x-1}{2} = \frac{26}{4}$:

1. Multiply the numerator of $\frac{x-1}{2}$ by the denominator of $\frac{26}{4}$.

$$\frac{x-1}{2} \searrow \frac{26}{4}$$
$$(x-1) \cdot 4$$

2. Multiply the denominator of $\frac{x-1}{2}$ by the numerator of $\frac{26}{4}$.

$$\frac{x-1}{2} \nearrow \frac{26}{4}$$
$$2 \cdot 26$$

3. Set the products equal to each other. $2 \cdot 26 = (x-1) \cdot 4$

4. Finish solving for x.

$$52 = 4x - 4$$

$$56 = 4x$$

$$14 = x$$

Sample Problems

1. Solve the equation $\frac{x-2}{4} + 3 = \frac{5}{6} - \frac{x}{6}$ for x. Determine if the solution is extraneous.

☐ a. Multiply both sides by the LCD of the fractions.

$$\underline{\quad}\left(\frac{x-2}{4} + 3\right) = \underline{\quad}\left(\frac{5}{6} - \frac{x}{6}\right)$$

a. *12, 12*

☐ b. Distribute the LCD and simplify.

$$\underline{\quad}\left(\frac{x-2}{4}\right) + \underline{\quad} \cdot 3 = \underline{\quad} \cdot \frac{5}{6} - \underline{\quad} \cdot \frac{x}{6}$$

$$\underline{\quad} + \underline{\quad} = 10 - 2x$$

b. *12, 12, 12, 12*

3(x − 2), 36

☐ c. Finish solving for x.

$$\underline{\qquad} = \underline{\qquad}$$
$$\underline{\qquad} = \underline{\qquad}$$
$$x = \underline{\quad}$$

c. *3x − 6 + 36 = 10 − 2x*

5x = −20

x = −4

☐ d. Substitute the solution for x in the original equation.

$$\underline{\qquad} = \underline{\qquad}$$

d. $\frac{-4-2}{4} + 3 = \frac{5}{6} - \frac{-4}{6}$

☐ e. Is the solution extraneous?

$$\underline{\qquad}$$

e. *No*

2. Solve the equation $\frac{2}{x} - \frac{1}{x+2} = \frac{2}{x(x+2)}$ for x. Determine if the solution is extraneous.

☑ a. Multiply both sides by the LCD of the fractions.

$$x(x+2)\left(\frac{2}{x} - \frac{1}{x+2}\right) = x(x+2) \cdot \frac{2}{x(x+2)}$$

☐ b. Distribute the LCD and simplify.

$$x(x+2)\left(\frac{2}{x}\right) - x(x+2)\left(\frac{1}{x+2}\right) = x(x+2)\frac{2}{x(x+2)}$$

$$\underline{\quad} - \underline{\quad} = \underline{\quad}$$

b. *2(x + 2), x, 2*

☐ c. Finish solving for x.

$$\underline{\qquad} = \underline{\quad}$$
$$x = \underline{\quad}$$

c. *2x + 4 − x, 2*

x = −2

☐ d. Substitute the solution for x in the original equation.

$$\underline{\qquad} = \underline{\qquad}$$

d. $\frac{2}{-2} - \frac{1}{-2+2} = \frac{2}{-2(-2+2)}$

☐ e. Is the solution extraneous?

$$\underline{\qquad}$$

e. *Yes*

3. Use cross multiplication to solve the proportion $\frac{7}{5x} = \frac{15}{10}$ for x.

☑ a. Multiply the numerator of $\frac{\mathbf{7}}{5x}$ by the denominator of $\frac{15}{\mathbf{10}}$.

$$\frac{\mathbf{7}}{5x} \searrow \frac{15}{\mathbf{10}}$$

$$7 \cdot 10$$

☐ b. Multiply the denominator of $\frac{7}{\mathbf{5x}}$ by the numerator of $\frac{\mathbf{15}}{10}$.

$$\frac{7}{\mathbf{5x}} \searrow \frac{\mathbf{15}}{10}$$

$$\underline{\quad} \cdot \underline{\quad}$$

b. *5x, 15 (in either order)*

☐ c. Set the products equal to each other.

$$\underline{\quad} = 70$$

c. *75x*

☐ d. Finish solving for x.

$$x = \underline{\quad}$$

d. $\frac{70}{75}$ *or* $\frac{14}{15}$

HOMEWORK

Homework Problems

Circle the homework problems assigned to you by the computer, then complete them below.

☀ Explain

Solving Equations with Rational Expressions

In problems 1 through 12 solve for the variable. Be sure to say if the solution is extraneous.

1. Solve for x: $\dfrac{5}{x} - \dfrac{2}{x} = 1$

2. Solve for y: $\dfrac{4}{7}y = -\dfrac{2}{7}$

3. Solve for x: $\dfrac{3}{x} + \dfrac{2}{x-2} = 1$

4. Solve for y: $\dfrac{y}{y-5} - \dfrac{2}{5} = \dfrac{5}{y-5}$

5. Solve for x: $\dfrac{3x+1}{11x-9} = \dfrac{1}{3}$

6. Solve for x: $\dfrac{2}{1-x} - \dfrac{1}{x} = \dfrac{7}{6}$

7. Solve for y: $\dfrac{4y-9}{8} = \dfrac{6-8y}{16}$

8. Solve for x: $\dfrac{3}{2x-3} - x = \dfrac{2}{4x-6}$

9. A person who weighs 100 pounds on Earth would weigh 38 pounds on Mars. Use the proportion below to figure out how much someone who weighs 160 pounds on Earth would weigh on Mars.

$$\frac{\text{weight on Mars}}{\text{weight on Earth}} = \frac{38}{100} = \frac{x}{160}$$

10. A person who weighs 100 pounds on Earth would weigh 234 pounds on Jupiter. Use the proportion below to figure out how much someone who weighs 160 pounds on Earth would weigh on Jupiter.

$$\frac{\text{weight on Jupiter}}{\text{weight on Earth}} = \frac{234}{100} = \frac{x}{160}$$

11. Solve for y: $\dfrac{6}{x-2} - \dfrac{3}{x} = \dfrac{-5}{x-4}$

12. Solve for x: $\dfrac{6}{x-2} - \dfrac{1}{3} = \dfrac{3x}{x-2}$

APPLY

Practice Problems

Here are some additional practice problems for you to try.

Solving Equations with Rational Expressions

1. Solve for x: $\dfrac{4}{x} + \dfrac{3}{x} = 1$

2. Solve for x: $\dfrac{8}{x} - \dfrac{4}{x} = 1$

3. Solve for x: $\dfrac{1}{x+1} + \dfrac{5}{x+1} = 2$

4. Solve for x: $\dfrac{10}{x-3} - \dfrac{4}{x-3} = -3$

5. Solve for x: $\dfrac{3}{x+2} + \dfrac{4}{x+2} = -1$

6. Solve for x: $\dfrac{x}{1-x} + \dfrac{3}{1-x} = -5$

7. Solve for x: $\dfrac{7}{x+3} - \dfrac{x}{x+3} = -3$

8. Solve for x: $\dfrac{4}{x-2} - \dfrac{x}{x-2} = -3$

9. Solve for x: $\dfrac{x-1}{4} + \dfrac{x}{3} = -2$

10. Solve for x: $\dfrac{x+2}{6} - \dfrac{x}{2} = 5$

11. Solve for x: $\dfrac{x+2}{3} + \dfrac{3x}{4} = 5$

12. Solve for x: $\dfrac{x-1}{5} + \dfrac{x}{3} = \dfrac{3x+2}{15}$

13. Solve for x: $\dfrac{x+3}{6} - \dfrac{x}{5} = \dfrac{9x-5}{30}$

14. Solve for x: $\dfrac{x+3}{5} + \dfrac{x}{4} = \dfrac{8x+4}{20}$

15. Solve for x: $\dfrac{4}{x} + \dfrac{2}{x+1} = 5$

16. Solve for x: $\dfrac{3}{x-2} - \dfrac{1}{x} = -4$

17. Solve for x: $\dfrac{2}{x} - \dfrac{1}{x-3} = 2$

18. Solve for x: $\dfrac{x}{5} + \dfrac{15}{3x} = \dfrac{x+3}{4}$

19. Solve for x: $\dfrac{24}{8x} - \dfrac{x}{3} = \dfrac{3-x}{6}$

20. Solve for x: $\dfrac{x}{6} + \dfrac{12}{2x} = \dfrac{x-2}{4}$

21. Solve for x: $\dfrac{x+1}{5} - \dfrac{3}{x} = \dfrac{x-2}{x}$

22. Solve for x: $\dfrac{3}{x} - \dfrac{x-2}{3} = \dfrac{x+1}{2x}$

23. Solve for x: $\dfrac{x-2}{7} - \dfrac{5}{x} = \dfrac{x+5}{x}$

24. Solve for x: $\dfrac{2}{3(x+4)} - \dfrac{4}{3} = \dfrac{2}{x}$

25. Solve for x: $\dfrac{6}{4(x-2)} + \dfrac{1}{4} = \dfrac{4}{x}$

26. Solve for x: $\dfrac{8}{5(x-3)} + \dfrac{1}{5} = \dfrac{5}{x}$

27. Solve for x: $5 - \dfrac{3}{x+4} = \dfrac{5x+20}{x+4}$

28. Solve for x: $\dfrac{2}{x-2} + 3 = \dfrac{3x-6}{x-2}$

Practice Test

Take this practice test to be sure that you are prepared for the final quiz in Evaluate.

1. Solve $\frac{2}{3x} - \frac{1}{x} = \frac{1}{15}$ for x. Is the solution extraneous?

2. Solve $\frac{4y}{y+3} - \frac{1}{2} = \frac{9}{y+3}$ for y. Is the solution extraneous?

3. The volume V of a right circular cone is $V = \frac{1}{3}\pi r^2 h$, where r is the radius, and h is the height. Solve this formula for h.

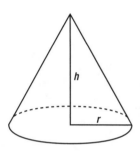

4. Solve this proportion for x: $\frac{5x-8}{2x+1} = \frac{4}{3}$

5. Solve $\frac{5}{4y} - \frac{2}{2y} = \frac{1}{16}$ for y. Is the solution extraneous?

6. Solve $\frac{x}{x+5} - \frac{1}{5} = \frac{3}{x+5}$ for x. Is the solution extraneous?

7. The surface area, S, of a right circular cylinder is

   $S = 2\pi rh + 2\pi r^2$, where r is the radius, and h is the height. Solve this formula for h.

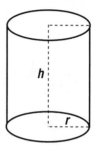

8. Solve this proportion for y: $\frac{6y-4}{6y+6} = \frac{4}{9}$

LESSON 8.4 — PROBLEM SOLVING

OVERVIEW

Here is what you'll learn in this lesson:

Rational Expressions

a. Ratio and proportion

b. Distance problems

c. Work problems

d. Variation

Even a day at the beach can involve algebra. For example, have you ever wondered how much faster you could get across a cove by swimming instead of jogging along the shore? Or how bad a sunburn you'd get if you used SPF 6 instead of SPF 15 sunscreen? Or how much faster you could set up the volleyball court if two people worked together?

In this lesson, you'll apply what you've learned about rational expressions to solve a variety of problems like these.

 **EXPLAIN**

RATIONAL EXPRESSIONS

Summary

Solving Word Problems Involving Rational Expressions

In this lesson you will learn how to solve word problems when the equations contain rational expressions. Using equations with rational expressions allows you to solve a wide variety of problems.

To solve word problems using equations that contain rational expressions, you can often use this process:

1. Carefully read the problem and find the important information.

2. Determine what the problem is asking for and assign a variable to this unknown.

3. Set up an equation.

4. Solve the equation.

5. Check your answer in the original problem.

Below are some examples.

Example 1 Working alone, it would take Michael 2 hours to paint a fence. If Beth worked alone, it would take her 3 hours to paint the fence. How long will it take them to paint the fence together?

1. Find the important information.

 Michael can paint the fence in 2 hours

 Beth can paint the fence in 3 hours

2. Assign the variable.

 Let t = the amount of time it will take them to paint the fence together

3. Set up an equation.

$$\underset{\substack{\text{Michael does}}}{\text{fraction of the job}} + \underset{\substack{\text{Beth does}}}{\text{fraction of the job}} = 1 \text{ complete job}$$

$$\frac{t}{2} + \frac{t}{3} = 1$$

4. Solve the equation.

$$6\left(\frac{t}{2} + \frac{t}{3}\right) = 6(1)$$

$$3t + 2t = 6$$

$$5t = 6$$

$$t = \frac{6}{5}$$

Where did the equation come from?

Well, Beth can paint the entire fence in 3 hours, so

in 1 hour she can paint $\frac{1}{3}$ of the fence.

in 2 hours she can paint $\frac{2}{3}$ of the fence.

in t hours she can paint $\frac{t}{3}$ of the fence.

Similarly, Michael can paint the entire

fence in 2 hours, so

in 1 hour he can paint $\frac{1}{2}$ of the fence.

in t hours he can paint $\frac{t}{2}$ of the fence.

5. Check your answer in the original problem.

Is $\dfrac{\text{fraction Michael paints}}{} + \dfrac{\text{fraction Beth paints}}{} = 1?$

Is $\dfrac{\frac{6}{5}}{2} + \dfrac{\frac{6}{5}}{3} = 1?$

Is $\dfrac{6}{5} \cdot \dfrac{1}{2} + \dfrac{6}{5} \cdot \dfrac{1}{3} = 1?$

Is $\dfrac{3}{5} + \dfrac{2}{5} = 1?$

Is $1 = 1?$ Yes.

So, working together it would take them $\frac{6}{5}$ of an hour, or 1 hour and 12 minutes, to paint the fence.

Example 2 Gretchen can run as fast as Petra. Last Saturday, Gretchen ran 4 miles farther than Petra ran. If Gretchen ran for 3 hours and Petra ran for 2 hours, how far did Petra go?

1. Find the important information.

Gretchen and Petra run the same speed
Petra ran for 2 hours
Gretchen ran for 3 hours
Gretchen ran 4 miles farther than Petra

2. Assign the variable.

Let $d =$ the distance Petra ran

If you solve the equation $d = r \cdot t$ for r you get $r = \frac{d}{t}$. You know you want to solve for r since the rates are equal.

3. Set up an equation.

Petra's rate = Gretchen's rate

$\dfrac{\text{distance Petra ran}}{\text{time Petra ran}} = \dfrac{\text{distance Gretchen ran}}{\text{time Gretchen ran}}$

$\dfrac{d}{2} = \dfrac{d + 4}{3}$

4. Solve the equation.

$\dfrac{d}{2} = \dfrac{d + 4}{3}$

$3d = 2(d + 4)$

$3d = 2d + 8$

$d = 8$

5. Check your answer in the original problem.

Is Petra's rate = Gretchen's rate ?

Is $\dfrac{\text{distance Petra ran}}{\text{time Petra ran}} = \dfrac{\text{distance Gretchen ran}}{\text{time Gretchen ran}}$?

Is $\dfrac{8}{2} = \dfrac{12}{3}$?

Is $4 = 4$? Yes.

So, Petra ran 8 miles.

Sample Problems

1. It would take Ernesto 2 hours to rake the yard by himself. It would take David 5 hours to rake the yard by himself. How long will it take them working together?

   ☑ a. Find the important information.

   Ernesto takes 2 hours to rake
   David takes 5 hours to rake

   ☑ b. Assign the variable.

   Let t = total time to rake the yard together

   ☐ c. Set up an equation. _____

   ☐ d. Solve the equation. $t =$ _____

   c. $\dfrac{t}{2} + \dfrac{t}{5} = 1$

   d. Below is one way to solve the equation.

   $$10\left(\dfrac{t}{2} + \dfrac{t}{5}\right) = 10(1)$$

   $$5t + 2t = 10$$

   $$7t = 10$$

   $$t = \dfrac{10}{7}$$

   So it would take them $\dfrac{10}{7}$ hours, or about 1 hour and 26 minutes, to rake the yard working together.

   ☐ e. Check your answer in the original problem.

   Is _____ = _____?

   Is _____ = _____?

   Is _____ = _____? _____

   e. Is $\dfrac{\frac{10}{7}}{2} + \dfrac{\frac{10}{7}}{5} = 1$?

   Is $\dfrac{10}{7} \cdot \dfrac{1}{2} + \dfrac{10}{7} \cdot \dfrac{1}{5} = 1$?

   Is $\dfrac{5}{7} + \dfrac{2}{7} = 1$?

   Is $1 = 1$? Yes.

2. In octane, the ratio of hydrogen atoms to carbon atoms is 9 to 4. If there are 846 atoms of hydrogen in 47 molecules of octane, how many atoms of carbon are in these 47 molecules of octane?

   ☑ a. Find the important information.

   The ratio of hydrogen atoms to carbon atoms is 9 to 4.
   47 molecules of octane contain 846 hydrogen atoms.

   ☑ b. Assign the variable.

   Let x = the number of carbon atoms in 47 molecules of octane

   ☑ c. Set up an equation.

   $\dfrac{\text{hydrogen atoms}}{\text{carbon atoms}} = \dfrac{9}{4} = \dfrac{846}{x}$

   ☐ d. Solve the equation. $x =$ _____

   ☐ e. Check your answer in the original problem.

   Is _____ = _____?

   Is _____ = _____?

   Is _____ = _____? _____

   d. Below is one way to solve the equation.

   $$9 \cdot x = 4 \cdot 846$$

   $$9x = 3384$$

   $$x = 376$$

   So, there are 376 carbon atoms in 47 molecules of octane.

   e. Is $\dfrac{9}{4} = \dfrac{846}{376}$?

   Is $\dfrac{9}{4} = \dfrac{2 \cdot 3 \cdot 3 \cdot \overset{1}{\cancel{47}}}{2 \cdot 2 \cdot 2 \cdot \underset{1}{\cancel{47}}}$?

   Is $\dfrac{9}{4} = \dfrac{9}{4}$? Yes.

3. The area of a triangle varies jointly with its base and its height. If a triangle of area 50 inches^2 has a base of 25 inches and a height of 4 inches, what is the base of a triangle whose height is 8 inches and whose area is 40 inches^2?

☑ a. Find the important information.

    area varies jointly with base and height

    when the base is 25 inches and the height is 4 inches, the area is 50 inches^2

☑ b. Assign the variable.

    Let x = base of the triangle

    Let y = area of the triangle

    Let z = height of the triangle

    Let k = the constant of variation

☐ c. Set up an equation.

    $y = kxz$

    _____

c. $50 = k(4)(25)$

☐ d. First find k, then substitute this value back into the equation $y = kxz$ to find the base of the triangle.

    $k =$ _____

    base = _____

d. $50 = k(100)$

$\dfrac{50}{100} = k$

$k = \dfrac{1}{2}$

$40 = \dfrac{1}{2}(x)8$

$40 = 4x$

$10 = x$

So, the base is 10 inches.

e. Is $\dfrac{1}{2} \cdot 10 \cdot 8 = 40?$

Is $\dfrac{1}{2} \cdot 80 = 40?$

Is $\quad 40 \quad = 40?$ Yes.

☐ e. Check your answer in the original problem.

Is _____ = _____?

Is _____ = _____?

Is _____ = _____? _____

HOMEWORK

Homework Problems

Circle the homework problems assigned to you by the computer, then complete them below.

☀ Explain
Rational Expressions

1. To fill up their swimming pool, the Johnsons decided to use both their high volume hose and their neighbor's regular garden hose. If they had used only their hose, it would have taken them 12 hours to fill the pool, but using both hoses it took them only 7 hours. How long would it have taken them to fill the pool using only their neighbor's hose?

2. On her camping trip, Li spent as much time hiking as she did rafting. She traveled 2.5 miles per hour when she was rafting and 3 miles per hour when she was hiking. If she went 3 miles more hiking than she did rafting, how far did she hike?

3. The ratio of jellybeans to gummy bears in a bag of candy is 7 to 2. If there are 459 pieces of candy in the bag, how many jellybeans are there?

4. The mass of an object varies jointly with its density and its volume. If 156 grams of iron has a volume of 20 cm^3 and a density of 7.8 $\frac{grams}{cm^3}$, what is the volume of 312 grams of iron with the same density?

5. To empty their swimming pool, the Johnsons decided to use both the regular drain and a pump. If it takes 15 hours for the pool to empty using the drain alone and 7 hours for the pool to empty using the pump alone, how long will it take for the pool to empty using both the drain and the pump?

6. One cyclist can ride 2 miles per hour faster than another cyclist. If it takes the first cyclist 2 hours and 20 minutes to ride as far as the second cyclist rides in 2 hours, how fast can each go?

7. The ratio of the length of a rectangle to its width is 3 to 1. If the perimeter of the rectangle is 32 inches, what are its dimensions?

8. The amount of energy that can be derived from particles varies directly with their mass. If $8.184 \cdot 10^{-14}$ Nm of energy is obtained from a particle whose mass is $9.1066 \cdot 10^{-31}$ kg, how much energy can be derived from particles whose mass is 0.001 kg?

9. Melanie and Alex have to prune all of the trees in their yard. Working alone, it would take Melanie 7 hours to do all of the pruning. It would take Alex 11 hours to do all of the pruning by himself. How long will it take them working together?

10. A bicyclist and a horseback rider are going the same speed. The rider stops after 11.1 miles. The bicyclist goes for another hour and travels a total of 18.5 miles. How fast is each one going?

11. Fish and game wardens can estimate the population of fish in a lake if they take a sample of fish, tag them, return them to the lake, take another sample of fish, and look at the ratio of tagged fish to untagged fish. If a warden tags 117 fish in the first sample, and then finds 13 out of 642 fish have tags in the second sample, how many fish were in the lake?

12. The height of a pyramid of constant volume is inversely proportional to the area of its base. If a pyramid of volume 300 meters^3 has a base area of 90 meters^2 and a height of 10 meters, what is the height of a pyramid whose base area is 100 meters^2?

 **APPLY**

Practice Problems

Here are some additional practice problems for you to try.

Rational Expressions

1. Working alone it would take Josie 4 hours to paint a room. It would take Curtis 5 hours to paint the same room by himself. How long would it take them to paint the room if they work together?

2. Before the library is remodeled, all of the books must be packed in boxes. Working alone, it would take Gail 15 workdays to do the packing. It would take Rob 18 workdays. How long will it take them working together?

3. Two computers are available to process a batch of data. The faster computer can process the batch in 36 minutes. If both computers run at the same time, they can process the batch in 20 minutes. How long would it take the slower computer to process the batch alone?

4. Two copy machines are available to print final exams. The faster copy machine can do the whole job in 75 minutes. If both machines print at the same time, they can do the whole job in 50 minutes. How long would it take the slower machine to do the whole job alone?

5. Two tomato harvesters are available to harvest a field of tomatoes. The slower harvester can harvest the whole field in 7 hours. If both machines harvest at the same time, they can harvest the whole field in 3 hours. How long would it take the faster machine to harvest the whole field by itself?

6. There are two overflow pipes at a dam. The larger overflow pipe can lower the level of the water in the reservoir by 1 foot in 45 minutes. The smaller one lowers the level of water by 1 foot in 2 hours 15 minutes. If both overflow pipes are open at the same time, how long will it take them to lower the level of water by 1 foot?

7. Two fire hoses are being used to flood the skating rink at the park. The larger hose alone can flood the park in 50 minutes. The smaller hose alone can flood the park in 1 hour and 15 minutes. If both hoses run at the same time, how long will it take them to flood the park?

8. Used by itself, the cold water faucet can fill a bathtub in 12 minutes. It takes 15 minutes for the hot water faucet to fill the bathtub. If both faucets are on, how long will it take to fill the bathtub?

9. A box of chocolates contains caramel chocolates and nougat chocolates. The ratio of the number of caramels in the box to the number of nougats in the box is 4 to 3. There are a total of 42 chocolates in the box. How many caramel chocolates are in the box? How many nougat chocolates are in the box?

10. A fast food stand sells muffins and cookies. Last Monday, the ratio of the number of muffins sold to the number of cookies sold was 16 to 13. A total of 145 muffins and cookies were sold. How many muffins were sold? How many cookies were sold?

11. At a certain animal shelter, the ratio of puppies to adult dogs is 7 to 4. This week, there are a total of 55 dogs in the shelter. How many puppies are in the shelter this week? How many adult dogs are in the shelter this week?

12. In a certain cookie recipe, the ratio of cups of flour to cups of sugar is 3 to 1. If the recipe uses $2\frac{1}{4}$ cups of flour, how much sugar does it use?

13. In one multivitamin pill, the ratio of the number of units of Vitamin C to the number of units of Vitamin E is 40 to 13. If the pill contains 200 units of Vitamin C, how many units of vitamin E does it contain?

14. The ratio of the amount of caffeine, in milligrams, in a 12-ounce serving of coffee to the amount of caffeine, in milligrams, in a 12-ounce serving of cola is 25 to 9. If a 12-ounce serving of cola contains 72 milligrams of caffeine, how much caffeine does a 12-ounce serving of coffee contain?

15. Jayme can ride his bike as fast as Terry. Each day, Jayme rides his bike for one hour and 20 minutes. Each day, Terry rides his bike for two hours and rides 15 miles further than Jayme. How far does each ride?

16. Saskia runs as fast as Tanya. Each day, Tanya runs for 40 minutes. Each day, Saskia runs for one hour and runs 2 miles farther than Tanya. How far does each run?

17. Leroy rows a boat as fast as Sasha rows a boat. If Leroy rows for 30 minutes, he travels 1 mile farther than Sasha when she rows for 20 minutes. How far does each row?

18. Pietro and Maria spend the same amount of time driving to school. Pietro averages 50 miles per hour and Maria averages 30 miles per hour. Pietro drives 10 miles farther than Maria. How far does each drive to school?

19. Ranji and Paula spend the same amount of time driving to work. Ranji averages 60 miles per hour and Paula averages 40 miles per hour. Ranji drives 15 miles farther than Paula. How far does each drive to work?

20. A car averages 55 miles per hour and an airplane averages 75 miles per hour. If the airplane and the car travel for the same amount of time, the airplane travels 100 miles farther than the car. How far does each travel?

21. The accuracy of a car's speedometer varies directly with the actual speed of the car. A car's speedometer reads 24 miles per hour when the car is actually traveling at 32 miles per hour. When the speedometer reads 51 miles per hour, how fast is the car actually going?

22. The force needed to stretch a spring a certain distance varies directly with the distance. An 8 pound force stretches a spring 3.5 inches. How much force is needed to stretch the spring 12 inches?

23. A person's weight on the moon varies directly as the person's weight on Earth. A person weighing 144 pounds on Earth weighs only 24 pounds on the moon. How much does a person weigh on Earth who weighs 30 pounds on the moon?

24. The current, i, in an electrical circuit with constant voltage varies inversely as the resistance, r, of the circuit. The current in a circuit with constant voltage is 5 amperes when the resistance is 8 ohms. What is the current in the circuit if the resistance is increased to 10 ohms?

25. For storage boxes with the same volume, the area of the bottom of the box varies inversely with the height of the box. The area of the bottom of the box is 108 square inches when the height is 20 inches. What is the area when the height is 16 inches?

26. The time it takes a car to travel a fixed distance varies inversely with the rate at which it travels. It takes the car 4 hours to travel a fixed distance when it travels at a rate of 50 miles per hour. How fast does the car have to travel to cover the same distance in $2\frac{1}{2}$ hours?

27. The volume of a gas is directly proportional to the temperature of the gas and inversely proportional to the pressure exerted on the gas. Write a formula expressing this property. Use V for volume, T for temperature, and P for pressure.

28. The resistance of an electric wire is directly proportional to the length of the wire and inversely proportional to the square of its diameter. Write a formula expressing this property using R for resistance, L for length, and D for diameter.

EVALUATE

Practice Test

Take this practice test to be sure that you are prepared for the final quiz in Evaluate.

1. Caleb and Daria are going to wash windows. Working alone, it would take Daria 4 hours to wash the windows. It would take Caleb 3 hours to wash the windows by himself. How long will it take them to wash the windows working together?

2. Trisha ran to the park and then walked home. It took her $\frac{1}{2}$ hour to get to the park and 1 hour and 20 minutes to get home. If she runs 5 miles an hour faster than she walks, how far does she live from the park?

3. The ratio of raisins to peanuts in a bag of party mix is 5 to 6. If the bag contains 462 items, how many peanuts are there?

4. The area of a kite varies jointly with the lengths of its two diagonals. If a kite with area 30 inches^2 has one diagonal of length 10 inches and the other diagonal of length 6 inches, what is the area of a kite with diagonals of length 8 inches and 13 inches?

5. Marta is helping Ned wash dishes after a big party. If Ned could do all of the dishes by himself in 60 minutes and Marta could do all of the dishes by herself in 90 minutes, how long will it take them to do the dishes working together?

6. A harpy eagle can fly 35 kilometers per hour faster than a ruby topaz hummingbird. In the same amount of time, an eagle can fly 8.5 kilometers and a hummingbird can fly 5 kilometers. How fast can each bird fly?

7. The ratio of roses to carnations that a florist ordered was 3 to 4. If the florist received a total of 441 flowers, how many of those were roses?

8. The speed of a wave varies jointly with the wavelength and the frequency of the wave. If the speed of a wave is 20 feet per second, its wavelength is 50 feet and its frequency is 0.4 waves per second. What is the speed of a wave whose wavelength is 1 foot and whose frequency is 8 waves per second?

TOPIC 8 CUMULATIVE ACTIVITIES

CUMULATIVE REVIEW PROBLEMS

These problems combine all of the material you have covered so far in this course. You may want to test your understanding of this material before you move on to the next topic. Or you may wish to do these problems to review for a test.

1. Solve $-13 \leq 5x - 3 < 4$ for x.

2. Find: $a^2b^3c \cdot ab^2c^3$

3. Factor: $a^2 - b^2$

4. Circle the true statements.

$$\frac{1}{2} + \frac{1}{3} = \frac{2}{5}$$

$$|19 + 4| = |19| + |4|$$

If $R = \{1, 2, 3\}$ and $S = \{1, 2, 3, 4, 5\}$, then $R \subset S$.

$$7 + 3 \cdot 6 = 60$$

$$\frac{56}{63} = \frac{8}{9}$$

The GCF and LCM of two numbers is usually the same.

5. Use the Pythagorean Theorem to find the distance between the points $(2, -5)$ and $(-3, 7)$. See Figure 8.1.

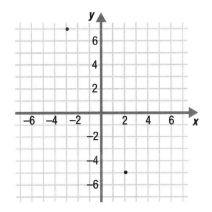

Figure 8.1

6. Factor: $18ab^4c^3 + 9a^4b^3c^2 + 12ab^2c^5$

7. Find the equation of the line parallel to the line $y = 3x + 2$ that passes through the point $(-2, 5)$.

8. Find: $(4a^2b + 3a - 9b) - (7a + 2b - 8a^2b)$

9. Solve $\frac{5}{x+3} - \frac{3}{x+3} = 1$ for x.

10. Write in scientific notation:

    a) 42,789,400

    b) 0.0025815

11. Find the slope and y-intercept of this line: $9x + 5y = 11$

12. Find:

    a. $3^4 \cdot 3$

    b. $\frac{a}{a^9}$

    c. $(x^7y^0)^5$

13. It would take Kendra 4 hours to type a report. It would take Gerri $2\frac{1}{2}$ hours to type the same report. How long would it take them to type the report working together?

14. Write the equation of the circle with radius 3 whose center is at $(1, 5)$.

15. Find: $\frac{3}{x-2} + \frac{1}{x-3}$

16. Factor: $6ab - 10a + 9b - 15$

17. Factor: $x^4 - y^4$

18. Evaluate the expression $5x^2 - 6xy^4 - 4 + 7y$ when $x = 3$ and $y = 0$.

19. Solve $2(3 + y) = 5(\frac{2}{5}y + 1)$ for y.

20. Solve for x: $\frac{2}{x} - \frac{1}{x-1} = \frac{5}{2x}$

21. Use the distance formula to find the square of the distance between the points $(1, 1)$ and $(7, -4)$.

22. For what values of x is the expression $\frac{7}{x^2 - 9}$ undefined?

23. Find: $(9a^8b^3c - 12a^4b^3c^6) \div 3a^7b^3c^4$

Use Figure 8.2 to answer questions 24 through 26.

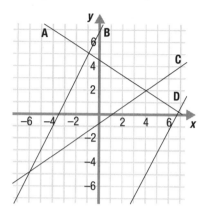

Figure 8.2

24. Which two lines form a system that has a solution of (4, 2)?

25. Which two lines form a system that has a solution of (−1, 5)?

26. Which two lines form a system that has no solution?

27. Graph the inequality $\frac{6}{5}x + 2y \geq 1$.

28. The ratio of dark brown candies to light brown candies in a bag is 3 to 1. If there are 53 light brown candies in the bag, how many are dark brown?

29. Find the slope of the line through the points (6, −2) and (4, −11).

30. Find: $-2s^3t(7r^2st^5 - r^3t)$

31. Factor: $6a^3b^2 - 24a^2b^3 + 24ab^2$

32. Graph the system of inequalities below to find its solution.

$$y \leq \frac{2}{3}x + 3$$

$$y > -3x - 4$$

33. Find: $\frac{x^2 - x - 6}{x^2 - 25} \cdot \frac{x^2 - 5x}{x^2 - x - 6}$

34. Factor: $8x^2y - 6xy^2 + 12x - 9y$

35. Solve $2 - \frac{x}{x+3} = \frac{-3}{x+3}$ for x.

36. Find the slope of the line that is perpendicular to the line that passes through the points (8, −9) and (−3, 11).

37. Find the equation of the line through the point (−2, 6) that has slope $\frac{4}{3}$:

    a. in point-slope form.

    b. in slope-intercept form.

    c. in standard form.

38. Emily withdrew $985 in 5-dollar and 20-dollar bills from her savings account. If she had 65 bills altogether, how many of each did she have?

39. Find:

    a. $2x^0 - 3x^0 + 4x^0$

    b. $(x^0 \cdot x^0 \cdot x^0)^2$

    c. $\frac{a^2 \cdot c}{a^5 \cdot b^6 \cdot c^4}$

40. Factor: $2x^2 + xy - 3y^2$

41. Find the radius and the center of the circle whose equation is:

$$(x - 2)^2 + (y + 6)^2 = 25$$

42. Find: $\dfrac{23}{\frac{9}{4^3} + \frac{5}{6^2}}$

43. Factor: $49x^2 - 14x + 1$

44. Find: $(a^2b^2 - 4a^2b - 4ab^2 + 16ab + 2b - 8) \div (b - 4)$

45. Solve $3(x + 2) = 3x + 6$ for x.

46. Graph the line $x = -2.5$.

47. Find the equation of the line perpendicular to the line $y = 2x - 5$ that passes through the point (6, −1).

48. Find the slope of the line through the points (6, −14) and (22, 17).

49. Factor: $x^2 - 4x + 3$

50. Write in expanded form:

    a. $7.1047 \cdot 10^{12}$

    b. $4.294036 \cdot 10^{-8}$

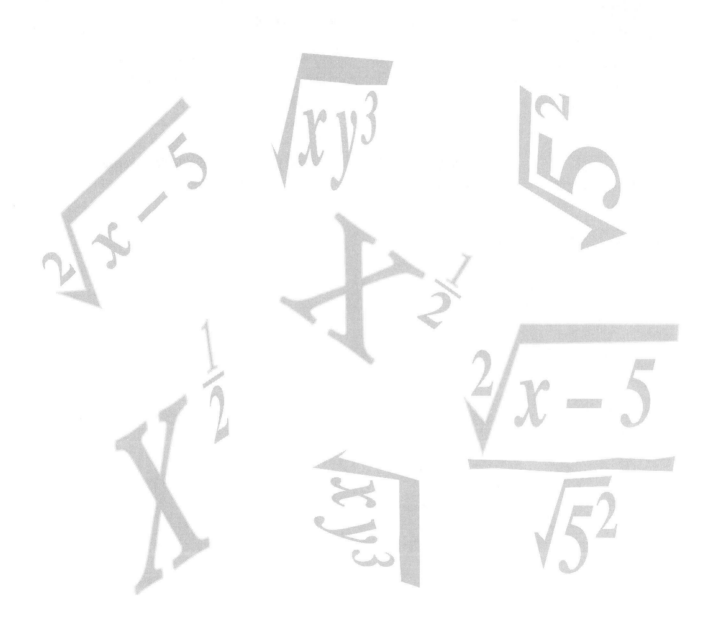

 **OVERVIEW**

Here's what you'll learn in this lesson:

Roots and Exponents

a. The nth root of a number

b. Definition of $a^{\frac{1}{n}}$ and $a^{\frac{m}{n}}$

c. Properties of rational exponents

Simplifying Radicals

a. Simplifying radicals

Operations on Radicals

a. Adding and subtracting radical expressions

b. Multiplying radical expressions

c. Dividing radical expressions

A farmer is experimenting with different fertilizers and varieties of corn in order to find ways to boost crop production. A researcher is studying the effects of pollutants and disease on fish populations around the world. A student volunteer is analyzing surveys to help increase donation levels for his organization.

Each of these people—Emerson Sarawop, the farmer; Sharon Ming, the researcher; and Vince Poloncic, the student volunteer—does work for the Center for World Hunger. And, each works every day with equations that involve radicals.

In this lesson, you will learn about radicals. You will learn how to simplify expressions that contain radicals. You will also learn how to add, subtract, multiply, and divide such expressions.

 EXPLAIN

ROOTS AND EXPONENTS

Summary

Square Roots

When you square the square root of a number, you get back the original number.

For example:

$$\left(\sqrt{16}\right)^2 = 16$$

Every positive integer has both a positive and a negative square root. The symbol $\sqrt{a}$ denotes the positive square root of a. The symbol $-\sqrt{a}$ denotes the negative square root of a.

For example:

$\sqrt{25} = 5$	$-\sqrt{25} = -5$
$\sqrt{16} = 4$	$-\sqrt{16} = -4$
$\sqrt{81} = 9$	$-\sqrt{81} = -9$

You can't take the square root of a negative number and get a real number because no real number times itself equals a negative number.

Cube Roots

When you cube the cube root of a number, you get back the original number.

For example:

$$\left(\sqrt[3]{64}\right)^3 = 64$$

Both positive and negative numbers have real cube roots.

For example:

$\sqrt[3]{125} = 5$ because $5 \cdot 5 \cdot 5 = 125$

$\sqrt[3]{-125} = -5$ because $(-5) \cdot (-5) \cdot (-5) = -125$

The square root of 16 is written as $\sqrt{16}$.

The positive square root is called the principal square root.

The cube root of 64 is written as $\sqrt[3]{64}$.

*n*th Roots

Numbers also have 4th roots, 5th roots, 6th roots, and so on.

For example:

$$\sqrt[4]{81} = 3 \qquad \text{because } 3 \cdot 3 \cdot 3 \cdot 3 = 81$$

$$\sqrt[5]{-1} = -1 \qquad \text{because } (-1) \cdot (-1) \cdot (-1) \cdot (-1) \cdot (-1) = -1$$

$$\sqrt[6]{64} = 2 \qquad \text{because } 2 \cdot 2 \cdot 2 \cdot 2 \cdot 2 \cdot 2 = 64$$

When you raise the *n*th root of a number to the *n*th power, you get back the original number.

For example:

$$\left(\sqrt[n]{10}\right)^n = 10$$

In general, the *n*th root of a number *a* is written

$$\sqrt[n]{a}$$

where *n* is a positive integer.

index

$$\sqrt[n]{a} \longleftarrow \textit{radicand}$$

When there isn't any number written as the index, it is understood to be 2. So $\sqrt{a}$ is the same as $\sqrt[2]{a}$.

Here, *a* is called the radicand and *n* is called the index. The index is the number of times that the root has to be multiplied in order to get the radicand.

When finding real roots:

• If *n* is odd, then $\sqrt[n]{a}$ is a real number.

• If *n* is even, then $\sqrt[n]{a}$ is a real number if $a \geq 0$.

For example:

If the index is even, then the radicand must be positive in order to get a real number. This is because there is no real number that, when multiplied by itself an even number of times, gives a negative number.

$$\sqrt[3]{8} = 2 \qquad \text{because } 2^3 = 2 \cdot 2 \cdot 2 = 8$$

$$\sqrt[3]{-8} = -2 \qquad \text{because } (-2)^3 = (-2) \cdot (-2) \cdot (-2) = -8$$

$$\sqrt{36} = 6 \qquad \text{because } 6^2 = 6 \cdot 6 = 36$$

$$\sqrt{-36} \text{ is not a real number}$$

Rational Exponents

All roots can also be written as rational, or fractional, exponents.

In general:

You may find it easier to solve problems if you first rewrite the exponent with a radical sign. For example,

$$16^{\frac{1}{2}} = \sqrt{16} = 4$$

$$\sqrt[n]{a} = a^{\frac{1}{n}}$$

For example:

$$\sqrt[3]{8} = 8^{\frac{1}{3}}$$

Since you can rewrite rational exponents as roots, the same rules that apply to roots also apply to rational exponents:

- If n is odd, then $a^{\frac{1}{n}}$ is a real number.

- If n is even, then $a^{\frac{1}{n}}$ is a real number when $a \geq 0$.

If the numerator of the rational exponent is not equal to 1, you can still rewrite the problem using radicals.

In general:
$$a^{\frac{m}{n}} = \left(\sqrt[n]{a}\right)^m = \sqrt[n]{a^m}$$

To simplify an expression when the rational exponent is not equal to 1:

1. Rewrite the problem using radicals.

2. Take the appropriate root.

3. Raise the result to the correct power.

For example, to find $32^{\frac{2}{5}}$:

1. Rewrite the problem using radicals. $\qquad = \left(\sqrt[5]{32}\right)^2$

2. Take the 5th root. $\qquad = 2^2$

3. Simplify. $\qquad = 4$

Always reduce a rational exponent to lowest terms or you may get the wrong answer.

For example:

$(-16)^{\frac{1}{2}} = \sqrt{-16}$, which is not a real number

$(-16)^{\frac{2}{4}} \neq \sqrt[4]{(-16)^2} \neq \sqrt[4]{256} \neq 4$

Since $\frac{2}{4}$ is not reduced to lowest terms, the answer, 4, is incorrect.

The basic properties for integer exponents also hold for rational exponents as long as the expressions represent real numbers.

Property of Exponents	Integer Exponents	Rational Exponents
Multiplication	$7^3 \cdot 7^5 = 7^{3+5} = 7^8$	$7^{\frac{1}{2}} \cdot 7^{\frac{1}{4}} = 7^{\frac{1}{2} + \frac{1}{4}} = 7^{\frac{3}{4}}$
Division	$\dfrac{3^2}{3^6} = 3^{2-6} = 3^{-4}$	$\dfrac{3^{\frac{1}{2}}}{3^{\frac{1}{4}}} = 3^{\frac{1}{2} - \frac{1}{4}} = 3^{\frac{1}{4}}$
Power of a Power	$(2^3)^4 = 2^{3 \cdot 4} = 2^{12}$	$\left(5^{\frac{1}{2}}\right)^{\frac{1}{3}} = 5^{\frac{1}{2} \cdot \frac{1}{3}} = 5^{\frac{1}{6}}$
Power of a Product	$(5 \cdot 7)^2 = 5^2 \cdot 7^2$	$(5 \cdot 7)^{\frac{2}{9}} = 5^{\frac{2}{9}} \cdot 7^{\frac{2}{9}}$
Power of a Quotient	$\left(\dfrac{3}{8}\right)^4 = \dfrac{3^4}{8^4}$	$\left(\dfrac{3}{8}\right)^{\frac{1}{4}} = \dfrac{3^{\frac{1}{4}}}{8^{\frac{1}{4}}}$

Notice that you get the same answer whether you first take the root of the number and then raise it to the appropriate power, or whether you first raise the radicand to the appropriate power and then take the root.

When dealing with large numbers, you may find it easier to first take the root of the number and then raise it to the correct power.

The properties of exponents can help you simplify some expressions.

For example, to simplify $(8 \cdot 27)^{\frac{1}{3}}$:

1. Apply the power of a product property.	$= (8)^{\frac{1}{3}} \cdot (27)^{\frac{1}{3}}$
2. Rewrite the problem using radicals.	$= \sqrt[3]{8} \cdot \sqrt[3]{27}$
3. Take the cube roots.	$= 2 \cdot 3$
4. Simplify.	$= 6$

Sample Problems

1. Find: $\sqrt[3]{-1331}$

   ☐ a. Simplify $\sqrt[3]{-1331}$. _____

2. Rewrite as a radical and evaluate: $\left(\dfrac{81}{100}\right)^{\frac{2}{4}}$

   ☑ a. Reduce the exponent to lowest terms. $= \left(\dfrac{81}{100}\right)^{\frac{1}{2}}$

   ☑ b. Apply the power of a quotient property of exponents. $= \dfrac{81^{\frac{1}{2}}}{100^{\frac{1}{2}}}$

   ☐ c. Rewrite as radicals. $= \underline{\hspace{1cm}}$

   ☐ d. Take the square root of the numerator and the denominator. $= \underline{\hspace{1cm}}$

3. Evaluate: $(8 \cdot 125)^{\frac{1}{3}}$

   ☑ a. Raise each term to the $\dfrac{1}{3}$ power. $8^{\frac{1}{3}} \cdot 125^{\frac{1}{3}}$

   ☑ b. Express exponents as radicals. $= \sqrt[3]{8} \cdot \sqrt[3]{125}$

   ☑ c. Simplify the radicals. $= 2 \cdot 5$

   ☐ d. Simplify. $= \underline{\hspace{1cm}}$

SIMPLIFYING RADICALS

Summary

Equations often contain radical expressions. In order to simplify these expressions, you have to know how to simplify radicals.

The Multiplication Property of Radicals

The rule for multiplying square roots is:

The square root of a product = the product of the square roots.

For example:

$$\sqrt{144 \cdot 121}$$

$$= \sqrt{144} \cdot \sqrt{121}$$

$$= 12 \cdot 11$$

$$= 132$$

In general:

$$\sqrt[n]{a \cdot b} = \sqrt[n]{a} \cdot \sqrt[n]{b}$$

Here, a and b are real numbers, $\sqrt[n]{a}$ and $\sqrt[n]{b}$ are real numbers, and n is a positive integer.

Division Property of Radicals

The rule for dividing square roots is:

The square root of a quotient = the quotient of the square roots.

For example:

$$\sqrt{\frac{16}{169}}$$

$$= \frac{\sqrt{16}}{\sqrt{169}}$$

$$= \frac{4}{13}$$

In general:

$$\sqrt[n]{\frac{a}{b}} = \frac{\sqrt[n]{a}}{\sqrt[n]{b}}$$

Here, a and b are real numbers, $\sqrt[n]{a}$ and $\sqrt[n]{b}$ are real numbers, and n is a positive integer.

Sums and Differences of Roots

The nth root of a sum is not equal to the sum of the nth roots.

For example:

Is $\sqrt{9 + 16}$ = $\sqrt{9} + \sqrt{16}$?

Is $\sqrt{25}$ = 3 + 4   ?

Is   5   =   7   ?   No.

Similarly, the nth root of a difference is not equal to the difference of the nth roots.

The Relationship Between Powers and Roots

If you start with a number, cube it, then take its cube root, you end up with the same number that you started with.

For example,

$$\sqrt[3]{8^3} = 8$$

However, if you start with a number, square it, then take its square root, you only get back the original number if the original number is greater than or equal to 0. If the original number is less than 0, taking the root will give you (–1) times the number.

For example:

$$\sqrt{9^2} = 9$$

$$\sqrt{(-9)^2} = 9 = (-1) \cdot (-9)$$

When taking roots:

- If the radicand is positive, $\sqrt[n]{a^n} = a$

- If a is negative and n is odd, $\sqrt[n]{a^n} = a$

- If a is negative and n is even, $\sqrt[n]{a^n} = -a$

Simplifying Radicals

A radical expression is in simplest terms if it meets the following conditions:

- In the expression $\sqrt[n]{a^n}$, a contains no factors (other than 1) that are raised to a power of n.

- There are no fractions under the radical sign.

- There are no radicals in the denominator of the expression.

To simplify a radical expression that contains factors which are powers of the index, n:

1. Write the radicand as a product of its prime factors.

2. Rewrite the factors using exponents.

3. Where possible, rewrite factors as a product having the index, n, as an exponent.

4. Bring all possible factors outside the radical.

5. Simplify.

For example, to simplify $\sqrt[3]{80}$:

1. Write 80 as a product of its prime factors. $\quad = \sqrt[3]{2 \cdot 2 \cdot 2 \cdot 2 \cdot 5}$

2. Rewrite the factors using exponents. $\quad = \sqrt[3]{2^4 \cdot 5}$

3. Rewrite 2^4 as a product including 2^3. $\quad = \sqrt[3]{2^3 \cdot 2 \cdot 5}$

4. Bring $\sqrt[3]{2^3}$ outside the radical. $\quad = 2\sqrt[3]{2 \cdot 5}$

5. Simplify. $\quad = 2\sqrt[3]{10}$

To simplify a radical expression that has a fraction under the radical sign:

1. Rewrite the fraction with two radical signs—one in the numerator and one in the denominator.

2. Multiply the numerator and denominator of the fraction by the same number to eliminate the radical in the denominator of the fraction.

3. Simplify.

For example, to simplify $\sqrt{\dfrac{2}{3}}$:

1. Rewrite the fraction with two radical signs. $\quad = \dfrac{\sqrt{2}}{\sqrt{3}}$

2. Multiply the numerator and denominator by $\sqrt{3}$. $\quad = \dfrac{\sqrt{2}}{\sqrt{3}} \cdot \dfrac{\sqrt{3}}{\sqrt{3}}$

3. Simplify. $\quad = \dfrac{\sqrt{2} \cdot \sqrt{3}}{\sqrt{3} \cdot \sqrt{3}}$

$\quad = \dfrac{\sqrt{6}}{3}$

To simplify a radical expression that has a radical in the denominator:

1. Multiply the numerator and denominator of the fraction by the same number to eliminate the radical in the denominator of the fraction.

2. Simplify.

When you multiply the numerator and denominator of a fraction by the same number, it is the same as multiplying the expression by 1, so the value of the rational expression doesn't change.

For example, to simplify $\frac{7}{\sqrt[3]{5}}$:

1. Multiply the numerator
    and denominator by $\sqrt[3]{5^2}$.

    $= \dfrac{7}{\sqrt[3]{5}} \cdot \dfrac{\sqrt[3]{5^2}}{\sqrt[3]{5^2}}$

2. Simplify.

    $= \dfrac{7\sqrt[3]{5^2}}{\sqrt[3]{5 \cdot 5^2}}$

    $= \dfrac{7\sqrt[3]{5^2}}{\sqrt[3]{5^3}}$

    $= \dfrac{7\sqrt[3]{25}}{5}$

When simplifying radicals, it is helpful to recognize some perfect squares and perfect cubes. You may want to remember the numbers in this table:

Number (n)	Square (n^2)	Cube (n^3)
1	1	1
2	4	8
3	9	27
4	16	64
5	25	125
6	36	216
7	49	343
8	64	512
9	81	729
10	100	1000

**Why do you multiply $\sqrt[3]{5}$ by $\sqrt[3]{5^2}$?
Because this gives $\sqrt[3]{5^3}$, which equals 5.**

Answers to Sample Problems

Sample Problems

1. Simplify: $\sqrt{\dfrac{49}{64}}$

    ☑ a. Rewrite the fraction using
       two radical signs.

       $= \dfrac{\sqrt{49}}{\sqrt{64}}$

    ☐ b. Simplify the square roots.

       $=$ _____

       b. $\dfrac{7}{8}$

2. Simplify: $\sqrt[3]{-125x^6 y^5}$

    ☑ a. Rewrite the radicand as a
       product of its prime factors.

       $= \sqrt{(-5)(-5)(-5)x^6 y^5}$

    ☑ b. Rewrite the factors using
       cubes, where possible.

       $= \sqrt[3]{(-5)^3 (x^2)^3 y^3 y^2}$

    ☐ c. Bring all perfect cubes
       outside the radical.

       $=$ _____

       c. $-5x^2 y \sqrt[3]{y^2}$

OPERATIONS ON RADICALS

Summary

Identifying Like Radical Terms

To add or subtract radical expressions or to eliminate a radical sign in the denominator of a fraction, you will need to identify similar, or like, radical terms.
Similar, or like, radical terms have the same index and the same radicand.

For example, here are two terms that are like terms:

$\sqrt[3]{7}$ index: 3; radicand: 7

$4\sqrt[3]{7}$ index: 3; radicand: 7

Here are two terms that are not like terms:

$2\sqrt[4]{5}$ index: 4; radicand: 5

$\sqrt{5}$ index: 2; radicand: 5

Here are two more terms that are not like terms:

$3\sqrt[5]{9}$ index: 5; radicand: 9

$6\sqrt[5]{8}$ index: 5; radicand: 8

Adding and Subtracting Radical Expressions

Now that you can identify like terms you can add and subtract radical expressions.

For example, to find $5\sqrt[3]{54} + 8\sqrt[3]{2} - 5\sqrt[3]{250}$:

1. Factor the radicands into their prime factors. $= 5\sqrt[3]{2 \cdot 3 \cdot 3 \cdot 3} + 8\sqrt[3]{2} - 5\sqrt[3]{2 \cdot 5 \cdot 5 \cdot 5}$

2. Rewrite the factors using cubes, where possible. $= 5\sqrt[3]{2 \cdot 3^3} + 8\sqrt[3]{2} - 5\sqrt[3]{2 \cdot 5^3}$

3. "Undo" the perfect cubes. $= \left(3 \cdot 5\sqrt[3]{2}\right) + 8\sqrt[3]{2} - \left(5 \cdot 5\sqrt[3]{2}\right)$

4. Simplify. $= 15\sqrt[3]{2} + 8\sqrt[3]{2} - 25\sqrt[3]{2}$

5. Combine like terms. $= -2\sqrt[3]{2}$

All the radicals in step (3) are like terms because they have the same index, 3, and the same radicand, 2.

Multiplying Radical Expressions

You can use the multiplication property of radicals to simplify complex radical expressions.

For example, to simplify $3\sqrt[4]{8} \cdot 5\sqrt[4]{4}$:

1. Apply the multiplication property of radicals. $= 3 \cdot 5\sqrt[4]{8 \cdot 4}$

2. Factor the radicands into their prime factors.

$$= 3 \cdot 5\sqrt[4]{2 \cdot 2 \cdot 2 \cdot 2 \cdot 2}$$

3. Write the 2's as a product involving a factor of 2^4.

$$= 15\sqrt[4]{2^4 \cdot 2}$$

4. Bring $\sqrt[4]{2^4}$ outside the radical.

$$= 15 \cdot 2\sqrt[4]{2}$$

5. Simplify.

$$= 30\sqrt[4]{2}$$

Sometimes when you multiply polynomials, you use the distributive property. This property is also useful when you multiply radicals.

For example, to simplify $\sqrt{8}\left(4\sqrt{2} + 3\sqrt{24}\right)$:

1. Apply the distributive property.

$$= \sqrt{8}(4\sqrt{2}) + \sqrt{8}\left(3\sqrt{24}\right)$$

2. Apply the multiplication property of radicals.

$$= 4\sqrt{8 \cdot 2} + 3\sqrt{8 \cdot 24}$$

3. Simplify.

$$= 4\sqrt{16} + 3\sqrt{192}$$
$$= 4\sqrt{4 \cdot 4} + 3\sqrt{2 \cdot 2 \cdot 2 \cdot 2 \cdot 2 \cdot 2 \cdot 3}$$
$$= 4\sqrt{4^2} + 3\sqrt{(2^3)^2 \cdot 3}$$
$$= 4\sqrt{4^2} + 3\sqrt{8^2 \cdot 3}$$
$$= (4 \cdot 4) + \left(3 \cdot 8\sqrt{3}\right)$$
$$= 16 + 24\sqrt{3}$$

Remember FOIL?

(x + 1)(x + 2)

$= (x \cdot x) + (x \cdot 2) + (1 \cdot x) + (1 \cdot 2)$

$= x^2 + 2x + x + 2$

$= x^2 + 3x + 2$

You can also use the FOIL method to multiply radicals.

For example, to find $(\sqrt{x} + 2)(3\sqrt{x} - 7)$:

1. Find the sum of the products of the first terms, the outer terms, the inner terms, and the last terms.

$$= \left(\sqrt{x} \cdot 3\sqrt{x}\right) - \left(\sqrt{x} \cdot 7\right) + \left(2 \cdot 3\sqrt{x}\right) - (2 \cdot 7)$$

2. Simplify.

$$= 3x - 7\sqrt{x} + 6\sqrt{x} - 14$$
$$= 3x - \sqrt{x} - 14$$

Conjugates

Sometimes when you multiply two irrational numbers you end up with a rational number.

For example, to find $\left(2 + \sqrt{7}\right)\left(2 - \sqrt{7}\right)$:

1. Find the sum of the products of the first terms, the outer terms, the inner terms, and the last terms.

$$= (2)^2 - 2\sqrt{7} + 2\sqrt{7} - \left(\sqrt{7}\right)^2$$

$$= 4 - 7$$

$$= -3$$

The expressions $2 + \sqrt{7}$ and $2 - \sqrt{7}$ are called conjugates of each other. When conjugates are multiplied, the result is a rational number.

In general, these expressions are conjugates of one another:

$$\left(a + \sqrt{b}\right) \text{ and } \left(a - \sqrt{b}\right)$$

$$\left(\sqrt{a} + \sqrt{b}\right) \text{ and } \left(\sqrt{a} - \sqrt{b}\right)$$

Here, $\sqrt{a}$ and $\sqrt{b}$ are real numbers.

When you multiply conjugates, here's what happens:

$$\left(\sqrt{a} + \sqrt{b}\right)\left(\sqrt{a} - \sqrt{b}\right)$$

$$= \left(\sqrt{a} \cdot \sqrt{a}\right) - \left(\sqrt{a} \cdot \sqrt{b}\right) + \left(\sqrt{b} \cdot \sqrt{a}\right) - \left(\sqrt{b} \cdot \sqrt{b}\right)$$

$$= \left(\sqrt{a}\right)^2 - \left(\sqrt{b}\right)^2$$

$$= a - b$$

As another example, to find $\left(\sqrt{6} + \sqrt{14}\right)\left(\sqrt{6} - \sqrt{14}\right)$:

1. Find the sum of the products of the first, outer, inner, and last terms. $\qquad = \left(\sqrt{6}\right)^2 - \left(\sqrt{14}\right)^2$

2. Simplify. $\qquad = 6 - 14$

$$= -8$$

Dividing Radical Expressions

You can use the division property of radicals to simplify radical expressions.

For example, to simplify $\sqrt{\dfrac{7}{2y}}$:

1. Apply the division property of radicals. $\qquad = \dfrac{\sqrt{7}}{\sqrt{2y}}$

2. Rationalize the denominator. $\qquad = \dfrac{\sqrt{7}}{\sqrt{2y}} \cdot \dfrac{\sqrt{2y}}{\sqrt{2y}}$

3. Perform the multiplication. $\qquad = \dfrac{\sqrt{7 \cdot 2y}}{2y}$

4. Simplify. $\qquad = \dfrac{\sqrt{14y}}{2y}$

Since the expression in the denominator is a square root, to eliminate it you must multiply it by itself one time (so there are a total of two factors of 2y under the square root sign). If the expression in the denominator had been $\sqrt[3]{2y}$, to eliminate it you would have had to multiply it by itself two times (so there would be a total of three factors of 2y under the cube root sign). In general, you need n factors to clear an nth root.

The process of eliminating a root in the denominator is called rationalizing the denominator.

As another example, to simplify $\sqrt[4]{\dfrac{2x}{5}}$:

1. Apply the division property of radicals.

$$= \dfrac{\sqrt[4]{2x}}{\sqrt[4]{5}}$$

2. Rationalize the denominator.

$$= \dfrac{\sqrt[4]{2x}}{\sqrt[4]{5}} \cdot \dfrac{\sqrt[4]{5^3}}{\sqrt[4]{5^3}}$$

3. Perform the multiplication.

$$= \dfrac{\sqrt[4]{5^3 \cdot 2x}}{\sqrt[4]{5^4}}$$

4. Simplify.

$$= \dfrac{\sqrt[4]{125 \cdot 2x}}{5}$$

$$= \dfrac{\sqrt[4]{250x}}{5}$$

When there is a sum or difference involving roots in the denominator of a radical expression, you can often simplify the expression by multiplying the numerator and denominator by the conjugate of the denominator.

For example, to simplify $\dfrac{3}{\sqrt{x}+5}$:

1. Multiply the numerator and denominator by the conjugate of $\left(\sqrt{x}+5\right)$, $\left(\sqrt{x}-5\right)$.

$$= \dfrac{3}{\sqrt{x}+5} \cdot \dfrac{\left(\sqrt{x}-5\right)}{\left(\sqrt{x}-5\right)}$$

$$= \dfrac{3\left(\sqrt{x}-5\right)}{\left(\sqrt{x}+5\right)\left(\sqrt{x}-5\right)}$$

2. Simplify.

$$= \dfrac{3 \cdot \sqrt{x} - 3 \cdot 5}{\left(\sqrt{x}\right)^2 - (5)^2}$$

$$= \dfrac{3\sqrt{x}-15}{x-25}$$

Sample Problems

1. Find: $5\sqrt{49} + 6\sqrt{3} + 15\sqrt{3} + 2\sqrt{48}$

   ☑ a. Factor the radicals into their prime factors.
   $= 5\sqrt{7 \cdot 7} + 6\sqrt{3} + 15\sqrt{3} + 2\sqrt{3 \cdot 4 \cdot 4}$

   ☑ b. Where possible, rewrite factors as perfect squares.
   $= 5\sqrt{7^2} + 6\sqrt{3} + 15\sqrt{3} + 2\sqrt{3 \cdot 4^2}$

   ☐ c. Take perfect squares out from under the radical signs.
   $=$ _____

   ☐ d. Simplify.
   $=$ _____

   ☐ e. Combine like terms.
   $=$ _____

2. Find: $(\sqrt{x} + 5)(6\sqrt{x} - 8)$

   ☐ a. Find the sum of the products of the first terms, the outer terms, the inner terms, and the last terms.
   $= 6x - 8\sqrt{x} +$ _____ $-$ _____

   ☐ b. Combine like terms.
   $=$ _____

3. Simplify: $\dfrac{5}{\sqrt[3]{x} \cdot \sqrt{y}}$

   ☑ a. Multiply the numerator and denominator by $\left(\sqrt[3]{x^2}\right) \cdot \left(\sqrt{y}\right)$.
   $= \dfrac{5}{\sqrt[3]{x} \cdot \sqrt{y}} \cdot \dfrac{\left(\sqrt[3]{x^2}\right) \cdot \left(\sqrt{y}\right)}{\left(\sqrt[3]{x^2}\right) \cdot \left(\sqrt{y}\right)}$

   ☐ b. Simplify the radicals.
   $=$ _____

HOMEWORK

Homework Problems

Circle the homework problems assigned to you by the computer, then complete them below.

☀ Explain
Roots and Exponents

1. Rewrite using a radical, then evaluate: $8^{\frac{1}{3}}$

2. Evaluate: $\sqrt[3]{-216}$

3. Evaluate: $2^{\frac{1}{4}} \cdot 2^{\frac{2}{5}}$

4. Rewrite using a radical, then evaluate: $4^{\frac{3}{2}}$

5. Evaluate: $\sqrt[5]{1024}$

6. Simplify the expression below. Write your answer using only positive exponents.

   $$\left(x^{\frac{1}{3}} \cdot y^{\frac{1}{2}} \right)^{-2}$$

7. Evaluate: $\sqrt[4]{-81}$

8. Simplify: $x^{\frac{4}{3}} \cdot x^{\frac{1}{3}}$

9. The number of cells of one type of bacteria doubles every 5 hours according to the formula $n_f = n_i \cdot 2^{\frac{t}{5}}$ where n_f is the final number of cells, n_i is the initial number of cells, and t is the initial number of hours since the growth began. If a biologist starts with a single cell of the bacteria, how many cells will she have after 50 hours?

10. Alan invests $100 in a savings account. How much money would he have after a year if the interest rate for this account is 3% compounded every 4 months?

    The amount A in a savings account can be expressed as

    $$A = P\left(1 + \frac{r}{n}\right)^{nt}$$

    where P is the amount of money initially invested, t is the number of years the money has been invested, r is the annual rate of interest, and n is the number of times the interest is compounded each year.

11. Evaluate the expression below. Express your answer using only positive exponents.

    $$\left(x^{\frac{5}{7}} \cdot x^{-\frac{3}{4}} \cdot y^{\frac{5}{4}} \right)^{-4}$$

12. Evaluate the expression below. Express your answer using only positive exponents.

    $$\left(\sqrt[3]{-1331} \right)\left(x^{\frac{2}{9}} \right)^{-3}\left(\frac{1}{x^{-\frac{1}{3}}} \right)$$

Simplifying Radicals

Simplify the expressions in problems (13)–(20). Assume x, y, and z are positive numbers.

13. $\sqrt[3]{\dfrac{54}{250}}$

14. $\sqrt{\dfrac{49x}{4}}$

15. $\dfrac{\sqrt{5x^3}}{\sqrt{5x^2}}$

16. $\sqrt{\dfrac{-1048}{775}}$

17. $\sqrt[3]{-27x^6y^3}$

18. $\dfrac{\sqrt{x}}{\sqrt{y^4z^{12}}}$

19. $\sqrt[3]{16x^9y^4z^2}$

20. $\sqrt{\dfrac{242xy^{12}}{288y^2z^4}}$

21. One of the three unsolved problems of antiquity was to "double a cube"—that is, to construct a cube with twice the volume of a given cube. What would be the length of a side of a cube with twice the volume of $1\ m^3$? (Hint: The volume, V, of a cube with sides of length L is $V = L \cdot L \cdot L = L^3$.)

22. In a cube with surface area, A, the length, s, of each side is given by this formula:

$$s = \sqrt{\frac{A}{6}}$$

The volume, V, of the cube is:

$$V = s^3$$

What is the volume of a cube with a surface area of 48 ft²?

Simplify the expressions in problems (23) and (24). Assume x, y, and z are positive numbers.

23. $\sqrt{4x^5y^7z^2}$

24. $\dfrac{\sqrt[3]{-8x^6y^3z^3}}{\sqrt{36x^2y^6z^2}}$

Operations on Radicals

25. Circle the like terms:

$7\sqrt[4]{360}$

$72\sqrt{360}$

$\dfrac{\sqrt[4]{360}}{72}$

$360\sqrt[4]{72}$

$-22\sqrt[3]{360}$

$-\dfrac{11}{4}\sqrt[4]{360}$

$\sqrt[4]{36}$

26. Simplify: $7\sqrt{125} + 2\sqrt{500} - \dfrac{3}{2}\sqrt{20} - 2\sqrt{10}$

27. Simplify: $5\sqrt{12}\left(6\sqrt{3} - 7\sqrt{27}\right)$

28. Circle the like terms:

$79 \qquad\qquad \sqrt{79}$

$\sqrt{79^3} \qquad\qquad \sqrt[2]{79}$

$\dfrac{1}{44}\sqrt{799} \qquad -\sqrt{79}$

$2\sqrt{3} \cdot \sqrt{79} \qquad -\sqrt[2]{79^2}$

29. Simplify: $8\sqrt[3]{24} - \dfrac{\sqrt{16}}{2} + \dfrac{1}{4}\sqrt[3]{2} - \sqrt[3]{-3^3}$

30. Simplify: $\left(7\sqrt{2} - 8\sqrt{3}\right)\left(5\sqrt{2} + 6\sqrt{3}\right)$

31. Circle the like terms:

$\sqrt[3]{24}$

$2\sqrt[3]{3}$

$\sqrt[3]{8} \cdot \sqrt[3]{3}$

$\sqrt{(-3)^2}$

$\dfrac{\sqrt[3]{3}}{27}$

3^{-3}

$3^{\frac{1}{3}}$

32. Simplify: $\dfrac{1 - \sqrt{5}}{\sqrt{5} - 9}$

33. The period of a simple pendulum is given by the formula $t = 2\pi\sqrt{\dfrac{L}{32}}$ where t is the period of the pendulum in seconds, and L is the length of the pendulum in feet. What is the period of a 16 foot pendulum?

34. The Pythagorean Theorem, $a^2 + b^2 = c^2$, gives the relationship between the lengths of the two legs of a right triangle, a and b, and the length of the hypotenuse of the triangle, c. If the lengths of the legs of a right triangle are $\sqrt{2}$ cm and $\sqrt{6}$ cm, how long is the hypotenuse?

35. Simplify: $\dfrac{\sqrt[3]{22}}{\sqrt[3]{77}}$

36. Simplify: $\dfrac{6\sqrt[3]{2} - 2\sqrt[3]{4}\left(2\sqrt[3]{32} - 2\sqrt[3]{4}\right)}{2\sqrt[3]{2} - \sqrt[3]{2} \cdot \sqrt[3]{2}}$

APPLY

Practice Problems

Here are some additional practice problems for you to try.

Roots and Exponents

1. Rewrite using a radical, then evaluate: $9^{\frac{5}{2}}$

2. Rewrite using a radical, then evaluate: $16^{\frac{3}{2}}$

3. Rewrite using a radical, then evaluate: $27^{\frac{2}{3}}$

4. Rewrite using a radical, then evaluate: $32^{\frac{4}{5}}$

5. Rewrite using a radical, then evaluate: $81^{\frac{3}{4}}$

6. Evaluate: $\sqrt[4]{625}$

7. Evaluate: $\sqrt[5]{7776}$

8. Evaluate: $\sqrt[5]{1024}$

9. Evaluate: $-\sqrt[4]{81}$

10. Evaluate: $\sqrt[5]{-32}$

11. Evaluate: $\sqrt[3]{-216}$

12. Rewrite using rational exponents: $\sqrt[4]{245^3}$

13. Rewrite using rational exponents: $\sqrt[5]{312^4}$

14. Rewrite using rational exponents: $\sqrt[3]{315^2}$

15. Rewrite using rational exponents: $\sqrt[7]{200^5}$

16. Rewrite using rational exponents: $\sqrt[8]{400^3}$

17. Find: $y^{\frac{2}{3}} \cdot y^{\frac{1}{4}}$

18. Find: $x^{\frac{1}{3}} \cdot z^{\frac{2}{5}}$

19. Find: $x^{\frac{1}{6}} \cdot x^{\frac{1}{5}}$

20. Find: $x^{\frac{1}{7}} \cdot x^{\frac{3}{7}} \cdot x^{\frac{2}{7}}$

21. Find: $x^{\frac{2}{9}} \cdot x^{\frac{5}{9}} \cdot x^{\frac{2}{9}}$

22. Find: $x^{\frac{3}{4}} \cdot x^{\frac{1}{2}} \cdot x^{\frac{3}{4}}$

23. Evaluate the expression below. Express your answer using only positive exponents.
$$\left(\frac{3a^{\frac{3}{4}}}{2b^2} \right)^{-4}$$

24. Evaluate the expression below. Express your answer using only positive exponents.
$$\left(\frac{x^{-\frac{4}{5}}}{2y} \right)^5$$

25. Evaluate the expression below. Express your answer using only positive exponents.
$$\left(\frac{4x^{-\frac{2}{3}}}{3y} \right)^3$$

26. Evaluate the expression below. Express your answer using only positive exponents.
$$\left(a^{\frac{3}{7}} \cdot b^{-\frac{2}{5}} \right)^3$$

27. Evaluate the expression below. Express your answer using only positive exponents.
$$\left(x^{-\frac{4}{9}} \cdot y^{\frac{6}{11}} \right)^2$$

28. Evaluate the expression below. Express your answer using only positive exponents.
$$\left(x^{-\frac{2}{3}} \cdot y^{\frac{3}{5}} \cdot z^{-\frac{4}{7}} \right)^3$$

Simplifying Radicals

29. Simplify: $\sqrt{\frac{121}{64}}$

30. Simplify: $\sqrt{\frac{289}{361}}$

31. Simplify: $\sqrt{\frac{169}{576}}$

32. Simplify: $\sqrt[3]{\frac{27}{8}}$

33. Simplify: $\sqrt[3]{\frac{-64}{125}}$

34. Simplify: $\sqrt[3]{\dfrac{-343}{27}}$

35. Simplify: $\sqrt[5]{\dfrac{-32}{243}}$

36. Simplify: $\sqrt[4]{\dfrac{625}{1296}}$

37. Simplify: $\sqrt[6]{\dfrac{729}{64}}$

38. Calculate: $\sqrt{(-35^2)}$

39. Calculate: $\sqrt{(-56)^2}$

40. Calculate: $\sqrt[3]{(13^3)}$

41. Calculate: $\sqrt[5]{(-47^5)}$

42. Calculate: $\sqrt[3]{(-29)^3}$

43. Which of the radical expressions below is in simplified form?

$\dfrac{\sqrt{81}}{x}$ $\sqrt{\dfrac{25}{49}}$ $\dfrac{6}{\sqrt{30}}$ $\dfrac{\sqrt[4]{7}}{x}$

44. Which of the radical expressions below is in simplified form?

$\dfrac{\sqrt[3]{5}}{x}$ $\dfrac{4}{\sqrt{20}}$ $\sqrt{\dfrac{16}{9}}$ $\dfrac{\sqrt{49}}{x}$

45. Simplify: $\sqrt{36a^2b^6}$

46. Simplify: $\sqrt{100m^6n^4}$

47. Simplify: $\sqrt{64x^4y^6z^{10}}$

48. Simplify: $\sqrt{54a^3b^8}$

49. Simplify: $\sqrt{108m^5n^9}$

50. Simplify: $\sqrt{72x^4y^7}$

51. Simplify: $\sqrt[3]{192a^3b^5c^9}$

52. Simplify: $\sqrt[3]{-250x^4y^6z^8}$

53. Simplify: $\sqrt[5]{160m^2n^7p^{12}}$

54. Simplify: $\dfrac{\sqrt{49a^3b^8}}{\sqrt{7ab^7}}$

55. Simplify: $\dfrac{\sqrt[3]{64m^7n^5}}{\sqrt[3]{2mn^3}}$

56. Simplify: $\dfrac{\sqrt[4]{81x^9y^6}}{\sqrt[4]{3xy^4}}$

Operations on Radicals

57. Circle the like terms:

$\dfrac{\sqrt{-5}}{4}$

$\sqrt{5}$

$\dfrac{1}{3}\sqrt{5}$

$-9\sqrt{50}$

$\dfrac{7\sqrt{5}}{8}$

$\dfrac{6\sqrt[5]{50}}{13}$

$-\sqrt[5]{2}$

58. Circle the like terms:

$\dfrac{5}{2}\sqrt{3}$

$-\sqrt[3]{2}$

$\dfrac{\sqrt{-3}}{3}$

$-6\sqrt{30}$

$\dfrac{4\sqrt[3]{30}}{15}$

$\sqrt{3}$

$\dfrac{6\sqrt{3}}{5}$

59. Simplify: $7\sqrt{5} + \sqrt{20} - 3\sqrt{80}$

60. Simplify: $10\sqrt{2} - \sqrt{128} + 3\sqrt{32}$

61. Simplify: $8\sqrt{3} + \sqrt{12} - 4\sqrt{27}$

62. Simplify: $\sqrt{40} + 3\sqrt{10} - \sqrt{18}$

63. Simplify: $4\sqrt{50} - 5\sqrt{27} + 2\sqrt{75}$

64. Simplify: $\sqrt{20} - 2\sqrt{18} + \sqrt{8}$

65. Simplify: $\sqrt[3]{250x^2} + 3\sqrt[3]{16x^5} - 3\sqrt[3]{432x^2}$

66. Simplify: $5\sqrt[4]{32y} + \sqrt[4]{162y^5} - \sqrt[4]{1250y}$

67. Simplify: $\sqrt[3]{128x} + 2\sqrt[3]{16x^4} - \sqrt[3]{54x}$

68. Simplify: $5\sqrt{6}\left(3\sqrt{8} - 9\sqrt{21}\right)$

69. Simplify: $2\sqrt[3]{9}\left(5\sqrt[3]{3} - 7\sqrt[3]{5}\right)$

70. Simplify: $3\sqrt[4]{8}\left(4\sqrt[2]{2} + 6\sqrt[4]{3}\right)$

71. Simplify: $3\sqrt{2y}\left(7\sqrt{10y} + 4\sqrt{3}\right)$

72. Simplify: $6\sqrt{3z}\left(2\sqrt{6z} - 3\sqrt{5}\right)$

73. Simplify: $2\sqrt[3]{4z}\left(5\sqrt[3]{2z^2} - 7\sqrt[3]{11z}\right)$

74. Simplify: $\left(\sqrt{5} + \sqrt{3}\right)\left(\sqrt{6} + \sqrt{11}\right)$

75. Simplify: $\left(\sqrt{5} - \sqrt{10}\right)\left(\sqrt{2} - \sqrt{15}\right)$

76. Simplify: $\left(\sqrt{6} + \sqrt{5}\right)\left(\sqrt{3} - \sqrt{10}\right)$

77. Simplify: $\left(3\sqrt{5z} + \sqrt{6}\right)\left(3\sqrt{5z} - \sqrt{6}\right)$

78. Simplify: $\left(2\sqrt{3y} + \sqrt{7}\right)\left(2\sqrt{3y} - \sqrt{7}\right)$

79. Simplify: $\left(5\sqrt{2y} - \sqrt{3x}\right)\left(5\sqrt{2y} + \sqrt{3x}\right)$

80. Simplify: $\dfrac{3\sqrt{y}}{\sqrt{6y}}$

81. Simplify: $\dfrac{2\sqrt{x}}{\sqrt{2x}}$

82. Simplify: $\dfrac{3\sqrt{5}}{x + \sqrt{5}}$

83. Simplify: $\dfrac{5\sqrt{2}}{x - \sqrt{2}}$

84. Simplify: $\dfrac{x - \sqrt{3}}{x + \sqrt{3}}$

EVALUATE

Practice Test

Take this practice test to be sure that you are prepared for the final quiz in Evaluate.

Assume that x, y, and z are positive numbers.

1. Simplify: $\sqrt[5]{x} \cdot \sqrt{x}$

2. Rewrite the expression using rational exponents.

   $\sqrt[5]{243^3}$

3. Circle the real number(s) in the list below:

   $\sqrt{-100}$

   $\sqrt[3]{-125}$

   $\sqrt[4]{-16}$

   $\sqrt[6]{-729}$

4. Simplify: $\left(\dfrac{8y^{-\frac{1}{2}}}{7^{\frac{3}{2}}x}\right)^2$

5. Simplify: $\sqrt{\dfrac{169}{225}}$

6. Calculate: $\sqrt{(-29)^2}$

7. Which of the radical expressions below is simplified?

   $\sqrt{\dfrac{3}{16}}$

   $\dfrac{xy}{\sqrt{8}}$

   $\dfrac{\sqrt{25}}{y}$

   $\dfrac{\sqrt[3]{3}}{2}$

8. Simplify: $\sqrt{\dfrac{81x^2y^2}{121z}}$

9. Simplify: $6\sqrt{5x} + 3\sqrt{125x} - 3$

10. Find: $(3\sqrt{5} - 8)(3\sqrt{5} + 8)$

11. Find: $(3\sqrt{2} + 3)(2\sqrt{2} - 6)$

12. Find: $\dfrac{\sqrt{y}}{\sqrt[3]{y}}$

CUMULATIVE REVIEW PROBLEMS

These problems combine all of the material you have covered so far in this course. You may want to test your understanding of this material before you move on to the next topic. Or you may wish to do these problems to review for a test.

1. Find: $(x^2 + 12x)(x + 3y^2 + 1)$

2. Solve for x: $\dfrac{2x + 5}{2 - x} = 3$

3. Solve for y: $\dfrac{1}{y} - \dfrac{2}{3} = \dfrac{y}{3}$

4. Find:

    a. $(125)^{\frac{1}{3}}(16)^{\frac{3}{4}}$

    b. $(x^3y)^{\frac{1}{3}}$

    c. $\dfrac{a^{\frac{1}{2}} b^2 a^{\frac{1}{3}}}{b^{-2}}$

5. Last year Scott earned 5% in interest on his savings account and 13% in interest on his money market account. If he had $14,125 in the bank and earned a total of $1706.25 in interest, how much did he have in each account?

6. Graph the line that passes through the point (0, –3) with slope 2.

7. For what values is the rational expression $\dfrac{x^3 - 3x + 29}{x^2 + 13x + 36}$ undefined?

8. Solve $-10 < 9x - 7 < 11$ for x.

9. Solve this system of equations:

    $$y = -\frac{2}{7}x + 3$$
    $$14y + 4x = 14$$

10. Factor: $2ab + 14a + 5b + 35$

11. Angela and Casey were asked to clean their classroom. Working alone, Angela could clean the room in 20 minutes. It would take Casey 25 minutes to clean the room by herself. How long would it take them to clean the room together?

12. Simplify: $\left(\dfrac{27}{x}\right)^{-\frac{1}{3}}\left(\sqrt{x}\right)^{\frac{4}{3}}$

13. Find the equation of the line that passes through the point (–7, 3) and has slope $\dfrac{5}{3}$. Write your answer in point-slope form, in slope-intercept form, and in standard form.

14. Simplify this expression: $2r^2s + 3t + 4s^2 - 5r^2s - 6s^2 + 7t$

15. Find:

    $(3a^2b^2 + 2a^2b - 7ab + a) - (a^2b^2 - 12ab + 2a^2b + b)$

16. Simplify: $\dfrac{2 + \sqrt{2}}{\sqrt{2} + \sqrt{6}}$

17. Find:

    a. $\dfrac{3^0 \cdot 10^2}{2^3}$

    b. $(-3a^2)^3$

    c. $[(x^3y^2)^2 z]^4$

18. Graph the inequality $2y - 10x \le 32$.

19. Solve for x: $3(x + 2) - x = 3x - 8$

20. Graph the line $y - \dfrac{1}{2} = \dfrac{1}{6}(x + 2)$.

21. Factor: $-16y^2 + 24y - 9$

22. Rewrite using radicals, then simplify: $\dfrac{\left(24^{\frac{1}{3}} + 100^{\frac{1}{2}}\right)}{16^{\frac{1}{4}}}$

23. Circle the true statements.

$2^2 + 3^2 = 5^2$

$|3 - 4| = |3| - |4|$

The GCF of 52 and 100 is 4.

$\frac{9}{25} = \frac{3}{5}$

The LCM of 30 and 36 is 180.

24. Find the slope of the line that is perpendicular to the line that passes through the points (8, 2) and (−4, 9).

25. Factor: $x^2 + 3x - 130$

26. In a bin, the ratio of red apples to green apples is 10 to 3. If there is a total of 15 green apples, how many red ones are there?

27. Find the slope and y-intercept of this line: $4y + 3x = -18$

28. Graph the system of linear inequalities below to find its solution.

$3y - 5x < 3$

$5y + 3x > -10$

29. Simplify: $\sqrt{\dfrac{72x^3y^2}{(2y^2)^2}}$

30. Find the slope of the line through the points $\left(\frac{3}{4}, 7\right)$ and $\left(\frac{1}{4}, -\frac{1}{2}\right)$.

31. Simplify: $\dfrac{8}{3 + \sqrt{11}}$

32. Graph the line $2y + 1 = 1 - 3x$.

33. Find: $(2x^3 + 21x^2 - 27x + 8) \div (2x - 1)$

34. Evaluate the expression $3a^2 + ab - b^2$ when $a = 2$ and $b = 8$.

35. Factor: $7x^2y^2 + 14xy^2 + 7y^2$

36. Factor: $5x^2 - 80$

37. Solve for y: $-10 < 5 - 3y \le 2$

38. Rewrite using only positive exponents: $\dfrac{a^3b^{-2}}{(c^{-1})^{-2}}$

39. Find:

a. $\sqrt{20} + \sqrt{80}$

b. $\sqrt{6}\left(\sqrt{6x^2} + \sqrt{3x^2}\right)$

c. $\left(a + \sqrt{b}\right)\left(a - \sqrt{b}\right)$

40. Factor: $a^2 + 6a + 6b + ab$

41. A juggler has 10 more balls than juggling pins. If the number of balls is 1 more than twice the number of pins, how many pins and balls are there?

42. Find: $\dfrac{x^2 - 9}{x^2 + 3x} \cdot \dfrac{x^2 - 7x}{x^2 - 10x + 21}$

43. Simplify: $\dfrac{x - y}{\sqrt{x} - \sqrt{y}}$

44. Solve for y: $3\left[7y + 5(1 - 2y)\right] = -27$

45. Factor: $8x^3 - 1$

46. Solve for x: $\frac{1}{2}(x + 7) = 12$

47. Solve for a: $\dfrac{a}{a - 5} + 1 = \dfrac{a - 3}{a - 5}$

48. Evaluate the expression $a^3b + 3 - ab^3 + 2ab$ when $a = -2$ and $b = 4$.

49. Simplify: $\left(\dfrac{2y^3}{3x^4}\right)^2$

50. Factor: $2x^2 - 40x + 198$

LESSON 10.1 — QUADRATIC EQUATIONS I

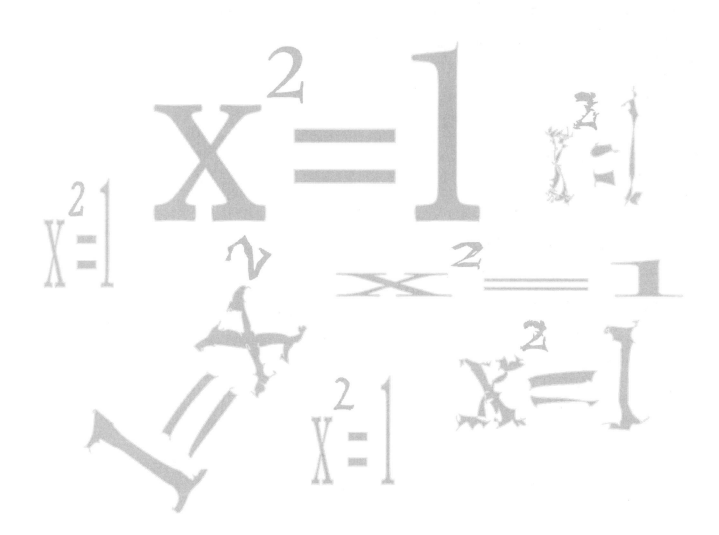

OVERVIEW

Here's what you'll learn in this lesson:

Solving by Factoring

a. The standard form of a quadratic equation

b. Putting a quadratic equation into standard form: $ax^2 + bx + c = 0$

c. Solving quadratic equations of the form $ax^2 + bx = 0$ by factoring

d. Solving quadratic equations of the form $ax^2 + bx + c = 0$ by factoring

Solving by Square Roots

a. Finding square roots

b. Solving quadratic equations of the form $ax^2 = b$

c. Solving quadratic equations of the form $(ax + b)^2 = c$

A baseball coach analyzes the path of a baseball to help his players with their batting. A building contractor is insulating the walls of an A-frame and needs to calculate the amount of insulation required to make the house more energy efficient. A police officer is measuring the length of skidmarks to determine how fast a driver was going.

Each of these people can find the information they are looking for by solving equations known as quadratic equations, or equations in one variable of degree two.

In this lesson you will learn how to solve quadratic equations by factoring and by using square roots.

 **EXPLAIN**

SOLVING BY FACTORING

Summary

Quadratic Equations

A quadratic equation in one variable is an equation that can be written in the form $ax^2 + bx + c = 0$ where a, b, and c are any real numbers, and $a \neq 0$.

Every quadratic equation has two solutions. If the two solutions are equal, the quadratic equation is said to have a solution of multiplicity two.

A quadratic equation written in the form $ax^2 + bx + c = 0$ is said to be in standard form. This means the right side is 0, and the terms on the left side are written in descending order.

The variable in a quadratic equation can be any letter, not just x. For example: $3y^2 + 5y - 9 = 0$ is a quadratic equation.

Quadratic equations are also called second degree equations or equations of degree 2.

Solving Quadratic Equations by Factoring

You have learned how to solve linear equations like $2x - 7 = 0$. You have also learned how to factor polynomials. You can combine these skills to solve some quadratic equations by factoring. However, not all quadratic equations can be solved by factoring.

When you solve quadratic equations by factoring you use the zero product property. This property states that if the product of two numbers (or polynomials) is equal to 0, then one (or both) factors must be equal to 0.

That is, if $P \cdot Q = 0$, then $P = 0$ or $Q = 0$ (or both P and $Q = 0$).

For example: if $3 \cdot x = 0$, then $x = 0$

if $a \cdot b = 0$, then $a = 0$ or $b = 0$

if $y(y - 3) = 0$, then $y = 0$ or $y - 3 = 0$

To solve a quadratic equation by factoring:

1. Write the quadratic equation in standard form.

2. Factor the left side.

3. Use the zero product property to set each factor equal to 0.

4. Finish solving for x.

If the left side can't be factored when the equation is in standard form, then the equation can't be solved by factoring.

For example, to solve $4x^2 = 3x$ by factoring:

1. Write the equation in standard form. $\qquad\qquad 4x^2 - 3x = 0$

2. Factor the left side. $\qquad\qquad\qquad\qquad\quad x(4x - 3) = 0$

3. Use the zero product property to set $\qquad x = 0 \quad$ or $\quad 4x - 3 = 0$
   each factor equal to 0.

4. Finish solving for x. $\qquad\qquad\qquad\qquad x = 0$ or $4x = 3$
$$x = \frac{3}{4}$$

So the two solutions of the equation $4x^2 = 3x$ are $x = 0$ or $x = \frac{3}{4}$.

You can check these solutions by substituting them into the original equation.

Check $x = 0$: $\qquad\qquad\qquad\qquad\qquad$ Check $x = \frac{3}{4}$:

Is $4(0)^2 = 3(0)$? $\qquad\qquad\qquad\qquad$ Is $4\left(\frac{3}{4}\right)^2 = 3\left(\frac{3}{4}\right)$?

Is $\quad 0 = 0 \quad$? Yes. $\qquad\qquad\qquad$ Is $4\left(\frac{9}{16}\right) = \frac{9}{4} \quad$?

$\qquad\qquad\qquad\qquad\qquad\qquad\qquad\qquad$ Is $\quad \frac{9}{4} = \frac{9}{4} \quad$? Yes.

As another example, to solve $x^2 + 4x = -4$ by factoring:

1. Write the equation in standard form. $\qquad\qquad x^2 + 4x + 4 = 0$

2. Factor the left side. $\qquad\qquad\qquad\qquad (x + 2)(x + 2) = 0$

3. Use the zero product property $\qquad x + 2 = 0 \quad$ or $\quad x + 2 = 0$
   to set each factor equal to 0.

4. Finish solving for x. $\qquad\qquad\qquad\qquad x = -2$ or $\qquad x = -2$

So the solution of the equation $x^2 + 4x = -4$ is $x = -2$. This is a solution of multiplicity two.

Check the solution by substituting it into the original equation.

Is $(-2)^2 + 4(-2) = -4$?

Is $\qquad 4 - 8 = -4$?

Is $\qquad\qquad -4 = -4$? Yes.

Sample Problems

1. Solve $x^2 - 6x = 0$ for x:

   ☑ a. Write the equation in standard form.　　　$x^2 - 6x = 0$

   ☐ b. Factor the left side.　　　_____ = 0

   ☐ c. Use the zero product property
      to set each factor equal to 0.　　____ = 0 or ____ = 0

   ☐ d. Finish solving for x.　　　$x = 0$ or $x =$ ____

2. Solve $3x^2 + 10x = 8$ by factoring:

   ☐ a. Write the equation in standard form.　　_____ = 0

   ☐ b. Factor the left side.　　(____)(____) = 0

   ☐ c. Use the zero product property to　(____) = 0 or (____) = 0
      set each factor equal to 0.

   ☐ d. Finish solving for x.　　$x =$ ____ or $x =$ ____

3. Solve $10x(x - 3) - x - 8 = 4 - 5x$ by factoring:

   ☐ a. Write the equation in standard form.　　_____ = 0

   ☐ b. Factor the left side.　　2(_____) = 0
   　　　　　　　　　　　　　　　　2(____)(____) = 0

   ☐ c. Use the zero product property to
      set each factor equal to 0.　(____) = 0 or (____) = 0

   ☐ d. Finish solving for x.　　$x =$ ____ or $x =$ ____

b. x(x – 6)

c. x, (x – 6) (in either order)

d. 6

a. 3x² + 10x – 8

b. 3x – 2, x + 4 (in either order)

c. 3x – 2, x + 4 (in either order)

d. $\frac{2}{3}$, –4

a. 10x² – 26x – 12

b. 5x² – 13x – 6
   5x + 2, x – 3 (in either order)

c. 5x + 2, x – 3 (in either order)

d. $\frac{-2}{5}$, 3 (in either order)

SOLVING BY SQUARE ROOTS

Summary

Square Roots

Another way to solve some quadratic equations is to use the square root property.

Here are some facts about square roots that you will need to know before you can solve quadratic equations.

1. Every positive number b has two square roots, a positive number a and a negative number $-a$, where $a^2 = b$ and $(-a)^2 = b$.

   For example, the square roots of 81 are 9 and -9 because $9^2 = 81$ and $(-9)^2 = 81$.

2. The radical symbol, $\sqrt{}$, is used to represent the positive square root: $\sqrt{81} = 9$. A negative sign is included to represent the negative square root: $-\sqrt{81} = -9$.

   You can use the radical symbol to represent the square roots of every nonnegative number.

   For example, the square roots of 17 are $\sqrt{17}$ and $-\sqrt{17}$ because $\left(\sqrt{17}\right)^2 = 17$ and $\left(-\sqrt{17}\right)^2 = 17$.

3. The square root of a product is the product of the square roots. That is, $\sqrt{a \cdot b} = \sqrt{a} \cdot \sqrt{b}$ where a and b are nonnegative real numbers.

   For example: $\sqrt{16 \cdot 9} = \sqrt{16} \cdot \sqrt{9} = 4 \cdot 3 = 12$.

4. The square root of a quotient is the quotient of the square roots. That is, $\sqrt{\dfrac{a}{b}} = \dfrac{\sqrt{a}}{\sqrt{b}}$ where a and b are nonnegative real numbers and $b \neq 0$.

   For example: $\sqrt{\dfrac{225}{25}} = \dfrac{\sqrt{225}}{\sqrt{25}} = \dfrac{15}{5} = 3$

Note that these properties can be used to simplify radicals.

For example $\sqrt{45}$ can be simplified by writing it as:

$$\sqrt{45} = \sqrt{9 \cdot 5} = \sqrt{9} \cdot \sqrt{5} = 3\sqrt{5}$$

and $\sqrt{\dfrac{108}{4}}$ can be simplified by writing it as:

$$\frac{\sqrt{108}}{\sqrt{4}} = \frac{\sqrt{108}}{2} = \frac{\sqrt{36} \cdot \sqrt{3}}{2} = \frac{6\sqrt{3}}{2} = 3\sqrt{3}$$

Solving Quadratic Equations by Square Roots

The square root property states that if $x^2 = a$ then $x = \sqrt{a}$ or $x = -\sqrt{a}$.

To solve a quadratic equation using the square root property:

1. Write the quadratic equation in the form $x^2 = a$.

2. Use the square root property: If $x^2 = a$ then $x = \sqrt{a}$ or $-\sqrt{a}$.

3. Finish solving for x.

For example, to solve $x^2 - 49 = 0$ for x:

1. Write the equation in the form $x^2 = a$. $x^2 = 49$

2. Use the square root property. $x = \sqrt{49}$ or $x = -\sqrt{49}$

3. Finish solving for x. $x = 7$ or $x = -7$

As another example, to solve $z^2 + 6z + 9 = 24$ for z:

1. Write the equation in the form $x^2 = a$. $(z + 3)^2 = 24$

2. Use the square root property. $z + 3 = \sqrt{24}$ or $z + 3 = -\sqrt{24}$

3. Finish solving for x. $z + 3 = 2\sqrt{6}$ or $z + 3 = -2\sqrt{6}$

                                       $z = -3 + 2\sqrt{6}$ or $z = -3 - 2\sqrt{6}$

Sample Problems

1. Solve for b: $b^2 = 32$

   ☑ a. Write the equation in the form $x^2 = a$. $b^2 = 32$

   ☐ b. Use the square root property. $b = $ _____ or $b = $ _____   *b. $\sqrt{32}, -\sqrt{32}$ (in either order)*

   ☐ c. Finish solving for b. $b = $ _____ or $b = $ _____   *c. $4\sqrt{2}, -4\sqrt{2}$ (in either order)*

2. Solve for x: $8x^2 = 128$

   ☐ a. Write the equation in the form $x^2 = a$. $x^2 = $ _____   *a. 16 or $\frac{128}{8}$*

   ☐ b. Use the square root property. $x = $ _____ or $x = $ _____   *b. $\sqrt{16}, -\sqrt{16}$*

   ☐ c. Finish solving for x. $x = $ _____ or $x = $ _____   *c. 4, –4*

3. Solve for s: $5(3s - 8)^2 = 10$

   ☐ a. Write the equation in the form $x^2 = a$. $(3s - 8)^2 = $ _____   *a. 2*

   ☐ b. Use the square root property.   _____ $= \sqrt{2}$ or   _____ $= -\sqrt{2}$ *b. $3s - 8, 3s - 8$*

   ☐ c. Finish solving for s. $s = $ _____ or $s = $ _____   *c. $\frac{8 + \sqrt{2}}{3}, \frac{8 - \sqrt{2}}{3}$*

HOMEWORK

Homework Problems

Circle the homework problems assigned to you by the computer, then complete them below.

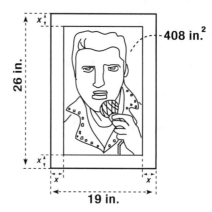

Explain
Solving by Factoring

1. Write $x(2x - 3) = 5$ in standard form.

2. Solve $x^2 - 5x = 0$ by factoring.

3. Solve $2y^2 + 8y = 0$ by factoring.

4. Write $2x - 3x(x - 5) + 1 = 7x - 10$ in standard form.

5. Circle the equations below that are quadratic.

   $-11x^2 = 0$

   $2a(a + 5) = 4$

   $x(7x^2 - 2x + 1) = 0$

   $10x^2 + 3x - 6 = 5x(2x + 7)$

   $6x - 9x^2 = 8$

6. Solve $x^2 + x = 12$ by factoring.

7. Solve $z^2 - 25 = 0$ by factoring.

8. Solve $4b^2 - 12b + 9 = 0$ by factoring.

9. The average depth of the Huang's rectangular swimming pool is 2 meters. If the pool holds 520 m^3 of water, and the length of one side is 6 meters less than 2 times the length of the other side, what are the dimensions of the swimming pool?

   Hint: Volume = depth · length · width

   $$520 = 2 \cdot x \cdot (2x - 6)$$

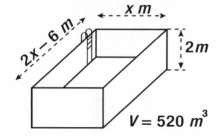

10. Lucy just put a mat around her new Elvis poster. If the matted poster is 19-by-26 inches, and there is 408 in.^{2} of the poster showing, how wide is the mat?

    Hint: Find x in this equation: $(19 - 2x)(26 - 2x) = 408$

11. Solve $3x^2 + 4x = 5 + 2x$ by factoring.

12. Solve $3r(4r - 7) - 2 = 4$ by factoring.

Solving by Square Roots

13. Solve $a^2 = 100$ using the square root property.

14. Simplify:

    a. $\sqrt{\dfrac{108}{16}}$

    b. $\sqrt{675}$

    c. $\sqrt{49 + 16}$

15. Solve $3x^2 = 108$ using the square root property.

16. Solve $c^2 - 112 = 0$ using the square root property.

17. Solve $2x^2 - 162 = 0$ using the square root property.

18. Solve $(x - 6)^2 = 36$ using the square root property.

19. Solve $x^2 - 2x + 1 = 75$ using the square root property.

20. Solve $9y^2 = 7$ using the square root property.

21. A tree is hit by lightning. The trunk of the tree breaks and the top of the tree touches the ground 20 ft. from the base of the tree. If the top part of the tree is twice as long as the bottom part, approximately how tall was the tree before it was hit by lightning?

    (Hint: In a right triangle, $a^2 + b^2 = c^2$, where a and b are the legs of the triangle, and c is the hypotenuse. So $20^2 + x^2 = (2x)^2$.)

    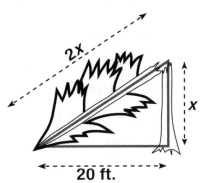

22. Malia is making a cake and she is supposed to pour the batter into a circular pan whose diameter is 10 inches. She doesn't have a circular pan, but she knows that one of her square pans has the same base area as a 10-inch circular pan. What is the length of a side of the square pan?

    Area is 25π square inches   Area is x^2 square inches

23. Solve $4x^2 - 36x + 81 = 5$ using the square root property.

24. Solve $4(z + 13)^2 = 17$ using the square root property.

APPLY

Practice Problems

Here are some additional practice problems for you to try.

Solving by Factoring

1. Solve $x^2 + 7x = 0$ by factoring.

2. Solve $x^2 + 13x = 0$ by factoring.

3. Solve $6x^2 - 24x = 0$ by factoring.

4. Solve $5x^2 + 35x = 0$ by factoring.

5. Solve $8x^2 + 40x = 0$ by factoring.

6. Solve $10x^2 - 70x = 0$ by factoring.

7. Solve $7x^2 - 42x = 0$ by factoring.

8. Solve $x^2 + 8x + 7 = 0$ by factoring.

9. Solve $x^2 - 7x + 6 = 0$ by factoring

10. Solve $x^2 + 12x + 11 = 0$ by factoring.

11. Solve $x^2 + 12x + 35 = 0$ by factoring.

12. Solve $x^2 - 17x + 66 = 0$ by factoring.

13. Solve $x^2 + 9x + 18 = 0$ by factoring.

14. Solve $x^2 - 3x - 40 = 0$ by factoring.

15. Solve $x^2 - 7x - 18 = 0$ by factoring.

16. Solve $x^2 - 4x - 21 = 0$ by factoring.

17. Solve $x^2 - 5x = 150$ by factoring.

18. Solve $x^2 - 8x = 33$ by factoring.

19. Solve $x^2 - 3x = 10$ by factoring.

20. Solve $x^2 + 7x = 44$ by factoring.

21. Solve $x^2 + 2x = 120$ by factoring.

22. Solve $x^2 + 2x = 24$ by factoring.

23. Solve $x^2 - 3x = 54$ by factoring.

24. Solve $x^2 + 2x = 99$ by factoring.

25. Solve $x^2 - x = 30$ by factoring.

26. Solve $x^2 = 5x + 66$ by factoring.

27. Solve $x^2 = 3x + 180$ by factoring.

28. Solve $x^2 = 4x + 32$ by factoring.

Solving by Square Roots

29. Solve $x^2 = 100$ using the square root property.

30. Solve $x^2 = 81$ using the square root property.

31. Solve $x^2 = 256$ using the square root property.

32. Solve $x^2 = 144$ using the square root property.

33. Solve $x^2 = 48$ using the square root property.

34. Solve $x^2 = 50$ using the square root property.

35. Solve $x^2 = 32$ using the square root property.

36. Solve $5x^2 = 245$ using the square root property.

37. Solve $4x^2 = 324$ using the square root property.

38. Solve $3x^2 = 108$ using the square root property.

39. Solve $7x^2 = 126$ using the square root property.

40. Solve $2x^2 = 90$ using the square root property.

41. Solve $5x^2 = 60$ using the square root property.

42. Solve $5x^2 - 180 = 0$ using the square root property.

43. Solve $2x^2 - 162 = 0$ using the square root property.

44. Solve $3x^2 - 147 = 0$ using the square root property.

45. Solve $(x + 5)^2 = 49$ using the square root property.

46. Solve $(x - 4)^2 = 225$ using the square root property.

47. Solve $(x + 9)^2 = 81$ using the square root property.

48. Solve $(x + 8)^2 = 10$ using the square root property.

49. Solve $(x - 3)^2 = 13$ using the square root property.

50. Solve $(x - 2)^2 = 7$ using the square root property.

51. Solve $x^2 + 6x + 9 = 64$ using the square root property.

52. Solve $x^2 - 10x + 25 = 121$ using the square root property.

53. Solve $x^2 + 4x + 4 = 49$ using the square root property.

54. Solve $4x^2 + 28x + 49 = 32$ using the square root property.

55. Solve $25x^2 - 40x + 16 = 75$ using the square root property.

56. Solve $9x^2 - 30x + 25 = 18$ using the square root property.

 **EVALUATE**

Practice Test

Take this practice test to be sure that you are prepared for the final quiz in Evaluate.

1. Write this quadratic equation in standard form and identify a, b, and c.

$$1 + 2x(x - 8) = x + 3$$

2. Solve the equation $6x^2 - 24x = 0$ by factoring.

3. Circle the quadratic equations.

$$x = 22 + 1$$

$$2 = (x - 3)^2$$

$$x^2 = x^2 + \frac{3}{x^2} + 8x$$

$$x(x + 9) = 4$$

$$x^2 - 9 = 7x + 2$$

4. Solve $2x^2 - x - 15 = 0$.

5. Circle the expressions below that are equal to 8.

$$-\sqrt{64} \qquad \frac{\sqrt{256}}{\sqrt{4}}$$

$$\sqrt{-8^2} \qquad \frac{\sqrt{192}}{3}$$

$$\sqrt{9 + 16} \qquad \sqrt{\frac{64}{5}} \cdot \sqrt{5}$$

6. Solve $x^2 = 343$ using the square root property.

7. Solve this equation for x:

$$x = \frac{\frac{\sqrt{20}}{\sqrt{3}}}{\frac{\sqrt{8}}{\sqrt{3}}}$$

8. Solve $(x - 5)^2 = 164$ using the square root property.

LESSON 10.2 – QUADRATIC EQUATIONS II

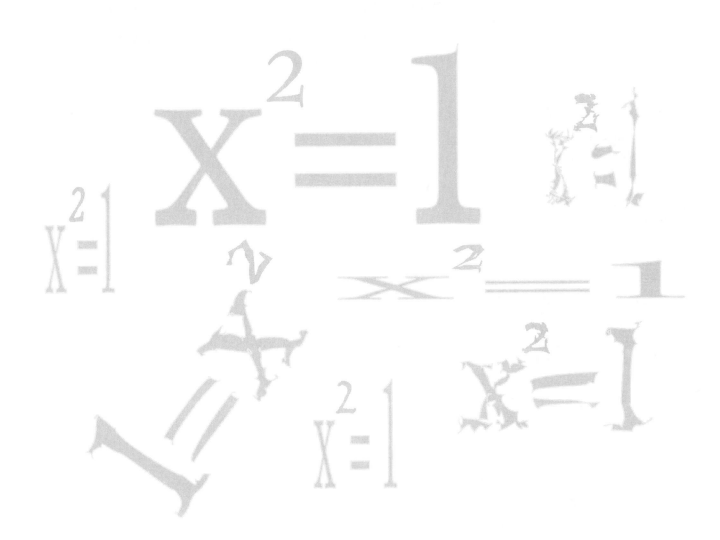

OVERVIEW

Here's what you'll learn in this lesson:

Completing the Square

a. Solving quadratic equations of the form $x^2 + bx + c = 0$ by completing the square

b. Solving quadratic equations of the form $ax^2 + bx + c = 0$, $a \neq 1$, by completing the square

The Quadratic Formula

a. Introduce the quadratic formula

b. Using the quadratic formula to solve quadratic equations of the form $ax^2 + bx + c = 0$

c. Using the discriminant of a quadratic equation to determine the nature of the solutions of the equation

An astronomer is measuring the distance to a star. A family wants to determine the best investments to help pay for their child's education. A paperclip manufacturer is trying to figure out the selling price that will produce the greatest revenue for the company.

Each of these people can find the information they are looking for by solving equations known as quadratic equations, or equations in one variable of degree two.

In this lesson you will learn how to solve quadratic equations by completing the square and by using the quadratic formula.

 **EXPLAIN**

COMPLETING THE SQUARE

Summary

You can solve some quadratic equations by factoring and you can solve some using the square root property. What do you do when you have an equation that can't be solved using either of these methods? You use a method called completing the square. In this method, you write the equation in the form $x^2 = a$, then you use the square root property. Completing the square can be used to solve any quadratic equation.

Completing the Square

To write an equation in the form $x^2 = a$, you must write the left side of the equation as a perfect square. You can learn how to do this with the example $x^2 + 6x = 0$. It sometimes helps to use algebra tiles.

Here's a model of the left side, $x^2 + 6x$:

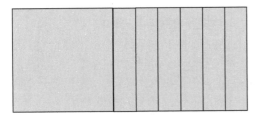

Try to rearrange the tiles to form a square by moving half the x tiles:

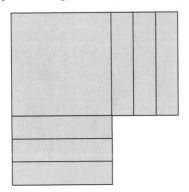

You see that the square is not complete.

To complete the square, you need to add 9 unit tiles.

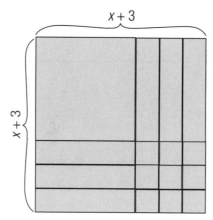

$x + 3$

$x + 3$

Notice that the area of the square is $(x + 3)(x + 3)$.

By adding 9 to both sides of the equation $x^2 + 6x = 0$, you have formed a square.

$$x^2 + 6x + 9 = 9$$
$$(x + 3)(x + 3) = 9$$
$$(x + 3)^2 = 9$$

The equation is now in the form $x^2 = a$.

What if you were trying to complete the square without using tiles. How would you do it?

1. Make sure the coefficient of x^2 is 1, then multiply the coefficient of x by $\frac{1}{2}$.

2. Square the result.

3. Add this number to both sides of the equation.

So, for example, to complete the square given $x^2 + 6x = 0$:

1. Make sure the coefficient of x^2 is 1, then multiply the coefficient of x by $\frac{1}{2}$. $\qquad 6 \cdot \frac{1}{2} = 3$

2. Square the result. $\qquad 3^2 = 9$

3. Add this number to both sides. $\qquad x^2 + 6x + 9 = 9$

The left side of the equation is a perfect square: $(x + 3)^2$.

Here's another example. Complete the square given $x^2 + 5x = 0$:

1. Make sure the coefficient of x^2 is 1, then multiply the coefficient of x by $\frac{1}{2}$. $\qquad 5 \cdot \frac{1}{2} = \frac{5}{2}$

2. Square the result. $\qquad \left(\frac{5}{2}\right)^2 = \frac{25}{4}$

3. Add this number to both sides. $\qquad x^2 + 5x + \frac{25}{4} = \frac{25}{4}$

The left side of the equation is a perfect square: $\left(x + \frac{5}{2}\right)^2$.

As a final example, complete the square given $4x^2 + 16x = 0$:

1. Make sure the coefficient of x^2 is 1, then multiply the coefficient of x by $\frac{1}{2}$.

$$x^2 + 4x = 0$$

$$4 \cdot \frac{1}{2} = 2$$

2. Square the result.

$$2^2 = 4$$

3. Add this number to both sides.

$$x^2 + 4x + 4 = 4$$

The left side of the equation is a perfect square: $(x + 2)^2$.

Now that you know how to complete the square, you can write any equation in the form $x^2 = a$, and then use the square root property to find the solutions.

Solving Quadratic Equations by Completing the Square

To solve a quadratic equation by completing the square:

1. Isolate the x^2- and x-terms on the left side of the equation.

2. Make sure the coefficient of x^2 is 1.
   (You may have to divide both sides by the coefficient of x^2.)

3. Complete the square.

   a. Multiply the coefficient of x by $\frac{1}{2}$.

   b. Square the result.

   c. Add this number to both sides of the equation.

4. Factor the left side of the equation. (It will factor as a perfect square.)

5. Finish solving for the variable.

For example, to solve $x^2 + 10x = 9$ for x:

1. Isolate the x^2- and x-terms on the left.   They are already isolated.

2. Make sure the coefficient of x^2 is 1.   The coefficient of x^2 is 1.

3. Complete the square.

   a. Multiply the coefficient of x by $\frac{1}{2}$. $10 \cdot \frac{1}{2} = 5$

   b. Square the result. $5^2 = 25$

   c. Add this number to both sides. $x^2 + 10x + 25 = 9 + 25$

4. Factor the left side of the equation. $(x + 5)^2 = 34$

5. Finish solving using the square root property.

$$x + 5 = \sqrt{34} \text{ or } x + 5 = -\sqrt{34}$$
$$x = -5 + \sqrt{34} \text{ or } x = -5 - \sqrt{34}$$

You can use a shortcut and write these two solutions as $x = -5 \pm \sqrt{34}$.

The "$\pm$" symbol is read as "plus or minus".

Some quadratic equations don't have real solutions.

For example, try to solve $2x^2 + 3x + 4 = 0$ by completing the square:

1. Isolate the x^2- and x-terms on the left.

$$2x^2 + 3x = -4$$

2. Make sure the coefficient of x^2 is 1.
   (Divide both sides by 2.)

$$x^2 + \frac{3}{2}x = -2$$

3. Complete the square.

   a. Multiply the coefficient of x by $\frac{1}{2}$.

   $$\frac{3}{2} \cdot \frac{1}{2} = \frac{3}{4}$$

   b. Square the result.

   $$\left(\frac{3}{4}\right)^2 = \frac{9}{16}$$

   c. Add this number to both sides.

   $$x^2 + \frac{3}{2}x + \frac{9}{16} = -2 + \frac{9}{16}$$

4. Factor the left side of the equation.

$$\left(x + \frac{3}{4}\right)^2 = -\frac{23}{16}$$

5. Finish solving using the square root property.

STOP!

To use the square root property you would have to take the square root of a negative number. This doesn't make sense for real numbers, so this quadratic equation has no real solutions.

Sample Problems

1. Solve $x^2 + 6x - 13$ for x:

   ☐ a. Isolate the x^2- and x-terms on the left. _____ = ____

   ☑ b. Make sure the coefficient of x^2 is 1. The coefficient of x^2 is 1.

   ☐ c. Complete the square.

   Multiply the coefficient of x by $\frac{1}{2}$. ____ $\cdot \frac{1}{2}$ = ____

   Square the result. $(\underline{\quad})^2 =$ ____

   Add this number to both sides. _____ $= 13 +$ ____

   ☐ d. Factor the left side of the equation. $(x + \underline{\quad})^2 =$ ____

   ☐ e. Finish solving for x. $x + \underline{\quad} = \underline{\quad}$ or $x + \underline{\quad} = \underline{\quad}$

   So, the solutions are: $x = \underline{\quad}$

2. Solve $3x^2 + 7x + 3 = 0$ by completing the square:

   ☑ a. Isolate the x^2- and x-terms on the left. $3x^2 + 7x = -3$

   ☑ b. Make sure the coefficient of x^2 is 1. $x^2 + \frac{7}{3}x = -1$
   (Divide both sides by 3.)

   ☐ c. Complete the square.

   Multiply the coefficient of x by $\frac{1}{2}$. ____ $\cdot \frac{1}{2}$ = ____

   Square the result. $(\underline{\quad})^2 =$ ____

   Add this number to both sides. $x^2 + \frac{7}{3}x +$ ____ $= -1 +$ ____

   ☐ d. Factor the left side of the equation. $(x + \underline{\quad})^2 =$ ____

   ☐ e. Finish solving for x. $x + \underline{\quad} = \underline{\quad}$ or $x + \underline{\quad} = \underline{\quad}$

   So, the solutions are: $x = \underline{\quad}$

Answers to Sample Problems

a. $x^2 + 6x$, 13

c. 6, 3

   3, 9

   $x^2 + 6x + 9$, 9

d. 3, 22

e. 3, $\sqrt{22}$, 3, $-\sqrt{22}$

   $-3 \pm \sqrt{22}$

c. $\frac{7}{3}, \frac{7}{6}$

   $\frac{7}{6}, \frac{49}{36}$

   $\frac{49}{36}, \frac{49}{36}$

d. $\frac{7}{6}, \frac{13}{36}$

e. $\frac{7}{6}, \sqrt{\frac{13}{36}}, \frac{7}{6}, -\sqrt{\frac{13}{36}}$

   $-\frac{7}{6} \pm \frac{\sqrt{13}}{6}$

THE QUADRATIC FORMULA

Summary

If you start with the quadratic equation $ax^2 + bx + c = 0$ and complete the square, you will get a general formula that will solve any quadratic equation. This formula is called the quadratic formula.

The quadratic formula is:
$$x = \frac{-b \pm \sqrt{b^2 - 4ac}}{2a}$$

So to find the solutions of any quadratic equation, you only have to substitute the values of a, b, and c into the quadratic formula.

Solving Quadratic Equations using the Quadratic Formula

To solve a quadratic equation using the quadratic formula:

1. Write the equation in standard form.

2. Identify the values of a, b, and c.

3. Substitute these values into the quadratic formula.

4. Simplify.

For example, to solve $3x^2 + 2x - 8 = 0$:

1. Write the equation in standard form.

    $3x^2 + 2x - 8 = 0$

Notice c = −8, not 8.

2. Identify the values of a, b, and c.

    $a = 3,\ b = 2,\ c = -8$

3. Substitute these values into the quadratic formula.

    $x = \dfrac{-b \pm \sqrt{b^2 - 4ac}}{2a}$

    $x = \dfrac{-2 \pm \sqrt{2^2 - 4(3)(-8)}}{2(3)}$

4. Simplify.

    $x = \dfrac{-2 \pm \sqrt{4 + 96}}{6}$

    $x = \dfrac{-2 \pm \sqrt{100}}{6}$

    $x = \dfrac{-2 \pm 10}{6}$

    $x = \dfrac{-2 + 10}{6}$ or $x = \dfrac{-2 - 10}{6}$

    $x = \dfrac{8}{6}$ or $x = \dfrac{-12}{6}$

    $x = \dfrac{4}{3}$ or $x = -2$

As another example, to solve $x^2 - 5x = 11$:

1. Write the equation in standard form.

    $x^2 - 5x - 11 = 0$

2. Identify the values of a, b, and c.

    $a = 1$, $b = -5$, $c = -11$

3. Substitute these values into the quadratic formula.

    $$x = \frac{-b \pm \sqrt{b^2 - 4ac}}{2a}$$

    $$x = \frac{-(-5) \pm \sqrt{(-5)^2 - 4(1)(-11)}}{2(1)}$$

4. Simplify.

    $$x = \frac{5 \pm \sqrt{25 + 44}}{2}$$

    $$x = \frac{5 \pm \sqrt{69}}{2}$$

As a third example, to solve $25x^2 + 9 = 30x$:

1. Write the equation in standard form.

    $25x^2 - 30x + 9 = 0$

2. Identify the values of a, b, and c.

    $a = 25$, $b = -30$, $c = 9$

3. Substitute these values into the quadratic formula.

    $$x = \frac{-b \pm \sqrt{b^2 - 4ac}}{2a}$$

    $$x = \frac{-(-30) \pm \sqrt{(-30)^2 - 4(25)(9)}}{2(25)}$$

4. Simplify.

    $$x = \frac{30 \pm \sqrt{900 - 900}}{50}$$

    $$x = \frac{30 \pm \sqrt{0}}{50}$$

    $$x = \frac{3}{5} \quad \text{or} \quad x = \frac{3}{5}$$

This equation has a solution of multiplicity two.

As a final example, to solve $x^2 + 3x + 9 = 0$:

1. Write the equation in standard form.

    $x^2 + 3x + 9 = 0$

2. Identify the values of a, b, and c.

    $a = 1$, $b = 3$, $c = 9$

3. Substitute these values into the quadratic formula.

    $$x = \frac{-b \pm \sqrt{b^2 - 4ac}}{2a}$$

    $$x = \frac{-3 \pm \sqrt{3^2 - 4(1)(9)}}{2(1)}$$

4. Simplify.

    $$x = \frac{-3 \pm \sqrt{9 - 36}}{2}$$

    $$x = \frac{-3 \pm \sqrt{-27}}{2}$$

Taking the square root of a negative number does not make sense for real numbers. So, this equation has no real solutions.

Discriminants

The equations above have solutions that look very different. You can tell what the solutions will look like without solving the entire quadratic formula by looking at the discriminant. The discriminant is $b^2 - 4ac$, the part of the quadratic formula under the square root sign.

Discriminant	Quadratic Formula	Solutions
$b^2 - 4ac > 0$	$x = \dfrac{-b \pm \sqrt{positive\ number}}{2a}$	two distinct real
$b^2 - 4ac < 0$	$x = \dfrac{-b \pm \sqrt{negative\ number}}{2a}$	no real
$b^2 - 4ac = 0$	$x = \dfrac{-b \pm \sqrt{0}}{2a}$ or $x = \dfrac{-b}{2a}$	two equal real

b. -5

c. $2, 3, -5$

d. $2, 64,$

   $2, 8$

   $1; -\dfrac{5}{3}$ or $\dfrac{-10}{6}$

b. $1, -12$

c. top: $-12, -12, 1$

   bottom: 1

d. $6 + \sqrt{33}, 6 - \sqrt{33}$

   Here is one way to simplify:

   $x = \dfrac{12 \pm \sqrt{132}}{2}$

   $x = \dfrac{12 \pm \sqrt{4} \cdot \sqrt{33}}{2}$

   $x = \dfrac{12 \pm 2\sqrt{33}}{2}$

   $x = 6 + \sqrt{33}$ or $x = 6 - \sqrt{33}$

Sample Problems

1. Solve $3x^2 + 2x - 5 = 0$:

   ☑ a. Write the equation in standard form. $3x^2 + 2x - 5 = 0$

   ☐ b. Identify the values of a, b, and c. $a = 3, b = 2, c = $ _____

   ☐ c. Substitute these values into the quadratic formula. $x = \dfrac{-b \pm \sqrt{b^2 - 4ac}}{2a}$

   $$x = \dfrac{-\underline{\ \ } \pm \sqrt{2^2 - 4(\underline{\ \ })(\underline{\ \ })}}{2(3)}$$

   ☐ d. Simplify. $x = \dfrac{-\underline{\ \ } \pm \sqrt{\underline{\ \ }}}{6}$

   $$x = \dfrac{-\underline{\ \ } \pm \underline{\ \ }}{6}$$

   $x = $ _____ or $x = $ _____

2. Solve $x^2 - 12x + 3 = 0$:

   ☑ a. Write the equation in standard form. $x^2 - 12x + 3 = 0$

   ☐ b. Identify the values of a, b, and c. $a = $ ___, $b = $ ___, $c = 3$

   ☐ c. Substitute these values into the quadratic formula. $x = \dfrac{-b \pm \sqrt{b^2 - 4ac}}{2a}$

   $$x = \dfrac{-\underline{\ \ } \pm \sqrt{(\underline{\ \ })^2 - 4(\underline{\ \ })(3)}}{2(\underline{\ \ })}$$

   ☐ d. Simplify. $x = $ _____ or $x = $ _____

3. Solve $4x^2 + 28x = -49$:

   ☑ a. Write the equation in standard form. $4x^2 + 28x + 49 = 0$

   ☑ b. Identify the values of a, b, and c. $a = 4$, $b = 28$, $c = 49$

   ☐ c. Substitute these values into
       the quadratic formula.

   $$x = \frac{-b \pm \sqrt{b^2 - 4ac}}{2a}$$

   $$x = \frac{-\underline{} \pm \sqrt{\underline{}^2 - 4(4)(49)}}{2(\underline{})}$$

   ☐ d. Simplify. $x = \underline{}$ or $x = \underline{}$

4. Solve $2x^2 + 4 = -2x$:

   ☑ a. Write the equation in standard form. $2x^2 + 2x + 4 = 0$

   ☐ b. Identify the values of a, b, and c. $a = \underline{}$, $b = \underline{}$, $c = \underline{}$

   ☐ c. Substitute these values into the
       quadratic formula.

   $$x = \frac{-b \pm \sqrt{b^2 - 4ac}}{2a}$$

   $$x = \frac{-2 \pm \sqrt{2^2 - 4(2)(4)}}{4}$$

   ☐ d. Simplify. $x = \underline{}$ or $x = \underline{}$

---

c. top: 28, 28
    bottom: 4

d. $-\dfrac{7}{2}$, $-\dfrac{7}{2}$

Here is one way to simplify:

$$x = \frac{-28 \pm \sqrt{0}}{8}$$

$$x = \frac{-28 \pm 0}{8}$$

$$x = -\frac{7}{2} \text{ or } x = -\frac{7}{2}$$

This quadratic equation has a solution of multiplicity two.

b. 2, 2, 4

c. 2, 2, 2, 4

d. $\dfrac{-2 + \sqrt{-28}}{4}$, $\dfrac{-2 - \sqrt{-28}}{4}$

This quadratic equation has no real solutions.

 **EXPLORE**

Sample Problems

On the computer you examined the solutions of quadratic equations and discriminants. Below are some additional exploration problems.

1. The solutions of a quadratic equation are $\frac{7}{3}$ and -2. Work backwards to find a quadratic equation with these solutions.

☑ a. Start with the solutions. $\qquad\qquad x = \frac{7}{3} \qquad\qquad x = -2$

☑ b. Write the equation in the form $\qquad 3x = 7$
   $ax + b = 0$. $\qquad\qquad\qquad 3x - 7 = 0 \qquad x + 2 = 0$

☑ c. Multiply. $\qquad\qquad\qquad\qquad\qquad (3x - 7)(x + 2) = 0$

d. −7x, −14

☐ d. Use the FOIL method. $\qquad\qquad 3x^2 + 6x \underline{\quad} + \underline{\quad} = 0$

e. $3x^2 - x - 14$

☐ e. Simplify. $\qquad\qquad\qquad\qquad \underline{\qquad\qquad\qquad} = 0$

2. The sum of the solutions of a quadratic equation is $-\frac{b}{a}$. The product of the solutions of a quadratic equation is $\frac{c}{a}$.

If you know the solutions of a quadratic equation you can use their sum and product to find an equation with those solutions.

The solutions of a quadratic equation are $-\frac{1}{2}$ and -3. Use their sum and product to find an equation with those solutions.

☑ a. Add the solutions. $\qquad\qquad\qquad -\frac{1}{2} + -3 = -\frac{7}{2} = -\frac{b}{a}$

☑ b. Multiply the solutions. $\qquad\qquad -\frac{1}{2} \cdot -3 = \frac{3}{2} = \frac{c}{a}$

c. 7

☐ c. Find possible values of a, b, and c. $\qquad a = 2, b = \underline{\quad}, c = 3$

d. 2, 7, 3

☐ d. Write a quadratic equation using these values. $\qquad \underline{\quad}x^2 + \underline{\quad}x + \underline{\quad} = 0$

3. The quadratic equation $2x^2 - 7x + c = 0$ has a discriminant of 9. What is the value of c?

☑ a. Set the discriminant equal to 9. $\qquad$ $b^2 - 4ac = 9$

☐ b. Substitute the values of a and b. $\qquad$ $(\underline{})^2 - 4(\underline{})(c) = 9$

☐ c. Simplify. $\qquad$ $\underline{} - \underline{}c = 9$

$\qquad\qquad\qquad\qquad\qquad\qquad\quad \underline{}c = \underline{}$

$\qquad\qquad\qquad\qquad\qquad\qquad\qquad\quad c = \underline{}$

4. The quadratic equation $2x^2 - 7x + c = 0$ has a discriminant of 9. What are the solutions?

☑ a. Substitute values into the quadratic equation. $\qquad x = \dfrac{-b \pm \sqrt{b^2 - 4ac}}{2a} = \dfrac{-(-7) \pm \sqrt{9}}{2(2)}$

☑ b. Simplify. $\qquad\qquad\qquad\qquad\qquad x = \dfrac{7 \pm 3}{4}$

$\qquad\qquad\qquad\qquad\qquad\qquad\qquad x = \dfrac{10}{4}$ or $x = \dfrac{4}{4}$

$\qquad\qquad\qquad\qquad\qquad\qquad\qquad x = \dfrac{5}{2}$ or $x = 1$

HOMEWORK

Homework Problems

Circle the homework problems assigned to you by the computer, then complete them below.

☼ **Explain**

Completing the Square

1. Complete the square: $x^2 + 13x$. What is the area of the completed square?

2. Solve $x^2 + 2x = 8$ by completing the square.

3. Solve $x^2 - 4x = 1$ by completing the square.

4. Solve $x^2 + 9x = -2$ by completing the square.

5. Solve $x^2 - 36x = 40$ by completing the square.

6. Solve $x^2 + 3x - 7 = 0$ by completing the square.

7. Solve $2x^2 + 6x = 2$ by completing the square.

8. Solve $3x^2 - 5x - 9 = 0$ by completing the square.

9. Seana is competing in the bicycle Race Across America. She rode 62 miles before lunch and 69 miles after lunch. She rode for one hour more after lunch than before lunch, but her speed after lunch was 2 mph slower. What were her speeds before and after lunch?

   Hint: Let t be her time spent riding before lunch. Then $t + 1$ is her time spent riding after lunch.

   Since speed $= \frac{\text{distance}}{\text{time}}$, we have $\frac{62}{t} = \frac{69}{t+1} + 2$.
   Now, solve for t.

10. Clair takes 1 hour longer than Jenna to mow the lawn. If they can mow the lawn together in 5 hours, how long would it take each of them to mow the lawn alone?

    (Hint: Let t be the time in hours it takes Jenna to mow the lawn. Then Clair can mow the lawn in $t + 1$ hours.

    To find t, solve $\frac{5}{t} + \frac{5}{t+1} = 1$.)

11. Solve $9x^2 - 15x = 32$ by completing the square.

12. Solve $5x^2 + 7x + 13 = 0$ by completing the square.

The Quadratic Formula

13. Solve $x^2 + 10x + 25 = 0$ using the quadratic formula.

14. Solve $x^2 - 6x - 16 = 0$ using the quadratic formula.

15. Solve $2x^2 + 3x + 1 = 0$ using the quadratic formula.

16. Solve $20x^2 - 42x - 26 = 0$ using the quadratic formula.

17. Solve $x^2 + 8x + 3 = 0$ using the quadratic formula.

18. Solve $x^2 - 5x = 9$ using the quadratic formula.

19. Solve $3x^2 - 15x = 20$ using the quadratic formula.

20. For each quadratic equation in the list below, calculate the discriminant, then write the letter of the statement that best describes its solutions.

    a. Two unequal real solutions

    b. Two equal real solutions

    c. No real solutions

Equation	Discriminant	Solutions
$x^2 - 2x + 8 = 0$	_____	_____
$x^2 - 5x - 16 = 0$	_____	_____
$49x^2 + 70x + 25 = 0$	_____	_____
$4x^2 - 6x - 9 = 0$	_____	_____
$-2x^2 - 7x + 10 = 0$	_____	_____

21. Pediatricians use formulas to convert adult dosages for medication to child dosages. Most pediatricians use formulas based on the child's weight. However, some use one of the formulas below, where a is the age of the child and d is the adult dosage.

$$\text{child's dosage} = \frac{a}{a+12}\,d \qquad \text{child's dosage} = \frac{a+1}{24}\,d$$

At approximately what age(s) do these two formulas yield the same child dosage?

(Hint: Begin by setting the expressions equal to each other:
$$\frac{a}{a+12}\,d = \frac{a+1}{24}\,d$$
Then divide by d: $\quad \frac{a}{a+12} = \frac{a+1}{24}$

Now find the LCD of the denominators, multiply by the LCD, and solve for a.)

22. Joe has a rectangular deck in his backyard. Its length measures 1 foot more than its width. He is planning to extend the length of the deck by 3 additional feet. If the new deck would have an area on 165 square feet, what is the width of the deck?

23. Solve $4x(x+1) - 9 = 6x^2 - x - 18$ using the quadratic formula.

24. Solve $x^2 + 5 = 0$ using the quadratic formula.

 Explore

25. The solutions of a quadratic equation are $-\frac{4}{3}$ and 5. What is an equation with these solutions?

26. The solutions of a quadratic equation are 3 and −2. Use their sum and product to find the equation.

27. The quadratic equation $3x^2 + 4x + c = 0$ has a discriminant of 196. What is the value of c? What are the solutions of the equation?

28. The solutions of a quadratic equation are $\frac{5 \pm 3\sqrt{13}}{2}$. What is an equation with these solutions?

29. The solutions of a quadratic equation are $\frac{5}{6}$ and $-\frac{2}{3}$. Use their sum and product to find an equation with these solutions.

30. The quadratic equation $ax^2 - 5x + 2 = 0$ has a discriminant of 17. What is the value of a? What are the solutions of the equation?

APPLY

Practice Problems

Here are some additional practice problems for you to try.

Completing the Square

1. Solve $x^2 - 6x = 27$ by completing the square.

2. Solve $x^2 + 2x = 15$ by completing the square.

3. Solve $x^2 - 4x = 45$ by completing the square.

4. Solve $x^2 + 10x = 56$ by completing the square.

5. Solve $x^2 + 8x = 20$ by completing the square.

6. Solve $x^2 + 12x = -11$ by completing the square.

7. Solve $x^2 - 8x = -7$ by completing the square.

8. Solve $x^2 + 6x = -5$ by completing the square.

9. Solve $x^2 + 6x = 12$ by completing the square.

10. Solve $x^2 - 16x = 13$ by completing the square.

11. Solve $x^2 + 4x = 21$ by completing the square.

12. Solve $x^2 + 12x = -14$ by completing the square.

13. Solve $x^2 - 18x = -57$ by completing the square.

14. Solve $x^2 - 8x = -5$ by completing the square.

15. Solve $x^2 + 3x = 16$ by completing the square.

16. Solve $x^2 + 9x = 15$ by completing the square.

17. Solve $x^2 + 7x = 9$ by completing the square.

18. Solve $x^2 - 7x = 3$ by completing the square.

19. Solve $x^2 - x = 14$ by completing the square.

20. Solve $x^2 - 5x = 2$ by completing the square.

21. Solve $4x^2 + 16x = 84$ by completing the square.

22. Solve $2x^2 + 16x = 40$ by completing the square.

23. Solve $3x^2 + 12x = 36$ by completing the square.

24. Solve $5x^2 - 30x = 200$ by completing the square.

25. Solve $3x^2 - 30x = 432$ by completing the square.

26. Solve $2x^2 - 2x = 112$ by completing the square.

27. Solve $3x^2 - 5x = 7$ by completing the square.

28. Solve $4x^2 + 3x = 6$ by completing the square.

The Quadratic Formula

29. Solve $4x^2 - 5x + 1 = 0$ using the quadratic formula.

30. Solve $3x^2 - 5x - 2 = 0$ using the quadratic formula.

31. Solve $x^2 - 4x + 1 = 0$ using the quadratic formula.

32. Solve $x^2 + 8x - 5 = 0$ using the quadratic formula.

33. Solve $x^2 - 6x + 4 = 0$ using the quadratic formula.

34. Solve $2x^2 - 7x + 2 = 0$ using the quadratic formula.

35. Solve $5x^2 + 3x - 4 = 0$ using the quadratic formula.

36. Solve $3x^2 + 7x - 7 = 0$ using the quadratic formula.

37. Solve $x^2 - 2x = 7$ using the quadratic formula.

38. Solve $x^2 + 8x = 5$ using the quadratic formula.

39. Solve $x^2 + 3x = 8$ using the quadratic formula.

40. Solve $4x^2 + 19x = -17$ using the quadratic formula.

41. Solve $5x^2 - 46x = -48$ using the quadratic formula.

42. Solve $3x^2 - 25x = -28$ using the quadratic formula.

43. Solve $x^2 = 3x + 7$ using the quadratic formula.

44. Solve $x^2 = -3x + 5$ using the quadratic formula.

45. Solve $x^2 = x + 1$ using the quadratic formula.

46. Solve $2x^2 = 5x - 3$ using the quadratic formula.

47. Solve $4x^2 = 9x - 3$ using the quadratic formula.

48. Solve $3x^2 = -2x + 7$ using the quadratic formula.

49. Calculate the discriminant of the quadratic equation $5x^2 + 8x - 9 = 0$ and determine the nature of the solutions of the equation.

50. Calculate the discriminant of the quadratic equation $3x^2 + 2x + 5 = 0$ and determine the nature of the solutions of the equation.

51. Calculate the discriminant of the quadratic equation $4x^2 - 12x + 9 = 0$ and determine the nature of the solutions of the equation.

52. Calculate the discriminant of the quadratic equation $x^2 + 7x - 6 = 0$ and determine the nature of the solutions of the equation.

53. Calculate the discriminant of the quadratic equation $9x^2 + 30x + 25 = 0$ and determine the nature of the solutions of the equation.

54. Calculate the discriminant of the quadratic equation $x^2 + 3x + 4 = 0$ and determine the nature of the solutions of the equation.

55. Calculate the discriminant of the quadratic equation $7x^2 - 6x + 5 = 0$ and determine the nature of the solutions of the equation.

56. Calculate the discriminant of the quadratic equation $4x^2 + 8x - 5 = 0$ and determine the nature of the solutions of the equation.

EVALUATE

Practice Test

Take this practice test to be sure that you are prepared for the final quiz in Evaluate.

1. Complete the square for this expression.

   $$x^2 + 9x + ?$$

   What is the perfect square?

2. Solve $4x^2 + 8x = 152$ by completing the square.

3. After completing the square by adding 16 to both sides, the result is $(x + 4)^2 = 2$. What was the original equation?

4. Solve $4x^2 - 5x + 1 = 0$ by completing the square.

5. Circle the equation below that has the solution $x = \dfrac{-2 \pm 3\sqrt{2}}{2}$.

   $$x^2 + 4x - 7 = 0$$

   $$2x^2 + 4x - 7 = 0$$

   $$2x^2 + 4x + 7 = 0$$

   $$x^2 - 4x - 7 = 0$$

6. Use the quadratic formula to solve this quadratic equation:

   $$6x = 1 - 5x^2$$

7. Circle the quadratic equations below that have no real solutions.

   $$x^2 + 5x - 9 = 0$$

   $$x^2 + 4x + 11 = 0$$

   $$x^2 - x + 1 = 0$$

   $$4x^2 + 5x + 1 = 0$$

   $$2x^2 - 10x - 3 = 0$$

   $$x^2 + 2x + 5 = 0$$

8. Find the two values for b for which the quadratic equation $9x^2 + bx + 36 = 0$ has two equal real solutions.

9. The quadratic equation $x^2 - 7x + c = 0$ has a discriminant of 45. What is the value of c? What are the solutions of the equation?

10. The sum of the solutions of a quadratic equation is $\dfrac{3}{2}$. The product of its solutions is 3. What is the equation?

11. Find a quadratic equation whose two solutions are -3 and $\dfrac{1}{5}$.

12. Find the greatest possible value of c in the quadratic equation $2x^2 - 7x + c = 0$ for which there are two real solutions.

LESSON 10.3 – COMPLEX NUMBERS

OVERVIEW

Here's what you'll learn in this lesson:

Complex Number System

a. *Definition of complex numbers*

b. *Powers of i*

c. *Operations on complex numbers*

Prior to the 19th century, mathematicians dismissed the idea that the square root of a negative number had any meaning—one mathematician even claimed that thinking about these quantities involved "putting aside…mental tortures." However, people eventually began accepting these values, which they called complex numbers, as important. Today, for example, engineers use complex numbers to describe certain properties of electrical circuits.

In this lesson, you will learn about complex numbers. You will learn how to add, subtract, multiply, and divide complex numbers. You will also see how you can use the discriminant of a quadratic equation to determine if the equation has complex solutions.

 **EXPLAIN**

COMPLEX NUMBER SYSTEM

Summary

The Number i

Sometimes when solving a quadratic equation such as $x^2 + 1 = 0$, you cannot find a real number x which solves the equation.

To solve such an equation, mathematicians defined i as:

$$i = \sqrt{-1}$$

The number i is called an imaginary number. If you were to square both sides, you would get $i^2 = -1$.

You can use i to help you rewrite negative roots.

For example:

$$\sqrt{-4} = \sqrt{(-1)(4)} = \sqrt{4}\,i = 2i$$

$$\sqrt{-9} = \sqrt{(-1)(9)} = \sqrt{9}\,i = 3i$$

$$\sqrt{-16} = \sqrt{(-1)(16)} = \sqrt{16}\,i = 4i$$

In general, if k is a positive number:

$$\sqrt{-k} = \sqrt{(-1)k} = \sqrt{k}\,i = i\sqrt{k}$$

You can use the imaginary number i to solve a quadratic equation that cannot be solved using only real numbers.

For example, to solve $x^2 = -4$:

1. Use the square root property. $\qquad\qquad x = \pm\sqrt{-4}$

2. Use $i = \sqrt{-1}$ to simplify $\qquad\qquad = \pm\sqrt{(-1)(4)}$
    the answer. $\qquad\qquad\qquad\qquad\quad = \pm 2\sqrt{-1}$
    $\qquad\qquad\qquad\qquad\qquad\qquad\quad = \pm 2i$

So $x = \pm 2i$ are the solutions of $x^2 = -4$.

Notice that in the term $\sqrt{k}\,i$, the i is outside the radical sign.

Here is another example.

To find $\sqrt{-9} \cdot \sqrt{-16}$:

1.	Rewrite both radicals in terms of i.	$= 3i \cdot 4i$
		$= 12i^2$
2.	Rewrite i^2 as -1.	$= 12(-1)$
		$= -12$

So $\sqrt{-9} \cdot \sqrt{-16} = -12$.

Powers of i

Look at the powers of i below:

$$i^1 = i$$
$$i^2 = -1$$
$$i^3 = i^2 \cdot i = (-1) \cdot i = -i$$
$$i^4 = i^2 \cdot i^2 = (-1)(-1) = 1$$
$$i^5 = i^4 \cdot i = 1 \cdot i = i$$

This pattern, i, -1, $-i$, 1, i, ..., continues. You can use this to rewrite powers of i.

To simplify powers of i when the power is greater than or equal to 4:

1. Rewrite the power as a product using as many factors of i^4 as possible.

2. Rewrite i^4 as 1.

3. Simplify.

For example, to simplify i^{33}:

1.	Rewrite i^{33} as a product using factors of i^4.	$= (i^4)^8 \cdot i$
2.	Rewrite i^4 as 1.	$= (1)^8 \cdot i$
3.	Simplify.	$= 1 \cdot i$
		$= i$

So $i^{33} = i$.

Complex Numbers

A complex number is a number that can be written in the form:

$$a + bi$$

where a and b are real numbers. The real number a is called the real part of the complex number. The real number b is called the imaginary part of the complex number.

For example, these are complex numbers:

$3 + 4i$	$a = 3, b = 4$
$2 + 17i$	$a = 2, b = 17$
$-6 + 3i$	$a = -6, b = 3$
$4 - 9i$	$a = 4, b = -9$
$-5 - 11i$	$a = -5, b = -11$

Do you see why b = –9 in this example? The expression 4 – 9i can be rewritten as 4 + (–9)i .

A complex number whose imaginary part is 0 is a real number.

For example, these are real numbers:

$3 = 3 + 0i$	$a = 3, b = 0$
$-7 = -7 + 0i$	$a = -7, b = 0$
$\frac{5}{8} = \frac{5}{8} + 0i$	$a = \frac{5}{8}, b = 0$

Notice that b = 0 in each of these examples.

A complex number whose real part is 0 is called a pure imaginary number.

For example, these are pure imaginary numbers:

$6i = 0 + 6i$	$a = 0, b = 6$
$19i = 0 + 19i$	$a = 0, b = 19$
$-2i = 0 + (-2i)$	$a = 0, b = -2$

Notice that a = 0 in each of these examples.

Two complex numbers are equal if their real parts are equal and their imaginary parts are equal.

For example,

$$5 + (8 - 2)i = [4 - (-1)] + 6i$$

Here, the real parts are equal. $5 = [4 - (-1)]$

And the imaginary parts are equal. $(8 - 2) = 6$

As another example,

$$(3 + 4) - 2i \neq 7 + 2i$$

Here the real parts are equal. $(3 + 4) = 7$

But the imaginary parts are **not** equal. $-2 \neq 2$

Remember that for two complex numbers to be equal, **both** the real parts and the imaginary parts of the complex numbers must be equal.

Complex Conjugates

When the real parts of two complex numbers are equal and the imaginary parts only differ in sign, the complex numbers are called complex conjugates.

For example, here are some pairs of complex numbers that are complex conjugates:

$$7 + 2i \quad \text{and} \quad 7 - 2i$$

$$-3 + 6i \quad \text{and} \quad -3 - 6i$$

$$12 - 8i \quad \text{and} \quad 12 + 8i$$

And here are some pairs of complex numbers that are **not** complex conjugates:

$$4 + 7i \quad \text{and} \quad 7 - 4i$$

$$3 + 5i \quad \text{and} \quad -3 + 5i$$

$$7 + 2i \quad \text{and} \quad -7 - 2i$$

In general, the pairs of complex numbers

$$a + bi \quad \text{and} \quad a - bi$$

are called complex conjugates.

Adding and Subtracting Complex Numbers

To add two complex numbers, just add their real parts and add their imaginary parts.

In general,

$$(a + bi) + (c + di) = (a + c) + (b + d)i$$

For example, to find $(2 + 7i) + (6 + 3i)$:

1. Add the real parts and $= (2 + 6) + (7 + 3)i$
   add the imaginary parts. $= 8 + 10i$

So $(2 + 7i) + (6 + 3i) = 8 + 10i$.

As another example, to find $(9 - 4i) + (-3 + 5i)$:

1. Add the real parts and $= [9 + (-3)] + [(-4) + 5]i$
   add the imaginary parts. $= 6 + i$

So $(9 - 4i) + (-3 + 5i) = 6 + i$.

To subtract two complex numbers, just subtract their real parts and subtract their imaginary parts as indicated.

In general,

$$(a + bi) - (c + di) = (a - c) + (b - d)i$$

For example, to find $(8 + 2i) - (5 + 4i)$:

1. Subtract the real parts and subtract the imaginary parts.

$$= (8 - 5) + (2 - 4)i$$
$$= 3 - 2i$$

So $(8 + 2i) - (5 + 4i) = 3 - 2i$.

As another example, to find $(11 - 7i) - (3 - 13i)$:

1. Subtract the real parts and subtract the imaginary parts.

$$= (11 - 3) + [-7 - (-13)]i$$
$$= 8 + 6i$$

So $(11 - 7i) - (3 - 13i) = 8 + 6i$.

Multiplying Complex Numbers

You multiply complex numbers in much the same way as you multiply polynomials.

For example, to find $2i \cdot 7i$:

1. Multiply the real parts. Multiply the imaginary parts.

$$= 2 \cdot i \cdot 7 \cdot i$$
$$= 2 \cdot 7 \cdot i \cdot i$$
$$= 14 \cdot i^2$$
$$= 14 \cdot (-1)$$
$$= -14$$

So $2i \cdot 7i = -14$.

As another example, to find $4i(9 - 5i)$:

1. Use the distributive property, then multiply.

$$= 4i \cdot 9 - 4i \cdot 5i$$
$$= 36i - 20i^2$$
$$= 36i - 20(-1)$$
$$= 36i + 20$$
$$= 20 + 36i$$

So $4i(9 - 5i) = 20 + 36i$.

As a third example, to find $(8 + 3i)(5 - 4i)$:

1. Use the FOIL method and simplify.

$$= 8 \cdot 5 - 8 \cdot 4i + 3i \cdot 5 - 3i \cdot 4i$$
$$= 40 - 32i + 15i - 12i^2$$
$$= 40 - 17i - 12(-1)$$
$$= 40 - 17i + 12$$
$$= 52 - 17i$$

So $(8 + 3i)(5 - 4i) = 52 - 17i$.

When subtracting complex numbers, make sure you subtract both the real and imaginary parts.

Don't forget that $i^2 = -1$.

In general,

$$(a + bi)(c + di) = ac + adi + bci + bdi^2$$

$$= ac + adi + bci + bd(-1)$$

$$= ac + adi + bci - bd$$

$$= ac - bd + adi + bci$$

$$= (ac - bd) + (ad + bc)i$$

Dividing Complex Numbers

You can also divide complex numbers.

To divide complex numbers:

1. Rewrite the division problem as a fraction.

2. Multiply the numerator and denominator of the fraction by the complex conjugate of the denominator.

3. Simplify.

For example, to find $(7 + 5i) \div (3 + 2i)$:

1. Rewrite as a fraction.

$$\frac{7 + 5i}{3 + 2i}$$

2. Multiply by the complex conjugate.

$$= \frac{7 + 5i}{3 + 2i} \cdot \frac{3 - 2i}{3 - 2i}$$

$$= \frac{(7 + 5i)(3 - 2i)}{(3 + 2i)(3 - 2i)}$$

3. Simplify.

$$= \frac{21 - 14i + 15i - 10i^2}{9 - 4i^2}$$

$$= \frac{21 + i - 10(-1)}{9 - 4(-1)}$$

$$= \frac{21 + i + 10}{9 + 4}$$

$$= \frac{31 + i}{13}$$

$$= \frac{31}{13} + \frac{1}{13}i$$

So $(7 + 5i) \div (3 + 2i) = \frac{31 + i}{13} = \frac{31}{13} + \frac{1}{13}i$.

In general,

$$(a + bi) \div (c + di) = \frac{a + bi}{c + di}$$

$$= \frac{a + bi}{c + di} \cdot \frac{c - di}{c - di}$$

$$= \frac{(a + bi)(c - di)}{(c + di)(c - di)}$$

$$= \frac{(a + bi)(c - di)}{c^2 + d^2}$$

Quadratic Equations

When you used the quadratic formula to find solutions of a quadratic equation, you learned that if the discriminant is negative, the equation has no real solutions. But now you can find solutions that are complex numbers.

For example, to solve $x^2 + 2x + 5 = 0$:

1. Find the values for a, b, and c.

    $a = 1, b = 2, c = 5$

2. Substitute these values into the quadratic formula.

    $$x = \frac{-2 \pm \sqrt{2^2 - 4(1)(5)}}{2(1)}$$

    $$= \frac{-2 \pm \sqrt{4 - 20}}{2}$$

    $$= \frac{-2 \pm \sqrt{-16}}{2}$$

    $$= \frac{-2 \pm 4i}{2}$$

    $$= -1 \pm 2i$$

So the solutions of $x^2 + 2x + 5 = 0$ are $x = -1 \pm 2i$.

Notice that these two solutions, $x = -1 + 2i$ and $x = -1 - 2i$, are complex conjugates. If one solution of a quadratic equation $ax^2 + bx + c = 0$ (a, b, and c real numbers) is a complex number, the other solution is its complex conjugate.

You can solve the quadratic equation $ax^2 + bx + c = 0$ using the quadratic formula, $x = \dfrac{-b \pm \sqrt{b^2 - 4ac}}{2a}$. You can use the sign of the discriminant, $b^2 - 4ac$, to determine the nature of the solutions.

Since the discriminant of $x^2 + 2x + 5 = 0$ is −16, the equation has two imaginary solutions.

Sample Problems

1. Find: $(2 + 7i) + (12 - 4i)$

    ☐ a. Add the real parts and add   = _____
the imaginary parts.

    ☐ b. Simplify.   = _____

2. Find: $(8 - 3i) - (5 - 2i)$

    ☐ a. Subtract the real parts and   = _____
subtract the imaginary parts.

    ☐ b. Simplify.   = _____

3. Find: $(7 + 4i)(3 - 6i)$

    ☐ a. Multiply using FOIL.   = _____

    ☐ b. Simplify.   = _____

4. Find: $(3 + 9i) \div (1 + 2i)$

    ☑ a. Rewrite as a fraction. $= \dfrac{3 + 9i}{1 + 2i}$

    ☑ b. Multiply the numerator and $= \dfrac{3 + 9i}{1 + 2i} \cdot \dfrac{1 - 2i}{1 - 2i}$
denominator by $(1 - 2i)$.

    ☐ c. Simplify.   = _____

5. Solve for x: $2x^2 + 3x + 7 = 0$

    ☐ a. Find the values for a, b, and c. $a =$ _____

    $b =$ _____

    $c =$ _____

    ☐ b. Substitute the values for a, b, and c $x = \dfrac{-b \pm \sqrt{b^2 - 4ac}}{2a}$
into the quadratic formula,

and simplify. $x =$ _____

    ☐ c. Write the solutions. $x =$ _____ or $x =$ _____

 **HOMEWORK**

Homework Problems

Circle the homework problems assigned to you by the computer, then complete them below.

 Explain
Complex Number System

1. Circle the complex numbers below that are equal to $7 - 4i$.

    $4i - 7$ $7 - (2 - 6)i$

    $-4i + 7$ $7 - 3i - i$

    $6 + 1 - 4i$

2. Circle the pairs below that are complex conjugates of each other.

    $2 + 3i$ and $2 - 3i$ $9 + 7i$ and $7 + 9i$

    $5 + 3i$ and $3 - 5i$ $4 + 6i$ and $4 - 6i$

    $8 + i$ and $-8 - i$

3. Simplify each expression below.

    a. $\sqrt{-16}$

    b. $\sqrt{-25}$

    c. $\sqrt{-16} + \sqrt{-25}$

    d. $\sqrt{-16} \cdot \sqrt{-25}$

    e. i^{20}

    f. i^{47}

In problems (4) – (12), perform the indicated operation.

4a. $(5 + 7i) + (2 + 4i)$

  b. $(6 + i) + (8 + 3i)$

5a. $(10 + 6i) + (7 - 4i)$

  b. $(8 + 9i) - (5 + 2i)$

  c. $3i(11 - 4i)$

6a. $(2 + 7i)(14 - i)$

  b. $(4 - 12i)(5 - 3i)$

  c. $(5 + 6i) \div (2 + 7i)$

7a. $(6 + 11i)(6 - 11i)$

  b. $(7 + 12i) \div (9 - 4i)$

  c. Solve for x: $x^2 + 4x + 8 = 0$

8a. $(3 + 10i) \div (8 - 5i)$

  b. $(9 + 5i) \div (7 - 2i)$

  c. Solve for x: $x^2 + 6x + 15 = 0$

9a. $(2 + 4i)(5 - 6i)$

  b. $(1 + 2i) \div (1 - 2i)$

10a. Solve for x: $x^2 - 4 = 0$

  b. Solve for x: $x^2 + 4 = 0$

  c. $(1 - 2i) \div (1 + 2i)$

11a. $(2i + 7)(3 - 8i)$

  b. $(6 + 5i) \div (1 - 4i)$

  c. Solve for x: $x^2 + 2x + 9 = 0$

12a. $(4 + 7i) \div (3 + 5i)$

  b. Solve for x: $x^2 - 4x + 11 = 0$

Practice Problems

Here are some additional practice problems for you to try.

Complex Number System

1. Simplify: $\sqrt{-169}$

2. Simplify: $-\sqrt{-144}$

3. Simplify: $\sqrt{16} - \sqrt{-25}$

4. Simplify: $-\sqrt{9} + \sqrt{-16}$

5. Simplify: i^{125}

6. Simplify: i^{234}

7. Simplify: i^{100}

8. Simplify: $\sqrt{-7} \cdot \sqrt{-16}$

9. Simplify: $\sqrt{-5} \cdot \sqrt{-10}$

10. Simplify: $\sqrt{-6} \cdot \sqrt{-4}$

11. Find: $(7 + \sqrt{-9}) + (12 - \sqrt{-36})$

12. Find: $(10 - \sqrt{-121}) - (3 + \sqrt{-25})$

13. Find: $(5 + \sqrt{-4}) + (8 - \sqrt{-49})$

14. Find: $5(3 - 6i) + 3(7 + 2i)$

15. Find: $4(8 + 3i) - 2(7 + 9i)$

16. Find: $3(4 + 3i) - 6(1 + 2i)$

17. Find: $(3 + 7i)(2 + 3i)$

18. Find: $(4 - 7i)(6 - 9i)$

19. Find: $(2 - 5i)(6 + 4i)$

20. $(4 - 7i)(4 + 7i)$

21. $(8 + 3i)(8 - 3i)$

22. Find: $(5 - 3i)(5 + 3i)$

23. Find: $(4 + 5i) \div (2 + 7i)$

24. Find: $(3 - 2i) \div (4 + i)$

25. Find: $(5 - 2i) \div (3 + 4i)$

26. Solve for x: $x^2 - 5x + 8 = 0$

27. Solve for x: $x^2 + 8x + 18 = 0$

28. Solve for x: $x^2 + 6x + 11 = 0$

EVALUATE

Practice Test

Take this practice test to be sure that you are prepared for the final quiz in Evaluate.

1. Find:

    a. $(7 + 2i) + (4 + 6i)$ $11 + 8i$

    b. $(7 + 2i) - (4 + 6i)$ $3 - 8i$

2. Circle the expressions below that are equal to $8 + 4i$.

    $(2i)(4i + 2)$

    $3i + 6i - i$

    $-8i^2 - 4i^3$

    $\sqrt{-64}$

3. Find: $(3 + \sqrt{-16}) + (2 + \sqrt{-9})$

4. Find: $(4 + 2i)(3 + 5i)$

5. Find: $(5 + 3i)(5 - 3i)$

6. Find: $2 \div (4 + 7i)$

7. Circle the expressions below that are equal to i.

    i^4

    i^{37}

    $(i^5)^{10}$

    $-i^3$

8. Use the quadratic formula to find the solutions of each equation below.

    a. $x^2 + 3x + 7 = 0$

    b. $x^2 - 5x + 9 = 0$

    c. $3x^2 + 2x + 1 = 0$

TOPIC 10 CUMULATIVE ACTIVITIES

CUMULATIVE REVIEW PROBLEMS

These problems combine all of the material you have covered so far in this course. You may want to test your understanding of this material before you move on to the next topic, or you may wish to do these problems to review for a test.

1. Factor: $3b + 6bx + 10x + 5$

2. Solve for x: $|3x + 4| = 10$

3. Find: $(xy^3 - 5x^2y + 11xy - 1) + (4xy^3 - 7 - x^2y + 3xy)$

4. Solve $x^2 - 11x + 30 = 0$ for x.

5. Graph the line that passes through the point $(-2, -4)$ with slope 1.

6. Factor: $5x^2 + 9xy - 2y^2$

7. Solve $\dfrac{1}{x-3} = \dfrac{x}{4}$ for x.

8. Simplify the expressions below.

    a. $\sqrt{\dfrac{81x^4}{16}}$

    b. $\left(9^{\frac{5}{2}}\right)(3^{-2})$

    c. $\dfrac{\sqrt[3]{x^6}}{5}$

9. Solve $y^2 + 1 = -3$ for y.

10. Find:

    a. $\dfrac{a^2 \cdot c}{a^5 \cdot b^6 \cdot c^4}$

    b. $6x^0 \cdot y^4$

    c. $(7a^0)^2$

11. Solve for x: $|x + 5| \le 7$

12. Factor: $x^2 - 14x + 13$

13. Find:

    a. $(7 + 4i)(1 - 5i)$

    b. $(6 - 2i)(6 + 2i)$

14. Solve $-1 < 2x - 3 \le 4$ for x, then graph its solution on the number line below.

15. Find the equation of the line that passes through the point $(11, -4)$ that has slope $-\dfrac{6}{5}$:

    a. in point-slope form.

    b. in slope-intercept form.

    c. in standard form.

16. Solve for x: $2(5 + x) = 2x - 4$

17. Evaluate the expression $5xy + 1 - 4xy^3 - x^2$ when $x = 3$ and $y = -2$.

18. Find:

    a. $3 \div (2 + 5i)$

    b. $(8 + 4i) \div (6 - 3i)$

19. Solve $2y^2 - 17y + 21 = 0$ for y.

20. Graph the line $x - \dfrac{4}{3}y = 6$.

21. Find: $\dfrac{1}{2^{-3} + 4^{-2}}$

22. Factor: $3x^2y + 36xy + 108y$

23. Simplify this expression: $w^2xy - w^2x + 4w^2xy + y - 5w^2x$

24a. Find the slope of the line through the points $(-14, 8)$ and $(-2, 5)$.

    b. Write the equation of the line passing through the point $(-4, 6)$ that is parallel to the line in (a).

25. Find the slope of the line through the points (28, 4) and (−19, −12).

26. Rewrite the expressions below using only positive exponents. Then simplify.

a. $\dfrac{3^{\frac{4}{3}}x^{-\frac{3}{2}}}{3^{-\frac{2}{3}}x^{-\frac{5}{2}}}$

b. $\dfrac{2^{\frac{1}{3}}x^{-\frac{4}{3}}}{2^{-\frac{1}{9}}x^{\frac{2}{3}}}$

27. Solve $\dfrac{5}{x} + \dfrac{7}{6} = \dfrac{4}{x-4}$ for x.

28. Solve $5y^2 - 16y + 4 = 0$ for y.

29. Solve for x: $|2x + 7| - 4 = 8$

30. Factor: $12xy + 18y - 2x - 3$

31. Find: $\left(\dfrac{2}{5}x^2 - 3x\right)\left(5x - \dfrac{7}{3}\right)$

32. For what values of x is the expression $\dfrac{2}{x^2 - 25}$ undefined?

33. Solve $3x^2 + 7x + 4 = 0$ for x.

34. Solve $\dfrac{3}{2x} + \dfrac{1}{6x} = \dfrac{4}{5}$ for x.

35. Factor: $25y^2 - 30y + 9$

36. Solve for x: $|2x - 3| > 6$

37. Circle the true statements.

$2\dfrac{6}{7} + 5\dfrac{1}{4} = 7\dfrac{7}{11}$

$|11 - 5| = 6$

$\dfrac{5}{3} \div \dfrac{2}{3} = \dfrac{10}{9}$

$8 + 4(3 + 6) = 12(3 + 6)$

The GCF of 64 and 81 is 1.

If $S = \{1, 2, 3, 4, 5, 6, 7, 8, 9, 10\}$, then $16 \notin S$.

38. Evaluate the expression $5a^3 + 4a^2b - b^2$ when $a = -3$ and $b = -2$.

39. Find: $(2a^4 + 5b^2)(2a^4 - 5b^2)$

40. Rewrite the expressions below without using radicals or exponents.

a. $\sqrt{-81}$

b. $\sqrt[3]{-64}$

c. $\sqrt{0.09}$

d. $4^{\frac{3}{2}}$

41. Solve $16x^2 - 16x + 1 = 0$ for x.

42. Factor: $5x^2 + 9xy - 2y^2$

43. Solve $-17 \leq 6y + 7 < 17$ for y.

44. Find:

a. $(3 + 5i) + (4 - 7i)$

b. $(8 - 2i) - (3 - 6i)$

45. Find the x- and y-intercepts of the line $4x - 3y = 5$.

46. Solve for a: $\dfrac{6}{a-2} + 1 = \dfrac{1}{a}$

47. Factor: $x^4 - y^2$

48. Find: $(6x + y)^2$

49. Solve $3x^2 - 5x - 22 = 0$ for x.

50. Find:

a. $7^3 \cdot 7^5$

b. $(5x^4)^3$

c. $\dfrac{b^{21}}{b^{24}}$

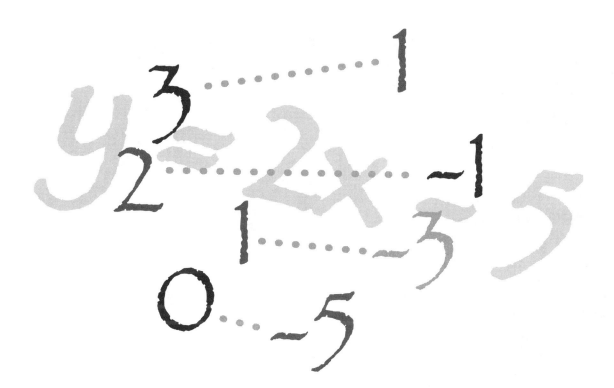

OVERVIEW

Here's what you'll learn in this lesson:

Functions and Graphs

a. *Definition of a function*

b. *Function as an ordered pair of numbers*

c. *Finding function values given a formula*

d. *Function notation: $y = f(x)$*

e. *Graphing simple functions*

f. *Domain and range of a function*

g. *The vertical line test*

Linear Functions

a. *Graphs of linear functions*

b. *Graphs of absolute value functions*

Quadratic Functions

a. *Graphs of quadratic functions*

b. *Intercepts of quadratic functions*

For centuries, artists from many cultures have used patterns to create beautiful mosaics, textiles, and stained glass windows. Many of these patterns can be described mathematically using functions.

In this lesson, you will learn about functions. You will learn how to graph a function, how to identify the domain and the range of a function, and how to use the notation for functions. In addition, you will study three types of functions that you will see again in your study of algebra: linear functions, absolute value functions, and quadratic functions.

 EXPLAIN

FUNCTIONS AND GRAPHS

Summary

Definition of a Function

Rules such as $y = 3x + 1$ and $y = 2x - 7$ are called functions. These rules assign to each real number x exactly one real number y.

Some examples of functions are:

$$y = x + 1 \qquad\qquad y = x^2$$

$$y = 4 - 2x \qquad\qquad y = \frac{1}{x}$$

Sometimes letters such as f, g, or h are used to denote functions. When x is the input number, the output number is written as $f(x)$, $g(x)$, or $h(x)$.

For example, you might write the above functions as:

$$f(x) = x + 1$$

$$g(x) = 4 - 2x$$

$$h(x) = x^2$$

$$k(x) = \frac{1}{x}$$

To find the value of a function at a given point, x:

1. Substitute the given x-value into the function.

2. Simplify to get the output value.

For example, to find the value of $f(x) = 3x - 7$ at $x = 1$:

1. Substitute 1 for x. $\qquad f(1) = 3(1) - 7$

2. Simplify. $\qquad\qquad\qquad = 3 - 7$

$$\qquad\qquad\qquad\qquad = -4$$

As another example, to find the value of $k(x) = \frac{1}{2x}$ at $x = -7$:

1. Substitute -7 for x. $\qquad f(-7) = \frac{1}{2(-7)}$

2. Simplify. $\qquad\qquad\qquad = \frac{1}{-14}$

$$\qquad\qquad\qquad\qquad = -\frac{1}{14}$$

You can think of a function as a machine that assigns to each input number, x, a unique output number, y.

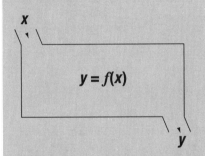

You read "f(x)" as "f of x."

Domain and Range of a Function

The domain is the x-values. The range is the y-values. Here's a way to help you remember which is which: **d**omain comes before **r**ange in alphabetical order just like **x** comes before **y**.

The **domain** of a function is all of the real numbers, x, for which the function is defined and produces a real number, y. The **range** of a function is all of the possible y-values you can get from the x-values.

The domain and range of a function don't always include all the real numbers.

- When a function includes a square root, the domain does not include any numbers that make the value under the square root negative.

For example, in the function $y = \sqrt{x - 1}$ you need to make sure that the $(x - 1)$ under the square root sign is never negative. So the domain includes all nonnegative real numbers greater than or equal to 1.

- When a function includes a fraction, the domain does not include any numbers that make the denominator of the fraction zero.

For example, in the function $y = \dfrac{1}{x + 2}$ you need to make sure that the $(x + 2)$ in the denominator is never zero. So the domain includes all real numbers except −2.

To find the domain and range of a function:

To help you find the domain, start with all real numbers and eliminate those numbers for which the function does not make sense.

1. Find all of the x-values for which the function is defined and produces a real number. This is the domain of the function.

2. Find all possible y-values for all x-values in the domain. This is the range of the function.

For example, to find the domain and range of the function $y = x - 1$:

The domain is all real numbers since $(x - 1)$ is defined for any real number. The range is all real numbers since $(x - 1)$ can equal any real number as x runs through all real numbers.

1. Find all the x-values for which the function is defined. All real numbers.

2. Find all possible y-values for all x-values in the domain. All real numbers.

As another example, to find the domain and range of the function $y = \dfrac{1}{x^2 - 4}$:

The domain can't include 2 or −2 because substituting either of these numbers makes the denominator equal to 0. The range can't include those real numbers which are greater than $-\dfrac{1}{4}$ and less than or equal to 0. There are no real numbers you can substitute for x that will make $\dfrac{1}{x^2 - 4}$ equal to numbers which are greater than $-\dfrac{1}{4}$ and less than or equal to 0.

1. Find all the x-values for which the function is defined. Any real number except $x = 2$ or $x = -2$.

2. Find all possible y-values for all x-values in the domain. All real numbers except those that are greater than $-\dfrac{1}{4}$ and less than or equal to 0.

Graphing Functions

When you are given a function, you can find ordered pairs of real numbers that satisfy the function and then use these ordered pairs to graph the function.

To find an ordered pair that satisfies a function $y = f(x)$:

1. Pick a value for x.

2. Substitute your chosen value, x, into the function to solve for the other variable, y.

3. Write the ordered pair as (x, y).

For example, to find an ordered pair that satisfies the function $y = 2x - 5$:

1. Pick a value for x. $x = 3$

2. Substitute $x = 3$ into $y = 2x - 5$
   $y = 2x - 5$ and solve for y. $y = 2(3) - 5$
                                       $y = 6 - 5$
                                       $y = 1$

3. Write the ordered pair as (x, y). $(3, 1)$

In general, there are infinitely many ordered pairs that satisfy a function. For the function $y = 2x - 5$, for any x you choose you can find a y. Once you have several ordered pairs that satisfy the function, you can plot the points on a grid to graph the function.

To graph a function:

1. Find several ordered pairs that satisfy the function.

2. Plot these points on a grid.

3. Use these points to sketch the graph of the function.

For example, to graph the function $y = 2x - 5$:

1. Find several ordered pairs that satisfy the function.

x	y
3	1
2	−1
1	−3
0	−5

2. Plot these points on a grid.

3. Use these points to sketch the graph of the function. See Figure 11.1.1.

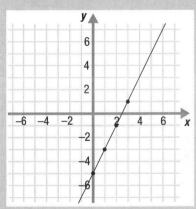

Figure 11.1.1

Vertical Line Test

Remember, a function is a relation that has a unique value, *y*, for any value of *x*. Therefore, you can tell if a graph is the graph of a function by using the vertical line test: if you draw a vertical line anywhere through the graph of a function, it will intersect the graph in only one place. Likewise, if a vertical line intersects a graph in more than one place then the graph is not the graph of a function.

For example, here are the graphs of some functions:

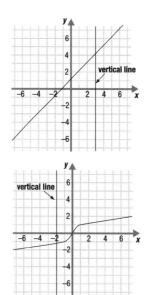 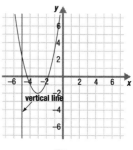

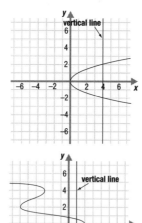 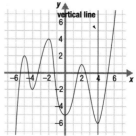

Any vertical line intersects each graph in one and only one place.

Here are some graphs that are not functions:

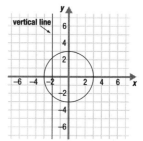

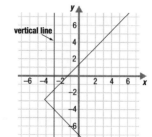

There is a vertical line that intersects these graphs in more than one place.

Sample Problems

1. For the function $g(x) = 4x - 5$, calculate:

   a. $g(0)$

   b. $g(2)$

   c. $g(-5)$

   ☑ a. Substitute 0 for x and simplify.

   $g(0) = 4(0) - 5$
   $= 0 - 5$
   $= -5$

   ☐ b. Substitute 2 for x and simplify.

   $g(2) = 4(2) - 5$
   $= \underline{\hspace{1cm}}$
   $= \underline{\hspace{1cm}}$

   b. $8 - 5$
   3

   ☐ c. Substitute -5 for x and simplify.

   $g(\underline{\hspace{0.5cm}}) = 4(\underline{\hspace{0.5cm}}) - 5$
   $= \underline{\hspace{1cm}}$
   $= \underline{\hspace{1cm}}$

   c. $-5, -5$
   $-20 - 5$
   -25

2. Find the domain and range for each of the functions below.

   a. $y = x - 7$

   b. $y = \dfrac{3}{2x}$

   c. $y = \sqrt{x - 2}$

   ☑ a. Find all of the x-values for which the function is defined.

   domain: all real numbers

   Find all possible y-values, given the possible x-values.

   range: all real numbers

   ☐ b. Find all of the x-values for which the function is defined.

   domain: $\underline{\hspace{2cm}}$

   b. all real numbers except $x = 0$
   $(x \neq 0)$

   Find all possible y-values, given the possible x-values.

   range: $\underline{\hspace{2cm}}$

   all real numbers except $y = 0$
   $(y \neq 0)$

   ☐ c. Find all of the x-values for which the function is defined.

   domain: $\underline{\hspace{2cm}}$

   c. all real numbers greater than or equal to 2 $(x \geq 2)$

   Find all possible y-values, given the possible x-values.

   range: $\underline{\hspace{2cm}}$

   all real numbers greater than or equal to 0 $(y \geq 0)$

a.

-1

0

1

2

b., c.

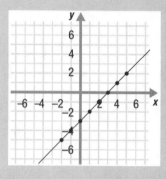

3. Complete the table below for the function $y = x - 3$. Then use the table to graph the function on the grid.

☐ a. Complete the table.

x	x − 3
−2	−5
−1	−4
0	−3
1	−2
2	___
3	___
4	___
5	___

☐ b. Plot the points.

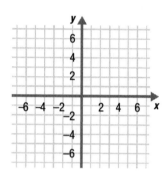

☐ c. Sketch the function.

LINEAR FUNCTIONS

Summary

Linear Functions

A linear function is a function whose graph is a straight line.

The equation of a linear function may be written in the form:

$$y = Ax + B \qquad \text{or} \qquad f(x) = Ax + B$$

For example, the following functions are linear functions:

$$y = x \qquad\qquad \text{or} \qquad f(x) = x$$

$$y = 3x + 1 \qquad \text{or} \qquad f(x) = 3x + 1$$

$$y = -2x + 5 \qquad \text{or} \qquad f(x) = -2x + 5$$

These functions are **not** linear functions:

$$y = x^2 + 6 \qquad \text{or} \qquad f(x) = x^2 + 6$$

$$y = \frac{1}{x} \qquad\qquad \text{or} \qquad f(x) = \frac{1}{x}$$

The output value, y, **depends** on the input value, x, so y is called the **dependent** variable. Because you pick the number x (**independent** of anything else) it is called the **independent** variable.

Domain and Range of a Linear Function

For any linear function $y = Ax + B$, $A \neq 0$:

- The domain is all real numbers.

- The range is all real numbers.

Graphs of Linear Functions

When you have two or more linear functions of the form $y = Ax + B$ with the same coefficient A of x, the graphs of the functions look almost the same. The lines have the same steepness but are shifted up or down depending on the value of the constant, B.

For example, look at the graphs of these functions in Figure 11.1.2:

$$y = 4x$$

$$y = 4x + 2$$

$$y = 4x - 3$$

Notice:

- The line $y = 4x + 2$ is the line $y = 4x$ shifted up 2 units.

- The line $y = 4x - 3$ is the line $y = 4x$ shifted down 3 units.

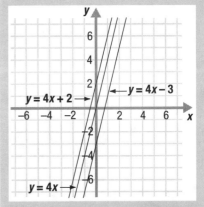

Figure 11.1.2

Here the coefficient of x in each function is 4.

Graphing Linear Functions

If you know the equation that describes a function, you can use the equation to graph the function.

To graph a function of the form $f(x) = Ax + B$:

1. Find several ordered pairs that satisfy the function.

2. Plot these points on a grid.

3. Use these points to sketch the graph of the function.

For example, to graph the function $y = f(x) = 2x + 1$:

1. Pick several values for x and find the corresponding values for $f(x)$.

x	$f(x)$
−3	−5
−1	−1
0	1
2	5

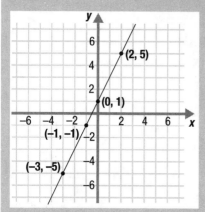

Figure 11.1.3

2. Plot the ordered pairs and draw a line through them. See Figure 11.1.3.

Absolute Value Functions

Here are four absolute value functions that are related to the linear function $y = Ax + B$:

$$y = |Ax| + B$$

$$y = -|Ax| + B$$

$$y = |Ax + B|$$

$$y = -|Ax + B|$$

If there is a negative sign in front of the absolute value then the "V" is inverted.

The graphs of these absolute value functions are in the shape of a "V" that is either upright or inverted.

Domain and Range of Absolute Value Functions

For an absolute value function based on a linear function:

• The domain is all real numbers.

• For functions of the form $y = |Ax| + B$ the range is real numbers $\geq B$.

• For functions of the form $y = -|Ax| + B$ the range is real numbers $\leq B$.

• For functions of the form $y = |Ax + B|$ the range is real numbers ≥ 0.

• For functions of the form $y = -|Ax + B|$ the range is real numbers ≤ 0.

For example, the domain and range of each of the functions below are:

$y = |3x| + 1$ domain: all real numbers
                         range: real numbers ≥ 1

$y = -|3x| + 1$ domain: all real numbers
                         range: real numbers ≤ 1

$y = |3x + 1|$ domain: all real numbers
                         range: real numbers ≥ 0

$y = -|3x + 1|$ domain: all real numbers
                         range: real numbers ≤ 0

Graphs of Absolute Value Functions

When you have two or more absolute value functions each based on a linear function and each having the same coefficient of x with no leading negative sign, their graphs look almost the same. Their only difference is the "V's" are shifted up or down depending on the constant term.

For example, look at the graphs of these absolute value functions in Figure 11.1.4:

$y = |x|$

$y = |x| + 2$

$y = |x| - 3$

Notice:

- The graph of $y = |x| + 2$ is the graph of $y = |x|$ shifted up 2 units.

- The graph of $y = |x| - 3$ is the graph of $y = |x|$ shifted down 3 units.

When there is a negative sign in front of the absolute value sign, the rules above still hold but the graphs of the functions are inverted "V's" rather than upright "V's."

For example, look at the graphs of the functions in Figure 11.1.5:

$y = -|x|$

$y = -|x| + 2$

$y = -|x| - 3$

Notice:

- The graph of $y = -|x| + 2$ is the graph of $y = -|x|$ shifted up 2 units.

- The graph of $y = -|x| - 3$ is the graph of $y = -|x|$ shifted down 3 units.

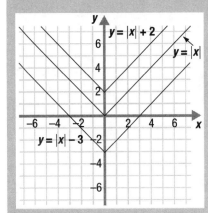

Figure 11.1.4

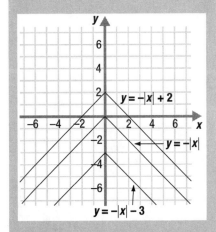

Figure 11.1.5

Sample Problems

1. Find the domain and range of each of the functions below.

   a. $y = x - 1$

   b. $y = |x + 2|$

   c. $y = 5 - 3x$

   ☑ a. The domain is all the possible values for x.   domain: all real numbers

   The range is all the possible values for y.   range: all real numbers

   ☐ b. The domain is all the possible values for x.   domain: _____

   The range is all the possible values for y.   range: _____

   ☐ c. The domain is all the possible values for x.   domain: _____

   The range is all the possible values for y.   range: _____

2. The graph of the function $y = -x$ is shown below. Determine the equations of the other lines shown.

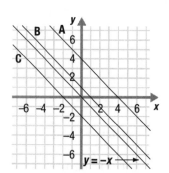

   ☑ a. Find the equation of line A.   This line is shifted **up** 4 units, so its equation is $y = -x + 4$.

   ☐ b. Find the equation of line B.   The equation is: _____

   ☐ c. Find the equation of line C.   The equation is: _____

QUADRATIC FUNCTIONS

Summary

Quadratic Functions

A quadratic function is a function whose graph is a parabola.

Equations of Quadratic Functions

The equation of a quadratic function may be written in the form:

$$y = Ax^2 + Bx + C$$

or $\quad f(x) = Ax^2 + Bx + C$

where $A \neq 0$.

For example, the following functions are quadratic functions:

$$y = 4x^2 + 5 \qquad\qquad y = 4x^2 + 0x + 5$$

$$f(x) = -3x^2 + 2x - 8 \qquad f(x) = -3x^2 + 2x + (-8)$$

These functions are **not** quadratic functions:

$$y = \frac{4}{x^2}$$

$$y = x + 7$$

$$y = x^3 + 5x^2 - 2$$

Notice that in each case the coefficient of the x^2-term is nonzero.

The Graph of $y = Ax^2 + Bx + C$ when $A > 0$

The graph of a quadratic function is a parabola that opens up or down depending on the coefficient of the x^2-term. If the coefficient of the x^2-term is positive, the parabola opens up.

For example, look at the graphs of these parabolas with positive x^2-terms. See Figure 11.1.6:

$$y = x^2$$

$$y = x^2 + 2$$

$$y = x^2 - 3$$

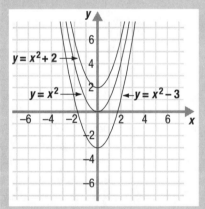

Figure 11.1.6

When the coefficient of the x^2-term is positive, the vertex of the parabola is the low point.

Notice:

- All of the graphs open up.

- The graph of $y = x^2 + 2$ is the graph of $y = x^2$ shifted up 2 units.

- The graph of $y = x^2 - 3$ is the graph of $y = x^2$ shifted down 3 units.

The Graph of $y = Ax^2 + Bx + C$ when $A < 0$

The graph of a quadratic function is a parabola that opens up or down depending on the coefficient of the x^2-term. If the coefficient of the x^2-term is negative, the parabola opens down.

For example, look at the graphs of these parabolas where the coefficient of x^2 is negative. See Figure 11.1.7:

$$y = -2x^2$$

$$y = -2x^2 + 2$$

$$y = -2x^2 - 3$$

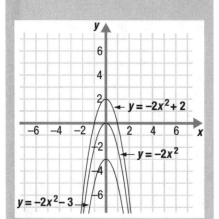

Figure 11.1.7

When the coefficient of the x^2-term is negative, the vertex of the parabola is the high point.

Notice:

- All of the graphs open down.

- The graph of $y = -2x^2 + 2$ is the graph of $y = -2x^2$ shifted up 2 units.

- The graph of $y = -2x^2 - 3$ is the graph of $y = -2x^2$ shifted down 3 units.

Domain and Range of Quadratic Functions

For a quadratic function $y = Ax^2 + Bx + C$:

- The domain is all real numbers.

- When $A > 0$, the range is all real numbers greater than or equal to the y-coordinate of the vertex of the parabola (the lowest point).

- When $A < 0$, the range is all real numbers less than or equal to the y-coordinate of the vertex of the parabola (the highest point).

For example, the domain and range of each of the functions below are:

$y = x^2$	domain: all real numbers range: all real numbers ≥ 0
$y = x^2 + 2$	domain: all real numbers range: all real numbers ≥ 2
$y = 2x^2 - 3$	domain: all real numbers range: all real numbers ≥ -3
$y = -5x^2$	domain: all real numbers range: all real numbers ≤ 0
$y = -x^2 + 2$	domain: all real numbers range: all real numbers ≤ 2
$y = -x^2 - 3$	domain: all real numbers range: all real numbers ≤ -3

Intercepts

The y-Intercept

The y-intercept of a function is the point where the graph crosses the y-axis; that is, the point where x is 0. A parabola of the form $y = Ax^2 + Bx + C$ always has one y-intercept.

To find the y-intercept of a function without a graph:

1. Set x equal to 0.

2. Solve for y.

3. Write the ordered pair as (0, y-value).

For example, to find the y-intercept of $y = x^2 - 3x - 10$:

1.	Set x equal to 0.	$y = (0)^2 - 3(0) - 10$
2.	Solve for y.	$y = 0 - 0 - 10$
		$y = -10$
3.	Write the ordered pair.	$(0, -10)$

So the y-intercept of $y = x^2 - 3x - 10$ is $(0, -10)$.

In general, if $y = Ax^2 + Bx + C$, then the y-intercept of the function is $(0, C)$.

The x-Intercept

The x-intercepts of a function are the points where the graph crosses the x-axis; that is, the points where y is 0. A parabola can have zero, one, or two x-intercepts (because a parabola can cross the x-axis in zero, one, or two places).

To find the x-intercept(s) of a function $y = f(x)$ without its graph:

1. Set y equal to 0.

2. Solve for x.

3. Write the ordered pair(s) as (x-value, 0).

For example, to find the x-intercepts of $y = x^2 - 3x - 10$:

1.	Set y equal to 0.	$0 = x^2 - 3x - 10$
2.	Solve for x.	$(x + 2)(x - 5) = 0$
		$x + 2 = 0$ or $x - 5 = 0$
		$x = -2$ or $x = 5$
3.	Write the ordered pairs.	$(-2, 0)$ and $(5, 0)$

So the two x-intercepts of $y = x^2 - 3x - 10$ are $(-2, 0)$ and $(5, 0)$.

As another example, to find the x-intercept of $y = x^2 + 2x + 1$:

1. Set y equal to 0.　　　　　　　　　　$0 = x^2 + 2x + 1$

2. Solve for x.　　　　　　　　　　　　$(x + 1)(x + 1) = 0$

　　　　　　　　　　　　　　　　　　　　　$x + 1 = 0$

　　　　　　　　　　　　　　　　　　　　　$x = -1$

3. Write the ordered pair.　　　　　　$(-1, 0)$

So the only x-intercept of $y = x^2 + 2x + 1$ is $(-1, 0)$.

As a third example, to find the x-intercepts of $y = x^2 + 9$:

1. Set y equal to 0.　　　　　　　　　　$0 = x^2 + 9$

2. Solve for x.　　　　　　　　　　　　$x^2 = -9$

　　　　　　　　　　　　　　　　　　　　　$x = \pm\sqrt{-9}$

　　　　　　　　　　　　　　　　　　　　　$x = \pm 3i$

This equation has two imaginary solutions, so the graph doesn't cross the x-axis and it has no x-intercepts.

The Vertex of a Parabola

When the parabola given by the quadratic function
$$y = Ax^2 + Bx + C$$
opens up ($A > 0$), its vertex is its low point.

When it opens down ($A < 0$), its vertex is its high point.

The x- coordinate of the vertex is given by:
$$x = -\frac{B}{2A}$$

To find the y-coordinate of the vertex, substitute $x = -\frac{B}{2A}$ into the equation $y = Ax^2 + Bx + C$.

For example, to find the vertex of the parabola given by the quadratic function
$$y = x^2 - 3x - 10 \quad \text{(Here, } A = 1, B = -3, C = -10\text{)}$$

1. Find the x-coordinate of the vertex

$$x = -\frac{B}{2A}$$

$$= \frac{-(-3)}{2(1)}$$

$$= \frac{3}{2}$$

2. Find the y-coordinate of the vertex

$$y = \left(\frac{3}{2}\right)^2 - 3\left(\frac{3}{2}\right) - 10$$

$$= \frac{9}{4} - \frac{9}{2} - 10$$

$$= -\frac{49}{4}$$

Notice that $\frac{3}{2}$, the x-coordinate of the vertex, lies halfway between the x-coordinates of the two x-intercepts, $x = -2$ and $x = 5$.

So the coordinates of the vertex are $\left(\frac{3}{2}, -\frac{49}{4}\right)$.

As another example, to find the vertex of the parabola given by the quadratic function

$y = x^2 + 2x + 1$ (Here, $A = 1$, $B = 2$, $C = 1$)

1. Find the x-coordinate of the vertex

$$x = \frac{-B}{2A}$$

$$= \frac{-2}{2(1)}$$

$$= -1$$

2. Find the y-coordinate of the vertex

$$y = (-1)^2 + 2(-1) + 1$$

$$= 1 - 2 + 1$$

$$= 0$$

So the coordinates of the vertex are $(-1, 0)$.

Sample Problems

1. Find the domain and range of each of the functions below.

a. $y = x^2$

b. $y = x^2 + 4$

c. $y = x^2 - 1$

☑ a. The domain is all the possible values for x.   domain: all real numbers

The range is all the possible values for y.   range: $y \geq 0$

☐ b. The domain is all the possible values for x.   domain: _____

The range is all the possible values for y.   range: _____

☐ c. The domain is all the possible values for x.   domain: _____

The range is all the possible values for y.   range: _____

Notice that the vertex, (–1, 0), is the same as the one and only one x-intercept.

Answers to Sample Problems

b. *all real numbers*

all real numbers greater than or equal to 4 (y ≥ 4)

c. *all real numbers*

all real numbers greater than or equal to –1 (y ≥ –1)

2. The graph of the function $y = x^2$ is shown below. Determine the equations of the other parabolas shown by finding the number of units each parabola is shifted up or down relative to $y = x^2$.

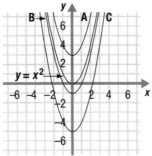

☑ a. Find the number of units parabola A is shifted up from $y = x^2$.

Shifted up: 3 units

Write the equation of parabola A.

The equation is: $y = x^2 + 3$.

☐ b. Find the number of units parabola B is shifted down from $y = x^2$.

Shifted down: 1 unit

b. $y = x^2 - 1$

Write the equation of parabola B.

The equation is: _____

☐ c. Find the number of units parabola C is shifted down from $y = x^2$.

Shifted down: _____

c. 5 units

Write the equation of parabola C.

The equation is: _____

$y = x^2 - 5$

3. Find the y-intercept of the parabola $y = x^2 + 6x + 8$.

☑ a. Set x equal to 0.

$y = (0)^2 + 6(0) + 8$

b. 8

☐ b. Solve for y.

$y = $ ____

c. (0, 8)

☐ c. Write the ordered pair.

(____ , ____)

4. Find the x-intercepts of the parabola $y = x^2 + 6x + 8$.

☑ a. Set y equal to 0.

$0 = x^2 + 6x + 8$

b. –2, –4

☐ b. Solve for x.

$x = $ ____ or $x = $ ____

c. (–2, 0), (–4, 0)

☐ c. Write the ordered pairs.

(____ , ____) or (____ , ____)

5. Find the vertex of the parabola $y = x^2 + 6x + 8$.

☑ a. Find the x-coordinate of the vertex.

$$x = -\frac{B}{2A}$$
$$= -\frac{6}{2(1)}$$
$$= -3$$

b. –1

☐ b. Find the y-coordinate of the vertex.

$y = $ ____

c. (–3, –1)

☐ c. The coordinates of the vertex are.

(____ , ____)

EXPLORE

Sample Problems

On the computer you used the Grapher to explore the relationship between a function and its graph.

1. Determine the domain and range of each function graphed below.

a.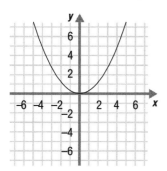

b.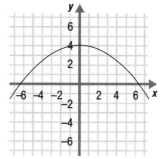

c.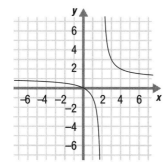

☑ a. The domain is all the possible values for *x*.   domain: all real numbers

   The range is all the possible values for *y*.   range: y ≥ 0

☐ b. The domain is all the possible values for *x*.   domain: _____

   The range is all the possible values for *y*.   range: _____

☐ c. The domain is all the possible values for *x*.   domain: _____

   The range is all the possible values for *y*.   range: _____

b. all real numbers

all real numbers less than or equal to

4 (y ≤ 4)

c. all real numbers except x = 2

all real numbers except y = 1

2a. If $f(x) = x + 8$, find $f(7)$.

b. If $f(x) = |x - 9|$, find $f(-3)$.

c. If $f(x) = 2x + 4$, find $f(5)$.

☑ a. Substitute $x = 7$ into $f(x) = x + 8$ and simplify.

$$f(x) = x + 8$$
$$f(7) = 7 + 8$$
$$= 15$$

☐ b. Substitute $x = -3$ into $f(x) = |x - 9|$ and simplify.

☐ c. Substitute $x = 5$ into $f(x) = 2x + 4$ and simplify.

b. $f(x) = |x - 9|$

$f(-3) = |-3 - 9|$

$= |-12|$

$= 12$

c. $f(x) = 2x + 4$

$f(5) = 2(5) + 4$

$= 10 + 4$

$= 14$

HOMEWORK

Homework Problems

Circle the homework problems assigned to you by the computer, then complete them below.

 Explain

Functions and Graphs

1. Complete the table below for the function $y = x + 7$.

x	y
−5	
−4	
−3	
−2	
−1	
0	
1	
2	

2. For the function $f(x) = 2x + 3$, calculate:

    a. $f(0)$

    b. $f(-4)$

    c. $f(2)$

    d. $f(-1)$

    e. $f(6)$

3. Find the domain and range of each of the functions below.

    a: $y = x - 8$

    b. $y = 4x + 7$

    c. $y = x^2$

    d. $y = x^2 + 1$

    e. $y = \dfrac{1}{x}$

4. Complete the table below for the function $y = x + 2$. Then use the table to graph the function.

x	y
−3	
−2	
−1	
0	
1	
2	
3	
4	

5. For the function $g(x) = x^2 - 5$, calculate:

    a. $g(0)$

    b. $g(4)$

    c. $g(7)$

    d. $g(-2)$

    e. $g(-6)$

6. Find the domain and range of each of the functions below.

    a. $y = 3x - 4$

    b. $y = \dfrac{x + 9}{7}$

    c. $y = \dfrac{7}{x + 9}$

    d. $y = x^3 + 5$

    e. $y = \dfrac{3}{x^2 - 4}$

7. For the function $h(x) = x^3 + 1$, calculate:

   a. $h(2)$

   b. $h(-1)$

   c. $h(0)$

   d. $h(-3)$

   e. $h(1)$

8. Make a table of at least seven ordered pairs that satisfy the equation $y = 2x - 5$. Then use the table to graph the function.

9. The world population increases at a net rate of approximately 3 people per second. This increase can be written as a function: $p(x) = 3x$, where x represents seconds. Find the number of people by which the world's population will increase in:

   a. 1 day (86,400 seconds)

   b. 1 week (604,800 seconds)

   c. 1 year (31,536,000 seconds)

10. America's population increases at a net rate of approximately 1 person every 14 seconds. This increase can be written as a function; $p(x) = \frac{x}{14}$, where x represents seconds. Find the number of people by which America's population will increase in:

    a. 1 day (86,400 seconds)

    b. 1 week (604,800 seconds)

    c. 1 year (31,536,000 seconds)

11. Find the domain and range for each of the functions below.

    a. $f(x) = \frac{x}{x-3}$

    b. $g(x) = \frac{2}{(x+1)(x+2)}$

    c. $h(z) = \sqrt{z^3 + 8}$

    d. $h(w) = \frac{(w+4)}{(w+4)(w-8)(w-11)}$

    e. $g(q) = \sqrt{4 - q^2}$

12. Make a table with at least seven ordered pairs that satisfy the equation $y = x^2 - 3$. (Use both positive and negative values.) Then use your table to graph the function.

Linear Functions

13. Circle the functions below that are linear.

    $y = 2x + 1$

    $y = x^3 + 4x^2 + 2x + 5$

    $y = 5 - 8x$

    $y = -7x^2 + 11x + 2$

    $y = 1 + 5x + 2x^4$

14. The graph of the function $y = |x|$ is shown on the grid in Figure 11.1.8. Determine the equations of the other lines shown.

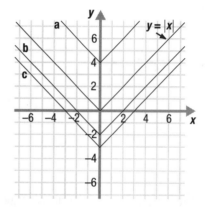

Figure 11.1.8

15. Find the domain and range of each of the functions below.

    a. $y = x + 3$

    b. $y = |x|$

    c. $y = 4x - 11$

    d. $y = |x - 4|$

16. The graph of the function $y = 2x$ is shown in Figure 11.1.9. Determine the equations of the other lines shown.

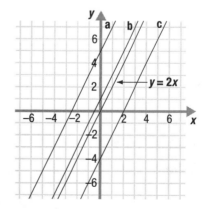

Figure 11.1.9

17. Find the domain and range of each of the functions below.

    a. $y = x - 5$

    b. $y = |x| - 5$

    c. $y = |x - 5|$

18. Graph the line $y = x - 2$.

19. The graph of the function $y = |x + 1|$ is shown in Figure 11.1.10. Determine the equations of the other functions shown.

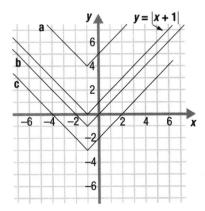

Figure 11.1.10

20. Find the domain and range of each of the functions below.

    a. $y = |4x + 3|$

    b. $y = |2 - x|$

    c. $y = 3 - 8x$

21. When Katy goes for a walk, she walks 2 miles then turns around and walks back home. The graph in Figure 11.1.11 shows her distance from home as she walks. Determine the domain and range of this function.

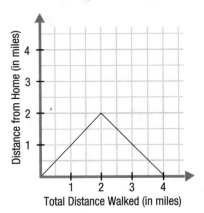

Figure 11.1.11

22. When Katy decides to go for a walk alone, she leaves immediately and walks at the rate of 4 miles an hour. When her dog Swirly comes, it takes them 15 minutes to get ready to leave, but they walk at the same speed. The graph in Figure 11.1.12 shows Katy's progress when she walks alone. On this grid, graph the line that shows her progress when she walks with her dog.

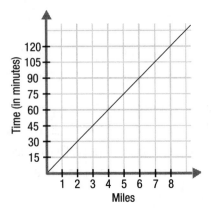

Figure 11.1.12

23. The graph of the line $y = -2x$ is shown on the grid in Figure 11.1.13. Graph the other lines whose equations are shown.

    a. $y = -2x + 5$

    b. $y = -2x + 1$

    c. $y = -2x - 4$

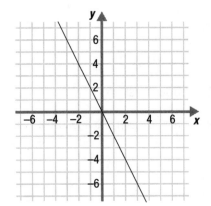

Figure 11.1.13

24. Find the domain and range of each of the functions below.

    a. $y = |2x| - 3$

    b. $y = 2|x| - 3$

    c. $y = |2x - 3|$

Quadratic Functions

25. Circle the functions below that are quadratic.

$y = 7 - x$

$y = 2x^3 + x^2 - 3x + 6$

$y = 4 - 2x + 6x^2$

$y = 8x^2 + 9x - 1$

$y = 4x - 11$

26. Determine whether each of the following parabolas opens up or down.

a. $y = x^2$

b. $y = 3x^2 - 5$

c. $y = -x^2 + 3x + 2$

d. $y = 4 - x^2$

e. $y = 2x^2 + 7x - 1$

f. $y = -5x^2 + 4x - 3$

27. Find the domain and range of each of the functions below.

a. $y = x^2 + 3$

b. $y = x^2 - 5$

c. $y = 3x^2$

d. $y = x^2 + 4x + 4$

28. The graph of the function $y = -x^2$ is shown on the grid in Figure 11.1.14. Determine the equations of the other parabolas shown.

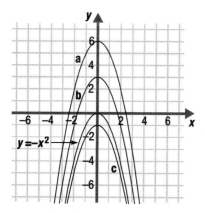

Figure 11.1.14

29. Find the x- and y-intercepts of each of the functions below.

a. $y = x^2$

b. $y = x^2 + 4x + 3$

c. $y = x^2 + 8x + 16$

30. Find the domain and range of each of the functions below.

a. $y = x^2 - 4$

b. $y = -x^2 - 4$

c. $y = x^2 + 2$

d. $y = -x^2 + 2$

31. The graph of the function $y = \frac{1}{2}x^2$ is shown on the grid in Figure 11.1.15. Determine the equations of the other parabolas shown.

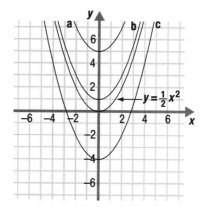

Figure 11.1.15

32. Find the vertex of the parabola given by each quadratic function below.

a. $y = x^2 + 3x - 10$

b. $y = x^2 - 2x + 1$

c. $y = x^2 + 4$

33. Duncan threw a ball in the air. The height of this ball over time is graphed on the grid in Figure 11.1.16. Use the graph to determine the domain (time) and range (height) of this function.

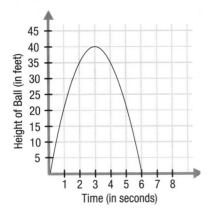

Figure 11.1.16

34. Duncan threw a ball in the air. The height of this ball over time is graphed on the grid in Figure 11.1.17. Next he climbed on top of his car (whose roof was 5 feet off the ground) and threw the ball the same way again. On this grid, draw the graph that shows the progress of the ball when it is thrown from the roof of the car.

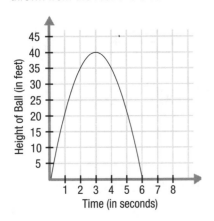

Figure 11.1.17

35. The graph of the function $y = x^2$ is shown on the grid in Figure 11.1.18. Graph the other parabolas whose equations are shown below.

a. $y = x^2 - 3$

b. $y = x^2 - 6$

c. $y = x^2 + 2$

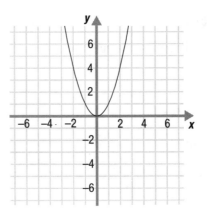

Figure 11.1.18

36. Find the domain and range of each of the functions below.

a. $y = x^2 - 2$

b. $y = (x - 2)^2$

 Explore

37. Determine which of the graphs below are functions by using the vertical line test.

a. 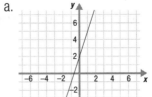   b.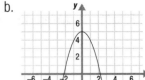

c. 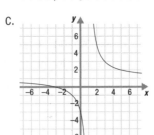   d.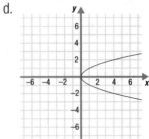

e.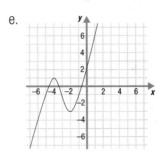

38a. If $f(x) = x$, find $f(6)$.

  b. If $f(x) = x + 2$, find $f(3)$.

  c. If $f(x) = 1 - x$, find $f(-2)$.

  d. If $f(x) = 2x + 7$, find $f(-4)$.

  e. If $f(x) = |x - 1|$, find $f(5)$.

39. Two reservoirs are filled at the rates shown in Figure 11.1.19. After 5 days, which reservoir has received the most water?

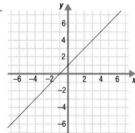

Figure 11.1.19

40. Determine the domain and range of each of the functions whose graphs are shown.

a.
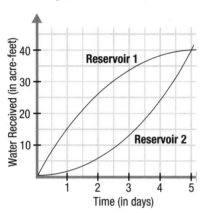

b.
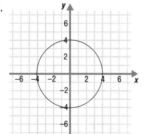

c.
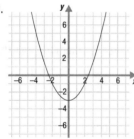

41a. If $f(x) = 3 - x$, find $f(3)$.

  b. If $f(x) = x^2 + 4$, find $f(1)$.

  c. If $f(x) = |5 - x|$, find $f(8)$.

  d. If $f(x) = 2 - x^3$, find $f(-2)$.

  e. If $f(x) = 2$, find $f(7)$.

42. If the two curves in Figure 11.1.20 represent the speed of two cars in the first minute of travel, where are the cars, relative to each other, after one minute?

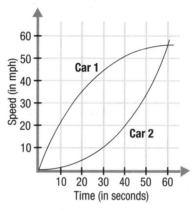

Figure 11.1.20

 **APPLY**

Practice Problems

Here are some additional practice problems for you to try.

Functions and Graphs

1. Which of the points below satisfy the function $y = 4x + 5$?

   (0, –5) (1, 9)

   (–1, –9) $\left(-\frac{5}{4}, 0\right)$

   (2, 13)

2. Which of the points below satisfy the function $y = 3x - 6$?

   (2, 0) (0, –6)

   (3, –3) (–1, –3)

   (1, –3)

3. Which of the points below satisfy the function $y = 6x - 2$?

   (0, 2) (1, 4)

   (–1, –4) $\left(\frac{1}{3}, 0\right)$

   (2, 10)

4. Which of the points below satisfy the equation $y = x^2 - 4x + 2$?

   (0, 2) (1, 1)

   (0, 0) (–1, 7)

   (2, –2)

5. Which of the points below satisfy the equation $y = -x^2 + 3x - 5$

   (0, 5) (1, –3)

   (–1, –3) (–1, –9)

   (2, –3)

6. Which of the points below satisfy the function $g(x) = x^2 - 2x + 3$?

   (0, 0) (1, 2)

   (–1, 4) (0, 3)

   (2, 11)

7. Make a table of at least three ordered pairs that satisfy the equation $y = 3x - 2$. Then use your table to graph the function.

8. Make a table of at least three ordered pairs that satisfy the equation $y = -x + 3$. Then use your table to graph the function.

9. Make a table of at least three ordered pairs that satisfy the equation $y = 2x - 4$. Then use your table to graph the function.

10. Make a table of at least seven ordered pairs that satisfy the equation $y = \frac{1}{3}x^2 - 3$. Then use your table to graph the function.

11. Make a table of at least seven ordered pairs that satisfy the equation $y = -x^2 + 4$. Then use your table to graph the function.

12. Make a table of at least seven ordered pairs that satisfy the equation $y = \frac{1}{2}x^2 - 2$. Then use your table to graph the function.

13. For the function $f(x) = 2x^2 - 3x$, calculate:

    a. $f(-2)$

    b. $f(-1)$

    c. $f(0)$

    d. $f(1)$

    e. $f(2)$

14. For the function $g(t) = -3t^2 + t - 4$, calculate:

   a. $g(-2)$

   b. $g(-1)$

   c. $g(0)$

   d. $g(1)$

   e. $g(2)$

15. For the function $g(t) = 2t^2 + 5t$, calculate:

   a. $g(-2)$

   b. $g(-1)$

   c. $g(0)$

   d. $g(1)$

   e. $g(2)$

16. For the function $h(s) = \sqrt{3s + 6}$, calculate:

   a. $h(-2)$

   b. $h(0)$

   c. $h(1)$

   d. $h(2)$

   e. $h(10)$

17. For the function $f(x) = \sqrt{2x + 4}$, calculate:

   a. $f(0)$

   b. $f(-1)$

   c. $f(-2)$

   d. $f(2)$

   e. $f(4)$

18. For the function $f(x) = \dfrac{2x - 1}{x + 3}$, calculate:

   a. $f\left(\dfrac{1}{2}\right)$

   b. $f(0)$

   c. $f(3)$

   d. $f(-1)$

   e. $f(-3)$

19. Use the graph of the function $y = 4x^2 - 16$, shown on the grid below, to find its domain and range.

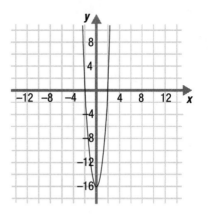

20. Use the graph of the function $y = -x^2 + 9$, shown on the grid below, to find its domain and range.

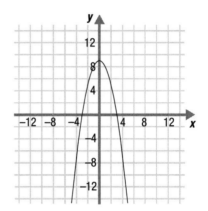

21. Use the graph of the function $y = 3x^2 + 9$, shown on the grid below, to find its domain and range.

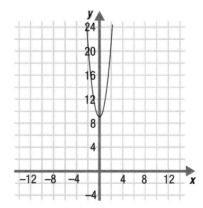

22. Use the graph of the function $y = \sqrt{2x - 3}$, shown on the grid below, to find its domain and range.

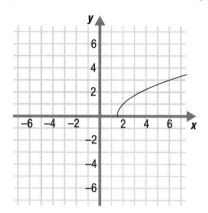

23. Use the graph of the function $y = \sqrt{3x} + 2$, shown on the grid below, to find its domain and range.

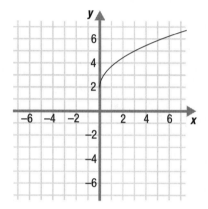

24. Use the graph of the function $y = \sqrt{2x} - 3$, shown on the grid below, to find its domain and range.

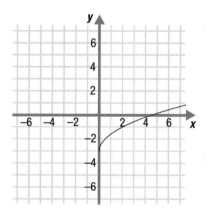

25. Use the graph of the function $y = -3x^2 + 3$, shown on the grid below, to find its domain and range.

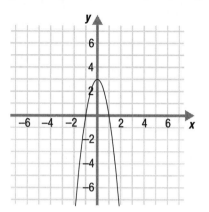

26. Use the graph of the function $y = -2x^2 + 4$, shown on the grid below, to find its domain and range.

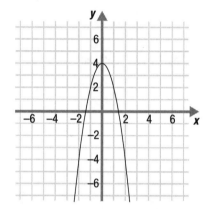

27. Use the graph of the function $y = \frac{3x - 2}{2x + 1}$, shown on the grid below, to find its domain and range.

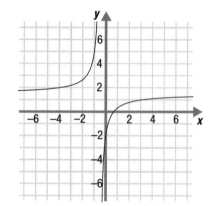

28. Use the graph of the function $y = \frac{2x + 1}{2x - 3}$, shown on the grid below, to find its domain and range.

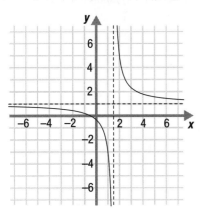

Linear Functions

29. Circle the functions below that are linear.

$f(x) = x^2 - 1$ $f(x) = 3x + 5$ $f(x) = 5^x$

$f(x) = \frac{x}{3} + 1$ $f(x) = 7 - \frac{3}{x}$

30. Circle the functions below that are linear.

$f(x) = 7 - 4x$ $f(x) = x^5 + 7x - 2$

$f(x) = 2^{x+1}$ $f(x) = x^{-1} + 3$ $f(x) = \frac{x}{5} - 3$

31. Circle the functions below that are linear.

$f(x) = 2x - 1$ $f(x) = x^4 - 15x^2 - 16$

$f(x) = \frac{x}{2}$ $f(x) = 10^x$ $f(x) = \frac{4}{x} + 3$

32. The graph of the function $y = \frac{1}{2}x$ is shown on the grid below.

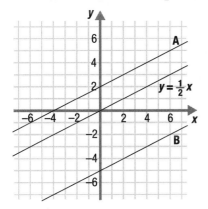

a. Find the equation of line A.

b. Find the equation of line B.

33. The graph of the function $y = -\frac{1}{3}x$ is shown on the grid below.

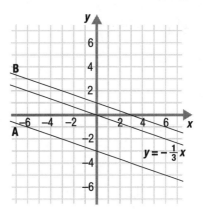

a. Find the equation of line A.

b. Find the equation of line B.

34. The graph of the function $y = |x - 3|$ is shown on the grid below.

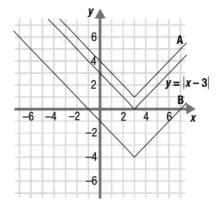

a. Find the equation of A.

b. Find the equation of B.

35. The graph of the function $y = |x + 2|$ is shown on the grid below.

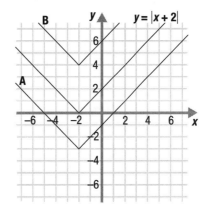

a. Find the equation of A.

b. Find the equation of B.

36. Graph the line $y = 2x + 4$.

37. Graph the line $y = -\frac{1}{3}x - 4$.

38. Graph the line $y = -\frac{1}{2}x + 3$.

39. Graph the function $y = |x| + 2$.

40. Graph the function $y = -|x| - 3$

41. Graph the function $y = -|x| + 1$.

42. Graph the function $y = |x + 1|$.

43. Graph the function $y = -|x - 2|$.

44. Graph the function $y = |x - 3|$.

45. Find the domain and range of the function $y = 7x + 9$.

46. Find the domain and range of the function $y = 10 - \frac{1}{5}x$.

47. Find the domain and range of the function $y = 5x - 2$.

48. Find the domain and range of the function $y = |x| + 5$.

49. Find the domain and range of the function $y = -|x + 4|$.

50. Find the domain and range of the function $y = -|x| - 1$.

51. The graph of the line $y = 2x$ is shown on the grid below. Graph the line whose equation is $y = 2x + 1$ on the same grid.

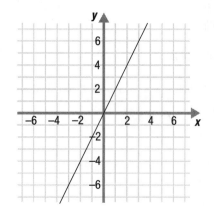

52. The graph of the line $y = -3x$ is shown on the grid below. Graph the line whose equation is $y = -3x + 2$ on the same grid.

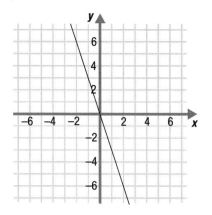

53. The graph of the line $y = \frac{1}{4}x$ is shown on the grid below. Graph the line whose equation is $y = \frac{1}{4}x - 5$ on the same grid.

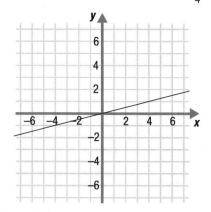

54. The graph of the function $y = |x + 1|$ is shown on the grid below. Graph the function $y = |x + 1| - 2$ on the same grid.

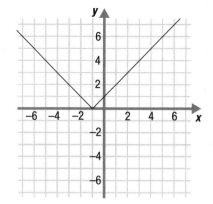

55. The graph of the function $y = |x - 2|$ is shown on the grid below. Graph the function $y = |x - 2| + 3$ on the same grid.

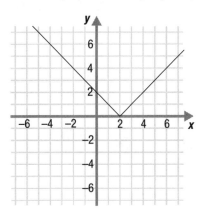

56. The graph of the function $y = -|x + 1|$ is shown on the grid below. Graph the function $y = -|x + 1| - 2$ on the same grid.

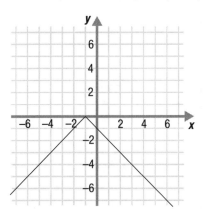

57. Circle the functions below that are quadratic.

$y = x^2 - 5$ $y = -x^2$

$y = 3x + 7$ $y = x^3 + 2x^2 - 6$

$y = 9 - x^2$

58. Circle the functions below that are quadratic.

$y = x + 10$ $y = -5x^2 + 3x - 7$

$y = 3x^2 + 2$ $y = 7 - 2x - 10x^2$

$y = 4 - x$

59. Circle the functions below that are quadratic.

$y = x^2 + 4$ $y = -5x^2 - 9$

$y = 2x - 6$ $y = 2x^3 + 4x^2 + x + 1$

$y = 3 - 2x - x^2$

60. Determine whether each of the following parabolas opens up or down.

a. $y = 3x^2$

b. $y = -\frac{1}{2}x^2$

c. $y = 8 - 3x + 7x^2$

d. $y = 25x^2 - 16$

e. $y = 12 - 3x^2$

61. Determine whether each of the following parabolas opens up or down.

a. $y = \frac{6}{7}x^2$

b. $y = 2.34x^2 - 3.7x$

c. $y = 9 + 4x - 2x^2$

d. $y = 49x^2 - 64$

e. $y = 32 - 2x^2$

62. Determine whether each of the following parabolas opens up or down.

a. $y = -2x^2$

b. $y = \frac{5}{7}x^2 - 3x + 7$

c. $y = 4 - 3x - 5x^2$

d. $y = 6x^2 - 9$

e. $y = 2x - x^2$

63. Use the graph of the function $y = x^2 - 6$, shown on the grid below, to find its domain and range.

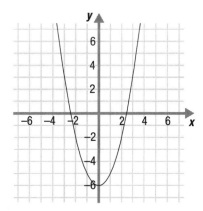

64. Use the graph of the function $y = 6 - x^2$, shown on the grid below, to find its domain and range.

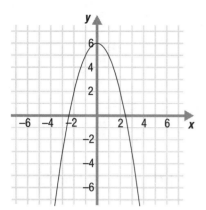

65. Use the graph of the function $y = x^2 + 9$, shown on the grid below, to find its domain and range.

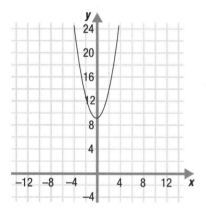

66. The graph of the function $y = x^2$ is shown on the grid below.

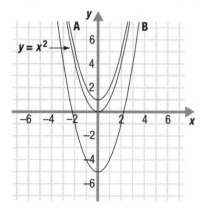

a. Find the equation of parabola A.

b. Find the equation of parabola B.

67. The graph of the function $y = -x^2$ is shown on the grid that follows.

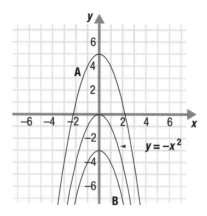

a. Find the equation of parabola A.

b. Find the equation of parabola B.

68. The graph of the function $y = (x + 3)^2$ is shown on the grid below.

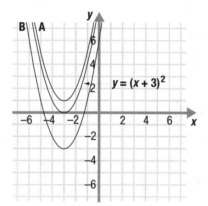

a. Find the equation of parabola A.

b. Find the equation of parabola B.

69. The graph of the function $y = (x + 2)^2$ is shown on the grid below.

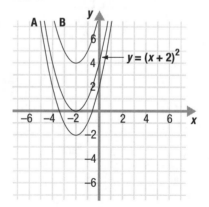

a. Find the equation of parabola A.

b. Find the equation of parabola B.

70. The graph of the function $y = x^2 - 2x$ is shown on the grid below. On the same grid, graph the parabola whose equation is $y = x^2 - 2x + 3$.

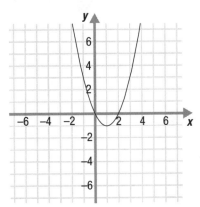

71. The graph of the function $y = -x^2 - 3x$ is shown on the grid below. On the same grid, graph the parabola whose equation is $y = -x^2 - 3x + 4$.

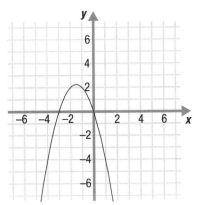

72. The graph of the function $y = -x^2 + 2x$ is shown on the grid below. On the same grid, graph the parabola whose equation is $y = -x^2 + 2x - 3$.

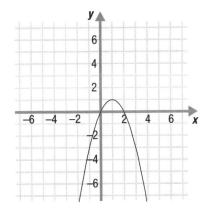

73. Find the x- and y-intercepts of the function $y = x^2 + 3x - 10$.

74. Find the x- and y-intercepts of the function $y = 2x^2 - 18$.

75. Find the x- and y-intercepts of the function $y = x^2 + 2x - 15$.

76. The graphs of the parabolas $y = x^2 + 1$ and $y = (x - 3)^2 + 1$ are shown on the grid below. Use these graphs to decide which of the statements below are true.

   a. Both functions have the same domain.

   b. Both functions have the same range.

   c. Both graphs have the same vertex.

   d. Both graphs have the same y-intercepts.

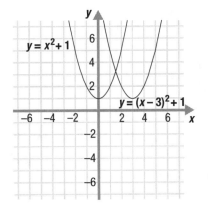

77. The graphs of the parabolas $y = x^2 - 1$ and $y = (x + 3)^2 - 1$ are shown on the grid below. Use these graphs to decide which of the statements below are true.

   a. Both functions have the same domain.

   b. Both functions have the same range.

   c. Both graphs have the same vertex.

   d. Goth graphs have the same y-intercepts.

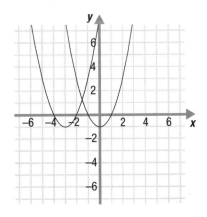

78. The graphs of the parabolas $y = -(x - 2)^2 - 1$ and $y = (x - 2)^2 - 1$ are shown on the grid below. Use these graphs to decide which of the statements below are true.

a. Both functions have the same domain.

b. Both functions have the same range.

c. Both graphs have the same vertex.

d. Both graphs have the same y-intercepts.

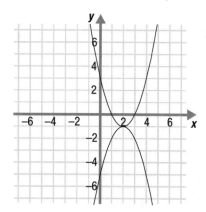

79. The graph of the parabola $y = x^2 - 6x + 8$ is shown on the grid below. If the graph is moved down 4 units, what is the vertex of the new parabola?

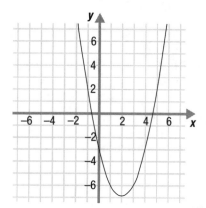

80. The graph of the parabola $y = -x^2 - 4x - 3$ is shown on the grid below. If the graph is moved up 3 units, what is the vertex of the new parabola?

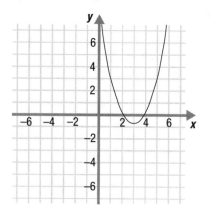

81. The graph of the parabola $y = x^2 - 8x + 12$ is shown on the grid below. If the graph is moved up 3 units, what is the vertex of the new parabola?

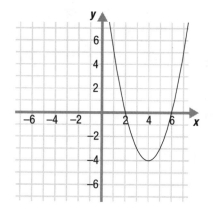

82. If the graph of the parabola $y = x^2 - 6x + 5$ is moved down 3 units, what is the equation of the new parabola?

83. If the graph of the parabola $y = x^2 + 4x + 3$ is moved up 3 units, what is the equation of the new parabola?

84. If the graph of the parabola $y = -x^2 + 6x - 5$ is moved down 2 units, what is the equation of the new parabola?

 EVALUATE

Practice Test

Take this practice test to be sure that you are prepared for the final quiz in Evaluate.

1. Which of the points below satisfy the function $y = 3x^2 + 1$?

    (0, 1)   (−2, 13)

    (3, 8)   (1, 4)

    (−4, 11)

2. Given the function $f(x) = x^2 + 3x$, find:

    a. $f(0)$

    b. $f(3)$

    c. $f(−2)$

    d. $f(−5)$

3. Use the graph of the function $y = x^2 − 2$, shown on the grid in Figure 11.1.21, to find its domain and range.

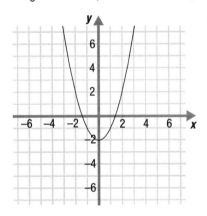

Figure 11.1.21

4. Given the functions $f(x) = 4x − 5$ and $g(x) = 1 − 3x$, find:

    a. $f(7)$

    b. $g(7)$

    c. $f(−2)$

    d. $g(−2)$

5. The graph of the function $y = 2x$ is shown on the grid in Figure 11.1.22.

    a. Find the equation of line A.

    b. Find the equation of line B.

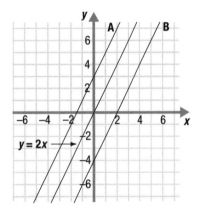

Figure 11.1.22

6. The graph of the function $y = |x − 5|$ is shown on the grid in Figure 11.1.23. Use this graph to find the domain and range of $y = |x − 5|$.

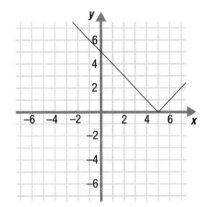

Figure 11.1.23

7. The graphs of the functions $y = |x| + 3$ and $y = -|x| + 3$ are shown on the grid in Figure 11.1.24. Use these graphs to decide which of the statements below are true.

Both functions have the same domain.

Both functions have the same range.

The point $(0, 3)$ satisfies both equations.

The point $(3, 0)$ satisfies both equations.

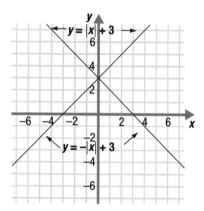

Figure 11.1.24

8. The graph of the function $y = x$ is shown on the grid in Figure 11.1.25. Circle the points below that lie on the graph of $y = x - 4$.

(5, 1)

(2, −2)

(1, 3)

(0, −4)

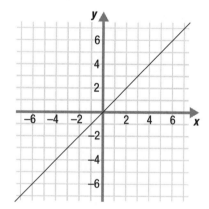

Figure 11.1.25

9. The graph of the parabola $y = x^2 - 4x - 5$ is shown on the grid in Figure 11.1.26. Use this graph to help you find the x-intercepts of the function.

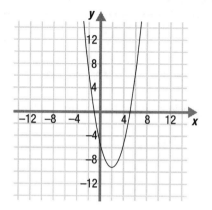

Figure 11.1.26

10. The graphs of the parabolas $y = x^2 - 4$ and $y = 2x^2 - 8$ are shown on the grid in Figure 11.1.27. Use these graphs to decide which of the statements below are true.

Both functions have the same domain.

Both functions have the same range.

Both graphs have the same vertex.

Both graphs have the same x-intercepts.

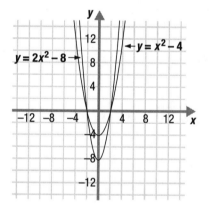

Figure 11.1.27

11. Circle the functions below that are **not** quadratic functions.

$y = 2x^2$

$y = 4x + 1$

$f(x) = -3x + 7$

$y = x^2 - 5x + 8$

$f(x) = -6 + x^2$

12. The graph of the parabola $y = x^2 + 4x + 3$ is shown on the grid in Figure 11.1.28. If the graph is moved up 3 units, what is the vertex of the new parabola?

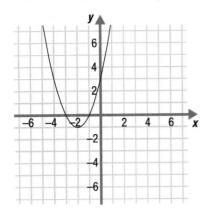

Figure 11.1.28

13. The graph of the function $f(x) = \frac{1}{4}x^3 - 2$ is shown on the grid in Figure 11.1.29. Use this graph to find the values of:

a. $f(2)$

b. $f(0)$

c. $f(-2)$

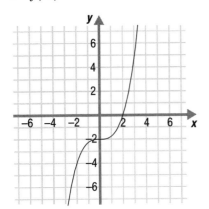

Figure 11.1.29

14. The graph of $y = 2|x - 1| + 4$ is shown on the grid in Figure 11.1.30. Use this graph to decide which of the statements below are true.

The domain of $y = 2|x - 1| + 4$ is all real numbers.

The range of $y = 2|x - 1| + 4$ is all real numbers.

For each input value, x, there is only one output value y.

For each output value, y, there is only one input value x.

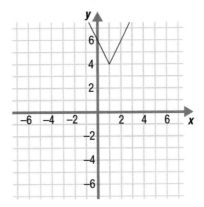

Figure 11.1.30

15. Determine which of the graphs below are functions by using the vertical line test.

a.
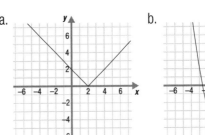

b.
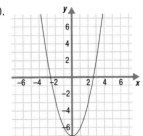

c.
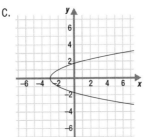

d.
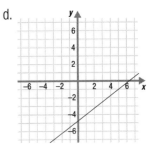

16. The graph of the function $y = -x^2 - 4x + 5$ is shown on the grid in Figure 11.1.31. Use this graph to:

a. Find the domain of the function $y = -x^2 - 4x + 5$.

b. Find the range of the function $y = -x^2 - 4x + 5$.

c. Find the vertex of the function $y = -x^2 - 4x + 5$.

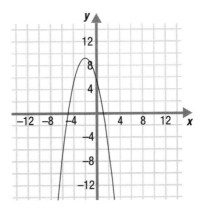

Figure 11.1.31

LESSON 11.2 – THE ALGEBRA OF FUNCTIONS

OVERVIEW

Here's what you'll learn in this lesson:

The Algebra of Functions

a. The sum and difference of functions

b. The product and quotient of functions

c. The composition of functions

Inverse Functions

a. Finding the inverse

b. One-to-one: checking whether a function has an inverse

c. Graphing inverse functions

Suppose you are a car salesman with a great idea for drawing people to your sales lot — you'll hire a balloonist to give free "tethered" balloon flights to anyone who test drives a new car!

In order to sell this idea to your boss, you've determined the cost of running the balloon promotion as a function of the number of days the promotion will be run. In addition, you've determined the cost of the liability insurance you'd have to pay. Again, this is a function of the number of days. In order to figure out the total cost of running the promotion, (so your boss can see what a good idea it really is), you add these two functions.

In this lesson, you will learn how to add functions. You will also learn how to subtract, multiply, and divide functions, and how to find the composition of two functions. In addition, you'll learn how to find and graph the inverse of a function.

 EXPLAIN

THE ALGEBRA OF FUNCTIONS

Summary

The Sum and Difference of Functions

In algebra, if you are given two functions f and g with the same domain, you can produce new functions called the sum $(f + g)$ or the difference $(f - g)$ of the two functions. The sum $(f + g)$ is defined as follows: $(f + g)(x) = f(x) + g(x)$. In other words, to get $(f + g)(x)$ you add the value $f(x)$ and the value $g(x)$. Similarly, the difference $(f - g)$ is defined as follows: $(f - g)(x) = f(x) - g(x)$. So to get $(f - g)(x)$ you subtract the value $g(x)$ from the value $f(x)$.

In general, to find the sum $(f + g)(x)$ of two functions $f(x)$ and $g(x)$ with the same domain:

1. Add the value $f(x)$ to the value $g(x)$.

2. Combine like terms, if any.

For example, if $f(x) = 2x^2 - 3x$ and $g(x) = x^2 + 2x$, to find $(f + g)(x)$:

1. Add the value $f(x)$ $(f + g)(x) = f(x) + g(x)$
    to the value $g(x)$. $= (2x^2 - 3x) + (x^2 + 2x)$

2. Combine like terms. $= 3x^2 - x$

So $(f + g)(x) = 3x^2 - x$.

The graphs of $y = f(x) = 2x^2 - 3x$, $y = g(x) = x^2 + 2x$, and $y = (f + g)(x) = 3x^2 - x$ are shown in Figure 11.2.1.

In general, to find the difference $(f - g)(x)$ of two functions $f(x)$ and $g(x)$ with the same domain:

1. Subtract the value $g(x)$ from the value $f(x)$.

2. Combine like terms, if any.

For example, if $f(x) = 2x^2 - 3x$ and $g(x) = x^2 + 2x$, to find $(f - g)(x)$:

1. Subtract the value $g(x)$ $(f - g)(x) = f(x) - g(x)$
    from the value $f(x)$. $= (2x^2 - 3x) - (x^2 + 2x)$
                                                    $= 2x^2 - 3x - x^2 - 2x$

2. Combine like terms. $= x^2 - 5x$

So $(f - g)(x) = x^2 - 5x$.

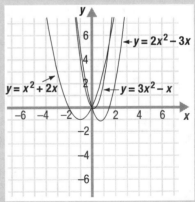

Figure 11.2.1

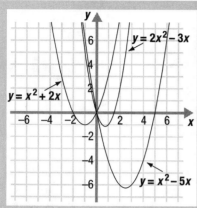

Figure 11.2.2

The graphs of $y = f(x) = 2x^2 - 3x$, $y = g(x) = x^2 + 2x$, and $y = (f - g)(x) = x^2 - 5x$ are shown in Figure 11.2.2.

Given two functions f and g and given a number $x = a$ in the domain of both f and g, you can evaluate the sum $(f + g)$ or the difference $(f - g)$ at $x = a$ using either one of two methods.

Method 1

To find the value $(f + g)(x)$ or $(f - g)(x)$ at $x = a$ using the first method:

1. Find $(f + g)(x)$ or $(f - g)(x)$.

2. Substitute a for x in $(f + g)(x)$ or $(f - g)(x)$.

3. Simplify.

For example, if $f(x) = 2x^2 - 3x$ and $g(x) = x^2 + 2x$, to find the value $(f + g)(x)$ at $x = 4$:

1. Find $(f + g)(x)$. $(f + g)(x) = 3x^2 - x$

2. Substitute 4 for x in $(f + g)(x)$. $(f + g)(4) = 3 \cdot 4^2 - 4$

3. Simplify. $= 48 - 4$

         $= 44$

So the value $(f + g)(x) = 3x^2 - x$ at $x = 4$ is 44.

As another example, if $f(x) = 2x^2 - 3x$ and $g(x) = x^2 + 2x$, to find the value $(f - g)(x)$ at $x = 4$:

1. Find $(f - g)(x)$. $(f - g)(x) = x^2 - 5x$

2. Substitute 4 for x in $(f - g)(x)$. $(f - g)(4) = 4^2 - 5 \cdot 4$

3. Simplify. $= 16 - 20$

         $= -4$

So the value $(f - g)(x) = x^2 - 5x$ at $x = 4$ is -4.

Method 2

To find the value $(f + g)(x)$ or $(f - g)(x)$ at $x = a$ using the second method:

1. Substitute a for x in $f(x)$ and a for x in $g(x)$.

2. Add or subtract the resulting values of the two functions.

You found $(f + g)(x)$ for these functions in an example above:

$(f + g)(x) = f(x) + g(x)$

$= (2x^2 - 3x) + (x^2 + 2x)$

$= 3x^2 - x$

You found $(f - g)(x)$ for these functions in an example above:

$(f - g)(x) = f(x) - g(x)$

$= (2x^2 - 3x) - (x^2 + 2x)$

$= 2x^2 - 3x - x^2 - 2x$

$= x^2 - 5x$

For example, if $f(x) = 2x^2 - 3x$ and $g(x) = x^2 + 2x$, to find the value $(f + g)(x)$ at $x = 4$:

1. Substitute 4 for x in both $f(x)$ and $g(x)$.

$$f(x) = 2x^2 - 3x \qquad\qquad g(x) = x^2 + 2x$$
$$f(4) = 2 \cdot 4^2 - 3 \cdot 4 \qquad g(4) = 4^2 + 2 \cdot 4$$
$$= 32 - 12 \qquad\qquad\qquad = 16 + 8$$
$$= 20 \qquad\qquad\qquad\qquad = 24$$

2. Add the resulting values.

$$f(4) + g(4) = 20 + 24$$
$$= 44$$

Again, the value $(f + g)(x) = f(x) + g(x)$ at $x = 4$ is 44.

As another example, if $f(x) = 2x^2 - 3x$ and $g(x) = x^2 + 2x$, to find the value $(f - g)(x)$ at $x = 4$:

1. Substitute 4 for x in both $f(x)$ and $g(x)$.

$$f(x) = 2x^2 - 3x \qquad\qquad g(x) = x^2 + 2x$$
$$f(4) = 2 \cdot 4^2 - 3 \cdot 4 \qquad g(4) = 4^2 + 2 \cdot 4$$
$$= 32 - 12 \qquad\qquad\qquad = 16 + 8$$
$$= 20 \qquad\qquad\qquad\qquad = 24$$

2. Subtract the resulting values.

$$f(4) - g(4) = 20 - 24$$
$$= -4$$

Again, the value $(f - g)(x) = f(x) - g(x)$ at $x = 4$ is -4.

The Product and Quotient of Functions

Given two functions f and g with the same domain, you can also produce new functions called the product $(f \cdot g)$ or quotient $\left(\dfrac{f}{g}\right)$ of the two functions. The product $(f \cdot g)$ is defined as follows: $(f \cdot g)(x) = f(x) \cdot g(x)$. The quotient $\left(\dfrac{f}{g}\right)$ is defined as follows: $\left(\dfrac{f}{g}\right)(x) = \dfrac{f(x)}{g(x)}$ (here, $g(x) \neq 0$).

In general, to find the product $(f \cdot g)(x)$ of two functions $f(x)$ and $g(x)$ with the same domain:

1. Multiply the values $f(x)$ and $g(x)$.

2. Combine like terms, if any.

For example, if $f(x) = x^2 - 2$ and $g(x) = x^2 + 1$, to find $(f \cdot g)(x)$:

1. Multiply the values $f(x)$ and $g(x)$.

$$(f \cdot g)(x) = f(x) \cdot g(x)$$
$$= (x^2 - 2) \cdot (x^2 + 1)$$
$$= x^2 \cdot x^2 + x^2 \cdot 1 - 2 \cdot x^2 - 2 \cdot 1$$
$$= x^4 + x^2 - 2x^2 - 2$$

2. Combine like terms.

$$= x^4 - x^2 - 2$$

So, $(f \cdot g)(x) = x^4 - x^2 - 2$.

Remember, you can use this pattern to multiply two binomials:

$$(a + b)(c + d) = ac + ad + bc + bd$$

In general, to find the quotient $\left(\frac{f}{g}\right)(x)$ of two functions $f(x)$ and $g(x)$ with the same domain, and with $g(x) \ne 0$:

1. Divide the value $f(x)$ by the value $g(x)$.

2. Simplify, if possible.

Here, we assume $g(x) = x + 2 \ne 0$. In other words, $x \ne -2$.

For example, to find $\left(\frac{f}{g}\right)(x)$ if $f(x) = x^2 - 4$ and $g(x) = x + 2$:

1. Divide the value $f(x)$ by the value $g(x)$.

$$\left(\frac{f}{g}\right)(x) = \frac{f(x)}{g(x)}$$

$$= \frac{x^2 - 4}{x + 2}$$

2. Simplify.

$$= \frac{(x - 2)(x + 2)}{x + 2}$$

$$= \frac{(x - 2)\overset{1}{\cancel{(x + 2)}}}{\underset{1}{\cancel{x + 2}}}$$

$$= x - 2$$

So, $\left(\frac{f}{g}\right)(x) = x - 2$.

Now you can evaluate the product $(f \cdot g)$ or quotient $\left(\frac{f}{g}\right)$ of two functions f and g at a given value $x = a$ using either one of two methods, just as you can when evaluating the sum or difference of two functions at $x = a$.

Method 1

To find the value $(f \cdot g)(x)$ or $\left(\frac{f}{g}\right)(x)$ at $x = a$ using the first method:

1. Find $(f \cdot g)(x)$ or $\left(\frac{f}{g}\right)(x)$.

2. Substitute a for x in $(f \cdot g)(x)$ or $\left(\frac{f}{g}\right)(x)$.

3. Simplify.

You found $(f \cdot g)(x)$ for these functions in an example above:

$(f \cdot g)(x) = f(x) \cdot g(x)$

$= (x^2 - 2) \cdot (x^2 + 1)$

$= x^2 \cdot x^2 + x^2 \cdot 1 - 2 \cdot x^2 - 2 \cdot 1$

$= x^4 + x^2 - 2x^2 - 2$

$= x^4 - x^2 - 2$

For example, if $f(x) = x^2 - 2$ and $g(x) = x^2 + 1$, to find the value $(f \cdot g)(x)$ at $x = -2$:

1. Find $(f \cdot g)(x)$.

$(f \cdot g)(x) = x^4 - x^2 - 2$

2. Substitute -2 for x in $(f \cdot g)(x)$.

$(f \cdot g)(-2) = (-2)^4 - (-2)^2 - 2$

3. Simplify.

$= 16 - 4 - 2$

$= 10$

So the value $(f \cdot g)(x) = x^4 - x^2 - 2$ at $x = -2$ is 10.

As another example, if $f(x) = x^2 - 4$ and $g(x) = x + 2$, to find the value $\left(\dfrac{f}{g}\right)(x)$ at $x = -3$:

1. Find $\left(\dfrac{f}{g}\right)(x)$. $\qquad\qquad\qquad\qquad$ $\left(\dfrac{f}{g}\right)(x) = x - 2$

2. Substitute -3 for x in $\left(\dfrac{f}{g}\right)(x)$. $\qquad$ $\left(\dfrac{f}{g}\right)(-3) = -3 - 2$

3. Simplify. $\qquad\qquad\qquad\qquad\qquad\qquad$ $= -5$

So the value $\left(\dfrac{f}{g}\right)(x) = x - 2$ at $x = -3$ is -5.

Method 2

To find the value $(f \cdot g)(x)$ or $\left(\dfrac{f}{g}\right)(x)$ at $x = a$ using the second method:

1. Substitute a for x in $f(x)$ and a for x in $g(x)$.

2. Multiply or divide the resulting values of the two functions.

For example, if $f(x) = x^2 - 2$ and $g(x) = x^2 + 1$, to find the value $(f \cdot g)(x)$ at $x = -2$:

1. Substitute -2 for x $\qquad$ $f(x) = x^2 - 2$ $\qquad$ $g(x) = x^2 + 1$
   in both $f(x)$ and $g(x)$. $\qquad$ $f(-2) = (-2)^2 - 2$ $\qquad$ $g(-2) = (-2)^2 + 1$
   $\qquad\qquad\qquad\qquad\qquad\qquad$ $= 4 - 2$ $\qquad\qquad\qquad$ $= 4 + 1$
   $\qquad\qquad\qquad\qquad\qquad\qquad$ $= 2$ $\qquad\qquad\qquad\quad$ $= 5$

2. Multiply the resulting values. $\qquad$ $f(-2) \cdot g(-2) = 2 \cdot 5$
   $\qquad\qquad\qquad\qquad\qquad\qquad\qquad\qquad$ $= 10$

Again, the value $(f \cdot g)(x) = f(x) \cdot g(x)$ at $x = -2$ is 10.

As another example, if $f(x) = x^2 - 4$ and $g(x) = x + 2$, to find the value $\left(\dfrac{f}{g}\right)(x)$ at $x = -3$:

1. Substitute -3 for x $\qquad$ $f(x) = x^2 - 4$ $\qquad$ $g(x) = x + 2$
   in both $f(x)$ and $g(x)$. $\qquad$ $f(-3) = (-3)^2 - 4$ $\qquad$ $g(-3) = -3 + 2$
   $\qquad\qquad\qquad\qquad\qquad\qquad$ $= 9 - 4$ $\qquad\qquad\qquad$ $= -1$
   $\qquad\qquad\qquad\qquad\qquad\qquad$ $= 5$

2. Divide the resulting values. $\qquad\qquad$ $\dfrac{f(-3)}{g(-3)} = \dfrac{5}{-1}$
   $\qquad\qquad\qquad\qquad\qquad\qquad\qquad\qquad$ $= -5$

So, again, the value $\left(\dfrac{f}{g}\right)(x) = \dfrac{f(x)}{g(x)}$ at $x = -3$ is -5.

The Composition of Functions

You have seen how to define addition, subtraction, multiplication, and division of two functions f and g to produce new functions $(f + g)$, $(f - g)$, $(f \cdot g)$, and $\left(\dfrac{f}{g}\right)$, respectively. You can also combine two functions f and g in yet another way. This new way is called the composition of f and g. It is denoted by $(f \circ g)$, and is defined as follows: $(f \circ g)(x) = f[g(x)]$. Observe that here the output of the function g is the input of the function f.

You found $\left(\dfrac{f}{g}\right)(x)$ for these functions in an example above:

$$\left(\dfrac{f}{g}\right)(x) = \dfrac{f(x)}{g(x)}$$
$$= \dfrac{x^2 - 4}{x + 2}$$
$$= \dfrac{(x - 2)(x + 2)}{x + 2}$$
$$= \dfrac{(x - 2)\overset{1}{\cancel{(x + 2)}}}{\underset{1}{\cancel{x + 2}}}$$
$$= x - 2$$

Here, we assume $x \neq 2$.

When you compose two functions f and g to form $f \circ g$, the range of g must be contained in the domain of f. In other words, all the possible outputs for g must be inputs for f.

In general, to find the composition $(f \circ g)(x)$ of two functions $f(x)$ and $g(x)$:

1. Use the output of the function g as the input of the function f. That is, evaluate the function f at the value $g(x)$.

2. Simplify, if possible.

For example, if $f(x) = x^2 - 2$ and $g(x) = 2x - 1$, to find $(f \circ g)(x)$:

1. Evaluate f at $g(x)$.
$$(f \circ g)(x) = f[g(x)]$$
$$= f(2x - 1)$$
$$= (2x - 1)^2 - 2$$

2. Simplify.
$$= (2x)^2 - 2(1)(2x) + (-1)^2 - 2$$
$$= 4x^2 - 4x + 1 - 2$$
$$= 4x^2 - 4x - 1$$

So, $(f \circ g)(x) = 4x^2 - 4x - 1$.

As another example, if $f(x) = 3x^2 - 4$ and $g(x) = x + 2$, to find $(f \circ g)(x)$:

1. Evaluate f at $g(x)$.
$$(f \circ g)(x) = f[g(x)]$$
$$= f(x + 2)$$
$$= 3(x + 2)^2 - 4$$

2. Simplify.
$$= 3[x^2 + 2(2)(x) + 2^2] - 4$$
$$= 3[x^2 + 4x + 4] - 4$$
$$= 3x^2 + 12x + 12 - 4$$
$$= 3x^2 + 12x + 8$$

So, $(f \circ g)(x) = 3x^2 + 12x + 8$.

You can evaluate the composition $(f \circ g)$ of two functions f and g at a given value $x = a$ using either one of two methods, just as you can when evaluating the sum or difference, or the product or quotient of two functions at $x = a$.

Method 1

To find the value $(f \circ g)(x)$ at $x = a$ using the first method:

1. Find $(f \circ g)(x)$.

2. Substitute a for x in $(f \circ g)(x)$.

3. Simplify.

Remember, this is the pattern for a perfect square trinomial:

$$(a - b)^2 = a^2 - 2ba + b^2$$

For example, if $f(x) = x^2 - 2$ and $g(x) = 2x - 1$, to find the value $(f \circ g)(x)$ at $x = -1$:

You found $(f \circ g)(x)$ for these functions in an example above:

$$(f \circ g)(x) = f[g(x)]$$
$$= f(2x - 1)$$
$$= (2x - 1)^2 - 2$$
$$= (2x)^2 - 2(1)(2x) + (-1)^2 - 2$$
$$= 4x^2 - 4x + 1 - 2$$
$$= 4x^2 - 4x - 1$$

1. Find $(f \circ g)(x)$. $\quad (f \circ g)(x) = 4x^2 - 4x - 1$

2. Substitute -1 for x in $(f \circ g)(x)$. $\quad (f \circ g)(-1) = 4(-1)^2 - 4(-1) - 1$

3. Simplify. $\quad = 4 + 4 - 1$

$$= 7$$

So the value $(f \circ g)(x) = 4x^2 - 4x - 1$ at $x = -1$ is 7.

Method 2

To find the value $(f \circ g)(x)$ at $x = a$ using the second method:

1. Substitute a for x in $g(x)$.

2. Evaluate f at $g(a)$.

For example, if $f(x) = x^2 - 2$ and $g(x) = 2x - 1$, to find the value $(f \circ g)(x)$ at $x = -1$:

1. Substitute -1 for x in $g(x)$. $\quad g(x) = 2x - 1$

$$g(-1) = 2(-1) - 1$$
$$= -2 - 1$$
$$= -3$$

2. Find $f[g(-1)]$. $\quad f[g(-1)] = f(-3)$

$$= (-3)^2 - 2$$
$$= 9 - 2$$
$$= 7$$

So, again, the value $(f \circ g)(x) = f[g(x)]$ at $x = -1$ is 7.

Sample Problems

1. Given $f(x) = 3x^2 + 4$ and $g(x) = -6x^2 - 5$, find $(f + g)(x)$.

   ☑ a. Add the value $f(x)$ to the value $g(x)$.

   $(f + g)(x) = f(x) + g(x)$
   $= (3x^2 + 4) + (-6x^2 - 5)$

   ☐ b. Combine like terms.

   $=$ _____

2. Given $f(x) = 2x^3$ and $g(x) = -x^2 + 5x$, find $(f \cdot g)(x)$.

   ☐ a. Multiply the values $f(x)$ and $g(x)$.

   $(f \cdot g)(x) = f(x) \cdot g(x)$
   $=$ _____

   ☐ b. Simplify.

   $=$ _____

3. Given $f(x) = 4x^2 - 9$ and $g(x) = 3 + 2x$, evaluate $\left(\dfrac{f}{g}\right)(x)$ at $x = 1$.

   ☑ a. Divide the value $f(x)$ by the value $g(x)$.

   $\left(\dfrac{f}{g}\right)(x) = \dfrac{f(x)}{g(x)}$

   $= \dfrac{4x^2 - 9}{3 + 2x}$

   ☐ b. Simplify.

   $=$ _____

   ☐ c. Find $\left(\dfrac{f}{g}\right)(1)$.

   $=$ _____

4. Given $f(x) = 4x^2 - 9$ and $g(x) = 3 + 2x$, evaluate $(f \circ g)(x)$ at $x = -2$.

   ☑ a. Substitute -2 for x in $g(x)$.

   $g(x) = 3 + 2x$
   $g(-2) = 3 + 2(-2)$
   $= 3 - 4$
   $= -1$

   ☐ b. Substitute -1 for x in $f(x)$ to find $f[g(-2)]$, and simplify.

   $f(x) = 4x^2 - 9$

   $f[g(-2)] = f(-1) = $ _____

INVERSE FUNCTIONS

Summary

Finding an Inverse Function and the Equation of an Inverse

Given two functions f and g with a common domain, you have seen how to produce new functions by combining these functions in various ways. Now you'll learn about another function, the inverse function.

Suppose you have a function $f(x) = 3x$ and a function $g(x) = \frac{1}{3}x$. The graphs of $y = f(x) = 3x$ and $y = g(x) = \frac{1}{3}x$ are shown in Figure 11.2.3. Notice that f and g are functions since for each function each input value is assigned to exactly one output value.

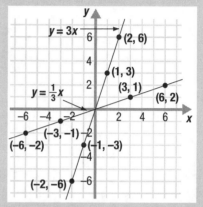

Figure 11.2.3

*Remember, another way to tell whether or not a graph is the graph of a function is to draw vertical lines anywhere on the graph. If any vertical line intersects the graph in more than one place, the graph is **not** the graph of a function.*

Here is a table of some of the ordered pairs that satisfy f:

Here is a table of some of the ordered pairs that satisfy g:

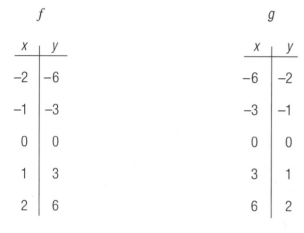

The y-values of f are the x-values of g, and the x-values of f are the y-values of g. Notice that f takes an input, x, and multiplies it by 3 to get the output, y; whereas $g(x)$ takes an input, x, and multiplies it by $\frac{1}{3}$ to get the output, y. So the function g "undoes" what the function f does—that is, the output of g is the original input of f. The function g is called the inverse of the function f. You can write this as $f^{-1}(x) = g(x) = \frac{1}{3}x$.

In general, when you find the inverse of a function f, if (x, y) is an ordered pair of f, then (y, x) is an ordered pair of f^{-1}.

To find the equation of the inverse, f^{-1}, of the function f:

1. Replace $f(x)$ with y.

2. In the equation, switch y and x.

3. Solve the equation for y.

4. Replace y with $f^{-1}(x)$.

*Be careful! Here the −1 in $f^{-1}(x)$ is just a notation. It is **not** an exponent, so $f^{-1}(x) \neq \frac{1}{f(x)}$.*

For example, to find the inverse of $f(x) = 5x - 3$:

1. Replace $f(x)$ with y. $y = 5x - 3$

2. Switch y and x. $x = 5y - 3$

3. Solve for y. $x + 3 = 5y$

$$\frac{x + 3}{5} = y$$

$$y = \frac{x + 3}{5}$$

4. Replace y with $f^{-1}(x)$. $f^{-1}(x) = \frac{x + 3}{5}$

So, if $f(x) = 5x - 3$, then $f^{-1}(x) = \frac{x + 3}{5}$.

Here's a slightly different example. If $f(x) = 5x - 3$, you now know that $f^{-1}(x) = \frac{x + 3}{5}$. So to find $(f^{-1} \circ f)(x)$:

1. Find f^{-1} at $f(x)$.

$$(f^{-1} \circ f)(x) = f^{-1}[f(x)]$$

$$= f^{-1}[5x - 3]$$

$$= \frac{(5x - 3) + 3}{5}$$

2. Simplify.

$$= \frac{5x}{5}$$

$$= x$$

So, if $f(x) = 5x - 3$, then $f^{-1}(x) = \frac{x + 3}{5}$, and $(f^{-1} \circ f)(x) = x$.

In general, if the inverse f^{-1} of a function f exists, then

$$(f^{-1} \circ f)(x) = f^{-1}[f(x)] = x$$

for every x in the domain of f and

$$(f \circ f^{-1})(x) = f[f^{-1}(x)] = x$$

for every x in the domain of f^{-1}.

That is, the inverse function f^{-1} "undoes" what the original function f does, and f "undoes" what the inverse function f^{-1} does.

One-to-One: Checking Whether a Function Has an Inverse

Some functions do not have inverses. As an example, if you look at the table of ordered pairs (x, y) for the function f below and interchange its x-values and y-values, you can get a new table whose ordered pairs do not represent a function.

Remember: A function is a rule that assigns to each input value exactly one output value.

Here is a table of some ordered pairs that satisfy $f(x) = 2x^2$:

x	y
−2	8
−1	2
0	0
1	2
2	8

Here is a table of these ordered pairs with the x- and y-values interchanged:

x	y
8	−2
2	−1
0	0
2	1
8	2

In the second table, when $x = 2$, $y = -1$ **and** $y = 1$. So the ordered pairs in the second table do not represent a function. If you look back at the table of ordered pairs for f, you see that when $y = 8$, $x = -2$ **and** $x = 2$. It turns out that for a function to have an inverse, each y-value of f must correspond to exactly one x-value. If this is true, the function is said to be one-to-one. The function $f(x) = 2x^2$ is not one-to-one, so it does not have an inverse.

A nice graphical way to tell if a function is one-to-one is to use the horizontal line test. This tests says that if every horizontal line crosses the graph of a function in at most one point, the function is one-to-one. So, for example, the function $y = f(x) = 2x^2$ shown in Figure 11.2.4 is **not** one-to-one since there is at least one horizontal line that crosses the graph in two places.

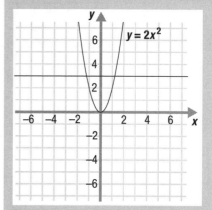

Figure 11.2.4

To determine if a function f is one-to-one, and hence, has an inverse:

1. Graph the function. Use a table of ordered pairs if necessary.

2. Use the horizontal line test.

For example, to determine whether or not the function $f(x) = x^2 - 1$ has an inverse:

1. Graph the function $y = f(x)$.   See the grid in Figure 11.2.5.

x	y
-2	3
-1	0
0	-1
1	0
2	3

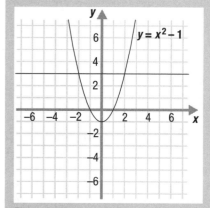

Figure 11.2.5

2. Use the horizontal line test.   See the grid in Figure 11.2.5.

Since you can find a horizontal line that crosses the graph in more than one place, the function $f(x) = x^2 - 1$ is **not** one-to-one, and hence does **not** have an inverse.

Graphing Inverse Functions

If a function f has an inverse f^{-1}, you can find that inverse by interchanging the x- and y-coordinates of f.

To graph the inverse of a function:

1. Graph the original function, $y = f(x)$.

2. Locate some points (a, b) on the graph of f, and then draw the points (b, a).

3. Connect the new points. This is the graph of $y = f^{-1}(x)$.

*If you make a table of values, you can also check the table to see if for every y-value in the table, there is exactly one corresponding x-value. As you can see in this table, when y = 3, x = −2 **and** x = 2, so this function is **not** one-to-one.*

If the graph of f is a line, then the graph of f⁻¹ is also a line and you only need two points to graph f⁻¹. But, if the graph of f is a more general curve, you will need more than two points to help you graph f⁻¹.

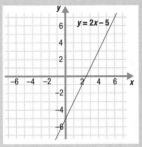

Figure 11.2.6

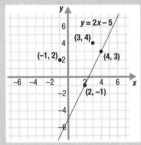

Figure 11.2.7

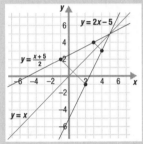

Figure 11.2.8

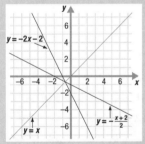

Figure 11.2.9

Answers to Sample Problems

b. $x = 3y + 4$

c. $y = \dfrac{x - 4}{3}$

d. $f^{-1}(x) = \dfrac{x - 4}{3}$

For example, to graph the inverse of $y = f(x) = 2x - 5$:

1. Graph the line $y = f(x) = 2x - 5$. See the grid in Figure 11.2.6.

2. Locate two points (a, b) on the graph of f, and two new points, (b, a). The points are shown in Figure 11.2.7.

3. Connect the new points. This is the graph of the inverse function, $y = f^{-1}(x) = \dfrac{x + 5}{2}$. See the grid in Figure 11.2.8.

Notice the line $y = x$ in Figure 11.2.8, and the line segments drawn from points (a, b) to points (b, a). The line $y = x$ is the perpendicular bisector of each segment that joins (a, b) to (b, a), and acts as a mirror for the line $y = f(x) = 2x - 5$. So, another way to graph the inverse f^{-1} of a function f is to reflect the graph of f about the line $y = x$.

For example, to graph the inverse of $y = f(x) = -2x - 2$:

1. Graph $y = f(x) = -2x - 2$. See the grid in Figure 11.2.9.

2. Reflect the graph of $f(x)$ about the line $y = x$. The result is the graph of $y = f^{-1}(x) = -\dfrac{x + 2}{2}$. See the grid in Figure 11.2.9.

Sample Problems

1. If $f(x) = 3x + 4$, find $f^{-1}(x)$.

    ☑ a. Replace $f(x)$ with y. $y = 3x + 4$

    ☐ b. Switch y and x. _____

    ☐ c. Solve for y. _____

    ☐ d. Replace y with $f^{-1}(x)$. _____

2. Does the function $f(x) = x^3 + 2$ have an inverse?

☑ a. Graph the function $y = f(x)$.

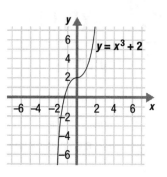

□ b. Use the horizontal line test. Does the function have an inverse? (Circle one)

The function has/does not have an inverse.

3. If $f(x) = 6x - 2$, find $f^{-1}(4)$.

□ a. Replace $f(x)$ with y.

_____

□ b. Switch y and x.

_____

□ c. Solve for y.

_____

□ d. Replace y with $f^{-1}(x)$.

_____

□ e. Find $f^{-1}(4)$.

$f^{-1}(4) = $ _____

4. Given $f(x) = 6x - 2$, graph $y = f^{-1}(x)$.

□ a. Graph $y = f(x)$.

□ b. Locate two points (a, b) on the graph, and two new points (b, a).

□ c. Connect the new points. This is the graph of $y = f^{-1}(x) = \dfrac{x + 2}{6}$.

Answers to Sample Problems

b. *Every horizontal line crosses the graph in at most one point, so the function **has** an inverse.*

a. $y = 6x - 2$

b. $x = 6y - 2$

c. $y = \dfrac{x + 2}{6}$

d. $f^{-1}(x) = \dfrac{x + 2}{6}$

e. 1

a.

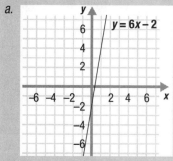

b. *Here are some possible points:*

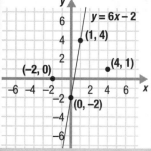

c.

Sample Problems

Answers to Sample Problems

On the computer, you used the Grapher to combine functions by adding, subtracting, multiplying, and dividing them. You also used the Grapher to learn more about the inverse of a function and the composition of two functions. Below are some additional exploration problems.

1. Let $r(x) = 4x + 2$, $c(x) = 2x + 50$, and $p(x) = (r + c)(x)$. Complete the following table.

x	r(x)	c(x)	p(x)
−3			
2		54	
		64	94

☑ a. Find $r(-3)$, $c(-3)$, and $p(-3)$ to complete the first row.

$r(-3) = 4(-3) + 2$
$\quad = -10$

$c(-3) = 2(-3) + 50$
$\quad = 44$

$p(-3) = (r + c)(-3)$

$\quad = r(-3) + c(-3)$
$\quad = -10 + 44$
$\quad = 34$

b. 10

☐ b. Find $r(2)$ and $p(2)$ to complete the second row.

$r(2) = \underline{\quad}$

64

$p(2) = \underline{\quad}$

c. 7

☐ c. Find x, then find $r(x)$ to complete the third row.

$x = \underline{\quad}$

7, 30

$r(\underline{\quad}) = \underline{\quad}$

2. The graphs of $y = f(x)$ and $y = g(x)$ are shown on the grid below. Use these graphs to find $(g \circ f)(1)$:

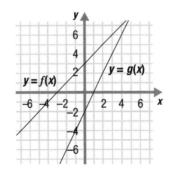

a. Find the *y*-coordinate of the point on the graph of *f* where $x = 1$.

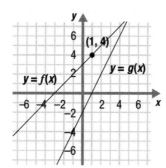

b. Use that *y*-value, 4, as an *x*-value for *g*. Find the *y*-coordinate of the point on the graph of *g* where $x = 4$.

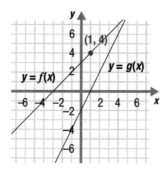

When $x = 4$, $y =$ _____.

b. 6

c. The value $(g \circ f)(1)$ is the *y*-value of the point found in (b).

$(g \circ f)(1) =$ _____

c. 6

3. Graph the function $y = f(x) = x^2 - 2$. If $g(x) = |x|$, sketch the graph of $(g \circ f)(x)$.

a. Graph $y = f(x) = x^2 - 2$.

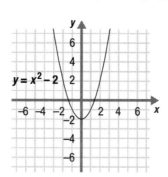

b. Find $(g \circ f)(x)$.

$(g \circ f)(x) = g[f(x)] =$ _____

b. $|x^2 - 2|$

c. Sketch the graph of $(g \circ f)$.

c.

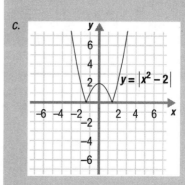

4. Let $f(x) = 3x + 0.5$, $g(x) = 2x - 8$, and $h(x) = (f \cdot g)(x)$. Complete the following table:

x	$f(x)$	$g(x)$	$h(x)$
-1		-10	
$\dfrac{1}{2}$			-14
3	9.5		

☐ a. Find $f(-1)$ and $h(-1)$ to complete the first row.

$f(-1) = $ _____

$h(-1) = $ _____

☐ b. Find $f\left(\dfrac{1}{2}\right)$ and $g\left(\dfrac{1}{2}\right)$ to complete the second row.

$f\left(\dfrac{1}{2}\right) = $ _____

$g\left(\dfrac{1}{2}\right) = $ _____

☐ c. Find $g(x)$ and $h(x)$ to complete the third row.

$g(x) = $ _____

$h(x) = $ _____

HOMEWORK

Homework Problems

Circle the homework problems assigned to you by the computer, then complete them below.

☀ Explain
The Algebra of Functions

1. Given $f(x) = 5x$ and $g(x) = -6x + 1$, find $(f + g)(x)$.

2. Given $f(x) = 5x$ and $g(x) = -6x + 1$, find $(f - g)(x)$.

3. Given $f(x) = x + 1$ and $h(x) = 3x$, find $(f \circ h)(x)$.

4. Given $f(x) = 5x$ and $g(x) = -6x + 1$, find $(f \cdot g)(x)$

5. Given $f(x) = 4x$ and $g(x) = -3x - 7$, evaluate $(f - g)(x)$ at $x = -3$.

6. Given $f(x) = 2x + 2$ and $h(x) = 3x$, evaluate $(f \circ h)(x)$ at $x = 2$.

7. Given $f(x) = 5x$ and $g(x) = -6x + 1$, evaluate $(f \cdot g)(x)$ at $x = -1$.

8. Given $f(x) = 4x^2$ and $g(x) = 12x$, find $\left(\dfrac{f}{g}\right)(x)$.

9. The function $h(C) = 331 + \dfrac{3}{5}C$ represents the speed of sound, h, in meters per second, where C is the temperature in degrees Celsius. The Celsius temperature C can also be expressed as a function of the Fahrenheit temperature F as follows: $C = g(F) = \dfrac{5}{9}(F - 32)$. Find $h[g(F)]$, the speed of sound expressed in terms of degrees Fahrenheit.

10. Suppose that the function $h(x) = x$ describes the number of days that your store runs a special promotion. And suppose that $g(x) = 4000x + 200$ describes the total cost of running this promotion for x days. Then the average daily cost of running the promotion is given by the quotient function $\left(\dfrac{g}{h}\right)(x)$, where $x \neq 0$. Find your average daily cost if you were to run the promotion for 5 days.

11. Given $f(x) = 1 - 4x^2$ and $g(x) = 1 + 2x$, evaluate $\left(\dfrac{f}{g}\right)(x)$ at $x = 2$.

12. Given $f(x) = x^2 - 2$ and $h(x) = \sqrt{x}$, find $(f \circ h)(2)$.

Inverse Functions

13. If $f(x) = 6x - 1$, find $f^{-1}(x)$.

14. Determine whether or not the function graphed in Figure 11.2.10 is one-to-one.

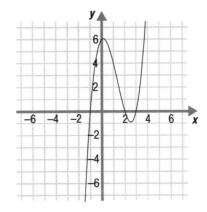

Figure 11.2.10

15. Graph the inverse of the function $y = f(x) = 2 - 3x$. Use the grid in Figure 11.2.11.

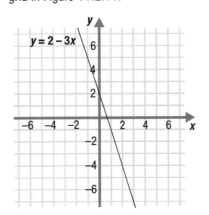

Figure 11.2.11

16. If $f(x) = 6x - 1$, find $(f^{-1} \circ f)(x)$.

17. Is the function $y = f(x) = x^2 - 3$ one-to-one?

18. Graph and find the equation of the inverse of the function $y = f(x) = 4x$. Use the grid in Figure 11.2.12.

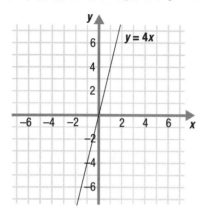

Figure 11.2.12

19. If $g(x) = \dfrac{2}{x} + 1$, find $g^{-1}(x)$.

20. If $y = f(x) = x^2 - 5$, does f^{-1} exist?

21. Julie parachutes from an airplane at 10,000 feet. She estimates her distance above the ground to be $f(x) = 10,000 - 20x$ at a given time x. Find $f^{-1}(x)$.

22. Rolph is in a hot air balloon at 20,000 feet that is descending. He estimates his distance above the ground to be $f(x) = 20,000 - 15x$ at a given time x. Find $f^{-1}(50)$.

23. If $y = g(x) = \dfrac{3}{x} - 1$, determine if g has an inverse, and if it does, find it. Use the graph of $y = g(x) = \dfrac{3}{x} - 1$ in Figure 11.2.13 to help you.

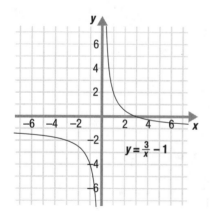

Figure 11.2.13

24. If f is a function that takes an input value x, increases it by 4, then multiplies that quantity by 3, find an equation for f. Then find $f^{-1}(-1)$.

 Explore

25. Let $r(x) = 4x$, $c(x) = 20x + 500$, and $p(x) = (r - c)(x)$. Complete the following table.

x	$r(x)$	$c(x)$	$p(x)$
5		600	
11	44		-676
20			

26. The graphs of $y = f(x)$ and $y = g(x)$ are shown on the grid in Figure 11.2.14. Use these graphs to find $(g \circ f)(-1)$.

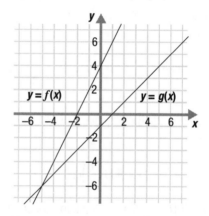

Figure 11.2.14

27. The graph of $y = f(x) = 2x^2 - 2$ is shown on the grid in Figure 11.2.15. If $g(x) = |x|$, which of the following is the graph of $y = (g \circ f)(x)$?

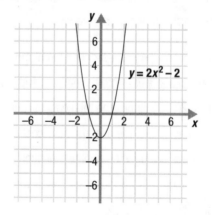

Figure 11.2.15

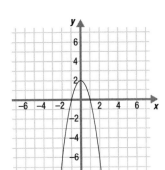

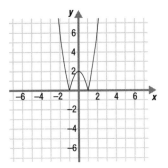

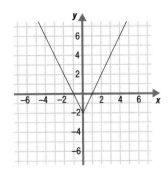

28. Let $f(x) = 3x$, $g(x) = 3x - 3$, and $h(x) = (f \cdot g)(x) + \left(\dfrac{f}{g}\right)(x)$. Complete the following table.

x	f(x)	g(x)	h(x)
−1	−3		
$\dfrac{1}{3}$		−2	
	12		$109\dfrac{1}{3}$

29. The graphs of $y = f(x) = x^2$ and $y = g(x) = 2x + 1$ are shown on the grid in Figure 11.2.16. Which of the following is the graph of $y = (f + g)(x)$?

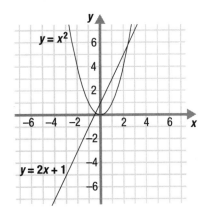

Figure 11.2.16

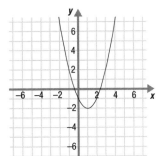

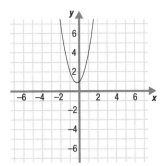

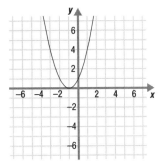

30. The graph of $y = f(x) = \sqrt{2x + 3}$ is shown on the grid in Figure 11.2.17. Which of the following is the graph of $y = f^{-1}(x)$?

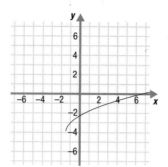

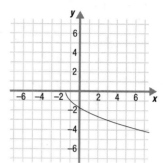

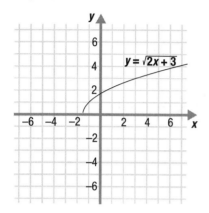

$y = \sqrt{2x + 3}$

Figure 11.2.17

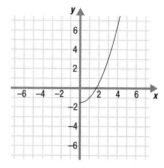

APPLY

Practice Problems

Here are some additional practice problems for you to try.

The Algebra of Functions

1. Given $g(x) = 3x + 1$ and $h(x) = x^2 - 7x + 4$, find $(g + h)(x)$.

2. Given $f(x) = x^2 - 9$ and $h(x) = x^2 + 3x + 5$, find $(f + h)(x)$.

3. Given $f(x) = 5x$ and $g(x) = 2x^2 - 8x + 7$, find $(f + g)(x)$.

4. Given $f(x) = 3x^2 + 7$ and $g(x) = 2x - 7$, find $(f - g)(x)$.

5. Given $g(x) = x^2 + 3x + 4$ and $h(x) = 3x^2 + 5x - 6$, find $(g - h)(x)$.

6. Given $f(x) = 5x^2 - 2$ and $h(x) = 7x + 2$, find $(f - h)(4)$.

7. Given $f(x) = 10x$ and $g(x) = 4x + 9$, find $(f \cdot g)(x)$.

8. Given $f(x) = x + 1$ and $g(x) = 3x - 5$, find $(f \cdot g)(x)$.

9. Given $g(x) = x^2 - 5$ and $h(x) = 4x + 7$, find $(g \cdot h)(x)$.

10. Given $f(x) = 3x + 5$ and $h(x) = 5x - 8$, find $(f \cdot h)(1)$.

11. Given $g(x) = -x^2 + 2x$ and $h(x) = 3x + 11$, find $(g \cdot h)(0)$.

12. Given $f(x) = -2x^2 - 3x$ and $g(x) = 6x + 2$, find $(f \cdot g)(-2)$.

13. Given $f(x) = 2x$ and $g(x) = x + 4$, find $\left(\dfrac{f}{g}\right)(x)$.

14. Given $f(x) = x^2 - 1$ and $h(x) = x + 3$, find $\left(\dfrac{f}{h}\right)(x)$.

15. Given $f(x) = x - 3$ and $g(x) = 4x^2 - 5$, find $\left(\dfrac{f}{g}\right)(x)$.

16. Given $f(x) = 2x$ and $h(x) = x^2 + 1$, find $(f \circ h)(x)$.

17. Given $g(x) = 3x + 7$ and $h(x) = 2x - 4$, find $(g \circ h)(x)$.

18. Given $f(x) = 4x - 3$ and $g(x) = 9x - 1$, find $(f \circ g)(x)$.

19. Given $f(x) = x^2$ and $g(x) = x + 2$, find $(f \circ g)(x)$.

20. Given $f(x) = 2x^2 + 1$ and $g(x) = x - 3$, find $(f \circ g)(x)$.

21. Given $f(x) = x^2 + x$ and $h(x) = x - 2$, find $(f \circ h)(x)$.

22. Given $f(x) = x^2$ and $g(x) = \sqrt{x + 1}$, find $(f \circ g)(3)$.

23. Given $g(x) = x^2 - 4$ and $h(x) = \sqrt{x - 2}$, find $(g \circ h)(3)$.

24. Given $f(x) = x^2 + 3$ and $g(x) = \sqrt{x - 5}$, find $(f \circ g)(7)$.

25. Given $g(x) = x^2 + 4$ and $h(x) = \sqrt{x + 5}$, find $(g \circ h)(4) - (h \circ g)(4)$.

26. Circle the pairs of functions f and g below for which the composition $(f \circ g)(x) = x$.

$f(x) = x + 7$ $f(x) = 3x$ $f(x) = 2x - 5$

$g(x) = x - 7$ $g(x) = -3x$ $g(x) = \dfrac{x}{2} + 5$

27. Circle the pairs of functions f and g below for which the composition $(f \circ g)(x) = x$.

$f(x) = 2x$ $f(x) = 3 - 6x$ $f(x) = 4x + 7$

$g(x) = \dfrac{x}{2}$ $g(x) = 2x - 1$ $g(x) = \dfrac{x - 7}{4}$

28. Circle the pairs of functions f and g below for which the composition $(f \circ g)(x) = x$.

$f(x) = 4 - x$ $f(x) = 6x$ $f(x) = 3x + 2$

$g(x) = \dfrac{x}{4} - 1$ $g(x) = \dfrac{x}{6}$ $g(x) = \dfrac{x - 2}{3}$

Inverse Functions

29. Given $f(x) = x + 8$, find $f^{-1}(x)$.

30. Given $f(x) = 3x - 2$, find $f^{-1}(x)$.

31. Given $f(x) = 4x - 1$, find $f^{-1}(x)$.

32. Given $g(x) = 7x + 8$, find $g^{-1}(x)$.

33. Given $f(x) = \dfrac{1}{x} + 2$, find $f^{-1}(x)$.

34. Given $f(x) = \dfrac{4}{5x} - 3$, find $f^{-1}(x)$.

35. Given $f(x) = \dfrac{2}{3x} + 1$, find $f^{-1}(x)$.

36. Use the graph of the function $y = f(x) = 3x - 5$ to determine if the function is one-to-one.

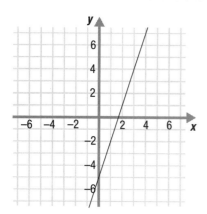

37. Use the graph of the function $y = f(x) = 7x + 1$ to determine if the function is one-to-one.

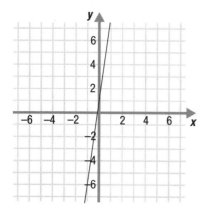

38. Use the graph of the function $y = f(x) = 2x + 1$ to determine if the function is one-to-one.

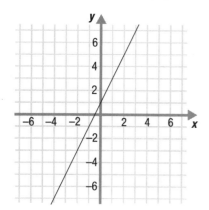

39. Use the graph of the function $y = f(x) = -(x + 2)^2 + 1$ to determine if the function is one-to-one.

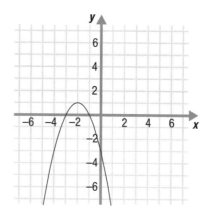

40. Use the graph of the function $y = f(x) = (x - 2)^2 - 3$ to determine if the function is one-to-one.

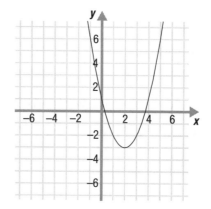

41. Use the graph of the function $y = f(x) = x^3 - 2$ to determine if the function is one-to-one.

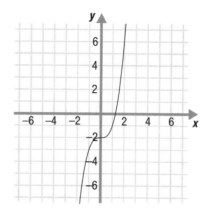

42. Given $f(x) = 8x - 4$, find $(f \circ f^{-1})(x)$.

43. Given $f(x) = 2x + 9$, find $(f^{-1} \circ f)(x)$.

44. Given $f(x) = 5x + 6$, find $(f \circ f^{-1})(x)$.

45. The graph of a function $y = f(x)$ is shown below. Fill in the table below for f^{-1}.

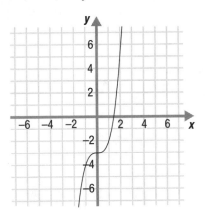

x	$y = f^{-1}(x)$
-4	
-3	
5	

47. The graph of a function $y = f(x)$ is shown below. Fill in the table below for f^{-1}.

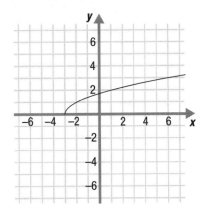

x	$y = f^{-1}(x)$
1	
2	
3	

46. The graph of a function $y = f(x)$ is shown below. Fill in the table below for f^{-1}.

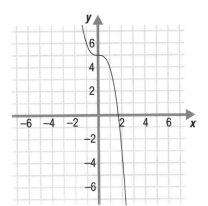

x	$f^{-1}(x)$
-3	
4	
6	

48. The graph of a function $y = f(x)$ is shown below. Fill in the table below for f^{-1}.

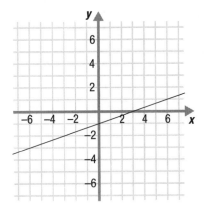

x	$y = f^{-1}(x)$
-3	
-1	
1	

49. The graph of a function $y = f(x)$ is shown below. Fill in the table below for f^{-1}.

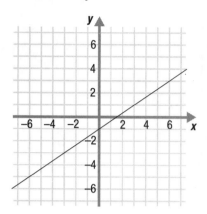

x	$y = f^{-1}(x)$
-3	
-1	
1	

50. The graph of a function $y = f(x)$ is shown below. Fill in the table below for f^{-1}.

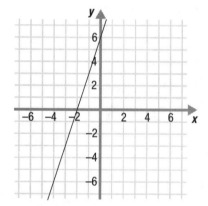

x	$y = f^{-1}(x)$
-3	
0	
6	

51. The graph of the function $y = f(x) = 2x + 4$ follows. Graph f^{-1}.

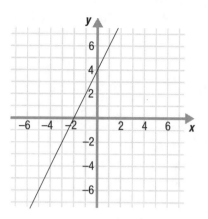

52. The graph of the function $y = f(x) = 3x - 2$ is shown below. Graph f^{-1}.

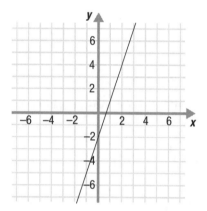

53. The graph of the function $y = f(x) = \sqrt{x}$ is shown below. Graph f^{-1}.

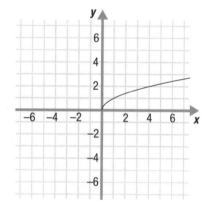

54. Given the function $y = f(x) = x + 5$, find the x-intercept and y-intercept of f^{-1}.

55. Given the function $y = f(x) = 3x + 12$, find the x-intercept and y-intercept of f^{-1}.

56. Given the function $y = f(x) = 2x - 10$, find the x-intercept and the y-intercept of f^{-1}.

EVALUATE

Practice Test

Take this practice test to be sure that you are prepared for the final quiz in Evaluate.

1. Given $h(x) = -7x$ and $f(x) = 7x$, find $(f - h)(-3)$.

2. Given $f(x) = 6x^4$ and $g(x) = 3x^3$, find $\left(\frac{f}{g}\right)(x) - (f \cdot g)(x)$.

3. Circle the pairs of functions f and g below for which the composition $(f \circ g)(x) = x$.

   $f(x) = 2x$ $f(x) = x^2$ $f(x) = 1 - x$ $f(x) = 2x + 1$

   $g(x) = \frac{x}{2}$ $g(x) = -x$ $g(x) = x + 1$ $g(x) = \frac{x}{2} - 1$

4. Given $f(x) = -6x + 5$ and $g(x) = -x^2 - 2$, evaluate $(f \circ g)(4)$.

5. The graph of a function, $y = f(x)$, is shown in Figure 11.2.20. Use the points shown on the graph to fill in the table below for f^{-1}:

   f^{-1}

x	y
-5	
-1	0
3	

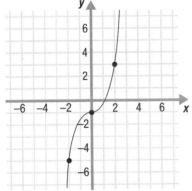

Figure 11.2.20

6. Given $h(x) = \frac{2x - 1}{5} + 2$, find $h^{-1}(x)$.

7. If $f(x) = 0.5x - 7$ and $g(x) = f^{-1}(x)$, find $(g \circ f)(x)$.

8. Which of the graphs below are graphs of functions that do **not** have inverses.

   a.
   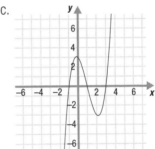

   b.
   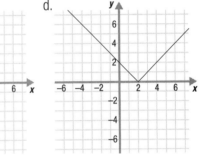

   c.
   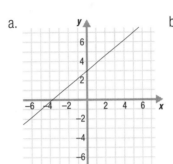

   d.
   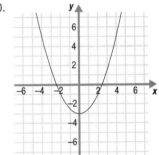

9. If $f(5) = 11$ and $g(5) = 3$, find $(f \cdot g)(5)$.

10. The graphs of two functions, $y = f(x)$ and $y = g(x)$, are shown on the grid in Figure 11.2.21. Use these graphs to find $(f - g)(2)$.

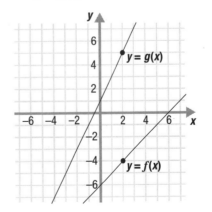

Figure 11.2.21

11. The graphs of $y = f(x)$ and $y = g(x)$ are shown on the grid in Figure 11.2.22. Use these graphs to find $(f \circ g)(2)$.

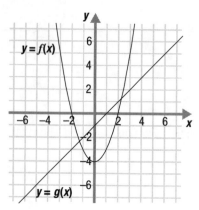

Figure 11.2.22

12. If $f(x) = 2x + 8$, circle the following statements that are true.

$$f^{-1}[f(x)] = x$$

$$f^{-1}(x) = \frac{1}{2x + 8}$$

$$(f \circ f)(x) = 4x + 24$$

$$(f \cdot f)(x) = (2x + 8)^2$$

TOPIC 11 CUMULATIVE ACTIVITIES

CUMULATIVE REVIEW PROBLEMS

These problems combine all of the material you have covered so far in this course. You may want to test your understanding of this material before you move on to the next topic, or you may wish to do these problems to review for a test.

1. Simplify this expression: $\sqrt{\dfrac{-225}{64}}$

2. Find:
$(12x^2 - 9xy + 5xy^2 + 2y^2 + 7) - (-4y^2 + 13 + 6x^2y - 11xy)$

3. Solve for x: $3x^2 - 7x + 4 = 0$

4. Find the domain and the range of each of the functions below.

   a. $y = 6x + 13$

   b. $y = x^2 - 3$

   c. $y = \dfrac{1}{x + 1}$

   d. $y = \sqrt{x^2 - 7}$

   e. $y = \dfrac{x + 8}{(x - 2)(x + 4)}$

5. If $f(x) = \dfrac{4}{2x - 4} + 7$, find $f^{-1}(x)$.

6. Find the domain and the range of each of the quadratic functions below.

   a. $y = x^2 + 9$

   b. $y = -x^2 + 3$

   c. $y = 6x2 + 12$

7. Simplify using the properties of exponents: $2a \cdot \dfrac{(3a)^3}{(3a^2)^2}$

8. Simplify the following expressions:

   a. $\sqrt{-9}$

   b. $\sqrt{-49}$

   c. $\sqrt{-9} \cdot \sqrt{-49}$

   d. $\sqrt{-9} + \sqrt{-49}$

   e. i^{42}

   f. i^{37}

9. Find: $(5\sqrt{3} + 2\sqrt{6})(7\sqrt{3} - \sqrt{6})$

10. Solve for x: $32 + 8x > 13(x + 2)$

11. Given $f(x) = 3x$ and $g(x) = -8x + 6$, find $(f + g)(x)$.

12. Determine whether each of the following parabolas opens up or down.

   a. $y = 7 - 2x^2$

   b. $y = 5x^2 + 8x - 5$

   c. $y = -x^2 + 14x - 11$

13. Solve for a: $\dfrac{2}{7}(a - 6) = \dfrac{2}{3}a + \dfrac{5}{7}$

14. If $g(x) = 7(x + 5)$, find $g^{-1}(-2)$.

15. Which of the following pairs are complex conjugates of each other?

   $4 + 6i$ and $6 - 4i$

   $2 + 9i$ and $2 - 9i$

   $5 + 5i$ and $-5 - 5i$

   $7 + 3i$ and $-3i + 7$

   $8 + 4i$ and $4i - 8$

16. Simplify the expression below. Assume x, y, and z are positive numbers.

   $\dfrac{(32)^{\frac{1}{5}} \cdot \sqrt[3]{125x^{27}}}{\sqrt{x^{18} \cdot z^{10} \cdot y^4}}$

17. Solve for x: $3x^2 - 4x - 7 = 0$

18. Solve for x: $\dfrac{5x}{3} + 4 = \dfrac{7}{8}$

19. Find the x- and y-intercepts of this line: $y + 6 = 2(x + 1)$

20. Which of the following functions are quadratic?

   $y = 7 - 2x$ $\qquad$ $y = 4 - 3x^2$

   $y = x^3 + 2x^2 - 6x + 4$ $\qquad$ $y = x + 3$

   $y = 7x^2 - 2x - 7$

21. Given $f(x) = 9x$ and $g(x) = -x + 1$, evaluate $(f \cdot g)(x)$ at $x = -2$.

22. Solve for x: $6 - 4|x + 3| \le -12$

23. Simplify the expression below. Leave your answer in exponential form.

$$\left(7^{\frac{1}{5}} \cdot 7^{\frac{-2}{3}}\right)^{-5}$$

24. Find the equation of the line through the point $(-4, 2)$ that is perpendicular to the line $y = -3x + 6$. Write your answer in slope-intercept form.

25. If $f(x) = 7x + 4$, find $f^{-1}(x)$.

26. Reduce to lowest terms: $\dfrac{3ba + 3b - 8a - 8}{ba + b + 2a + 2}$

27. Does the function f, shown in Figure 11.1, have an inverse?

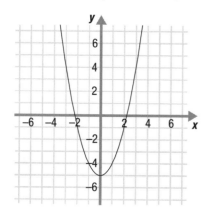

Figure 11.1

28. Which of the following functions are linear?

$y = 3x + 7$ $y = x^3 + x^2 + 4x + 9$

$y = x^2 - 2x + 2$ $y = \sqrt{x^3 - 1}$

$y = \dfrac{2}{3} - 5x$

29. Solve for x: $x^2 + 3x = 9$

30. Combine like terms: $8\sqrt{3y} + 3\sqrt{48y} - 9$

31. Solve for x: $2x^2 - 5x - 20 = -8x - 15$

32. Solve for x: $|x| \ge 7$

33. Find the x- and y-intercepts of each of the functions below.

   a. $y = 2x^2 - 3x - 9$

   b. $y = 9x^2 - x$

   c. $y = 4 - x^2$

34. Factor: $64a^3 + 40a^2 - 24a$

35. Simplify: $\left(a^{\frac{4}{5}} \cdot a^{\frac{-3}{2}} \cdot b^{\frac{3}{5}}\right)^{-5}$

36. Given $f(x) = 4x - 2$ and $g(x) = -2x - 7$, find $(f \circ g)(x)$.

37. If $f(x) = 3x - 8$, find $(f^{-1} \circ f)(x)$.

38. Find the distance d between the points $(3, -4)$ and $(-1, 5)$.

39. Solve for a: $a^2 - 64 = 0$

40. Given $f(x) = 5x^2$ and $g(x) = 15x$, find $\left(\dfrac{f}{g}\right)(x)$.

41. Find the domain and the range of each of the functions below.

   a. $y = 4x - 1$ b. $y = |x| + 6$

   c. $y = |x - 6|$

42. Solve for x: $(x - 4)^2 = 64$

43. Solve for x: $2x^2 - 3x + 5 = 0$

44. Find: $(8 + 9i) - (-3 + 7i)$

45. Given $g(x) = 2x^2 - 4$, calculate:

   a. $g(0)$ b. $g(-1)$

   c. $g(2)$ d. $g(-3)$

   e. $g(1)$

46. Given $f(x) = 4x + 1$, calculate:

   a. $f(0)$ b. $f(3)$

   c. $f(-1)$ d. $f\left(\dfrac{1}{2}\right)$

   e. $f(-5)$

47. Solve for x: $|5x + 2| = 9$

48. Find: $(12 + 4i) \div (9 - 6i)$

49. Combine like terms and simplify:

$$4\sqrt{313} + 7\sqrt{47} + \sqrt[3]{64} - 2\sqrt{313} - 5\sqrt[4]{16} - 3\sqrt{47}$$

50. Find: $(4 - 9i)(1 + i)$

LESSON 12.1 – EXPONENTIAL FUNCTIONS

OVERVIEW

Here's what you'll learn in this lesson:

The Exponential Function

a. Recognizing and graphing an exponential function

b. Applications of the exponential function

c. Solving exponential equations

Suppose you have one amoeba that splits into two. Then suppose that these two amoebas each split, producing four. If this process continues, you will have eight amoebas, then sixteen, then thirty-two, then sixty-four, and so on. The growth of your amoeba population can be described using a function called an exponential function. In this lesson, you'll learn about exponential functions.

 **EXPLAIN**

THE EXPONENTIAL FUNCTION

Summary

Recognizing and Graphing an Exponential Function

You have studied linear functions of the form $y = ax + b$, and quadratic functions of the form $y = ax^2 + bx + c$. Now you will learn about exponential functions.

An exponential function is a function of the form $y = b^x$. Here the number b is the base and x is the exponent. Also, $b > 0$ and $b \neq 1$. Examples of exponential functions are $y = 2^x$, $y = 3^x$, and $y = \left(\frac{1}{3}\right)^x$.

In general, to graph an exponential function $f(x) = b^x$, where $b > 0$ and $b \neq 1$:

1. Make a table of ordered pairs for $y = f(x) = b^x$ by substituting several values for x into $y = b^x$, and solving for y.

2. Plot these ordered pairs on a grid, and join them with a smooth curve.

For example, to graph $f(x) = 3^x$:

1. Substitute values for x into $y = f(x) = 3^x$, and make a table of ordered pairs.

Let $x = 2$, then:
$$y = 3^2$$
$$= 9$$

Let $x = \frac{3}{2}$, then:
$$y = 3^{\frac{3}{2}}$$
$$= \left(\sqrt{3}\right)^3$$
$$= 3\sqrt{3}$$
$$\approx 5.196$$

Let $x = 1$, then:
$$y = 3^1$$
$$= 3$$

Let $x = 0$, then:
$$y = 3^0$$
$$= 1$$

Let $x = -1$, then:
$$y = 3^{-1}$$
$$= \frac{1}{3}$$

The independent variable, x, doesn't have to be an integer. You can pick values for x that aren't integers, then approximate the y-value by using a calculator. You can use the symbol " ≈ " to mean "is approximately equal to."

Let $x = -\frac{3}{2}$, then:

$$y = 3^{-\frac{3}{2}}$$

$$= \frac{1}{3^{\frac{3}{2}}}$$

$$= \frac{1}{3\sqrt{3}}$$

$$= \frac{\sqrt{3}}{9}$$

$$\approx .1925$$

Let $x = -2$, then:

$$y = 3^{-2}$$

$$= \frac{1}{3^2}$$

$$= \frac{1}{9}$$

$$y = f(x) = 3^x$$

x	y
2	9
$\frac{3}{2}$	5.196
1	3
0	1
-1	$\frac{1}{3}$
$-\frac{3}{2}$	.1925
-2	$\frac{1}{9}$

2. Plot the ordered pairs, and join them with a smooth curve.

See the grid in Figure 12.1.1.

As x becomes large (see Figure 12.1.1), the graph rises rapidly. In other words, as the x-values increase (that is, as you move to the right along the x-axis), the y-values become large. Also, regardless of x, 3^x is never 0 or negative, so the graph never touches or crosses the x-axis. Notice that the y-intercept of the graph is the point (0, 1).

Here's another example. To graph $f(x) = \left(\frac{1}{3}\right)^x$:

1. Substitute values for x into $y = f(x) = \left(\frac{1}{3}\right)^x$, and make a table of ordered pairs.

Let $x = 2$, then:

$$y = \left(\frac{1}{3}\right)^2$$

$$= \frac{1}{9}$$

Let $x = 1$, then:

$$y = \left(\frac{1}{3}\right)^1$$

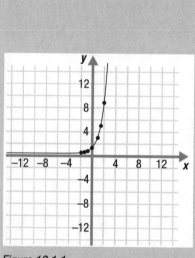

Figure 12.1.1

The domain of $f(x) = 3^x$ is all real numbers, since the input value, x, can be any number. The range is all positive real numbers, since the output value, y, is never 0 or negative.

$$= \frac{1}{3}$$

Let $x = 0$, then:

$$y = \left(\frac{1}{3}\right)^0$$

$$= 1$$

Let $x = -1$, then:

$$y = 3^{-(-1)}$$

$$= 3$$

Let $x = -2$, then:

$$y = 3^{-(-2)}$$

$$= 9$$

$$y = f(x) = \left(\frac{1}{3}\right)^x$$

x	y
2	$\frac{1}{9}$
1	$\frac{1}{3}$
0	1
−1	3
−2	9

2. Plot the ordered pairs, and join them with a smooth curve.

See the grid in Figure 12.1.2.

As x becomes large (see Figure 12.1.2), the graph decreases rapidly. In other words, as the x-values increase (as you move to the right along the x-axis), the y-values become small. Also, regardless of x, $\left(\frac{1}{3}\right)^x$ is never 0 or negative, so the graph never touches or crosses the x-axis. Notice that the y-intercept of this graph is also the point (0, 1).

Exponential Growth and Decay

You have learned that an exponential function with base b is a function of the form $y = f(x) = b^x$, where $b > 0$ and $b \neq 1$. Exponential functions have many applications in the natural and social sciences.

Look at the graph of the exponential function $y = f(x) = 3^x$ shown on the grid in Figure 12.1.3. Notice that as you move from left to right, the graph rises slowly, and then rises rapidly. This is typical of an exponential function where the base b is greater than 1, and is often called **exponential growth**. Populations often grow exponentially. Money invested where it is earning compounded interest also grows exponentially.

Remember: $\left(\frac{1}{3}\right)^x = \frac{1^x}{3^x} = \frac{1}{3^x} = 3^{-x}$. When you substitute negative values for x into $y = f(x) = \left(\frac{1}{3}\right)^x$, you might find it easier to substitute these values into $y = f(x) = 3^{-x}$.

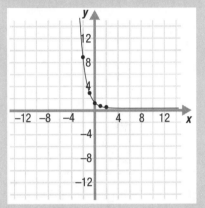

Figure 12.1.2

Once again, the domain of the exponential function $y = f(x) = \left(\frac{1}{3}\right)^x$ is all real numbers, since the input value, x, can be any number. The range of this exponential function is all positive real numbers, since the output value, y, is never 0 or negative.

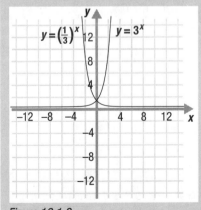

Figure 12.1.3

Now look at the graph of the exponential function $y = g(x) = \left(\dfrac{1}{3}\right)^x$ shown on the grid in Figure 12.1.3. Notice that as you move from left to right, the graph drops sharply and then levels off slowly. This is typical of an exponential function where the base b is between 0 and 1, and is called **exponential decay**. Radioactive substances decay exponentially. The temperature of a round object such as a potato that has been baked then left to cool declines exponentially as well.

The Exponential Function with Base e

There is a particular base that is very important for applications that involve exponential growth and decay. This base is the irrational number e, a number that lies between 2 and 3 on the number line. Since e is an irrational number like π or $\sqrt{2}$, its decimal representation never ends and never repeats. You will often approximate the value of e to be 2.718.

The exponential function with base e is of the form $y = f(x) = e^x$. This is called the natural exponential function and it arises "naturally" in many applications in the natural and social sciences. You can see the relationship between the graphs of $y = e^x$, $y = 2^x$, and $y = 3^x$ in Figure 12.1.4.

On some calculators there is an "e^x" key. So you can find the approximate value of e raised to a power of x by entering the power x first, then pressing the "e^x" key.

In general, to use a calculator to approximate e^x:

1. Enter the power (or exponent), x.

2. Press the "e^x" key, then round your answer to the desired number of decimal places.

For example, to find $e^{1.8}$ approximated to two decimal places:

    1. Enter 1.8 in your calculator.

    2. Press the "e^x" key, then   6.0496475
        round your answer.   6.05

So, $e^{1.8} \approx 6.05$.

As another example, to find $e^{-2.2}$ approximated to two decimal places:

    1. Enter −2.2 in your calculator.
        Use the "±" key.

    2. Press the "e^x" key, then   0.1108031
        round your answer.   0.11

So, $e^{-2.2} \approx 0.11$.

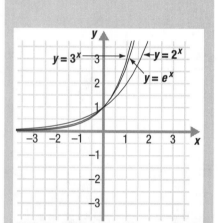

Figure 12.1.4

If x is a negative value, you may have to use the "±" key to get the negative sign.

To round 6.0496475 to two decimal places, look at the number in the third decimal place. It is 9, so round the 4 up to 5.

To round 0.1108031 to two decimal places, look at the number in the third decimal place. It is 0, so don't round the 1.

Applications of the Exponential Function $y = e^x$

There are many situations in the natural and social sciences that can be described using functions that are based on the natural exponential function $y = e^x$.

A Learning Curve

For example, you can use a function of the form $y = A + Be^{g(x)}$ to predict the number of items that a factory worker can produce on an assembly line. The function $g(x)$ depends upon the particular assembly line. As you might expect, a worker who is just learning how to use the machinery produces very few items. As the worker becomes more experienced with the operation of the machinery, she produces more and more items. But eventually, the number of items that the worker produces will reach a maximum number and level off.

Here's a function that describes this situation for a worker who has x days of experience on the assembly line and produces y items per day: $y = 45 - 30e^{-0.02x}$. Look at the value of this function for different input values of x.

- A new worker has 0 days of experience.

$$\begin{aligned} \text{Here, } x = 0, \text{ so } y &= 45 - 30e^{-0.02(0)} \\ &= 45 - 30e^0 \\ &= 45 - 30(1) \\ &= 45 - 30 \\ &= 15 \end{aligned}$$

So a new worker can expect to produce 15 items per day.

Look at the number of items a worker can produce for various levels of experience:

- A worker who has 3 days of experience.

$$\begin{aligned} \text{Here, } x = 3, \text{ so } y &= 45 - 30e^{-0.02(3)} \\ &= 45 - 30e^{-0.06} \\ &\approx 45 - 30(.942) \\ &= 45 - 28.26 \\ &= 16.74 \end{aligned}$$

So after 3 days, a worker can expect to produce almost 17 items per day.

- A worker who has 2 months (60 days) of experience.

$$\begin{aligned} \text{Here, } x = 60, \text{ so } y &= 45 - 30e^{-0.02(60)} \\ &= 45 - 30e^{-1.2} \\ &\approx 45 - 30(.301) \\ &= 45 - 9.03 \\ &= 35.97 \end{aligned}$$

So after 2 months, a worker can expect to produce almost 36 items per day, or almost twice as much as she was producing after 3 days.

- A worker who has 6 months (180 days) of experience.

Here, $x = 180$, so $y = 45 - 30e^{-0.02(180)}$

$= 45 - 30e^{-3.6}$

$\approx 45 - 30(.027)$

$= 45 - 0.81$

$= 44.19$

So after 6 months a worker can expect to produce about 44 items per day.

- A worker who has 1 year (365 days) of experience.

Here, $x = 365$, so $y = 45 - 30e^{-0.02(365)}$

$= 45 - 30e^{-7.3}$

$\approx 45 - 30(.001)$

$= 45 - 0.03$

$= 44.97$

So after 1 year, a worker can expect to produce almost 45 items per day, which is not much more than she could produce after 6 months on the job.

Now to graph the function $y = 45 - 30e^{-0.02x}$:

You already found several ordered pairs for the table, but you will be able to graph the curve more easily if you find a few more.

1. Substitute values for x into $y = f(x) = 45 - 30e^{-0.02x}$, and make a table of ordered pairs.

Let $x = 20$, then:

$y = 45 - 30e^{-0.02(20)}$

$= 45 - 30e^{-0.4}$

$\approx 45 - 30(.670)$

$= 45 - 20.1$

$= 24.9$

Let $x = 100$, then:

$y = 45 - 30e^{-0.02(100)}$

$= 45 - 30e^{-2}$

$\approx 45 - 30(.135)$

$= 45 - 4.05$

$= 40.95$

Let $x = 275$, then:

$y = 45 - 30e^{-0.02(275)}$

$= 45 - 30e^{-5.5}$

$\approx 45 - 30(.004)$

$= 45 - 0.12$

$= 44.88$

$y = f(x) = 45 - 30e^{-0.02x}$

x	y
0	15
3	16.74
20	24.9
60	35.97
100	40.95
180	44.19
275	44.88
365	44.97

2. Plot the ordered pairs, and   See the grid in Figure 12.1.5.
   join them with a smooth curve.

Notice that the graph levels off at approximately $y = 45$. So you would expect that an experienced worker could produce about 45 items per day.

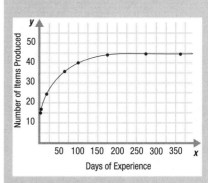

Figure 12.1.5

The Compound Interest Formula

Here's another application of exponential functions. Suppose you open a bank account by investing $100 for one year at an interest rate of 5%. Also, suppose that interest is compounded 2 times during the year. This means that after 6 months, the bank pays you $\frac{1}{2}$ of the 5% interest, or 2.5%, on the $100 you invested.

So at the end of 6 months you have: $100 + $2.50 = $102.50

original investment   interest

After 12 months, the bank pays you $\frac{1}{2}$ of the 5% interest, or 2.5%, on your original investment **and** on the $2.50 interest that you have earned.

So at the end of 12 months you have: $102.50 + .025(102.50) ≈ $105.06

original investment + $2.50 interest   interest

In general, if you invest P dollars at an interest rate r, and interest is compounded n times per year, the amount of money in the account after one year is given by the compound interest formula: $A = P \cdot \left(1 + \frac{r}{n}\right)^n$. For this exponential function, the base is $1 + \frac{r}{n}$ and the exponent is n.

The number e arises naturally from a special case of the compound interest formula. Suppose you invest $1.00 at a rate of 100%, or 1.00 per year. This is a fantastic rate! If you compound this investment n times per year, the interest rate for each individual period is $\frac{1.00}{n}$ or $\frac{1}{n}$.

So after 1 year, your $1.00 investment will grow to $A = \left(1 + \frac{1}{n}\right)^n$ dollars.

Look at how much the investment will grow in a year for different numbers of compounding periods:

- Interest compounded just 1 time per year.

$$\text{Here, } n = 1, \text{ so } A = \left(1 + \frac{1}{1}\right)^1$$
$$= 2^1$$
$$= \$2.00$$

So you will have $2.00 at the end of the year if the interest is compounded once a year.

- Interest compounded 2 times per year.

$$\text{Here, } n = 2, \text{ so } A = \left(1 + \frac{1}{2}\right)^2$$
$$= \left(1\frac{1}{2}\right)^2$$
$$= \left(\frac{3}{2}\right)^2$$
$$= \frac{9}{4}$$
$$= \$2.25$$

So you will have $2.25 at the end of the year if the interest is compounded twice a year.

- Interest compounded 4 times per year (that is, every three months).

$$\text{Here, } n = 4, \text{ so } A = \left(1 + \frac{1}{4}\right)^4$$
$$= \left(1\frac{1}{4}\right)^4$$
$$= \left(\frac{5}{4}\right)^4$$
$$= \frac{625}{256}$$
$$\approx \$2.44$$

So you will have $2.44 at the end of the year if the interest is compounded four times a year.

From these values, you can see that the more times the bank compounds the interest earned on your investment, the more money you have in your account at the end of the year. So it seems that if the bank compounds your interest many times, your $1.00 investment will grow very large. To see what really happens, look at the table of increasing values of n and the corresponding values of $\left(1 + \frac{1}{n}\right)^n$ that appears on the facing page.

Notice that as n gets larger, the value of $\left(1 + \frac{1}{n}\right)^n$ increases quickly for a time. Then, as n gets very large, the value of $\left(1 + \frac{1}{n}\right)^n$ levels

off to approximately 2.718, which is also the approximation of the number e. It turns out that if you deposit P dollars in an account for one year at an interest rate r, and the bank compounds the interest as frequently as possible (this is called continuous compounding), your investment will grow to $A = P \cdot e^r$ dollars.

Interest Compounded	n	$\left(1 + \dfrac{1}{n}\right)^n$ = total amount
once a year	1	$\left(1 + \dfrac{1}{1}\right)^1 = 2.00$
twice a year	2	$\left(1 + \dfrac{1}{2}\right)^2 = 2.25$
four times a year	4	$\left(1 + \dfrac{1}{4}\right)^4 \approx 2.4414$
every month	12	$\left(1 + \dfrac{1}{12}\right)^{12} \approx 2.6130$
every day	365	$\left(1 + \dfrac{1}{365}\right)^{365} \approx 2.71457$
every hour	8760	$\left(1 + \dfrac{1}{8760}\right)^{8760} \approx 2.718127$
every minute	525,600	$\left(1 + \dfrac{1}{525600}\right)^{525,600} \approx 2.718283$

Solving Exponential Equations

One place you will use what you have learned about exponents is when solving exponential equations. These are equations in which the variable is an exponent.

In order to solve some exponential equations, you can use the fact that the exponential function is one-to-one. In other words, if two exponential values are the same, they must have come from the same input value. This can be expressed with the following property:

If $b^m = b^n$, then $m = n$ (here $b > 0$ and $b \neq 1$).

In general, to solve an exponential equation that can be written in the form $b^m = b^n$:

1. Rewrite the equation so that it is in the form $b^m = b^n$.

2. Use the property "If $b^m = b^n$, then $m = n$" to set the exponents equal to each other, then solve.

For example, to solve $2^{4t} = 64$ for t:

1. Rewrite the equation. $\qquad\qquad\qquad\qquad 2^{4t} = 2^6$

2. Set the exponents equal to each other, then solve for t.
$$4t = 6$$
$$t = \frac{6}{4}$$
$$= \frac{3}{2}$$

So if $2^{4t} = 64$, then $t = \dfrac{3}{2}$.

Here's another example. To solve $27^{x-1} = 81^{2x}$ for x:

1. Rewrite the equation.
$$\left(3^3\right)^{x-1} = \left(3^4\right)^{2x}$$
$$3^{3(x-1)} = 3^{8x}$$

> Remember, a function f is one-to-one if each *y*-value corresponds to one and only one *x*-value.

2. Set the exponents equal to each other, then solve for x.

$$3(x - 1) = 8x$$
$$3x - 3 = 8x$$
$$-5x = 3$$
$$x = -\frac{3}{5}$$

So if $27^{x-1} = 81^{2x}$, then $x = -\frac{3}{5}$.

In a later lesson you will learn how to solve exponential equations like $2^t = 5$, where you cannot rewrite the equation in the form $b^m = b^n$.

Sample Problems

1. Given $y = f(x) = \left(\frac{3}{4}\right)^x$, complete the table of ordered pairs below (to one decimal place), then sketch the graph of the function on the grid.

☑ a. To find the y-value for $x = 2$, substitute 2 into $y = \left(\frac{3}{4}\right)^x$, then round to one decimal place.

$$y = \left(\frac{3}{4}\right)^x$$
$$= \left(\frac{3}{4}\right)^2$$
$$= \frac{9}{16}$$
$$\approx .5625$$
$$\approx .6$$

☐ b. To find the y-value for $x = 0$, substitute 0 into $y = \left(\frac{3}{4}\right)^x$, then round to one decimal place.

$y = $ ___

☐ c. To find the y-value for $x = -3$, substitute −3 into $y = \left(\frac{3}{4}\right)^x$, then round to one decimal place.

$y \approx$ ___

☐ d. Complete the table.

x	y
3	.4
2	.6
1	.8
0	___
−1	1.3
−2	1.8
−3	___

□ e. Plot the ordered pairs, and
   join them with a smooth curve.

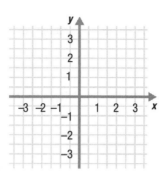

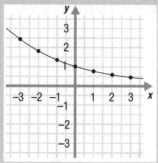

2. Use your calculator to find $e^{-7.8}$ approximated to five decimal places.

   □ a. Enter 7.8 on your calculator, then press the "±" key.

   □ b. Press the "e^x" key and write the result. _____

   □ c. Round the result of (b) to five decimal places. _____

3. Solve for x: $3^{4-2x} = 9^{8x}$

   □ a. Rewrite the equation in the
      form $b^m = b^n$, and simplify.

   $3^{4-2x} = 9^{8x}$

   $3^{4-2x} = 3^{2(8x)}$

   _____

   □ b. Set the exponents equal to
      each other, then solve for x.

   _____

   $x = $ _____

4. The half-life of radium is approximately 1600 years. This means that approximately half of an original amount of radium remains after 1600 years. Use the formula $A = A_0\left(\dfrac{1}{2}\right)^{\left(\frac{1}{1600}\right)t}$, where A is the amount of radium left after time t and A_0 is the original amount of radium, to find how long it would take for 80mg of radium to disintegrate to 20mg of radium.

   ☑ a. Substitute the values for
      A_0 and A into the formula.

   $A = A_0\left(\dfrac{1}{2}\right)^{\left(\frac{1}{1600}\right)t}$

   $20 = 80\left(\dfrac{1}{2}\right)^{\left(\frac{1}{1600}\right)t}$

   ☑ b. Simplify.

   $\dfrac{20}{80} = \left(\dfrac{1}{2}\right)^{\left(\frac{1}{1600}\right)t}$

   $\dfrac{1}{4} = \left(\dfrac{1}{2}\right)^{\left(\frac{1}{1600}\right)t}$

   □ c. Rewrite the equation in the form $b^m = b^n$. _____

   □ d. Set the exponents equal to each other,
      then solve for t.

   $2 = \dfrac{1}{1600}t$

   $t = $ _____ years

HOMEWORK

HOMEWORK PROBLEMS

Circle the homework problems assigned to you by the computer, then complete them below.

 Explain
The Exponential Function

1. Given $y = f(x) = 2^x$, complete this table of ordered pairs.

x	y
3	
2	
1	2
0	1
−1	$\frac{1}{2}$
−2	
−3	$\frac{1}{8}$

2. The graph of the function $y = f(x) = 5^x$ is shown on the grid in Figure 12.1.6. As the value of x increases, does the value of y increase or decrease?

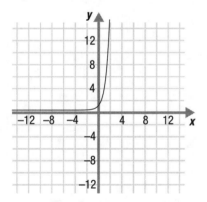

Figure 12.1.6

3. Find $e^{-2.6}$ approximated to two decimal places. (Hint: Use your calculator.)

4. If the graph of $f(x) = b^x$ passes through the point (2, 9), find $f(4)$.

5. The graphs of $y = f(x) = b^x$ and $y = g(x) = a^x$ are shown on the grid in Figure 12.1.7. If $b > 1$ and $0 < a < 1$, which graph is the graph of $y = f(x)$, and which graph is the graph of $y = g(x)$?

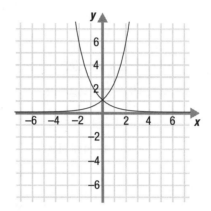

Figure 12.1.7

6. Solve for t: $4^{2t} = 64$

7. The point (2, 100) lies on the graph of an exponential function $y = b^x$. Find the base b of this exponential function.

8. Graph the function $y = f(x) = \left(\frac{2}{3}\right)^x$. (Hint: Use your calculator and round the y-values to the first decimal place.)

9. Pam's biology class is doing an experiment to study the growth of a certain type of bacteria. Suppose that this bacteria grows in such a way that $N = N_0 e^{0.04t}$, where N is the number of bacteria present at the end of t hours, and N_0 is the number of bacteria present at the start of the experiment. If Pam starts with 12 bacteria in a culture, approximately how many bacteria will be present at the end of 2 days? Round your answer to the nearest whole number. (Hint: 2 days = 48 hours)

10. Use the formula $A = P \cdot e^r$ and a calculator to find the amount A you would have one year after you invested a principal $P = \$1000$ at the rate of $r = 0.03$, compounded continuously.

11. What is the base b of the exponential function $f(x) = b^x$ if $f(-2) = \frac{1}{64}$?

12. Solve for x: $e^{3x+1} = \frac{1}{e^{x-2}}$

 APPLY

Practice Problems

Here are some additional practice problems for you to try.

The Exponetial Function

1. Circle the ordered pairs below that satisfy the function $y = \left(\frac{1}{4}\right)^x$.

    (0, 1) (1, 0)

    (–1, 4) (–1, –4)

2. Circle the ordered pairs below that satisfy the function $y = \left(\frac{2}{5}\right)^x$.

    (5, 2) (0, 1)

    (–1, 2.5) (–1, –2.5)

3. Circle the ordered pairs below that satisfy the exponential function $y = \left(\frac{2}{3}\right)^x$.

    (0, 1) (1, 0)

    (–1, –1.5) (–1, 1.5)

4. Find $e^{3.4}$ approximated to two decimal places. (Hint: Use your calculator.)

5. Find $e^{1.2}$ approximated to two decimal places. (Hint: Use your calculator.)

6. Find $e^{-2.1}$ approximated to two decimal places. (Hint: Use your calculator.)

7. Graph the function $y = f(x) = 4^x$.

8. Graph the function $y = f(x) = \left(\frac{1}{5}\right)^x$

9. Graph the function $y = f(x) = \left(\frac{1}{3}\right)^x$

10. If the graph of $f(x) = b^x$ passes through the point (2, 64), find $f(-1)$.

11. If the graph of $f(x) = b^x$ passes through the point (–3, 8), find $f(2)$.

12. If the graph of $f(x) = b^x$ passes through the point $\left(3, \frac{125}{8}\right)$, find $f(-1)$.

13. If the graph of $f(x) = b^x$ passes through the point $\left(-2, \frac{1}{16}\right)$, find $f\left(\frac{1}{2}\right)$.

14. The point (3, 27) lies on the graph of the exponential function $y = b^x$. Find the base, b, of this exponential function.

15. The point (–2, 9) lies on the graph of an exponential function $y = b^x$. Find the base, b, of this exponential function.

16. The point $\left(-2, \frac{9}{16}\right)$ lies on the graph of the exponential function $y = b^x$. Find the base, b, of this exponential function.

17. Solve for x: $9^x = 27$

18. Solve for x: $8^x = 16$

19. Solve for x: $125^x = \frac{1}{25}$

20. Solve for t: $4^{3t} = 64$

21. Solve for t: $9^{4t} = \frac{1}{81}$

22. Solve for t: $9^{5t} = \frac{1}{243}$

23. Solve for x: $e^{x+1} = e^{3x-1}$

24. Solve for x: $e^{2x+3} = \frac{1}{e^{x-9}}$

25. Solve for x: $e^{x-1} = \frac{1}{e^{2x-5}}$

26. What is the base, b, of the exponential function $f(x) = b^x$ if $f(3) = 64$?

27. What is the base, b, of the exponential function $f(x) = b^x$ if $f(2) = \frac{1}{49}$?

28. What is the base, b, of the exponential function $f(x) = b^x$ if $f(-2) = \frac{25}{64}$?

EVALUATE

Practice Test

Take this practice test to be sure that you are prepared for the final quiz in Evaluate.

1. Circle the ordered pairs below that satisfy the exponential function $y = 2^x$.

   (1, 2)   (0, 1)

   (−4, −8) $\left(-2, \dfrac{1}{4}\right)$

2. Given $f(x) = \left(\dfrac{1}{2}\right)^x$,
   complete the table of ordered pairs below. Round your answers to four decimal places.

x	y
2	$\dfrac{1}{4}$
1.5	
1	
0	1
−1	2
−1.5	
−2	

3. Circle the statements below that are true for an exponential function of the form $y = f(x) = b^x$.

   The base b can be negative.

   The domain of f is all positive real numbers.

   The range of f includes y-values that are negative.

   The function f is one-to-one.

4. Find the approximate value of each of the following to 2 decimal places. (Hint: You may want to use a calculator.)

   a. e^3 b. e^{-1}

5. The compound interest formula is given by $A = P \cdot \left(1 + \dfrac{r}{n}\right)^n$, where A is the amount of money in an account after one year, P is the amount invested, n is the number of times the investment is compounded in a year, and r is the rate at which it is compounded. Determine the amount of money in an account at the end of one year if the original investment is $200 compounded every 4 months at a rate of 3%. (Hint: Since the money is compounded every 4 months, $n = 3$. Also: 3% = .03)

6. A radioactive isotope is said to decay over time. That is, after t years, the original amount of an isotope, N_0 grams, decays until the amount is N grams, where N is defined as $N = N_0\left(\dfrac{1}{2}\right)^{\frac{t}{200}}$. How much of this isotope, in terms of N_0, remains at time $t = 600$?

7. Solve this exponential equation for x: $3^{2x + 1} = 27$

8. Solve this exponential equation for x: $e^{x - 2} = \dfrac{1}{e^{3x + 4}}$

LESSON 12.2 – LOGS AND THEIR PROPERTIES

$$Y = 3^X$$

$$\log_{10} X^2 = 2\log_{10} X$$

$$\log_6 36 = L$$

OVERVIEW

Here's what you'll learn in this lesson:

The Logarithm Function

a. Converting from exponents to logarithms and from logarithms to exponents

b. Graphing a logarithmic function

Logarithmic Properties

a. The algebra of logarithmic functions

Suppose you see a newspaper ad for a savings and loan that promises to multiply your savings by 10 in 30 years. Before you open an account, you want to know if the claim is true. Or, suppose a truck load of hazardous materials is accidentally spilled into a lake that supplies drinking water to your community. Local health officials say that the water will be safe to drink after 10 days, but you want to be sure that this is true. In both of these situations, you can use logarithms to help you find the answer you are looking for.

In this lesson, you'll learn about logarithms and how to graph logarithmic functions. In addition, you'll learn some properties of logarithms.

EXPLAIN

THE LOGARITHM FUNCTION

Summary

An Introduction to the Logarithmic Function

You have already worked with exponential functions, and you have learned many of the properties of exponential functions. You have also learned how to graph exponential functions. For example, Figure 12.2.1 shows the graph of the exponential function $y = 3^x$.

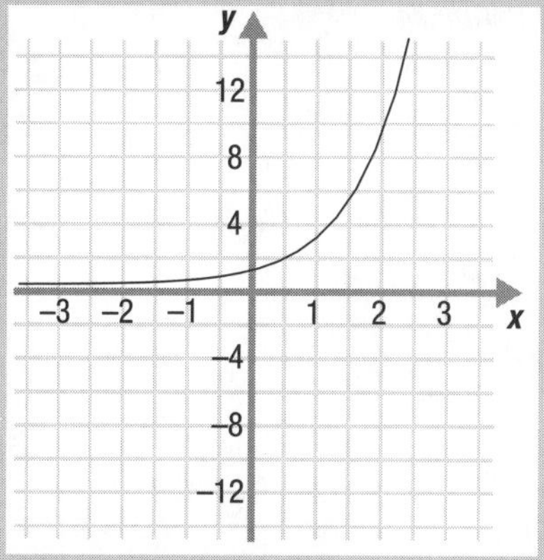

Figure 12.2.1

Here is a table of ordered pairs that satisfy $y = 3^x$:

x	y
0	1
1	3
2	9
−1	$\frac{1}{3}$
−2	$\frac{1}{9}$

When you graph $y = 3^x$, you begin by making a table of ordered pairs. To do this, you choose values for x, and then calculate their corresponding values for y.

For example, if you choose $x = 2$, then $y = 3^2 = 9$.

If you go in the reverse direction – that is, if you start with a value for y and calculate the corresponding value for x – then you obtain a new function called a logarithm. The logarithmic function is the inverse function of the exponential function. That is, the logarithmic function "undoes" what the exponential function does.

In the example above, if you choose $y = 9$, then $9 = 3^x$.

To solve for x, you can write this as a logarithm: $x = \log_3 9$.

Switching Between Exponential and Logarithmic Notation

So you now have two equivalent ways of writing the same information. Here is a statement in exponential form:

exponent ↙

$3^2 = 9$

↑ base

You can also write the same statement in logarithmic form:

In both cases the base is 3. In exponential form the number 2 is called the exponent. In logarithmic form it is called the logarithm.

Here's a way to generalize this:

$$b^L = x \text{ is the same as } \log_b x = L$$

These two statements are equivalent. They contain the same information written in different ways. In both cases, b is the base and is a positive number and $b \neq 1$. The x is also a positive number.

Here are some more examples:

Exponential Form	Logarithmic Form
$b^L = x$	$\log_b x = L$
$5^2 = 25$	$\log_5 25 = 2$
$\left(\dfrac{1}{3}\right)^4 = \dfrac{1}{81}$	$\log_{\frac{1}{3}}\left(\dfrac{1}{81}\right) = 4$

So any exponential statement can be written as a logarithmic statement. And any logarithmic statement can be written as an exponential statement. To do this:

1. Rewrite the statement, using the fact that $b^L = x$ is the same as $\log_b x = L$.

For example, to write $\left(\dfrac{1}{2}\right)^3 = 8$ in logarithmic form:

    1. Rewrite the statement, using the fact that $b^L = x$ is the same as $\log_b x = L$.
          $\log_{\frac{1}{2}} 8 = 3$

Similarly, to write $\log_{10} 10,000 = 4$ in exponential form:

    1. Rewrite the statement, using the fact that $b^L = x$ is the same as $\log_b x = L$.
          $10^4 = 10,000$

Finding Logarithms

By rewriting a logarithmic statement in exponential form, you can find the value of the logarithm. Here are the steps:

1. Set the logarithm equal to L to create an equation. The variable L is the unknown value of the logarithm.

2. Rewrite the equation in exponential form.

3. Rewrite the constant term using the base.

4. Solve for L.

You've already seen this idea of writing the same information in two different ways when you studied powers and roots. For example, the statements $3^2 = 9$ and $\sqrt{9} = 3$ express the same information.

For example, to find $\log_6 36$:

1. Set the logarithm equal to L to create an equation. $\log_6 36 = L$

2. Rewrite in exponential form. $6^L = 36$

3. Rewrite the constant term, 36, using the base, 6. $6^L = 6^2$

4. Solve for L. $L = 2$

So, $\log_6 36 = 2$.

Here's a similar example that uses fractions. To find $\log_3\left(\frac{1}{81}\right)$:

1. Set the logarithm equal to L to create an equation. $\log_3\left(\frac{1}{81}\right) = L$

2. Rewrite in exponential form. $3^L = \frac{1}{81}$

3. Rewrite the constant term, $\frac{1}{81}$, using the base, 3. $3^L = 3^{-4}$

4. Solve for L. $L = -4$

So, $\log_3\left(\frac{1}{81}\right) = -4$.

When the constant cannot be easily rewritten as a power of the base, you will need to use a calculator. You'll learn how to do this in another lesson.

Graphing Logarithmic Functions

You can graph a logarithmic function by using your knowledge of inverses and exponential functions. Here are the steps for graphing a logarithmic function:

1. Write the logarithmic function in the form $y = \log_b x$.

2. Switch x and y to get $x = \log_b y$.

3. Rewrite in exponential form, $y = b^x$.

4. Graph this exponential function, $y = b^x$.

5. Graph the original logarithmic function $y = \log_b x$ by reflecting the exponential function $y = b^x$ about the line $y = x$.

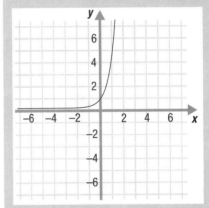

Figure 12.2.2

For example, to graph the logarithmic function $y = \log_5 x$:

1. The logarithmic function is in the form $y = \log_b x$. $y = \log_5 x$

2. Switch x and y. $x = \log_5 y$

3. Rewrite in exponential form, $y = b^x$. $y = 5^x$

4. Graph this exponential function, $y = 5^x$. See Figure 12.2.2.

5. Graph $y = \log_5 x$ by reflecting $y = 5^x$ about the line $y = x$. See Figure 12.2.3.

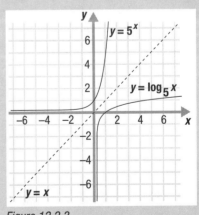

Figure 12.2.3

Sample Problems

1. Write in logarithmic form: $7^3 = 343$

    ☑ a. Rewrite the statement, using the fact that $b^L = x$ is the same as $\log_b x = L$. $\log_7 343 = 3$

2. Write in exponential form: $\log_{11} 200 = x$

    ☐ a. Rewrite the statement, using the fact that $b^L = x$ is the same as $\log_b x = L$.   _____

3. Find: $\log_4 1024$

    ☑ a. Set the logarithm equal to L to create an equation. $\log_4 1024 = L$

    ☐ b. Rewrite the equation in exponential form.   _____

    ☐ c. Rewrite 1024 using the base, 4. $4^L =$ _____

    ☐ d. Solve for L. $L =$ _____

4. Graph the function $y = \log_2 x$.

    ☑ a. Write the logarithmic function in the form $y = \log_b x$. $y = \log_2 x$

    ☑ b. Switch x and y. $x = \log_2 y$

    ☑ c. Rewrite in exponential form, $y = b^x$. $y = 2^x$

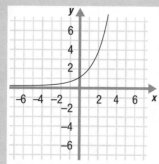

    ☐ d. Graph this exponential function, $y = 2^x$.

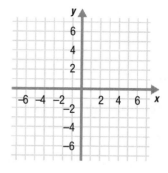

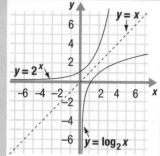

    ☐ e. Graph $y = \log_2 x$ by reflecting $y = 2^x$ about the line $y = x$.

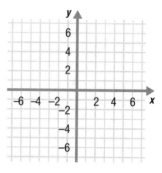

5. Graph the function $y = \log_{\frac{1}{3}} x$.

☑ a. Write the logarithmic function in the form $y = \log_b x$.

$y = \log_{\frac{1}{3}} x$

☑ b. Switch x and y.

$x = \log_{\frac{1}{3}} y$

☐ c. Rewrite in exponential form, $y = b^x$.

_____

☐ d. Graph this exponential function, $y = \left(\frac{1}{3}\right)^x$.

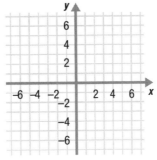

☐ e. Graph $y = \log_{\frac{1}{3}} x$ by reflecting

$y = \left(\frac{1}{3}\right)^x$ about the line $y = x$.

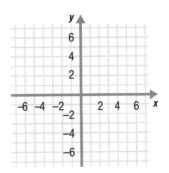

c. $y = \left(\frac{1}{3}\right)^x$

d.

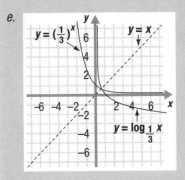

e.

$y = \left(\frac{1}{3}\right)^x$ $y = x$

$y = \log_{\frac{1}{3}} x$

LOGARITHMIC PROPERTIES

Summary

Logarithms can be rewritten using a variety of algebraic properties. These properties are like the properties that you already know for exponents.

Here are the fundamental properties of exponents.

Properties of Exponents	Examples
$b^1 = b$	$7^1 = 7$
$b^0 = 1$	$13^0 = 1$
$b^x b^y = b^{x+y}$	$11^5 11^8 = 11^{5+8} = 11^{13}$
$\dfrac{b^x}{b^y} = b^{x-y}$	$\dfrac{11^8}{11^5} = 11^{8-5} = 11^3$
$\left(b^x\right)^n = b^{nx}$	$\left(8^5\right)^7 = 8^{5 \cdot 7} = 8^{35}$

Here are the corresponding fundamental properties of logarithms, as well as several others.

Properties of Logarithms	Examples
$\log_b b = 1$	$\log_9 9 = 1$
$\log_b 1 = 0$	$\log_{17} 1 = 0$
Log of a Product: $\log_b uv = \log_b u + \log_b v$	$\log_{11} 17x = \log_{11} 17 + \log_{11} x$
Log of a Quotient: $\log_b \dfrac{u}{v} = \log_b u - \log_b v$	$\log_{11} \dfrac{x}{17} = \log_{11} x - \log_{11} 17$
Log of a Power: $\log_b u^n = n \cdot \log_b u$	$\log_8 13^{22} = 22 \cdot \log_8 13$
$b^{\log_b x} = x$	$5^{\log_5 23} = 23$
$\log_b \dfrac{1}{v} = -\log_b v$	$\log_6 \dfrac{1}{32} = -\log_6 32$
$\log_b b^n = n$	$\log_{14} 14^{29} = 29$

As with the properties of exponents, you can use one or several properties of logarithms to rewrite statements that contain logarithms. Here are some examples.

To find $\log_{17} 17^{2x-3}$:

    1. Use the property $\log_b b^n = n$. $\log_{17} 17^{2x-3} = 2x - 3$

So, $\log_{17} 17^{2x-3} = 2x - 3$.

To rewrite the expression $\log_{11}[x^3(5x + 17)]$:

1. Use the log of a product property.

$$\log_{11}[x^3(5x + 17)]$$
$$= \log_{11}x^3 + \log_{11}(5x + 17)$$

2. Use the log of a power property.

$$= 3 \cdot \log_{11}x + \log_{11}(5x + 17)$$

So, $\log_{11}[x^3(5x + 17)] = 3 \cdot \log_{11}x + \log_{11}(5x + 17)$.

To rewrite $3\log_b x - 5\log_b y + 7\log_b z$ as a single logarithm:

1. Use the log of a power property.

$$3\log_b x - 5\log_b y + 7\log_b z$$
$$= \log_b x^3 - \log_b y^5 + \log_b z^7$$

2. Use the log of a quotient property.

$$= \log_b \frac{x^3}{y^5} + \log_b z^7$$

3. Use the log of a product property.

$$= \log_b \frac{x^3 z^7}{y^5}$$

So, $3\log_b x - 5\log_b y + 7\log_b z = \log_b \frac{x^3 z^7}{y^5}$.

You have to work from left to right when you rewrite logarithms. In this example, you subtract the two logs, then you add.

Sample Problems

Answers to Sample Problems

1. Simplify: $\log_{3x - 7}(3x - 7)$

☑ a. Use the property $\log_b b = 1$.

$$\log_{3x - 7}(3x - 7) = 1$$

2. Simplify: $y^{\log_y(13z + 6x - 119)}$

☑ a. Use the property $b^{\log_b x} = x$.

$$y^{\log_y(13z + 6x - 119)}$$
$$= 13z + 6x - 119$$

3. Rewrite this expression: $\log_{37}\frac{3p}{qr}$

☑ a. Use the log of a quotient property.

$$\log_{37}\frac{3p}{qr}$$
$$= \log_{37}3p - \log_{37}qr$$

☐ b. Use the log of a product property.

$$= \underline{\hspace{4cm}}$$

b. $\log_{37}3 + \log_{37}p - (\log_{37}q + \log_{37}r)$

or $\log_{37}3 + \log_{37}p - \log_{37}q - \log_{37}r$

4. Write $\log_9 \frac{1}{2}$ using $\log_9 2$.

☐ a. Use the property $\log_b \frac{1}{v} = -\log_b v$.

$$\log_9 \frac{1}{2}$$
$$= \underline{\hspace{3cm}}$$

a. $-\log_9 2$

5. Write as a single logarithm not containing exponents: $\log_w f^7 + \log_w g^7$

☐ a. Use the log of a product property.

$$\log_w f^7 + \log_w g^7$$
$$= \underline{\hspace{3cm}}$$

a. $\log_w f^7 g^7$ or $\log_w (fg)^7$

☐ b. Use the log of a power property.

$$= \underline{\hspace{3cm}}$$

b. $7\log_w (fg)$

⬡ **EXPLORE**

Sample Problems

On the computer you used the grapher to graph various logarithmic functions and compared their behavior. You also explored the algebraic properties of logarithms. Below are some additional problems to explore.

1. Graph these logarithmic functions on the same set of axes, and answer the questions below.

$$y = \log_2 x$$

$$y = \log_3 x$$

$$y = \log_e x$$

☑ a. To graph the functions, start by finding the inverse function of each logarithmic function.

Function	Inverse
$y = \log_2 x$	$y = 2^x$
$y = \log_3 x$	$y = 3^x$
$y = \log_e x$	$y = e^x$

☑ b. Then graph each exponential function.

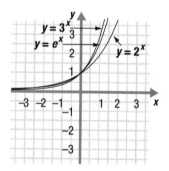

☐ c. Finally, reflect the graph of each exponential function about the line $y = x$ to graph each corresponding logarithmic function.

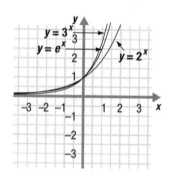

c.

☐ d. What is the domain of each of the logarithmic functions? _____

d. $x > 0$

Notice by looking at the graphs that they are not defined for $x \le 0$.

□ e. What is the range of each of the logarithmic functions? _____

□ f. Find the point which is common to all three logarithmic graphs. _____

□ g. For $x > 1$, list the three logarithmic graphs in order of their distance from the x-axis. Start with the closest one.
_____
_____
_____

□ h. For $x < 1$, list the three logarithmic graphs in order of their distance from the y-axis.
_____
_____
_____

Answers to Sample Problems

e. *All real numbers.*
   Notice by looking at the graphs that y-values can be positive or negative.

f. *(1, 0)*

g. $y = \log_3 x$
   $y = \log_e x$
   $y = \log_2 x$

h. $y = \log_3 x$
   $y = \log_e x$
   $y = \log_2 x$

2. Graph these logarithmic functions on the same set of axes, and answer the questions below.

$y = \log_2 x$

$y = \log_{\frac{1}{2}} x$

☑ a. To graph the functions, start by finding the inverse function of each logarithmic function.

Function	Inverse
$y = \log_2 x$	$y = 2^x$
$y = \log_{\frac{1}{2}} x$	$y = \frac{1}{2}^x$

☑ b. Then graph each exponential function.

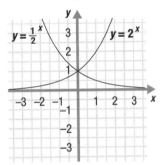

□ c. Finally, reflect the graph of each exponential function about the line $y = x$ to graph each corresponding logarithmic function.

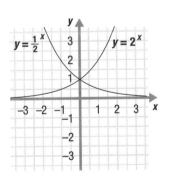

c.

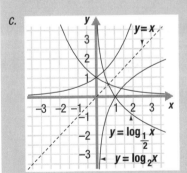

d. $x > 0$

Notice by looking at the graphs that they are not defined for $x \le 0$.

□ d. What is the domain of each of the logarithmic functions? _____

☐ e. What is the range of each
    of the logarithmic functions? _____

☐ f. Find the point which is common to _____
    both logarithmic graphs.

☐ g. Does the graph of $y = \log_2 x$ increase or decrease _____
    as you move from left to right along the x-axis?
    How about $y = \log_{\frac{1}{2}} x$? _____

3. This is a formula for political fund raising: $C = a + b\log_{10} \dfrac{M}{1000}$. Here, the variable M
   describes the number of funding requests that have been mailed, and the variable C
   describes the number of contributions received in response to the requests. The
   numbers a and b are constants which are determined by the nature of the request
   and the contents of each specific mailing.

   Here is a graph which illustrates the formula for a given a and b.

☑ a. Use the graph to find the number of   When $M = 1000$,
     contributions, C, corresponding $C = 5000$.
     to $M = 1000$ mailings.

☐ b. Use the graph to find the number of   When $M =$ _____,
     mailings, M, required to produce $C = 6000$.
     $C = 6000$ contributions.

☐ c. What happens to the contributions as as _____
     the number of mailings increase? _____

4. Calculate the following logarithms: $\log_2 4$, $\log_2 8$, $\log_2 16$, $\log_2 32$. Use your answers to illustrate the log of a product property, the log of a quotient property and the log of a power property, respectively.

☑ a. Calculate $\log_2 4$.

- Set $\log_2 4$ equal to x. $\log_2 4 = x$

- Rewrite the expression in exponential form. $2^x = 4$

- Solve for x. $x = 2$

☐ b. Calculate $\log_2 8$.

- Set $\log_2 8$ equal to x. _____

- Rewrite the expression in exponential form. _____

- Solve for x. $x =$ _____

b. $\log_2 8 = x$

  $2^x = 8$

  3

☐ c. Calculate $\log_2 16$.

- Set $\log_2 16$ equal to x. _____

- Rewrite the expression in exponential form. _____

- Solve for x. $x =$ _____

c. $\log_2 16 = x$

  $2^x = 16$

  4

☐ d. Calculate $\log_2 32$.

- Set $\log_2 32$ equal to x. _____

- Rewrite the expression in exponential form. _____

- Solve for x. $x =$ _____

d. $\log_2 32 = x$

  $2^x = 32$

  5

☑ e. Illustrate the log of a product property. Substitute the values you found in (a), (b), and (d) for $\log_2 4$, $\log_2 8$, and $\log_2 32$.

$\log_2 4 + \log_2 8 = \log_2(4 \cdot 8)$
$= \log_2 32$
$2 + 3 = 5$

☐ f. Illustrate the log of a quotient property. Substitute the values you found in (d), (b) and (a) for $\log_2 32$, $\log_2 8$, and $\log_2 4$.

$\log_2 32 - \log_2 8 = \log_2 \frac{32}{8}$
$= \log_2 4$

____ – ____ = ____

f. 5, 3, 2

☐ g. Illustrate the log of a power property. Substitute the values you found in (a) and (c) for $\log_2 4$ and $\log_2 16$.

$2 \cdot \log_2 4 = \log_2 4^2$
$= \log_2 16$

_____ = ____

g. 2 · 2, 4

HOMEWORK

Homework Problems

Circle the homework problems assigned to you by the computer, then complete them below.

☀ **Explain**

The Logarithm Function

1. Write this exponential statement in logarithmic form:
    $10^4 = 10000$

2. Find: $\log_5 125$

3. In order to graph the function $y = \log_7 x$, you can use its inverse function. What is this inverse function?

4. Write this logarithmic statement in exponential form:
    $\log_3 500 = x$

5. Find: $\log_2 \frac{1}{64}$

6. Graph the function $y = \log_2 x$.

7. Write this exponential statement in logarithmic form: $7^{12} = y$

8. Find: $\log_{3.2} 3.2$

9. Suppose that when you graduate from college, you deposit one dollar in a savings account for your retirement, t years later. The final amount in your savings account after t years is given by A. If you receive 7% interest compounded annually, the formula which tells you how many years, t it takes to accumulate the amount A is $t = \log_{1.07} A$. Graph this function.

10. The formula for the rate of decay of a radioactive chemical is given by $T = \log_B\left(\frac{R}{S}\right)$. Here, S is the starting amount of the chemical in grams and R is the amount of the chemical in grams remaining after T years. Write this formula in exponential form.

11. Find: $\log_{\frac{1}{4}} 16$

12. Graph the function $y = \log_{\frac{1}{3}} x$.

Logarithmic Properties

13. Find: $\log_{17} 17^y$

14. Rewrite as a single logarithm: $\log_d 3x + \log_d 4y$

15. Rewrite using the log of a power property: $14\log_{13} 12$

16. Simplify: $10^{\log_{10} 12abc}$

17. Rewrite using the log of a quotient property: $\log_2 \frac{3ab}{7cd}$

18. Rewrite using the log of a product property and the log of a power property: $\log_5 x^2 y^3$

19. Simplify: $\log_{3x-y} 3x - y$

20. Rewrite using the log of a product property to get an expression with four terms: $\log_B 7xyz$

21. The magnitude of an earthquake of intensity I as compared to one of minimum intensity M is measured on the Richter scale as $R = \log_{10} \frac{I}{M}$. Use a property of logarithms to rewrite this formula for R in terms of logarithms that do not contain fractions.

22. The pH of a particular fruit juice is given by the formula $x = -\log_{10}(1.56 \cdot 10^{-4})$. Find x to four decimal places, given that $\log_{10} 1.56 = 0.1931$. (Hint: Use properties of logarithms to get your answer.)

23. Find: $\log_x\left(\frac{1}{x}\right)$

24. Write as a single logarithm: $3\log_{16} u + 5\log_{16} v - 8\log_{16} w$

25. The following functions are graphed on the grid in Figure 12.2.4:

    $y = \log_3 x$

    $y = \log_{15} x$

    $y = \log_7 x$

    Label each graph with the appropriate function.

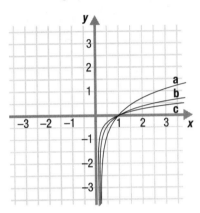

Figure 12.2.4

26. The following functions are graphed on the grid in Figure 12.2.8:

    $y = \log_5 x$

    $y = \log_{\frac{1}{5}} x$

    Label each graph with the appropriate function. About what line can you reflect one graph in order to get the other?

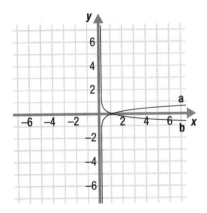

Figure 12.2.5

27. The log of a product property allows you to rewrite $\log_4 64$ as $\log_4 16 + \log_4 4$. Use numbers which are powers of 2 to:

    a. rewrite $\log_4 64$ in a different way using the Log of a Product Property.

    b. rewrite $\log_4 64$ using the Log of a Quotient Property.

    c. rewrite $\log_4 64$ using the Log of a Power Property.

28. Describe how the graph of the logarithm $y = \log_b x$ changes as you increase the value of the base b for $b > 1$. Use the graph of $y = \log_5 x$ shown in Figure 12.2.6 to help you.

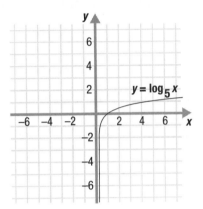

Figure 12.2.6

29. Compare a logarithmic graph $y = \log_b x$ with base $b < 1$ and another logarithmic graph with base $b > 1$. What is the same on both graphs?

30. Circle the expressions below that are equal.

    $\log_{\frac{1}{2}} 4$ -2 $-\log_2 4$ $\log_2 \frac{1}{4}$

 APPLY

Practice Problems

Here are some additional practice problems for you to try.

The Logarithm Function

1. Write this exponential statement in logarithmic form: $2^3 = 8$

2. Write this exponential statement in logarithmic form: $4^5 = 1024$

3. Write this exponential statement in logarithmic form: $5^4 = 625$

4. Write this exponential statement in logarithmic form: $x^2 = 225$

5. Write this exponential statement in logarithmic form: $3^x = a$

6. Write this exponential statement in logarithmic form: $b^x = 36$

7. Find: $\log_2 64$

8. Find: $\log_3 27$

9. Find: $\log_3 81$

10. Find: $\log_7 49$

11. Find: $\log_5 \dfrac{1}{125}$

12. Find: $\log_{\frac{1}{4}} 16$

13. What is the inverse of the function $f(x) = \log_{12} x$?

14. What is the inverse of the function $f(x) = \log_7 x$?

15. What is the inverse of the function $f(x) = \log_{11} x$?

16. What is the inverse of the function $f(x) = \log_{2.5} x$?

17. What is the inverse of the function $f(x) = \log_b x$?

18. What is the inverse of the function $f(x) = \log_a x$?

19. Write this logarithmic statement in exponential form: $\log_5 125 = 3$

20. Write this logarithmic statement in exponential form: $\log_4 \dfrac{1}{16} = -2$

21. Write this logarithmic statement in exponential form: $\log_8 512 = x$

22. Write this logarithmic statement in exponential form: $\log_b 81 = 4$

23. Write this logarithmic statement in exponential form: $\log_6 x = 5$

24. Write this logarithmic statement in exponential form: $\log_4 1024 = x$

25. Graph the function $y = \log_3 x$.

26. Graph the function $y = \log_2 x$.

27. Graph the function $y = \log_{\frac{1}{3}} x$.

28. Graph the function $y = \log_{\frac{1}{2}} x$.

Logarithmic Properties

29. Rewrite using the log of a power property: $\log_2 33^4$

30. Rewrite using the log of a power property: $a\log_b 8$

31. Rewrite using the log of a power property: $17\log_3 25$

32. Rewrite using the log of a quotient property: $\log_7 \dfrac{5}{3m}$

33. Rewrite using the log of a quotient property: $\log_b \dfrac{7x}{8y}$

34. Rewrite using the log of a quotient property: $\log_5 \dfrac{6y}{11z}$

35. Rewrite using the log of a product property to get an expression with two terms: $\log_3 5x$

36. Rewrite using the log of a product property to get an expression with three terms: $\log_b 15mn$

37. Rewrite using the log of a product property to get an expression with three terms: $\log_b 21xy$

38. Rewrite as a single logarithm: $\log_c 5 + \log_c 7$

39. Rewrite as a single logarithm: $\log_b 3 + \log_b 4$

40. Rewrite as a single logarithm: $\log_b 12 + \log_b z$

41. Rewrite as a single logarithm: $\log_5 4x^7 - \log_5 2x$

42. Rewrite as a single logarithm: $\log_2 9y^5 - \log_2 3y^2$

43. Rewrite as a single logarithm: $\log_7 6xy - \log_7 12xy$

44. Rewrite as a single logarithm: $2\log_y a + 7\log_y b - \log_y c$

45. Rewrite as a single logarithm: $3\log_x a + 5\log_x b - \log_x c$

46. Rewrite as a single logarithm: $a\log_3 x - b\log_3 y - c\log_3 z$

47. Simplify: $\log_3 3$

48. Simplify: $7^{\log_7 3}$

49. Simplify: $\log_{2xy}(2xy)^{-3}$

50. Simplify: $5^{\log_5 3}$

51. Simplify: $\log_{3z}(3z)^{-2}$

52. Simplify: $(2m)^{\log_{2m} -6}$

53. Rewrite using the log of a quotient property and the log of a power property: $\log_5 \dfrac{z}{y^4}$

54. Rewrite using the log of a product property and the log of a power property: $\log_7 xy^5$

55. Rewrite using the properties of logarithms to get an expression with three terms: $\log_3 \dfrac{x}{2y}$

56. Rewrite using the properties of logarithms to get an expression with three terms: $\log_{11} \dfrac{x^2 y^3}{z}$

PRACTICE TEST

Take this practice test to make sure you are ready for the final quiz in Evaluate.

1. Write this exponential statement in logarithmic form:
    $18^{12x} = yz$

2. Find the value of $\log_6 \frac{1}{216}$.

3. Graph the function $y = \log_5 x$.

4. The formula for the growth rate of an experimental bacterium is given by the exponential function $P = Ie^{kt}$. Here, I is the starting population number, P is the population after t years, and k is a proportionality constant determined by the laboratory conditions. Write this formula in logarithmic form.

5. Simplify: $15^{\log_{15} xyz}$

6. Using the log of a product property twice, rewrite:
    $\log_{29} 19AB$

7. Write as a single logarithm: $\log_w 11 - \log_w x - \log_w y$

8. Use properties of logarithms to rewrite as an expression with three terms:

    $\log_b \frac{17x^5}{y^{12}}$

9. Graph the functions $y = \log_a x$ and $y = \log_b x$, where you know that a and b are greater than 1 and $a > b$. Use the grid in Figure 12.2.7.

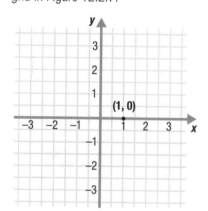

Figure 12.2.7

10. Consider the graphs shown in Figure 12.2.8.

    a. Identify which graph represents an exponential function with base less than 1.

    b. Identify which graph represents a logarithmic function with base greater than 1.

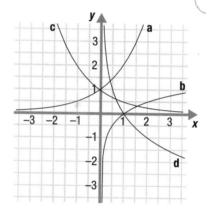

Figure 12.2.8

11. Use properties of logarithms to determine whether the following is true:

    $\log_5 25 = -\log_{\frac{1}{5}} 25$

12. Suppose that you have a formula to measure the loudness of sounds on a logarithmic scale. You do not know what base to use for your logarithms, but you have determined by experiment that $\log_b 2 = 3.4$ and that $\log_b 3 = 5.4$. Use this information and properties of logarithms to calculate $\log_b 24$.

$$\log_e X = \ln X$$

$$\log X = \log_{10} X$$

$$\log 2 + 3(\log X)$$

$$\log_x \frac{2}{3} = -1$$

OVERVIEW

Here's what you'll learn in this lesson:

Natural and Common Logs

a. *Base e and natural logarithms*

b. *Finding logs in base 10 and powers of 10 using a calculator*

c. *Finding logs in base e and powers of e using a calculator*

d. *Change of base formula*

Solving Equations

a. *Solving logarithmic equations*

b. *Solving exponential equations*

Have you ever wondered how to measure the loudness of a noise like an ambulance siren? Or pondered the difference between earthquakes of magnitudes 5 and 6 on the Richter scale? Or thought about how medical examiners can determine the time of someone's death? Well in this lesson, you'll learn more about logarithms, which can be used to investigate all of these things.

In particular, you'll learn about two kinds of logarithms that are useful in many applications: common logarithms and natural logarithms. In addition, you'll see how to use what you've already learned about logarithms to solve equations containing logarithms and to solve exponential equations.

 EXPLAIN

NATURAL AND COMMON LOGS

Summary

Logarithms were invented by the Scottish mathematician John Napier (1550 – 1617) in the sixteenth century to make multiplication and division of large numbers easier. Using properties of logarithms, a multiplication or division problem can be replaced with an easier addition or subtraction problem. Today you can often use a calculator or computer to solve problems with large numbers. But you can also use logarithms with base 10 or base e to help you solve problems in the fields of chemistry, earth science, physics, economics, and environmental studies.

Notation for Common Logarithms

You have learned about logarithms such as $\log_{10} 100$. Here, the base is 10. Logarithms with base 10 are called common logarithms and they were commonly used for computation in the base 10 number system.

Here are some examples of common logarithms:

$$\log_{10} 22 \qquad \log_{10} 311 \qquad \log_{10} 12 \qquad \log_{10} 1667$$

Common logarithms are often written without the base. For example, the logarithms above can be written respectively as:

$$\log 22 \qquad \log 311 \qquad \log 12 \qquad \log 1667$$

Sometimes, you can find the value of common logarithms without the use of a calculator by switching from logarithmic to exponential notation.

To find the value of a common logarithm $\log b$ when b is a power of 10:
1. Set the logarithm equal to x.
2. Rewrite the log in exponential form.
3. Rewrite the result of (2) in the form $10^x = 10^n$, then solve for x.

For example, find the value of $\log_{10} 1000$:

1. Set the logarithm equal to x. $\log_{10} 1000 = x$
2. Rewrite in exponential form. $10^x = 1000$
3. Rewrite in the form $10^x = 10^n$, $10^x = 10^3$
      then solve for x. $x = 3$

So $\log_{10} 1000 = 3$.

Remember, $\log_{10} 100 = 2$ because $10^2 = 100$.

In general, $\log x = \log_{10} x$. (Here, $x > 0$.)

Finding Logs in Base 10 and Powers of 10 Using a Calculator

You can find the value of a common logarithm when you can't rewrite the exponential form of the logarithm in the form $10^x = 10^n$ by using a calculator.

To find the approximate value of the common logarithm log a using a calculator:
1. Enter the number a in the calculator.
2. Press the "log" key.
3. Round the result of (2) to the desired number of decimal places.

For example, find the value of $\log_{10} 311$ rounded to three decimal places:

    1. Enter 311 in the calculator.

    2. Press the "log" key.   2.4927604

    3. Round to three decimal places.   2.493

So $\log_{10} 311 \approx 2.493$.

So $\log_{10} 311 \approx 2.493$.

Remember you use the symbol "$\approx$" to mean "is approximately equal to."

Sometimes, you are given the common logarithm of a number, and you want to find the number itself. You can work backwards to find this number using a calculator when necessary. In general, to find the number whose common logarithm is k:
1.   Let x be the number you are trying to find, and let $\log_{10} x = k$.
2.   Rewrite the equation in exponential form, $10^k = x$, and solve for x using a calculator if necessary.
3.   If necessary, round your answer.

For example, suppose you want to find the number whose common logarithm is 4.

    1. Substitute 4 for k in the $\log_{10} x = 4$
       equation $\log_{10} x = k$.

    2. Rewrite the equation in exponential $10^4 = x$
       and solve for x. $x = 10,000$

So 10,000 is the number whose common logarithm is 4.

Here's how to use your calculator to find $10^{-2.23}$ to two decimal places.

1. Enter 2.23, then press the "±" key.

2. Press the "10^x" key.

3. Round to two decimal places.
   So, $x \approx 0.01$.

Here's an example where you will need to use your calculator. To find an approximation of the number (to two decimal places) whose common logarithm is −2.23:

    1. Substitute −2.23 for k in the $\log_{10} x = -2.23$
       equation $\log_{10} x = k$.

    2. Rewrite the equation in exponential $10^{-2.23} = x$
       form and solve for x using $x \approx 0.0058884$
       your calculator.

    3. Round to two decimal places. $x \approx 0.01$

So the number whose common logarithm is −2.23 is approximately 0.01.

Properties of Common Logarithms

You are already familiar with some properties of logarithms. For example, $\log_b b = 1$ since $b^1 = b$. (Here, the base b is a positive number and $b \neq 1$.) When the base b is 10, this property states that $\log_{10} 10 = 1$ or $\log 10 = 1$ because $10^1 = 10$.

Here are some other properties of logarithms written for the case when the base b is 10. Here $x > 0$, $u > 0$, and $v > 0$.

Name of Property	Property
Log of a Product	$\log uv = \log u + \log v$
Log of a Quotient	$\log \dfrac{u}{v} = \log u - \log v$ When $u = 1$, you have this special case: $\log \dfrac{1}{v} = -\log v$
Log of a Power	$\log u^n = n \cdot \log u$ When $u = 10$, you have this special case: $\log 10^n = n$
Other Properties	$\log 10 = 1$ $\log 1 = 0$ $10^{\log x} = x$

Here's an example of how you can use these properties to write $\log 5 + 3[\log (x - 1)]$ as a single logarithm:

$$\log 5 + 3[\log (x - 1)]$$

1. Use the log of a power property. $\qquad = \log 5 + \log (x - 1)^3$

2. Use the log of a product property. $\qquad = \log[5(x - 1)^3]$

So $\log 5 + 3[\log (x - 1)]$ written as a single logarithm is $\log[5(x - 1)^3]$.

Notation for Natural Logarithms

You have learned about common logarithms (logarithms with base $b = 10$). Now you will learn about logarithms with base $b = e$. These logarithms are called natural logarithms. Natural logs have many applications such as describing the growth of populations and the growth of money invested at compound interest.

Here are some examples of natural logarithms:

$$\log_e 22 \qquad \log_e 311 \qquad \log_e 12 \qquad \log_e 1667$$

Natural logarithms are often written without the base, and abbreviated "ln." For example, the logarithms above can be written respectively as:

$$\ln 22 \qquad \ln 311 \qquad \ln 12 \qquad \ln 1667$$

Recall that e is an irrational number that lies between 2 and 3 and is approximately equal to 2.718.

Natural *logarithms get their name from the exponential function with base e,* $y = e^x$*, which is called the* ***natural*** *exponential function. This function arises* ***naturally*** *in many applications in the biological and social sciences.*

In general, $\ln x = \log_e x$*. (Here,* $x > 0$*.) Sometimes it may help to rewrite ln x as* $\log_e x$ *to remind yourself that you're working with base e, especially when switching to exponential form without the use of a calculator.*

Sometimes you can find the value of a natural logarithm without the use of a calculator by switching from logarithmic to exponential notation, just as you did with common logs.

To find the value of a natural logarithm $\ln b$ when b is a power of e:

1. Set the logarithm equal to x.

2. Rewrite the logarithm in exponential form $e^x = e^n$.

3. Solve for x.

For example, find the value of $\log_e e^3$ (which is the same as $\ln e^3$):

1. Set the logarithm equal to x. $\qquad\qquad\qquad\qquad \log_e e^3 = x$

2. Rewrite in exponential form, $\qquad\qquad\qquad\qquad e^x = e^3$
   $e^x = e^n$.

3. Solve for x. $\qquad\qquad\qquad\qquad\qquad\qquad\qquad x = 3$

So $\log_e e^3 = \ln e^3 = 3$.

Finding Logs in Base e and Powers of e Using a Calculator

What if you want to find the value of a natural logarithm and you can't rewrite the exponential form of the logarithm in the form $e^x = e^n$? For example, what if you want to find $\ln 35$? If you set $\ln 35 = x$ and switch to exponential form, you end up with $e^x = 35$. But 35 isn't a power of e, so it can't be written easily in the form e^n. So how do you find x? You can't find the exact value for x, but you can find the approximate value using a calculator.

To find the approximate value of the natural logarithm $x = \log_e a$ (which is the same as $\ln a$) using a calculator:

1. Enter the number a in the calculator.

2. Press the "ln" key.

3. Round the result of (2) to the desired number of decimal places.

For example, to find the value of $\ln 35$ (which is the same as $\log_e 35$) rounded to two decimal places:

1. Enter 35 in the calculator.

2. Press the "ln" key. $\qquad\qquad\qquad\qquad\qquad$ 3.5553481

3. Round to two decimal places. $\qquad\qquad\qquad\quad$ 3.56

So $\ln 35 \approx 3.56$.

Sometimes, you are given the natural logarithm of a number, and you want to find the number itself. You can work backwards to find the approximate value of this number using a calculator. In general, to find the number whose natural logarithm is k:

1. Let x be the number you are trying to find, and let $\log_e x = k$.

2. Rewrite the equation in exponential form $e^k = x$, and solve for x using a calculator as follows:
   - Enter k in your calculator.
   - Press the "e^x" key.

3. If necessary, round your answer.

For example, to find the number (rounded to two decimal places) whose natural logarithm is 3.

1. Substitute 3 for k in the equation $\log_e x = k$.

$$\log_e x = 3$$

2. Rewrite the equation in expontial form and solve for x, using a calculator as follows:
   - Enter 3 in your calculator.
   - Press the "e^x" key.

$$e^3 = x$$

$$x \approx 20.085537$$

3. Round your answer to two decimal places.

$$x \approx 20.09$$

So the number whose natural logarithm is 3 is approximately 20.09.

Properties of Natural Logarithms

You are already familiar with some properties of logarithms. For example, $\log_b b = 1$ since $b^1 = b$. (Here, the base b is a positive number and $b \neq 1$.) When the base b is e, this property states that $\log_e e = 1$ or $\ln e = 1$ because $e^1 = e$.

Here are some other properties of logarithms written for the case when the base b is e. Here $x > 0$, $u > 0$, and $v > 0$.

Name of Property	Property
Log of a Product	$\ln uv = \ln u + \ln v$
Log of a Quotient	$\ln \frac{u}{v} = \ln u - \ln v$ When $u = 1$, you have this special case: $\ln \frac{1}{v} = -\ln v$
Log of a Power	$\ln u^n = n \cdot \ln u$ When $u = e$, you have this special case: $\ln e^n = n$
Other Properties	$\ln e = 1$ $\ln 1 = 0$ $e^{\ln x} = x$

Here's an example of how you can use these properties to write $2\ln x + \ln (x + 1)$ as a single logarithm:

$$2\ln x + \ln (x + 1)$$

1. Use the log of a power property. $= \ln x^2 + \ln (x + 1)$

2. Use the log of a product property. $= \ln \left[x^2(x + 1)\right]$

3. Simplify. $= \ln \left(x^3 + x^2\right)$

So, $2\ln x + \ln (x + 1)$ written as a single logarithm is $\ln \left(x^3 + x^2\right)$.

Change of Base Formula

So far you have learned how to find the value of a logarithm with base 10 or base e, and you've seen how to use your calculator to do so. Now you will learn how to find the value of a logarithm with any base b (where $b > 0$ and $b \neq 1$) using a formula called the "change of base formula," which states the following:

$$\log_b x = \frac{\log_c x}{\log_c b}$$

Here, c is any number greater than 0 and not equal to 1.

Because you already know how to find natural logarithms with the "ln" key on your calculator, it is convenient to choose $c = e$. So the formula becomes:

$$\log_b x = \frac{\log_e x}{\log_e b} = \frac{\ln x}{\ln b}$$

In general, to find $\log_b x$ using the change of base formula:

1. Substitute values for b and x in the formula $\log_b x = \frac{\log_e x}{\log_e b} = \frac{\ln x}{\ln b}$.

2. Simplify using your calculator if necessary.

3. Round as required.

Here's an example. To use your calculator to find $\log_5 29$ using the change of base formula (round the answer to two decimal places):

1. Substitute 29 for x and 5 for b in the formula.

$$\log_b x = \frac{\log_e x}{\log_e b} = \frac{\ln x}{\ln b}$$

$$\log_5 29 = \frac{\log_e 29}{\log_e 5} = \frac{\ln 29}{\ln 5}$$

Here is a shortcut for doing this problem with your calculator:

- *Enter 29 in your calculator.*
- *Press the "ln" key.*
- *Press the "÷" key.*
- *Enter 5.*
- *Press the "ln" key.*
- *Press the "=" key*
- *Record the result.*
- *Round the result.*

2. Simplify using your calculator.

$$\approx \frac{3.3672958}{1.6094379}$$

$$\approx 2.0922185$$

3. Round the answer to two decimal places.

$$\approx 2.09$$

So, $\log_5 29$ is approximately 2.09.

Sample Problems

1. Use your calculator to find $\log_{10} 89$ approximated to four decimal places.

   ☐ a. Enter 89 in your calculator.

   ☐ b. Press the "log" key and write the result. _____

   ☐ c. Round the result of (b) to four decimal places. _____

2. Use your calculator to find an approximation of the number whose common logarithm is -0.03. Round your answer to three decimal places.

   ☑ a. Substitute -0.03 for k $\qquad$ $\log_{10} x = -0.03$
      in the equation $\log_{10} x = k$.

   ☑ b. Rewrite the equation in exponential form. $10^{-0.03} = x$

   ☐ c. Enter .03 in your calculator, and press the "±" key.

   ☐ d. Press the "10^x" key and write the result. $x \approx$ _____

   ☐ e. Round the result of (d) to three decimal places. $\approx$ _____

3. Use properties of logarithms to write $2\log x - 3(\log x + \log 3)$ as a single logarithm.

   ☑ a. Use the log of a product $\qquad$ $2\log x - 3(\log x + \log 3)$
      property. $\qquad\qquad\qquad\qquad$ $= 2\log x - 3(\log 3x)$

   ☐ b. Use the log of a power property. $=$ _____

   ☐ c. Use the log of a quotient property. $=$ _____

   ☐ d. Simplify. $=$ _____

4. Use your calculator to find ln 13 rounded to two decimal places.

   ☐ a. Enter 13 in your calculator.

   ☐ b. Press the "ln" key and write the result. _____

   ☐ c. Round the result of (b) to two decimal places. _____

5. Use your calculator to find $\log_9 146$ using the change of base formula. Round your answer to two decimal places.

   ☑ a. Substitute 146 for x and 9 $\qquad$ $\log_b x = \dfrac{\ln x}{\ln b}$
      for b in the formula.
      $\qquad\qquad\qquad\qquad\qquad$ $\log_9 146 =$ _____

   ☐ b. Simplify using your calculator. $\approx$ _____

   ☐ c. Round the result of (b) to two decimal places. $\approx$ _____

Answers to Sample Problems

b. 1.94939

c. 1.9494

d. 0.9332543

e. 0.933

b. $\log x^2 - \log(3x)^3$

c. $\log \dfrac{x^2}{(3x)^3}$

d. $\log \dfrac{1}{27x}$

b. 2.5649494

c. 2.56

a. $\dfrac{\ln 146}{\ln 9}$

b. 2.2681371

c. 2.27

SOLVING EQUATIONS

Summary

Solving Logarithmic Equations

You have studied logarithms and learned how to switch between logarithmic and exponential notation. You have also studied properties of common and natural logarithms and learned how to approximate the values of common and natural logarithms. Now you will learn how to solve equations that contain logarithms.

To solve some equations that contain logarithms you will need to switch from logarithmic to exponential notation. Recall that the following are equivalent:

logarithmic notation	exponential notation
$\log_b x = L$	$b^L = x$

Here, $b > 0$, $b \neq 1$, and $x > 0$.

For example, here's how to write $\log_5 25$ in logarithmic and exponential notation:

logarithmic notation	exponential notation
$\log_5 25 = 2$	$5^2 = 25$

In general, to solve a logarithmic equation that can be written in the form $\log_b x = L$, where $b > 0$, $b \neq 1$, and $x > 0$:

1. Rewrite the equation in exponential form $b^L = x$.

2. Solve for x.

3. Check your answer.

For example, to solve $\log_3 x = 4$ for x:

1. Rewrite the equation in exponential form.

$$3^4 = x$$

2. Solve for x.

$$81 = x$$
$$x = 81$$

3. Check.

The answer, 81, is positive, so it is an appropriate solution.

So if $\log_3 x = 4$, then $x = 81$.

Here's an equation where the variable, x, is the base of the logarithm. You will need to check the answer to make sure that $x \neq 1$ and $x > 0$.

To solve $\log_x 25 = 2$ for x:

1. Rewrite the equation in exponential form.

$$x^2 = 25$$

2. Solve for x. $x = 5$ or $x = -5$

3. Check.   Since x is the base of a logarithm, you need to check that $x > 0$ and $x \neq 1$. The number 5 satisfies these conditions but -5 does not.

So if $\log_x 25 = 2$, then $x = 5$.

Here's another equation where the variable, x, is the base of the logarithm. Again, you will need to check the answer to make sure that $x \neq 1$ and $x > 0$.

To solve $\log_x \frac{1}{9} = -1$ for x:

1. Rewrite the equation in exponential form. $x^{-1} = \frac{1}{9}$

2. Solve for x:

   • Substitute $\frac{1}{x}$ for x^{-1}. $\frac{1}{x} = \frac{1}{9}$

   • Multiply both sides by $9x$. $9x \cdot \frac{1}{x} = 9x \cdot \frac{1}{9}$

   • Cancel out the common factors. $9\overset{1}{x} \cdot \frac{1}{\underset{1}{x}} = \overset{1}{9}x \cdot \frac{1}{\underset{1}{9}}$

   • Simplify $9 = x$

   $x = 9$

3. Check.   Since x is the base of a logarithm, you need to check that $x > 0$ and $x \neq 1$. Since $9 > 0$ and $9 \neq 1$, it is an appropriate solution.

So if $\log_x \frac{1}{9} = -1$, then $x = 9$.

Using Properties of Logs to Solve Logarithmic Equations

So far, all of the equations you have solved contain only one logarithm. Now you will learn how to solve equations that contain more than one logarithm. You will sometimes find it easier to solve these equations if you use properties of logs to combine the logs into a single logarithm. Below is a chart you can use to review these properties. Here $b > 0$, $b \neq 1$, $u > 0$, and $v > 0$.

Name of Property	Property
Log of a Product	$\log_b uv = \log_b u + \log_b v$
Log of a Quotient	$\log_b \frac{u}{v} = \log_b u - \log_b v$
Log of a Power	$\log_b u^n = n \cdot \log_b u$
Other Properties	$\log_b b = 1$ $\log_b 1 = 0$

There is one more property of logs that will help you solve logarithmic equations. This property states that if the logs of two real numbers are equal, the real numbers must also be equal. That is, if $\log_b x = \log_b y$, then $x = y$. You can use this property to solve an equation in which one log is set equal to another log and both have the same base. For example, suppose you want to solve $\log_5 x = \log_5 9$ for x. Since the logs are equal and they have the same base, the quantities are equal. That is, $x = 9$.

Here's an example of how to solve a logarithmic equation using properties of logs. To solve $\ln(3x - 1) = \ln(x + 3)$ for x:

1. Since $\ln a$ is shorthand for $\log_e a$, rewrite the equation.

$\log_e(3x - 1) = \log_e(x + 3)$

2. Since the logs are equal and they have the same base, set the quantities to each other and solve for x.

$3x - 1 = x + 3$
$2x = 4$
$x = 2$

3. Check.

When $x = 2$, the expressions $3x - 1$ and $x + 3$ must be positive. Since these expressions are equal, you only need to check one of them. When you substitute 2 for x in $3x - 1$, you get $3 \cdot 2 - 1 = 6 - 1 = 5$, which is positive.

So $x = 2$ is an appropriate solution of $\ln(3x - 1) = \ln(x + 3)$.

Now here's an example where you use properties of logs to combine two logs into a single log.

To solve $\log_3 x + \log_3 2 = \log_3 6$ for x:

1. Use the log of a product property to rewrite the left side as a single log.

$\log_3 2x = \log_3 6$

2. Since the logs are equal and they have the same base, set the quantities equal to each other and solve for x.

$2x = 6$
$x = 3$

3. Check.

Since 3 is positive, the number $x = 3$ is an appropriate solution.

So if $\log_3 x + \log_3 2 = \log_3 6$, then $x = 3$.

In the next example you use properties of logs to combine two logs into a single log, then you rewrite the equation in exponential form.

To solve $\log x + \log(x - 21) = 2$ for x:

1. Use the log of a product property to rewrite the left side as a single log.

$\log[x(x - 21)] = 2$

2. Since log a is shorthand for $\log_{10} a$, rewrite the equation.

$$\log_{10}[x(x-21)] = 2$$

3. Rewrite the equation in exponential form.

$$10^2 = x(x-21)$$

4. Finish solving for x.

$$100 = x^2 - 21x$$
$$x^2 - 21x - 100 = 0$$
$$(x-25)(x+4) = 0$$
$$x - 25 = 0 \text{ or } x + 4 = 0$$
$$x = 25 \text{ or } \boxed{x = -4}$$

5. Check.

For log x and log $(x-21)$ to be defined, x and $x-21$ must be positive. So $x = 25$ checks, but $x = -4$ does not.

The numbers $x = -4$ and $x - 21 = -4 - 21 = -25$ are both negative, so the logarithm is not defined for $x = -4$.

So the only solution of log x + log $(x-21) = 2$ is $x = 25$.

In the next example, you start with one log term on the left side of the equation, and one log term on the right side of the equation. Since you want to write the equation using a single log term, you need to get both log terms on the left side of the equation.

To solve $2\log_3 x = 2 + \log_3 9$ for x:

1. Subtract $\log_3 9$ from both sides of the equation.

$$2\log_3 x - \log_3 9 = 2$$

2. Use the log of a power property to rewrite $2\log_3 x$.

$$\log_3 x^2 - \log_3 9 = 2$$

3. Use the log of a quotient property to rewrite the left side as a single log.

$$\log_3\left(\frac{x^2}{9}\right) = 2$$

4. Rewrite in exponential form.

$$3^2 = \frac{x^2}{9}$$
$$9 = \frac{x^2}{9}$$

5. Finish solving for x.

$$9 \cdot 9 = x^2$$
$$81 = x^2$$
$$x^2 = 81$$
$$x = 9 \text{ or } \boxed{x = -9}$$

6. Check.

For $\log_3 x$ to be defined, x must be positive. So $x = 9$ checks but $x = -9$ does not.

So the only solution of $2\log_3 x = 2 + \log_3 9$ is $x = 9$.

Solving Exponential Equations

You have solved equations that contain logarithms. Now you will use logarithms to solve equations where the variable appears in an exponent. These equations are called exponential equations. Here are some examples:

$$2^x = 5 \qquad\qquad 2^x + 9 = 15 \qquad\qquad 4e^{2x-1} = 6$$

To solve exponential equations, you will frequently use two properties of logs. The first property states that if two quantities are equal, their logs are equal. That is:

If $x = y$, then $\log_b x = \log_b y$.

The other property is the log of a power property:

$$\log_b u^n = n \cdot \log_b u \quad \text{(Here } b > 0, b \neq 1, \text{ and } u > 0\text{)}$$

In general, to solve an exponential equation:

1. Isolate the term that contains the exponent.

Remember, a natural log is just a logarithm with base e. So you can let $b = e$ and use the property "If $x = y$, then ln x = ln y."

2. Take the log of both sides of the equation. You may want to use natural logs since you can easily approximate them using a calculator.

3. Use the log of a power property to rewrite the term that contains the exponent. Then the variable will no longer be in the exponent.

4. Finish solving for x.

Once you find x, you can approximate your answer by using a calculator to compute the natural logs, and then round your answer.

Here's an example. To solve the exponential equation $2^x = 5$ for x:

1. The term that contains the exponent is isolated on the left side. $\qquad\qquad 2^x = 5$

2. Take the natural log (ln) of both sides of the equation. $\qquad \ln 2^x = \ln 5$

3. Use the log of a power property to get x out of the exponent. $\qquad x \cdot \ln 2 = \ln 5$

4. Finish solving for x by dividing both sides by ln 2. $\qquad x = \dfrac{\ln 5}{\ln 2}$

So if $2^x = 5$ then $x = \dfrac{\ln 5}{\ln 2}$. You can approximate this answer by using a calculator and round your answer. For example, $\dfrac{\ln 5}{\ln 2}$ rounded to two decimal places is approximately $\dfrac{1.6094379}{0.6931471} \approx 2.32$.

As another example, to solve $4e^{2x-1} = 5$ for x:

1. Isolate the term e^{2x-1} by dividing both sides by 4. $\qquad\qquad 4e^{2x-1} = 5$
$$e^{2x-1} = \frac{5}{4}$$

2. Take the natural log (ln) of both sides of the equation.

$$\ln e^{2x-1} = \ln \frac{5}{4}$$

3. Use the log of a power property to get $2x - 1$ out of the exponent.

$$(2x - 1) \cdot \ln e = \ln \frac{5}{4}$$

4. Finish solving for x. Recall that $\ln e = 1$.

$$(2x - 1) \cdot 1 = \ln \frac{5}{4}$$
$$2x - 1 = \ln 1.25$$
$$2x = \ln 1.25 + 1$$
$$x = \frac{\ln 1.25 + 1}{2}$$

So if $4e^{2x-1} = 5$ then $x = \frac{\ln 1.25 + 1}{2}$. You can approximate this answer using a calculator and rounding. For example, $\frac{\ln 1.25 + 1}{2}$ rounded to two decimal places is approximately $\frac{0.2231435 + 1}{2} \approx 0.61$.

Sample Problems

1. Solve for x: $\log_{10} x = 3$

   ☑ a. Rewrite the equation in expontential form.

   $10^3 = x$

   ☐ b. Solve for x.

   $x = \underline{\hspace{2cm}}$

   b. 1000

2. Solve for x: $\log_x 3 = \frac{1}{2}$

   ☐ a. Rewrite the equation in exponential form.

   $\underline{\hspace{3cm}}$

   a. $x^{\frac{1}{2}} = 3$

   ☑ b. Rewrite the equation using a radical sign.

   $\sqrt{x} = 3$

   ☐ c. Solve for x.

   $x = \underline{\hspace{2cm}}$

   c. 9

   ☑ d. Check.

   The number $x = 9$ is greater than 0 and not equal to 1 so the answer checks.

3. Solve for x: $\ln 2x + \ln \left(x + \frac{3}{2}\right) = \ln 5$

   ☐ a. Use the log of a product property to rewrite the left side as a single log.

   $\underline{\hspace{3cm}} = \ln 5$

   a. $\ln \left[2x\left(x + \frac{3}{2}\right)\right]$ or $\ln \left(2x^2 + 3x\right)$

   ☐ b. Since the two natural logs are equal, set the quantities equal to each other.

   $\underline{\hspace{3cm}}$

   b. $2x\left(x + \frac{3}{2}\right) = 5$ or $2x^2 + 3x = 5$

   ☐ c. Solve for x.

   $x = \underline{\hspace{1.5cm}}$ or $x = \underline{\hspace{1.5cm}}$

   c. $1, -\frac{5}{2}$

$\checkmark$ d. Check your answer.

The numbers $x = -\dfrac{5}{2}$,

$2x = 2\left(-\dfrac{5}{2}\right) = -5$,

and $\left(x + \dfrac{3}{2}\right) = \left(-\dfrac{5}{2} + \dfrac{3}{2}\right) = -1$

are all negative, so the logarithm

is not defined for $x = -\dfrac{5}{2}$.

The only solution is $x = 1$.

4. Solve for x: $2\log_2 x = 2 + \log_2 4$

b. $\log_2 x^2$

$\checkmark$ a. Subtract $\log_2 4$ from both sides. $\qquad$ $2\log_2 x - \log_2 4 = 2$

☐ b. Use the log of a power $\qquad$ _____ $- \log_2 4 = 2$
  property to rewrite $2\log_2 x$.

c. $\log_2\left(\dfrac{x^2}{4}\right)$

☐ c. Use the log of a quotient property $\qquad$ _____ $= 2$
  to rewrite the left side as a
  single log.

☐ d. Rewrite in exponential form, $\qquad$ $2^2 = \dfrac{x^2}{4}$
  and simplify.

d. $4 = \left(\dfrac{x^2}{4}\right)$

_____

e. $4, -4$

☐ e. Finish solving for x.

$x = $ _____ or $x = $ _____

$\checkmark$ f. Check your answer. The number $x = -4$ is negative,
so $x = 4$ is the only solution.

5. Solve for x: $2^x + 9 = 15$

$\checkmark$ a. Subtract 9 from both sides $\qquad$ $2^x = 6$
  to isolate 2^x.

b. $\ln 2^x = \ln 6$

☐ b. Take the natural log of both $\qquad$ _____
  sides of the equation.

☐ c. Use the log of a power property $\qquad$ _____
  to get x out of the exponent.

c. $x \cdot \ln 2 = \ln 6$

☐ d. Finish solving for x.

$x = $ _____

d. $\dfrac{\ln 6}{\ln 2}$

e. 2.59

☐ e. Use a calculator to compute $\qquad$ $x \approx $ _____
  the answer to two decimal places.

HOMEWORK

Homework Problems

Circle the homework problems assigned to you by the computer, then complete them below.

☼ Explain
Natural and Common Logs

1. Find: $\log_{10} 10000$

2. Find: $\ln 1 + \ln e$

3. Use properties of logarithms to write $\log 2 + 3(\log x)$ as a single logarithm.

4. Use your calculator to find $\log_{10} 62$ rounded to three decimal places.

5. Use your calculator to find $\ln 31$ rounded to two decimal places.

6. Use properties of logarithms to write $\log x - 2(\log x + \log x^2)$ as a single logarithm.

7. Find an approximation of the number whose common logarithm is -3.56. Round your answer to four decimal places.

8. Use properties of logarithms to write $3\ln x - \ln (x + 3)$ as a single logarithm.

9. Jack is an environmental scientist. A recent experiment of his resulted in some data that he represented with the following logarithmic expressions.

   Expression 1: $2\log(x + 1)$

   Expression 2: $2\log(2 + x)$

   Expression 3: $4 + \log 2 + 3\log x$

   Jack wants to include this data in a report using a single logarithm. Help Jack by writing the sum of the three logarithmic expressions as a single logarithm.

10. Jack, the environmental scientist, now wants to evaluate the sum of the following expressions for $x = 2$:

    Expression 1: $2\log(x + 1)$

    Expression 2: $2\log(2 + x)$

    Expression 3: $4 + \log 2 + 3\log x$

    Evaluate the total of these expressions for $x = 2$. The round your answer to two decimal places.

11. Use your calculator to find $\log_6 801$ using the change of base formula. Round your answer to three decimal places.

12. Use properties of logarithms to write $\frac{1}{2}\log x + 3\log \sqrt{x} - 2\log (x + 1)$ as a single logarithm.

Solving Equations

13. Solve for x: $\log_4 x = 3$

14. Solve for x: $\ln (x + 3) = \ln (2x - 1)$

15. Solve for x: $\log_4 2x - \log_4 2 = \log_4 5$

16. Solve for x: $\log_x 8 = 3$

17. Solve for x: $\ln x + \ln (x + 1) = \ln 2$

18. Solve for x: $2\log_7 x = 2 + \log_7 16$

19. Solve for x: $\log_x 7 = \frac{1}{2}$

20. Solve for x: $4^x + 5 = 12$

21. The number of bacteria, N, at a certain time t is described by the equation $N = N_0 e^{4t}$, where N_0 is the number of bacteria present at time $t = 0$. What is t when the number of bacteria has doubled from what it was when $t = 0$? Round your answer to two decimal places. (Hint: You want to know when $N = 2N_0$, so solve the equation $2N_0 = N_0 e^{4t}$, which is the same as $2 = e^{4t}$, for t.)

22. On the Richter scale the magnitude M of an earthquake of intensity I is given by the equation $M = \log \frac{I}{I_0}$, where I_0 is a minimum intensity measurement used for comparing earthquakes. If an earthquake has an intensity I that is $10^{6.2}$ times the minimum intensity I_0, what is its magnitude M on the Richter scale?

23. Solve for x: $5^x - 4 = 2$. Round your answer to two decimal places.

24. Solve for x: $2e^{2x+3} - 3 = 5$. Round your answer to three decimal places.

 **APPLY**

Practice Problems

Here are some additional practice problems for you to try.

Natural and Common Logs

1. Find: log 400 – log 40

2. Find: log 30 – log 3000

3. Find: log 50 – log 50,000

4. Find: $\ln e^3 + \ln e^7$

5. Find: $\ln e^2 + \ln e^5$

6. Find: $\ln e - \ln e^2$

7. Use properties of logarithms to write $\log x^3 + \log x^5$ as a single logarithm.

8. Use properties of logarithms to write $5\log x^4 + 6\log x^7$ as a single logarithm.

9. Use properties of logarithms to write $5(\log x^2 + \log x^{11})$ as a single logarithm.

10. Use properties of logarithms to write $4\ln x + \ln (x - 7)$ as a single logarithm.

11. Use properties of logarithms to write $7\ln x + \ln (2x + 3)$ as a single logarithm.

12. Use properties of logarithms to write $3\ln x^3 - \ln x^6 - \ln (3x + 2)$ as a single logarithm.

13. Use properties of logarithms to write $2\ln x^2 + \ln x - \ln (4x - 1)$ as a single logarithm.

14. Use your calculator to find log 35 rounded to two decimal places.

15. Use your calculator to find log 24.7 rounded to two decimal places.

16. Use your calculator to find log 28 rounded to two decimal places.

17. Use your calculator to find ln 90 rounded to two decimal places.

18. Use your calculator to find ln 83 rounded to two decimal places.

19. Use your calculator to find ln 54.9 rounded to two decimal places.

20. Use your calculator to find $\log_{11} 34$ rounded to two decimal places. (Round your answer at the end of your calculations.)

21. Use your calculator to find $\log_7 23.4$ rounded to two decimal places. (Round your answer at the end of your calculations.)

22. Use your calculator to find $\log_{12} 55$ rounded to two decimal places. (Round your answer at the end of your calculations.)

23. Find an approximation of the number whose common logarithm is 2.125. Round your answer to two decimal places.

24. Find an approximation of the number whose common logarithm is –0.005. Round your answer to two decimal places.

25. Find an approximation of the number whose common logarithm is 1.375. Round your answer to two decimal places.

26. Find an approximation of the number whose natural logarithm is 1.07. Round your answer to two decimal places.

27. Find an approximation of the number whose natural logarithm is –1.24. Round your answer to two decimal places.

28. Find an approximation of the number whose natural logarithm is –2.2. Round your answer to two decimal places.

Solving Equations

29. Solve for x: $\log_2 x = 4$

30. Solve for x: $\log_5 x = 3$

31. Solve for x: $\log_3 x = 5$

32. Solve for x: $\log_x 64 = 3$

33. Solve for x: $\log_x 125 = 3$

34. Solve for x: $\log_x 6 = \dfrac{1}{2}$

35. Solve for x: $\log_x 4 = \dfrac{1}{3}$

36. Solve for x: $\log_3 3x + \log_3 9x = 4$

37. Solve for x: $\log_{12} 6x + \log_{12} 4x = 2$

38. Solve for x: $\log 12x^2 - \log 6x = 2$

39. Solve for x: $\log_8 32 = 5\log_8 x$

40. Solve for x: $\log_{13} 125 = 3\log_{13} x$

41. Solve for x: $4\log_{15} x = \log_{15} 81$

42. Solve for x: $\log (x + 1) = 2$

43. Solve for x: $\log_6 (x + 2) - \log_6 5 = \log_6 (x - 2)$

44. Solve for x: $\log_5 (x + 1) - \log_5 7 = \log_5 (x - 1)$

45. Solve for x: $2\ln (x - 3) = \ln (30 - 2x)$

46. Solve for x: $2\ln (x + 2) = \ln (7x + 44)$

47. Solve for x: $\ln (x - 3) + \ln 5 = \ln (x + 17)$

48. Solve for x: $4^x + 7 = 31$

49. Solve for x: $5^x + 2 = 21$

50. Solve for x: $12^x - 8 = 32$

51. Solve for x: $8 + 3^{x-2} = 30$

52. Solve for x: $19 - 4^{x+1} = 7$

53. Solve for x: $20 = 14 + 7^{x-3}$

54. Solve for x: $8e^{3x+2} - 3 = 45$

55. Solve for x: $5e^{4x-1} + 12 = 37$

56. Solve for x: $6e^{5x-1} + 2 = 44$

EVALUATE

Practice Test

Take this practice test to be sure that you are prepared for the final quiz in Evaluate.

1. Use a calculator to approximate the values of x below to two decimal places.

   a. $\log_{10} 113 = x$

   b. $\log_{10} x = 0.34$

2. Use properties of logs to rewrite $\log 5 + \log 2x + 2\log(x-2)$ as a single logarithm.

3. Circle the statements below that are always true. Assume $x > 0$, $a > 0$, and $b > 0$.

   $\ln 1 = 0$

   $\log a + \log b = \log (a + b)$

   $\log_e 5 = \ln 5$

   $2\log x = \log x + \log x$

4. Use a calculator to approximate $\log_{13} 230$ to two decimal places.

5. Solve for x: $\log_x 36 = 2$

6. Solve the following equations for x:

   a. $\log_3 x = -2$

   b. $\log_x \frac{2}{3} = -1$

7. Solve for x: $\log_3 (2x - 2) - 1 = \log_3 8$

8. Solve for x: $6^x - 5 = 17$. Round your answer to two decimal places.

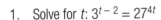

CUMULATIVE REVIEW PROBLEMS

These problems combine all of the material you have covered so far in this course. You may want to test your understanding of this material before you move on to the next topic, or you may wish to do these problems to review for a test.

1. Solve for t: $3^{t-2} = 27^{4t}$

2. Find: $(-3 + 4i) - (-2 + 9i)$

3. Find: $(4 + 3i) \div (11 - 6i)$

4. Find the domain and range of each of the functions below.

   a. $y = x - 8$

   b. $y = |x| - 8$

   c. $y = |x - 8|$

5. Rewrite each of the following using properties of logarithms:

   a. $\log_7(4^6)$

   b. $\log_2\left(\dfrac{1}{x^3}\right)$

   c. $\log_3(3^x)$

   d. $\log_5 100 - \log_5 10$

   e. $\log_{10}\sqrt{31}$

6. Given $f(x) = 14x$ and $g(x) = 3x - 5$, evaluate $\left(\dfrac{f}{g}\right)(x)$ at $x = 2$.

7. The graph of the function $y = -2x^2$ is show on the grid in Figure 12.1. Determine the equations of the other parabolas shown.

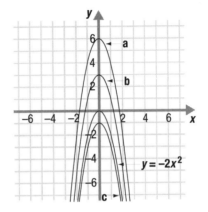

Figure 12.1

8. Find: $(x + 7)(4x - 1)$

9. Evaluate the following:

   a. $\log_8 8$

   b. $\log_{41} 1$

   c. $6^{\log_6 14}$

10. Solve for x: $2\ln x = \ln(x + 6)$

11. Write the equation $4(y + 2) = 3x + 4$ in slope-intercept form.

12. Solve for x: $\ln x + \ln 9 = \ln(2x + 7)$

13. Combine like terms and simplify:

$$5\sqrt{13} + 2\sqrt{41} - \sqrt[3]{-27} - 6\sqrt{41} + \sqrt{13} + 1$$

14. Rewrite the following in logarithmic form:

   a. $3^4 = 81$

   b. $10^6 = 1,000,000$

   c. $2^7 = 128$

   d. $\left(\dfrac{1}{4}\right)^3 = \dfrac{1}{64}$

   e. $49^{\frac{1}{2}} = 7$

15. Find the domain and range of each of the functions below.

   a. $f(x) = x^2 - 2$

   b. $f(x) = 21x + 13$

   c. $f(x) = \dfrac{14x}{(2x-1)(x+7)}$

   d. $f(x) = \sqrt{x^2 - 4}$

16. Rewrite each of the following using more than one logarithm. Use properties of logarithms.

   a. $\log_4 (8 \cdot 3)$

   b. $\log_{17}\left(\dfrac{m}{n}\right)$

   c. $\log_{10}[5(a+1)]$

   d. $\log_9\left(\dfrac{7}{8}\right)$

   e. $\log_{10}\left(\dfrac{5x}{x^2}\right)$

17. Rewrite using properties of logarithms: $\ln \dfrac{x^2}{x+2}$

18. Find the reciprocal: $6a \cdot \left(\dfrac{2b^2 + 9c}{a+c}\right)$

19. Solve for x: $x^2 - 70 = 11$

20. Find the equation of the line through the point $(-3, 2)$ that is parallel to the line $y + 4 = 2(x - 1)$. Write your answer in slope-intercept form.

21. Find the inverse of the function $y = 3^x$ in logarithmic form. Then graph $y = 3^x$ and its inverse.

22. Using your calculator and the change of base formula, find each of the following. Round your answers to two decimal places.

   a. $\log_4 72$

   b. $\log_8 16$

   c. $\log_{21} 11$

   d. $\log_5\left(\dfrac{7}{9}\right)$

23. Does $y = f(x) = x^3$, shown in Figure 12.2, have an inverse? If yes, find and graph the inverse $y = f^{-1}(x)$.

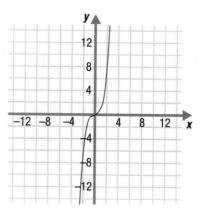

Figure 12.2

24. Find: $(11 + 2i)(2 + 11i)$

25. Solve for y: $5y^2 - 7y + 2 = 0$

26. Solve for y: $-8|y + 2| > 7$

27. Solve for x: $4e^{3x-6} + 2 = 11$

28. Find: $\dfrac{3z^2}{x^2} + \dfrac{10x}{xy}$

29. Using the compound interest formula, $A = P\left(1 + \dfrac{r}{n}\right)^n$, determine how much money, A, you would have after one year if you invested $P = \$500$ in an account that compounded interest quarterly ($n = 4$) at an interest rate of $r = 3\%$. Use a calculator.

30. Find: $\log_5 125$

31. Simplify the following expressions:

   a. $\sqrt{-81}$

   b. $\sqrt{25}$

   c. $\sqrt{-81} \cdot \sqrt{25}$

   d. $\sqrt{-81} \cdot \sqrt{-25}$

   e. i^{31}

   f. i^{50}

32. Simplify: $\dfrac{\sqrt{ab^2c}}{a^{\frac{1}{4}}}$

33. Solve for q: $2q^2 - 11q = 7 - 6q$

34. Find the x- and y-intercepts of each of the functions below.

   a. $y = x^2 - 10x - 6$

   b. $y = -8x^2$

   c. $y = x^2 - 3x - 4$

35. Complete the table of ordered pairs below for $y = 2^{x+1}$, then graph the function. As the value of x becomes smaller and smaller, what value does y approach?

x	y
0	
−1	
1	
−2	
2	

36. Solve for x: $\log_x 81 = 2$

37. Factor: $2x^2 - 6x + xy - 3y$

38. Solve for x: $3x(x - 2) - 2 = 4x^2 - 9x + 3$

39. Solve for d: $\frac{d-7}{3} = \frac{2d}{5} - 1$

40. Using your calculator, find x for each of the following to two decimal places.

   a. $\log 365 = x$ b. $\log 727 = x$

   c. $\log x = -0.78$ d. $\ln x = 3.14$

   e. $\ln 216 = x$

41. If $f(x) = 33x - 18$, find $f^{-1}(x)$.

42. Solve for z: $z^2 + 6z + 7 = 0$

43. Complete the table below for the function $y = x + 4$. Then use the table to graph the function.

x	y
−6	
−4	
−2	
0	
1	
3	

44. If $f(x) = \frac{3}{x + 9}$, find $f^{-1}(x)$.

45. Solve for m: $-4(2m - 5) \geq -3m + 27$

46. Which of the following functions are linear?

$$y = 4x + 3$$

$$y = x^2 + 2x - 6$$

$$y = \sqrt{x}$$

$$y = 244x$$

$$y = x^3 - 7x + 2$$

47. Find: $\left(6\sqrt{5} - 3\sqrt{6}\right)\left(5\sqrt{5} + 9\sqrt{6}\right)$

48. Solve for p: $11 - 2|3p + 1| = 9$

49. Given $f(x) = x^2 + 2x - 8$ and $g(x) = x + 1$, find $(f \circ g)(x)$.

50. Solve for x: $(2x - 6)^2 = 4$

LESSON 13.1 — NONLINEAR EQUATIONS

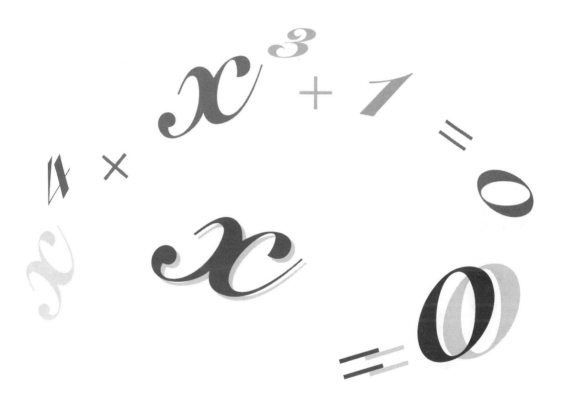

OVERVIEW

Here's what you'll learn in this lesson:

Solving Equations

a. Solving polynomial equations by factoring

b. Solving quadratic type equations by factoring or by substitution

Radical Equations

a. Solving $\sqrt{ax + b} = cx + d$

b. Solving $\sqrt{ax + b} + \sqrt{cx + d} = ex + f$

c. Solving $\sqrt[n]{ax + b} = \sqrt[n]{cx + d}$

d. Solving equations that contain rational exponents

Suppose you were the owner of a hot dog cart. Of course you would be interested in knowing how many hot dogs you would need to sell each day to make the most profit. One way to figure this out would be to solve a profit equation based on your costs and pricing strategy. Most likely, such an equation would be nonlinear.

In this lesson, you will learn some techniques for solving nonlinear equations. First, you will learn to solve a variety of polynomial equations by factoring. Then you will learn to solve some equations that resemble quadratic equations. Finally, you will learn to solve equations in which variables appear under radical signs.

SOLVING EQUATIONS

Summary

Solving Nonlinear Equations by Factoring

Sometimes you will need to solve nonlinear equations. You can solve some nonlinear equations by rewriting them so they look like polynomial equations, then factoring.

To solve such a nonlinear equation by factoring:

1. Write the equation as a polynomial equation in standard form.

2. Factor the left side.

3. Use the Zero Product Property to set each factor equal to 0.

4. Finish solving for the variable.

For example, to solve $x - 6 = -\frac{5}{x}$:

1.	Write the equation in standard form.	$x \cdot (x - 6) = x \cdot \left(-\frac{5}{x}\right)$
		$x^2 - 6x = -5$
		$x^2 - 6x + 5 = 0$
2.	Factor the left side.	$(x - 1)(x - 5) = 0$
3.	Use the Zero Product Property.	$x - 1 = 0$ or $x - 5 = 0$
4.	Finish solving for x.	$x = 1$ or $x = 5$

So the two solutions of $x - 6 = -\frac{5}{x}$ are $x = 1$ and $x = 5$.

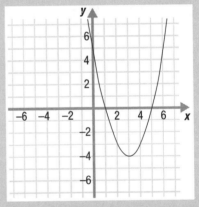

The graph of the function $y = x^2 - 6x + 5$ crosses the x-axis at $x = 1$ and $x = 5$, the two solutions of $x - 6 = -\frac{5}{x}$.

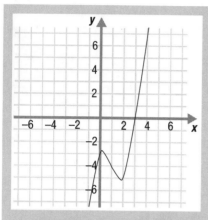

The graph of $y = x^3 - 3x^2 + x - 3$ only crosses the x-axis in one place because only one solution of the equation, $x^3 - 3x^2 + x = 3$, is a real number.

As another example, to solve $x^3 - 3x^2 + x = 3$:

1. Write the equation in standard form.

$$x^3 - 3x^2 + x - 3 = 0$$

2. Factor the left side.

$$(x^3 - 3x^2) + (x - 3) = 0$$
$$x^2(x - 3) + (x - 3) = 0$$
$$(x - 3)(x^2 + 1) = 0$$

3. Use the Zero Product Property.

$$x - 3 = 0 \text{ or } x^2 + 1 = 0$$

4. Finish solving for x.

$$x = 3 \text{ or } x^2 = -1$$
$$x = \pm\sqrt{-1}$$
$$x = \pm i$$

So the three solutions of $x^3 - 3x^2 + x = 3$ are $x = 3$, $x = -i$, and $x = i$.

Solving Nonlinear Equations by Substitution

The equations you looked at in the previous section could be solved by factoring. But sometimes when a nonlinear equation resembles a quadratic equation, it is easier to first use substitution to rewrite the problem before you factor it.

To solve such a polynomial equation using substitution:

1. Write the polynomial equation so it has the form of a quadratic equation.

2. Use an appropriate substitution to make the polynomial equation easier.

3. Factor the new equation.

4. Use the Zero Product Property to set each factor equal to 0.

5. Finish solving for the new variable.

6. Substitute the original expression back for the new variable.

7. Finish solving for the original variable.

For example, to solve $(x - 4)^2 + 6(x - 4) = -8$:

1. Write the polynomial so it is quadratic in form. $(x - 4)^2 + 6(x - 4) + 8 = 0$

2. Substitute u for $(x - 4)$. $u^2 + 6u + 8 = 0$

3. Factor. $(u + 4)(u + 2) = 0$

4. Use the Zero Product Property. $u + 4 = 0$ or $u + 2 = 0$

5. Finish solving for u. $u = -4$ or $u = -2$

6. Substitute $(x - 4)$ for u. $x - 4 = -4$ or $x - 4 = -2$

7. Finish solving for x. $x = 0$ or $x = 2$

The solutions, $x = 0$ and $x = 2$, both satisfy the original equation, $(x - 4)^2 + 6(x - 4) = -8$.

Sample Problems

1. Solve for x: $x + 7 + \dfrac{10}{x} = 0$.

   ☑ a. Write the equation in standard form.
   $$x + 7 + \frac{10}{x} = 0$$
   $$x \cdot \left(x + 7 + \frac{10}{x}\right) = x \cdot 0$$
   $$x^2 + 7x + 10 = 0$$

   ☐ b. Factor the left side. _____ $= 0$

   ☐ c. Use the Zero Product Property. _____ $= 0$ or _____ $= 0$

   ☐ d. Finish solving for x. $x =$ _____ or $x =$ _____

2. Use substitution to solve for y: $(y - 2)^2 + 10(y - 2) + 9 = 0$.

   ☑ a. Substitute u for $(y - 2)$. $u^2 + 10u + 9 = 0$

   ☐ b. Factor. _____ $= 0$

   ☐ c. Use the Zero Product Property. _____ $= 0$ or _____ $= 0$

   ☐ d. Finish solving for u. $u =$ _____ or $u =$ _____

   ☐ e. Substitute $(y - 2)$ for u. $y - 2 =$ ____ or $y - 2 =$ ____

   ☐ f. Finish solving for y. $y =$ _____ or $y =$ _____

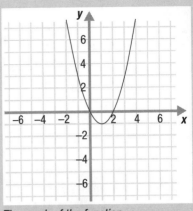

The graph of the function $y = (x - 4)^2 + 6(x - 4) + 8$ crosses the x-axis at $x = 0$ and $x = 2$, the two solutions of the equation, $(x - 4)^2 + 6(x - 4) = -8$.

Answers to Sample Problems

b. $(x + 2)(x + 5)$

c. $x + 2, x + 5$

d. $-2, -5$

b. $(u + 1)(u + 9)$

c. $u + 1, u + 9$

d. $-1, -9$

e. $-1, -9$

f. $1, -7$

RADICAL EQUATIONS

Summary

Equations with Radicals

Sometimes in an equation the variable is under a radical sign. When this is the case you need to get the variable out from under the radical sign in order to solve the equation.

To solve an equation in which the variable is under a radical sign:

If an equation has variables that appear under two radical signs, separate the radicals on opposite sides of the equation.

For example, in step (2), if the index of the radical is 2, square both sides of the equation. If the index of the radical is 3, cube both sides of the equation.

1. Isolate a radical on one side of the equation.

2. Raise both sides of the equation to the same power as the index of the isolated radical. Repeat steps (1) and (2) as often as necessary to eliminate all radicals.

3. Rewrite the resulting equation as a polynomial equation in standard form.

4. Factor.

5. Solve for the variable.

It is very important to check your answer in the original equation. It is possible to get answers that don't make sense.

6. Check your answer in the original equation.

For example, to solve $x + 1 = \sqrt{x + 3}$:

1. Isolate a radical on one side of the equation. $\sqrt{x+3}$ is already isolated

2. Square both sides of the equation.

$$(x + 1)^2 = \left(\sqrt{x + 3}\right)^2$$
$$x^2 + 2x + 1 = x + 3$$

3. Rewrite the equation in standard form.

$$x^2 + x - 2 = 0$$

4. Factor.

$$(x + 2)(x - 1) = 0$$

5. Solve for x.

$$x + 2 = 0 \quad \text{or} \quad x - 1 = 0$$
$$x = -2 \text{ or} \quad x = 1$$

6. Check your answer.

    Check $x = -2$:

    Is $(-2) + 1 = \sqrt{-2 + 3}$?

    Is $\quad -1 = \sqrt{1} \quad$?

    Is $\quad -1 = 1 \quad$? No.

    Check $x = 1$:

    Is $1 + 1 = \sqrt{1 + 3}$?

    Is $\quad 2 = \sqrt{4} \quad$?

    Is $\quad 2 = 2 \quad$? Yes.

$x = -2$ is an extraneous solution because it does not satisfy the original equation.

So the only solution of the equation $x + 1 = \sqrt{x + 3}$ is $x = 1$.

As another example, to solve $\sqrt{x+5} - \sqrt{x-3} = 2$:

1. Isolate a radical on one side of the equation.

$$\sqrt{x+5} = \sqrt{x-3} + 2$$

2. Square both sides of the equation.

$$\left(\sqrt{x+5}\right)^2 = \left(\sqrt{x-3} + 2\right)^2$$
$$x + 5 = (x-3) + 4\sqrt{x-3} + 4$$
$$x + 5 = x + 1 + 4\sqrt{x-3}$$

3. Isolate a radical on one side of the equation.

$$4 = 4\sqrt{x-3}$$

4. Square both sides of the equation.

$$4^2 = \left(4\sqrt{x-3}\right)^2$$

5. Rewrite the equation so it is a polynomial equation.

$$16 = 16(x-3)$$

6. Solve for x.

$$1 = x - 3$$
$$4 = x$$

7. Check your answer.

Check $x = 4$:

Is $\sqrt{4+5} - \sqrt{4-3} = 2$?

Is $\sqrt{9} - \sqrt{1} = 2$?

Is $3 - 1 = 2$?

Is $2 = 2$? Yes.

So $x = 4$ is the solution of the equation $\sqrt{x+5} - \sqrt{x-3} = 2$.

Equations with Rational Exponents

Equations that contain rational exponents are solved in much the same way as equations that contain radical expressions.

To solve an equation that contains a rational exponent:

1. Isolate the term that contains the rational exponent on one side of the equation.

2. Rewrite the equation using a radical.

3. Finish solving for the variable.

4. Check your answer in the original equation.

For example, to solve the equation $(x+1)^{\frac{3}{2}} - 5 = -1$:

1. Isolate the term with the rational exponent.

$$(x+1)^{\frac{3}{2}} = 4$$

2. Rewrite the equation using a radical.

$$\sqrt{(x+1)^3} = 4$$

3. Finish solving for x.

$$\left(\sqrt{(x+1)^3}\right)^2 = (4)^2$$

$$(x+1)^3 = 16$$

$$\sqrt[3]{(x+1)^3} = \sqrt[3]{16}$$

$$x + 1 = \sqrt[3]{2^3 \cdot 2}$$

$$x + 1 = 2\sqrt[3]{2}$$

$$x = 2\sqrt[3]{2} - 1$$

4. Check your answer.

Check $x = 2\sqrt[3]{2} - 1$:

Is $\left[(2\sqrt[3]{2} - 1) + 1\right]^{\frac{3}{2}} - 5 = -1$?

Is $\left(2\sqrt[3]{2}\right)^{\frac{3}{2}} - 5 = -1$?

Is $\left(2 \cdot 2^{\frac{1}{3}}\right)^{\frac{3}{2}} - 5 = -1$?

Is $\left(2^3 \cdot 2^{\frac{1}{3} \cdot 3}\right)^{\frac{1}{2}} - 5 = -1$?

Is $(8 \cdot 2)^{\frac{1}{2}} - 5 = -1$?

Is $(16)^{\frac{1}{2}} - 5 = -1$?

Is $\sqrt{16} - 5 = -1$?

Is $4 - 5 = -1$?

Is $-1 = -1$? Yes.

So $x = 2\sqrt[3]{2} - 1$ is the solution of the equation $(x+1)^{\frac{3}{2}} - 5 = -1$.

Sample Problems

1. Solve for x: $(x-61)^{\frac{3}{5}} + 34 = 7$.

   ☑ a. Isolate the term with the rational exponent.

   $$(x-61)^{\frac{3}{5}} = -27$$

   ☑ b. Rewrite the equation using a radical.

   $$\sqrt[5]{(x-61)^3} = -27$$

☐ c. Finish solving for x.

$$\left(\sqrt[5]{(x-61)^3}\right)^5 = (-27)^5$$

$$(x-61)^3 = (-27)^5$$

$$\sqrt[3]{(x-61)^3} = \left(\sqrt[3]{-27}\right)^5$$

$$x-61 = \left(\sqrt[3]{-27}\right)^5$$

$$x-61 = (-3)^5$$

$$x-61 = \underline{\hspace{1cm}}$$

$$x = \underline{\hspace{1cm}}$$

☐ d. Check your answer.

2. Solve for x: $(x+6)^{\frac{2}{5}} + 2 = 3$.

☑ a. Isolate the term with the rational exponent.

$$(x+6)^{\frac{2}{5}} = 1$$

☑ b. Rewrite the equation using a radical.

$$\sqrt[5]{(x+6)^2} = 1$$

☐ c. Finish solving for x.

$$\left(\sqrt[5]{(x+6)^2}\right)^5 = (1)^5$$

$$(x+6)^2 = 1$$

$$x+6 = \underline{\hspace{1cm}} \quad \text{or} \quad x+6 = \underline{\hspace{1cm}}$$

$$x = \underline{\hspace{1cm}} \quad \text{or} \quad x = \underline{\hspace{1cm}}$$

☐ d. Check your answer.

EXPLORE

Sample Problems

On the computer you used the Grapher to investigate the real solutions of nonlinear equations. Below are some additional exploration problems.

1. Graph the parabola $y = x^2 + 8x + 12$. Use your graph to find the solutions of the quadratic equation $x^2 + 8x + 12 = 0$.

☑ a. Graph the parabola
$y = x^2 + 8x + 12$.
Make a table of values.

x	y
−1	5
−2	0
−3	−3
−4	−4
−5	−3
−6	0

Plot these points on a grid and graph the parabola.

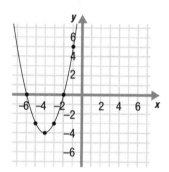

☑ b. Find the points where the graph of the parabola crosses the x-axis. The x-coordinates of these points are the solutions of the equation.

The graph crosses the x-axis at (−2, 0) and (−6, 0). So the solutions of $x^2 + 8x + 12 = 0$ are $x = -2$ and $x = -6$.

c. Here's one way to check:

Check x = −2:

Is $(-2)^2 + 8(-2) + 12 = 0$?

Is $\quad 4 - 16 + 12 = 0$?

Is $\quad\quad 0 = 0$? Yes.

Check x = −6:

Is $(-6)^2 + 8(-6) + 12 = 0$?

Is $\quad 36 - 48 + 12 = 0$?

Is $\quad\quad 0 = 0$? Yes.

☐ c. Check your answer using algebra.

2. Solve the equation $x^2 + 3x = 4$ by graphing $y = x^2 + 3x$ and $y = 4$.

☐ a. Graph the parabola
   $y = x^2 + 3x$.
   Make a table of values.

x	y
1	4
0	0
−1	−2
−1.5	−2.25
−2	−2
−3	0
−4	4

Plot these points on the grid
below and graph the parabola.

a.

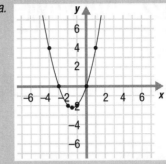

☐ b. Graph the line $y = 4$.
   Make a table of values.

x	y
−3	4
0	4
2	4

Plot these points on the
grid and graph the line.

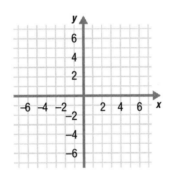

b.

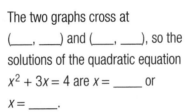

☐ c. Find where the graphs cross.
   The x-coordinates of the points
   of intersection are the solutions
   of the equation $x^2 + 3x = 4$.

The two graphs cross at
(___, ___) and (___, ___), so the
solutions of the quadratic equation
$x^2 + 3x = 4$ are $x =$ _____ or
$x =$ _____.

c. (−4, 4) and (1, 4)

   −4

   1

☐ d. Check your answer using
   algebra.

d. Here's one way to check:

   Check x = −4:
   Is $0 = (−4)^2 + 3(−4) − 4$?
   Is $0 = 16 − 12 − 4$?
   Is $0 = 0$? Yes.

   Check x = 1:
   Is $0 = 1^2 + 3(1) − 4$?
   Is $0 = 1 + 3 − 4$?
   Is $0 = 0$? Yes.

Homework Problems

Circle the homework problems assigned to you by the computer, then complete them below.

 Explain

Solving Equations

1. Solve for x: $x^4 - 7x^2 + 12 = 0$

2. Use substitution to solve for y: $y^4 - 7y^2 + 12 = 0$

3. Solve for w: $2w^3 - w^2 + 6w - 3 = 0$

4. Use substitution to solve for x: $x^{-2} + x^{-1} = 20$

5. Solve for a: $\dfrac{a}{2} - 3 = -\dfrac{4}{a}$

6. Use substitution to solve for y: $y^4 - 17y^2 + 72 = 0$

7. Solve for b: $b^4 - 8b^3 + 15b^2 = 0$

8. Use substitution to solve for
    z: $\left(z + \dfrac{6}{z}\right)^2 + 12\left(z + \dfrac{6}{z}\right) + 35 = 0$

9. Solve for x: $x - \dfrac{3}{2x} = \dfrac{1}{2}$

10. Use substitution to solve for w: $(w - 6)^4 - 81 = 0$

11. Solve for p: $p + \dfrac{2}{p} = \dfrac{19}{3}$

12. Use substitution to solve for r: $\left(r + \dfrac{4}{r}\right)^2 = 5\left(r + \dfrac{4}{r}\right)$

Radical Equations

13. Solve for x: $\sqrt{2x + 5} = x + 3$

14. Solve for x: $(x - 5)^{\frac{1}{2}} + 4 = 6$

15. Solve for x: $x + 6 = \sqrt{x + 6}$

16. Solve for x: $\sqrt{1 - x} = x + 5$

17. Solve for x: $(x + 9)^{\frac{3}{4}} + 2 = 10$

18. Solve for x: $x + 2 = \sqrt[3]{3x^2 + 9x + 7}$

19. Solve for x: $(x + 26)^{\frac{3}{5}} = 27$

20. Solve for x: $\sqrt{3x + 4} - \sqrt{2x + 1} = 1$

21. Solve for x: $\sqrt[4]{x^2 - 8x - 4} - 2 = 0$

22. Solve for x: $(x - 3)^{\frac{2}{3}} = 5$

23. Solve for x: $\sqrt{3(x + 4)} = 1 + \sqrt{2x + 7}$

24. Solve for x: $(x - 11)^{\frac{5}{2}} - 5 = 27$

 Explore

25. Graph the parabola $y = x^2 + 3x + 2$. Use your graph to find the solutions of the quadratic equation $x^2 + 3x + 2 = 0$.

26. For what value of c does the equation $x^2 + 12x + c = 0$ have only one distinct solution?

27. Solve the equation $x^2 + 5x = -6$ by graphing $y = x^2 + 5x$ and $y = -6$.

28. Graph the parabola $y = x^2 + 3x - 10$. Use your graph to find the solutions of the quadratic equation $x^2 + 3x - 10 = 0$.

29. For what values of c does the equation $x^2 + c = 0$ have at least one real solution?

30. Solve the equation $x^2 = x + 2$ by graphing $y = x^2$ and $y = x + 2$.

 APPLY

Practice Problems

Here are some additional practice problems for you to try.

1. Solve for w: $w^4 - 15w^2 - 16 = 0$

2. Solve for t: $t^4 + 15 = 8t^2$

3. Use substitution to solve for w: $w^{-2} + 5w^{-1} - 36 = 0$

4. Use substitution to solve for y: $(y + 5)^2 - 4(y + 5) - 12 = 0$

5. Solve for x by grouping: $x^3 - 5x^2 - 9x + 45 = 0$

6. Solve for r by grouping: $3r^3 - 2r^2 - 21r + 14 = 0$

7. Solve for z: $z - \dfrac{6}{5z} = -\dfrac{7}{5}$

8. Solve for q: $\dfrac{q}{3} - 3 = \dfrac{12}{q}$

9. Solve for x: $x^4 - 8x^3 - 9x^2 = 0$

10. Solve for t: $t^{-2} - t^{-1} - 42 = 0$

11. Solve for x: $\sqrt{4x + 21} = 3$

12. Solve for x: $\sqrt{9x - 2} = x + 2$

13. Solve for x: $\sqrt{x^2 + 16} - x = 2$

14. Solve for x: $x - 9 = \sqrt{x - 9}$

15. Solve for x: $(x - 1)^{\frac{3}{2}} - 1 = 7$

16. Solve for x: $(5x - 3)^{\frac{1}{3}} + 8 = 11$

17. Solve for x: $\sqrt[4]{x^2 - 4x + 11} + 6 = 8$

18. Solve for x: $\sqrt[3]{x^2 - 7x - 15} + 3 = 0$

19. Solve for x: $\sqrt{x + 10} - \sqrt{x + 5} = 1$

20. Solve for x: $\sqrt{5 - x} + 3 = \sqrt{3x + 4}$

Practice Test

Take this practice test to be sure that you are prepared for the final quiz in Evaluate.

1. Solve for x: $\dfrac{1}{x} - \dfrac{4}{3} = \dfrac{x}{4}$

2. What value of u can you substitute into the equation $5x^{-8} + 3x^{-4} - 2 = 0$ to get an equation in standard quadratic form?

3. Solve for y: $y^3 - 28 = 4y - 7y^2$

4. Solve for w: $\left(\dfrac{w-4}{w}\right)^2 = 5\left(\dfrac{w-4}{w}\right) - 6$

5. Solve for x: $\sqrt{x^2 + 6} - x = 3$

6. Solve for y: $\sqrt{y+8} - \sqrt{y+1} = 2$

7. Solve for x: $\sqrt[4]{x^3 + 54} - 2 = 1$

8. Solve for w: $(w+5)^{\frac{2}{3}} - 5 = -1$

9. Graph the parabola $y = x^2 - x - 2$. Use your graph to find the solutions of the quadratic equation $x^2 - x - 2 = 0$.

10. Graph the parabola $y = x^2 - 4x$ and the line $y = -3$. Use your graphs to find the solutions of the equation $x^2 - 4x = -3$.

11. The graphs of four parabolas are shown on the grid in Figure 13.1.1. Use these graphs to decide which of the equations below have one positive solution and one negative solution.

$0 = x^2 + \dfrac{9}{2}x + 5$ (use graph a)

$0 = \dfrac{1}{2}x^2 - \dfrac{5}{2}x - 3$ (use graph b)

$0 = \dfrac{1}{3}x^2 - \dfrac{4}{3}x + 1$ (use graph c)

$0 = x^2 - \dfrac{8}{3}x - 1$ (use graph d)

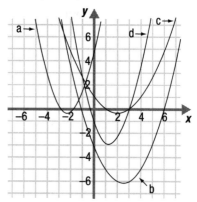

Figure 13.1.1

12. Graph the parabolas below. Use your graphs to determine for what values of c the equation $x^2 + 2x + c = 0$ has no solution.

$$y = x^2 + 2x - 3$$

$$y = x^2 + 2x + 1$$

$$y = x^2 + 2x + 4$$

LESSON 13.2 – NONLINEAR SYSTEMS

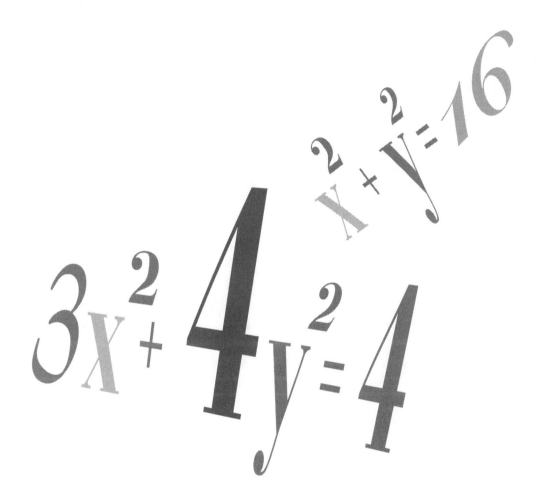

OVERVIEW

Here's what you'll learn in this lesson:

Solving Systems

a. *Recognizing the solution(s) of a nonlinear system, both graphically and algebraically.*

b. *Solving a nonlinear system using the substitution method.*

c. *Solving a nonlinear system using the elimination method.*

Suppose you wanted to use your carpentry skills to make some extra money selling bookcases to your friends. One of the first things you might be interested in knowing is how many bookcases you would have to sell before you started to make a profit.

If you knew how much it would cost you to make the bookcases, and how much you planned to charge for each one, you could figure out when you might start making money by setting up and solving a nonlinear system of equations.

In this lesson, you will learn about nonlinear systems of equations. You will learn how to solve such systems both by graphing and by using algebra.

 EXPLAIN

SOLVING SYSTEMS

Summary

Nonlinear Systems of Equations

A nonlinear system of equations is a system in which at least one of the equations is not the equation of a line.

You have learned how to solve systems of linear equations by graphing, by the substitution method, and by the elimination method. You can also use these methods to solve nonlinear systems of equations.

A linear system of equations can have either no solutions, one solution, or an infinite number of solutions. A nonlinear system can have no solutions, one solution, two solutions, three solutions, four solutions, five solutions, etc., up to an infinite number of solutions.

Solving Nonlinear Systems by Graphing

To solve a nonlinear system of equations by graphing:

1. Graph each equation in the system.

2. Find the points of intersection, if any. These are the real solutions of the system.

For example, to solve this nonlinear system:

$$x + y = 5$$
$$y = x^2 - 1$$

1. Graph the line $x + y = 5$ and graph the parabola $y = x^2 - 1$. See the grid in Figure 13.2.1.

2. Find the points of intersection: $(-3, 8)$ and $(2, 3)$. See the grid in Figure 13.2.2.

So this nonlinear system has two real solutions, $(x, y) = (-3, 8)$ or $(x, y) = (2, 3)$.

As another example, to solve this nonlinear system:

$$y = 5$$
$$x^2 + y^2 = 25$$

1. Graph the line $y = 5$ and graph the circle $x^2 + y^2 = 25$. See the grid in Figure 13.2.3.

2. Find the point of intersection: $(0, 5)$. See the grid in Figure 13.2.4.

So this nonlinear system has one solution, $(x, y) = (0, 5)$.

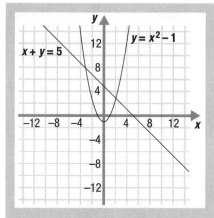

Figure 13.2.1

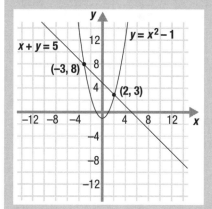

Figure 13.2.2

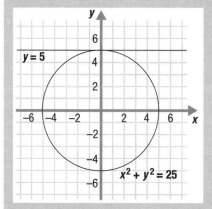

Figure 13.2.3

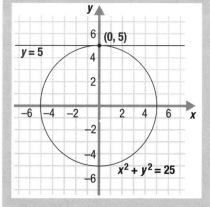

Figure 13.2.4

You have to be careful when you use graphing to solve a nonlinear system. If the solutions are fractions or irrational numbers, it may be difficult to read them accurately from the graph. If the solutions include very large numbers, you may not be able to see all of the solutions. Also, the points of intersection of the graphs represent only the **real** solutions of the system. To determine if the system has solutions that include imaginary numbers you will need to use one of the following algebraic methods.

Solving Nonlinear Systems by Substitution

To solve a nonlinear system of two equations by substitution:

1. Solve one equation for one variable in terms of the other variable.

2. Substitute into the other equation to obtain an equation in one variable.

3. Solve the resulting equation.

4. Complete the ordered pair(s) by substituting each solution into one of the original equations.

5. Check each ordered pair in both of the original equations to see if it is a solution of the system.

For example, to solve this nonlinear system by substitution:

$$y = x^2 - 1$$
$$x + y = 1$$

1. The first equation is already solved for y in terms of x. $\qquad y = x^2 - 1$

2. Substitute $y = x^2 - 1$ into the second equation. $\qquad x + (x^2 - 1) = 1$

3. Solve the resulting equation. $\qquad\qquad\qquad\qquad x^2 + x - 2 = 0$

$$(x - 1)(x + 2) = 0$$

$$x - 1 = 0 \quad \text{or} \quad x + 2 = 0$$

$$x = 1 \quad \text{or} \quad x = -2$$

4. Complete the ordered pairs.

$x + y = 1$	$x + y = 1$
$1 + y = 1$	$-2 + y = 1$
$y = 0$	$y = 3$

So the ordered pairs are (1, 0) and (−2, 3).

5. Check each ordered pair.

   Check (1, 0):

   $y = x^2 - 1$ $x + y = 1$

   Is $0 = 1^2 - 1$?   Is $1 + 0 = 1$?

   Is $0 = 1 - 1$?   Is $1 = 1$? Yes.

   Is $0 = 0$? Yes.

   Check (−2, 3):

   $y = x^2 - 1$ $x + y = 1$

   Is $3 = (-2)^2 - 1$?   Is $-2 + 3 = 1$?

   Is $3 = 4 - 1$?   Is $1 = 1$? Yes.

   Is $3 = 3$? Yes.

So the solutions of this nonlinear system are the points $(x, y) = (1, 0)$ or $(x, y) = (-2, 3)$.

Sometimes nonlinear systems have equations that are expressed in terms of x or y. So you can start by solving one of the equations for x or for y. However, it is often more convenient to solve one of the equations for something other than x or y.

For example, to solve this nonlinear system by substitution:

$$x^2 + y^2 + 2y = 3$$
$$x^2 - y = 5$$

1. Solve the second equation for x^2 in terms of y. $x^2 = y + 5$

2. Substitute this value for x^2 into the first $(y + 5) + y^2 + 2y = 3$
   equation.

3. Solve this quadratic equation. $y^2 + 3y + 2 = 0$

   $$(y + 2)(y + 1) = 0$$

   $$y + 2 = 0 \text{ or } y + 1 = 0$$

   $$y = -2 \text{ or } \quad y = -1$$

4. Complete the ordered pairs. $x^2 - y = 5$ $x^2 - y = 5$

   $$x^2 - (-2) = 5 \qquad x^2 - (-1) = 5$$

   $$x^2 + 2 = 5 \qquad x^2 + 1 = 5$$

   $$x^2 = 3 \qquad\qquad x^2 = 4$$

   $$x = \pm\sqrt{3} \qquad\qquad x = \pm 2$$

So the ordered pairs are $(\sqrt{3}, -2)$, $(-\sqrt{3}, -2)$, $(2, -1)$, and $(-2, -1)$.

5. Check each ordered pair.

Check $(\sqrt{3}, -2)$:

$$x^2 + y^2 + 2y = 3 \qquad\qquad x^2 - y = 5$$

Is $\left(\sqrt{3}\right)^2 + (-2)^2 + 2(-2) = 3$? $\qquad$ Is $\left(\sqrt{3}\right)^2 - (-2) = 5$?

Is $\quad 3 \quad + \quad 4 \quad - \quad 4 \quad = 3$? $\qquad$ Is $\quad 3 \quad + \quad 2 = 5$?

Is $\qquad\qquad\qquad 3 = 3$? Yes. $\qquad$ Is $\qquad\qquad 5 = 5$? Yes.

Check $(-\sqrt{3}, -2)$:

$$x^2 + y^2 + 2y = 3 \qquad\qquad x^2 - y = 5$$

Is $\left(-\sqrt{3}\right)^2 + (-2)^2 + 2(-2) = 3$? $\qquad$ Is $\left(-\sqrt{3}\right)^2 - (-2) = 5$?

Is $\quad 3 \quad + \quad 4 \quad - \quad 4 \quad = 3$? $\qquad$ Is $\quad 3 \quad + \quad 2 = 5$?

Is $\qquad\qquad\qquad 3 = 3$? Yes. $\qquad$ Is $\qquad\qquad 5 = 5$? Yes.

Check $(2, -1)$:

$$x^2 + y^2 + 2y = 3 \qquad\qquad x^2 - y = 5$$

Is $2^2 + (-1)^2 + 2(-1) = 3$? $\qquad$ Is $2^2 - (-1) = 5$?

Is $4 \; + \; 1 \; - \; 2 \quad = 3$? $\qquad$ Is $4 \; + \; 1 = 5$?

Is $\qquad\qquad\qquad 3 = 3$? Yes. $\qquad$ Is $\qquad 5 = 5$? Yes.

Check $(-2, -1)$:

$$x^2 + y^2 + 2y = 3 \qquad\qquad x^2 - y = 5$$

Is $(-2)^2 + (-1)^2 + 2(-1) = 3$? $\qquad$ Is $(-2)^2 - (-1) = 5$?

Is $4 \; + \; 1 \; - \; 2 \quad = 3$? $\qquad$ Is $4 \; + \; 1 = 5$?

Is $\qquad\qquad\qquad 3 = 3$? Yes. $\quad$ Is $\qquad 5 = 5$? Yes.

So this nonlinear system has four solutions: $\qquad (x, y) = (\sqrt{3}, -2)$

$(x, y) = (-\sqrt{3}, -2)$

$(x, y) = (2, -1)$

$(x, y) = (-2, -1)$

Solving Nonlinear Systems by Elimination

To solve a nonlinear system of two equations by elimination:

1. Eliminate a variable by multiplying the equations by appropriate numbers and adding them.

2. Solve for the other variable.

3. Complete the ordered pair(s) by substituting each solution into one of the original equations.

4. Check each ordered pair in both of the original equations to see if it is a solution of the system.

For example, to solve this nonlinear system by elimination:

$$x^2 - y^2 = 7$$
$$2x^2 + 3y^2 = 24$$

1. Eliminate the y^2-terms by multiplying the first equation by 3 and adding it to the other equation.

$$3(x^2 - y^2) = 3(7) \rightarrow \begin{array}{r} 3x^2 - 3y^2 = 21 \\ 2x^2 + 3y^2 = 24 \\ \hline 5x^2 = 45 \end{array}$$

2. Solve for x.

$$5x^2 = 45$$
$$x^2 = 9$$
$$x = \pm 3$$

3. Complete the ordered pairs.

$$\begin{array}{ll} x^2 - y^2 = 7 & x^2 - y^2 = 7 \\ (3)^2 - y^2 = 7 & (-3)^2 - y^2 = 7 \\ 9 - y^2 = 7 & 9 - y^2 = 7 \\ -y^2 = -2 & -y^2 = -2 \\ y = \pm\sqrt{2} & y = \pm\sqrt{2} \end{array}$$

So the ordered pairs are $(3, \sqrt{2})$, $(3, -\sqrt{2})$, $(-3, \sqrt{2})$, and $(-3, -\sqrt{2})$.

4. Check each ordered pair.

Check $(3, \sqrt{2})$:

$$x^2 - y^2 = 7 \qquad\qquad 2x^2 + 3y^2 = 24$$

Is $3^2 - \left(\sqrt{2}\right)^2 = 7$?   Is $2(3)^2 + 3\left(\sqrt{2}\right)^2 = 24$?

Is $9 - 2 = 7$?   Is $18 + 6 = 24$?

Is $7 = 7$? Yes.   Is $24 = 24$? Yes.

Check $(3, -\sqrt{2})$:

$$x^2 - y^2 = 7 \qquad\qquad 2x^2 + 3y^2 = 24$$

Is $3^2 - \left(-\sqrt{2}\right)^2 = 7$? $\qquad$ Is $2(3)^2 + 3\left(-\sqrt{2}\right)^2 = 24$?

Is $9 \quad - \quad 2 \quad\ = 7$? $\qquad\qquad$ Is $18 + \quad 6 \quad\ = 24$?

Is $\qquad\qquad 7 = 7$? Yes. $\qquad$ Is $\qquad\qquad 24 = 24$? Yes.

Check $(-3, \sqrt{2})$:

$$x^2 - y^2 = 7 \qquad\qquad 2x^2 + 3y^2 = 24$$

Is $(-3)^2 - \left(\sqrt{2}\right)^2 = 7$? $\qquad$ Is $2(-3)^2 + 3\left(\sqrt{2}\right)^2 = 24$?

Is $9 \ - \ 2 \quad\ = 7$? $\qquad\qquad$ Is $18 \ + \ 6 \quad = 24$?

Is $\qquad\qquad 7 = 7$? Yes. $\qquad$ Is $\qquad\qquad 24 = 24$? Yes.

Check $(-3, -\sqrt{2})$:

$$x^2 - y^2 = 7 \qquad\qquad 2x^2 + 3y^2 = 24$$

Is $(-3)^2 - \left(-\sqrt{2}\right)^2 = 7$? $\qquad$ Is $2(-3)^2 + 3\left(-\sqrt{2}\right)^2 = 24$?

Is $9 \ - \ 2 \quad\ = 7$? $\qquad\qquad$ Is $18 \ + \ 6 \quad = 24$?

Is $\qquad\qquad 7 = 7$? Yes. $\qquad$ Is $\qquad\qquad 24 = 24$? Yes.

So this system has four solutions: $(x, y) = (3, \sqrt{2})$

$$(x, y) = (3, -\sqrt{2})$$

$$(x, y) = (-3, \sqrt{2})$$

$$(x, y) = (-3, -\sqrt{2})$$

Sometimes nonlinear systems have no real solutions.

As an example, to solve this system by elimination:

$$x^2 - y = -3$$
$$-x^2 - y = -1$$

1. Eliminate the x^2-terms by adding the $\qquad$ $x^2 - y = -3$
   equations. $\qquad\qquad\qquad\qquad\qquad\qquad\underline{-x^2 - y = -1}$
   $\qquad\qquad\qquad\qquad\qquad\qquad\qquad\qquad -2y = -4$

2. Solve for y. $\qquad\qquad\qquad\qquad\qquad\qquad -2y = -4$

$$y = 2$$

3. Complete the ordered pairs.

$$x^2 - y = -3$$

$$x^2 - 2 = -3$$

$$x^2 = -1$$

$$x = \pm\sqrt{-1}$$

$$x = \pm i$$

So the ordered pairs are $(i, 2)$ and $(-i, 2)$.

4. Check each ordered pair.

Check $(i, 2)$:

$x^2 - y = -3$ $-x^2 - y = -1$

Is $i^2 - 2 = -3$?   Is $-(i)^2 - 2 = -1$?

Is $-1 - 2 = -3$?   Is $-(-1) - 2 = -1$?

Is $\quad -3 = -3$? Yes.   Is $\quad -1 = -1$? Yes.

Check $(-i, 2)$:

$x^2 - y = -3$ $-x^2 - y = -1$

Is $(-i)^2 - 2 = -3$?   Is $-(-i)^2 - 2 = -1$?

Is $-1 - 2 = -3$?   Is $-(-1) - 2 = -1$?

Is $\quad -3 = -3$? Yes.   Is $\quad -1 = -1$? Yes.

This system has no real solutions. It has two solutions that include imaginary numbers: $(x, y) = (i, 2)$ or $(x, y) = (-i, 2)$.

Sample Problems

1. Solve this nonlinear system by graphing:
   $$y = x^2$$
   $$y = 2x - 1$$

   ☑ a. Graph the parabola $y = x^2$.

   ☐ b. Graph the line $y = 2x - 1$.

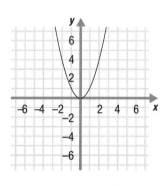

   ☐ c. Find the point of intersection.

   $(x, y) = ($____$,$ ____$)$

b.

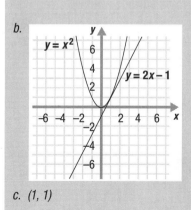

c. (1, 1)

2. Solve this nonlinear system by substitution:

$$x^2 + y = 5$$
$$y = 2x^2 + 4$$

☑ a. The second equation is already solved for y in terms of x.

$$y = 2x^2 + 4$$

☑ b. Substitute $y = 2x^2 + 4$ into the first equation.

$$x^2 + y = 5$$
$$x^2 + (2x^2 + 4) = 5$$

☐ c. Solve for x.

$$3x^2 - 1 = 0$$
$$x = \underline{\quad} \text{ or } x = \underline{\quad}$$

c. $x = \pm\sqrt{\dfrac{1}{3}}$ or $x = \pm\dfrac{\sqrt{3}}{3}$

☐ d. Complete the ordered pairs.

$$y = \underline{\quad} \qquad y = \underline{\quad}$$

d. $\dfrac{14}{3}, \dfrac{14}{3}$

e. Here's one way to do it.

Check $\left(-\sqrt{\dfrac{1}{3}}, \dfrac{14}{3}\right)$:

$$x^2 + y = 5$$

Is $\left(-\sqrt{\dfrac{1}{3}}\right)^2 + \dfrac{14}{3} = 5$? Yes.

$$y = 2x^2 + 4$$

Is $\dfrac{14}{3} = 2\left(-\sqrt{\dfrac{1}{3}}\right)^2 + 4$? Yes.

Check $\left(\sqrt{\dfrac{1}{3}}, \dfrac{14}{3}\right)$:

$$x^2 + y = 5$$

Is $\left(\sqrt{\dfrac{1}{3}}\right)^2 + \dfrac{14}{3} = 5$? Yes.

$$y = 2x^2 + 4$$

Is $\dfrac{14}{3} = 2\left(\sqrt{\dfrac{1}{3}}\right)^2 + 4$? Yes.

☐ e. Check each ordered pair.

f. $(x, y) = \left(-\sqrt{\dfrac{1}{3}}, \dfrac{14}{3}\right)$

$(x, y) = \left(\sqrt{\dfrac{1}{3}}, \dfrac{14}{3}\right)$

☐ f. Write the solution(s).

3. Solve this nonlinear system by elimination:

$$3x^2 + 4y^2 = 4$$
$$2x^2 - 3y^2 = -3$$

☑ a. Eliminate the x^2-terms by
   multiplying the first equation
   by 2 and the second equation
   by –3 and adding.

$$2(3x^2 + 4y^2) = 2(4) \longrightarrow 6x^2 + 8y^2 = 8$$
$$-3(2x^2 - 3y^2) = -3(-3) \longrightarrow \underline{-6x^2 + 9y^2 = 9}$$
$$17y^2 = 17$$

☐ b. Finish solving for y.

$$y = \underline{\quad} \text{ or } y = \underline{\quad}$$

☐ c. Complete the ordered pair(s).

$$x = \underline{\quad} \qquad x = \underline{\quad}$$

☐ d. Check each ordered pair.

☐ e. Write the solution(s).

Answers to Sample Problems

b. 1, –1 (in either order)

c. 0, 0

d. Check (0, 1):

$$3x^2 + 4y^2 = 4$$

Is $3(0)^2 + 4(1)^2 = 4$? Yes.

$$2x^2 - 3y^2 = -3$$

Is $2(0)^2 - 3(1)^2 = -3$? Yes.

Check (0, –1):

$$3x^2 + 4y^2 = 4$$

Is $3(0)^2 + 4(-1)^2 = 4$? Yes.

$$2x^2 - 3y^2 = -3$$

Is $2(0)^2 - 3(-1)^2 = 3$? Yes.

e. $(x, y) = (0, 1)$ or $(x, y) = (0, -1)$

Answers to Sample Problems

Sample Problems

On the computer, you used the Grapher to find the real solutions of nonlinear systems of equations by graphing the systems. Below are some additional exploration problems.

1. Graph each system below. Find the system that has only one real solution.

☑ a. circle: $x^2 + y^2 = 16$

parabola: $y = x^2 + 4x$

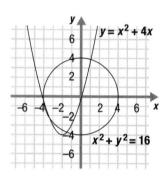

b.

☐ b. circle: $x^2 + y^2 = 16$

circle: $x^2 + y^2 = 9$

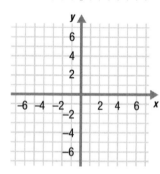

c.

☐ c. line: $y = -x + 3$

parabola: $y = x^2 - 5$

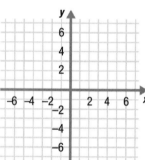

d.

☐ d. line: $y = 5$

parabola: $y = -x^2 + 5$

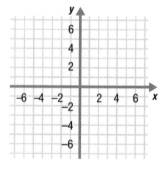

e. $y = 5$ and $y = -x^2 + 5$

☐ e. Write the equations of the system above whose graphs intersect in one and only one point.

2. Which of the following systems have a solution $(x, y) = (1, 1)$?

$y = x^2$ $y = x^2 - 1$ $x^2 + y^2 = 1$ $y = -x^2 + 2$

$y = x$ $y = 2x - 1$ $y = -x^2 + 1$ $y = -x + 2$

Substitute $(x, y) = (1, 1)$ into both equations in the first system.

$y = x^2$ $y = x$

Is $1 = 1^2$?   Is $1 = 1$? Yes.

Is $1 = 1$? Yes.

Is $(x, y) = (1, 1)$ a solution of this system? Yes.

☐ b. Substitute $(x, y) = (1, 1)$ into both equations in the second system.

$y = x^2 - 1$ $y = 2x - 1$

Is $(x, y) = (1, 1)$ a solution of this system? _____

☐ c. Substitute $(x, y) = (1, 1)$ into both equations in the third system.

$x^2 + y^2 = 1$ $y = -x^2 + 1$

Is $(x, y) = (1, 1)$ a solution of this system? _____

☐ d. Substitute $(x, y) = (1, 1)$ into both equations in the third system.

$y = -x^2 + 2$ $y = -x + 2$

Is $(x, y) = (1, 1)$ a solution of this system? _____

Answers to Sample Problems

b. Is $1 = 1^2 - 1$?   Is $1 = 2(1) - 1$?

Is $1 = 0$? No.   Is $1 = 1$? Yes.

No.

c. Is $1^2 + 1^2 = 1$?   Is $1 = -1^2 + 1$?

Is $2 = 1$? No.   Is $1 = -1 + 1$?

Is $1 = 0$? No.

No.

d. Is $1 = -(1)^2 + 2$?   Is $1 = -1 + 2$?

Is $1 = -1 + 2$?   Is $1 = 1$? Yes.

Is $1 = 1$? Yes.

Yes.

HOMEWORK

Homework Problems

Circle the homework problems assigned to you by the computer, then complete them below.

 Explain
Solving Systems

1. Solve the following nonlinear system by graphing:
$$y = x^2 + 2x + 1$$
$$y = x + 3$$

2. Solve the following nonlinear system by substitution:
$$y = x^2 - 4$$
$$y = 6x - 13$$

3. Solve the following nonlinear system by elimination:
$$3x^2 + 4y^2 = 16$$
$$x^2 - y^2 = 3$$

4. Solve the following nonlinear system by graphing:
$$x^2 + y^2 = 9$$
$$y = -\frac{1}{3}x^2 + 3$$

5. Solve the following nonlinear system by substitution:
$$x^2 + y^2 = 13$$
$$x + y = 5$$

6. Solve the following nonlinear system by elimination:
$$4x^2 + 3y^2 = 12$$
$$x^2 + 3y^2 = 12$$

7. Solve the following nonlinear system by graphing:
$$x^2 + y^2 = 26$$
$$y = -x + 8$$

8. Solve the following nonlinear system by substitution:
$$x^2 + y^2 = 16$$
$$x^2 - 2y = 8$$

9. The supply equation for Socorro Rodriguez's hand painted jackets is $y = \frac{x}{50} + 15$, where x is the number of jackets supplied at a price of y dollars each. The demand equation is $y = \frac{12500}{x}$, where x is the number of jackets demanded at a price of y dollars each. Use substitution to find the point (x, y) where supply equals demand. How many jackets are sold and for what price?

10. Chakotay is building a house with a gabled roof, as shown in Figure 13.2.5. The plans call for a vent in the shape of an isosceles triangle in each gable. The ratio of the height to the base of these triangles must be 1 to 4. If each vent has to provide a total venting area of 3500 square inches, what should the height and base of each vent be?

Hint: The area of a triangle is $\frac{1}{2}$base · height.

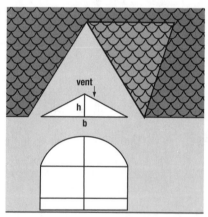

Figure 13.2.5

11. Solve the following nonlinear system by substitution:
$$x^2 + y^2 = 16$$
$$x + y = 4$$

12. Solve the following nonlinear system by elimination:
$$4x^2 - 9y^2 + 132 = 0$$
$$x^2 + 4y^2 - 67 = 0$$

Explore

13. The graph of the parabola $y = x^2 + 4x + 3$ is shown in Figure 13.2.6. What is the equation of the horizontal line that, with the parabola, forms a nonlinear system with exactly one solution?

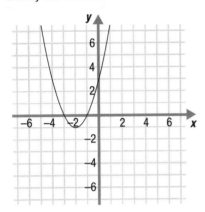

Figure 13.2.6

14. Graph this system of equations on the grid in Figure 13.2.7 to find its solutions:
$$y = x^2 - 4x + 2$$
$$y = x + 2$$

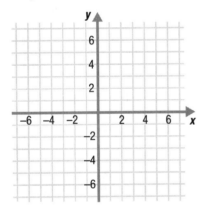

Figure 13.2.7

15. Which nonlinear system(s) below have a solution $(x, y) = (5, 0)$?

$x^2 + y^2 = 25$ $y = 5$
    $y = 0$ $y = x^2 - 4x - 5$

$y = -x^2 + 4x + 5$ $y = x^2 + 4x - 5$
$y = -x + 5$ $x^2 + y^2 = 25$

16. Which of the system(s) below have only one real solution?

$(x - 3)^2 + y^2 = 18$ $y = 3$
$y = -x^2 - 3x + 4$ $y = x^2 + 3x - 4$
$y = 2x - 7$ $y = x^2 - 6$

17. Graph this system of equations to find its solutions:
$$y = -x^2 - 3x + 4$$
$$y = x^2 + 3x - 4$$

18. Which nonlinear system(s) below have a solution $(x, y) = (1, -8)$?

$y = x^2 - 2x - 7$ $(x - 1)^2 + y^2 = 64$
$y = -3x - 5$ $y = -8$

$y = -x^2 + 2x - 9$ $y = -x^2 + 8x - 15$
$y = 2x^2 - 4x - 6$ $y = 2x - 10$

APPLY

Practice Problems

Here are some additional practice problems for you to try.

1. Use the graphs below to solve the following nonlinear system:
$$y = (x-2)^2$$
$$y = -(x-2)^2 + 2$$

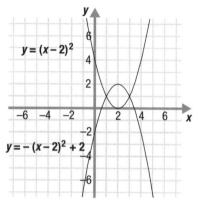

2. Use the graphs below to solve the following nonlinear system:
$$x^2 + y^2 = 4$$
$$y = x + 5$$

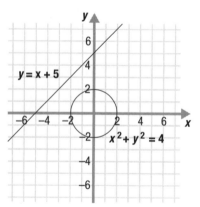

3. Use the graphs below to solve the following nonlinear system:
$$y = -x^2 + 1$$
$$y = x - 1$$

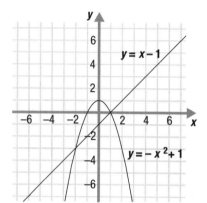

4. Solve the following nonlinear system by substitution:
$$x^2 + y^2 = 10$$
$$-3x + y = 0$$

5. Solve the following nonlinear system by substitution:
$$6x^2 - 2y = 4$$
$$3x^2 + y = 22$$

6. Solve the following nonlinear system by elimination:
$$7x^2 + 11y^2 = 28$$
$$x^2 + y^2 = 4$$

7. Solve the following nonlinear system by elimination:
$$10x^2 + y^2 = 14$$
$$2x^2 + y^2 = 6$$

8. Use the graphs below to find the number of real solutions of this nonlinear system:

$x^2 + y^2 = 18$

$y = x^2 - 5$

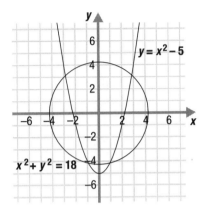

9. Use the graphs below to find the number of real solutions of this nonlinear system:

$y = x^2 + 3$

$2x + y = -6$

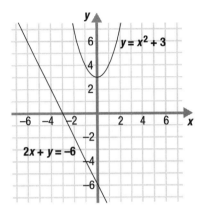

10. How many real solutions does this nonlinear system have?

$x^2 + y^2 = 16$

$y = x - 3$

Practice Test

Take this practice test to be sure that you are prepared for the final quiz in Evaluate.

1. Use the graphs in Figure 13.2.8 to find the number of solutions of this nonlinear system:

$$x^2 + y^2 = 13$$
$$y = 2x^2 - 5$$

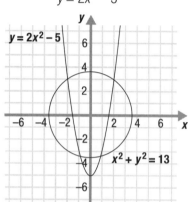

Figure 13.2.8

2. Use the substitution method to solve this nonlinear system:

$$y = x^2 - 2x + 1$$
$$y + x = 3$$

3. Use the elimination method to solve this nonlinear system:

$$x^2 - 4y = 8$$
$$x + 2y = 8$$

4. How many real solutions does this nonlinear system have?

$$x^2 + y^2 = 25$$
$$x + y = 9$$

5. The graph of the circle $x^2 + y^2 = 18$ is shown in Figure 13.2.9. Identify the equation of the line below that, with this circle, forms a nonlinear system with exactly two solutions.

$$y = 2x + 11$$
$$y = -x$$
$$y = 3\sqrt{2}$$
$$y = 9$$

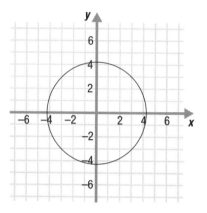

Figure 13.2.9

6. The graph of the nonlinear system below is shown in Figure 13.2.10.

$$x^2 + y^2 = 25$$
$$y = -\frac{1}{5}x^2 + 5$$

Find the solutions of this system that also satisfy the equation $y = -x + 5$.

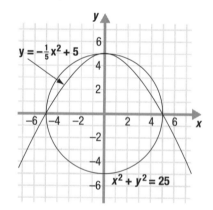

Figure 13.2.10

7. The graph of the parabola $y = x^2 - 5$ is shown in Figure 13.2.11. Circle the equation of the line below that, with the parabola, forms a nonlinear system with solutions $(x, y) = (-3, 4)$ and $(x, y) = (1, -4)$.

$y = -2x - 2$

$y = -2x$

$y = \dfrac{4}{3}x$

$y = 4$

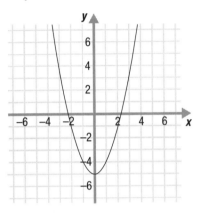

Figure 13.2.11

8. Which of the nonlinear systems below have $(x, y) = (2, -1)$ as one of their solutions?

$y = x^2 - 5$
$y = -2x + 3$

$x^2 + y^2 = 5$
$y = -x^2 + 3$

$y = x^2 - 4$
$y = -2x + 5$

$x^2 + y^2 = 5$
$y = 2$

LESSON 13.3 – INEQUALITIES

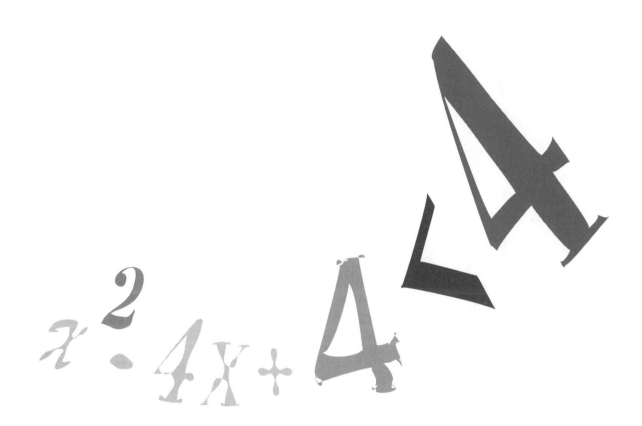

 **OVERVIEW**

Here's what you'll learn in this lesson:

Quadratic Inequalities

a. *Solving quadratic inequalities*

Rational Inequalities

a. *Definition of a rational inequality*

b. *Solving rational inequalities*

An ice cream company is considering producing several new special-edition flavors, and the members of the market research team are hard at work. They are analyzing each of the potential new flavors, and trying to determine how much of each the company would have to sell to make a profit. They are also trying to estimate the most economical number of production runs the company should make for each flavor.

To do these tasks, the members of the market research team are setting up and solving quadratic and rational inequalities. In this lesson, you will learn how to solve such inequalities.

 **EXPLAIN**

QUADRATIC INEQUALITIES

Summary

Graphing Method

When you graph the function $y = ax + b$, the x-coordinate of the point at which the line crosses the x-axis is the solution of the equation $ax + b = 0$. Similarly, when you graph the function $y = ax^2 + bx + c$, the x-coordinates of the points at which the parabola crosses the x-axis are the solutions of the equation $ax^2 + bx + c = 0$.

You can also use graphing to solve quadratic inequalities.

If you graph the function corresponding to...	The solutions are ...
$ax^2 + bx + c > 0$	the x-coordinates of all the points on the graph that lie **above** the x-axis.
$ax^2 + bx + c \geq 0$	the x-coordinates of all the points on the graph that lie **on and above** the x-axis.
$ax^2 + bx + c < 0$	the x-coordinates of all the points on the graph that lie **below** the x-axis.
$ax^2 + bx + c \leq 0$	the x-coordinates of all the points on the graph that lie **on and below** the x-axis.

To find the solutions of a quadratic inequality by graphing:

1. Graph the corresponding parabola, $y = ax^2 + bx + c$.

2. Find the x-coordinates of the appropriate points on the graph, as indicated in the table above.

For example, to solve the inequality $x^2 + 2x - 8 < 0$ by graphing:

1. Graph the parabola $y = x^2 + 2x - 8$. See Figure 13.3.1.

2. Find the x-coordinates of the points on the graph that lie **below** the x-axis. See Figure 13.3.2.

So $x^2 + 2x - 8 < 0$ when $-4 < x < 2$.

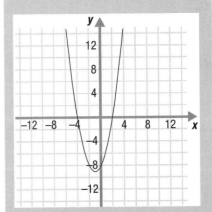

Figure 13.3.1

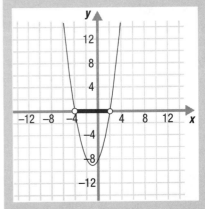

Figure 13.3.2

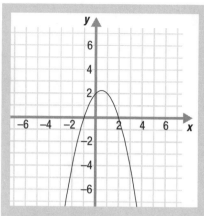

Figure 13.3.3

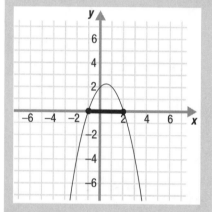

Figure 13.3.4

As another example, to solve the inequality $-x^2 + x + 2 \geq 0$ by graphing:

1. Graph the parabola $y = -x^2 + x + 2$. See Figure 13.3.3.

2. Find the x-coordinates of the points on the graph that lie **on and above** the x-axis. See Figure 13.3.4.

So $-x^2 + x + 2 \geq 0$ when $-1 \leq x \leq 2$.

Test Point Method

When you use graphing to solve a quadratic inequality, you look at the x-coordinates of the points on the graph of the corresponding function. You can also solve a quadratic inequality by finding the sign of the corresponding function at different intervals on the x-axis.

When a quadratic function crosses the x-axis, the value of the function either goes from positive to negative or from negative to positive. A function crosses the x-axis when it is equal to 0, that is, at the solutions of the corresponding equation. So the solutions of the corresponding equation are the dividing points where the function changes sign.

Since the value of a quadratic function is always positive or negative in an interval between dividing points, you can evaluate the function at any "test point" in the interval to find out whether all the values for points in that interval are positive or negative.

To find the solutions of a quadratic inequality without graphing:

1. Solve the corresponding equation, $ax^2 + bx + c = 0$.

2. Use the solutions to divide the number line into intervals.

3. Substitute a test point from each interval into $ax^2 + bx + c$ to determine its sign.

4. Find the intervals that satisfy the original inequality.

For example, to solve $x^2 - 4x + 3 < 0$ without graphing:

1. Solve the equation
    $x^2 - 4x + 3 = 0$.

    $x^2 - 4x + 3 = 0$
    $(x - 1)(x - 3) = 0$
    $x - 1 = 0$ or $x - 3 = 0$
    $x = 1$ or $\quad x = 3$

2. Divide the number line into intervals, using $x = 1$ and $x = 3$ as the dividing points.

3. Substitute a test point from each interval into $x^2 - 4x + 3$ to determine its sign; for example, $x = 0$, $x = 2$, and $x = 5$.

    Test $x = 0$:
    $(0)^2 - 4(0) + 3$
    $0 - 0 + 3 = 3$ is **positive**

Test $x = 2$:

$(2)^2 - 4(2) + 3$

$4 - 8 + 3 = -1$ is **negative**

Test $x = 5$:

$(5)^2 - 4(5) + 3$

$25 - 20 + 3 = 8$ is **positive**

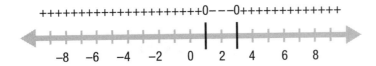

4. Find the intervals where $x^2 - 4x + 3 < 0$.

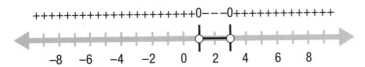

*You want the value of $x^2 - 4x + 3$ to be **less than 0**, that is, **negative**.*

So $x^2 - 4x + 3 < 0$ when $1 < x < 3$.

As another example, to solve $2x^2 + 7x - 15 \geq 0$ without graphing:

1. Solve the equation
   $2x^2 + 7x - 15 = 0$.

   $2x^2 + 7x - 15 = 0$

   $(x + 5)(2x - 3) = 0$

   $x + 5 = 0$ or $2x - 3 = 0$

   $x = -5$ or $\qquad x = \dfrac{3}{2}$

2. Divide the number line into intervals, using $x = -5$ and $x = \dfrac{3}{2}$ as the dividing points.

3. Substitute a test point from each interval into $2x^2 + 7x - 15$ to determine its sign; for example, $x = -7$, $x = 0$, and $x = 6$.

   Test $x = -7$:

   $2(-7)^2 + 7(-7) - 15$

   $98 - 49 - 15 = 34$ is **positive**

   Test $x = 0$:

   $2(0)^2 + 7(0) - 15$

   $0 + 0 - 15 = -15$ is **negative**

   Test $x = 6$:

   $2(6)^2 + 7(6) - 15$

   $72 + 42 - 15 = 99$ is **positive**

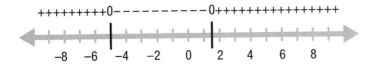

4. Find the intervals where $2x^2 + 7x - 15 \geq 0$.

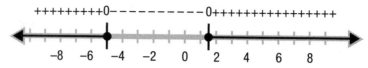

So $2x^2 + 7x - 15 \geq 0$ when $x \leq -5$ or $x \geq \frac{3}{2}$.

Although in the two preceding examples the number line was divided into three intervals, this is not always the case.

For example, to solve $x^2 - 4x + 4 < 0$ without graphing:

1. Solve the equation $x^2 - 4x + 4 = 0$.

$$x^2 - 4x + 4 = 0$$
$$(x - 2)(x - 2) = 0$$
$$x - 2 = 0 \text{ or } x - 2 = 0$$
$$x = 2 \text{ or } \quad x = 2$$

2. Divide the number line into intervals, using $x = 2$ as the dividing point.

3. Substitute a test point from each interval into $x^2 - 4x + 4$ to determine its sign; for example, $x = 0$ and $x = 3$.

    Test $x = 0$:
    $(0)^2 - 4(0) + 4$
    $0 - 0 + 4 = 4$ is **positive**

    Test $x = 3$:
    $(3)^2 - 4(3) + 4$
    $9 - 12 + 4 = 1$ is **positive**

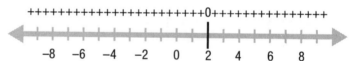

4. Find the intervals where $x^2 - 4x + 4 < 0$.

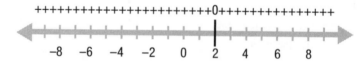

So $x^2 - 4x + 4 < 0$ is **never true** since $x^2 - 4x + 4$ is positive in both intervals, and 0 at $x = 2$.

Sample Problems

1. Find the solution of $x^2 - 2x - 3 \leq 0$ by graphing the equation $y = x^2 - 2x - 3$.

   ☑ a. Graph the equation
      $y = x^2 - 2x - 3$.

   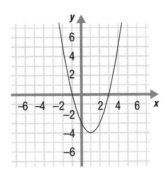

   ☐ b. Find the x-coordinates of
      the points on the graph that
      lie **on and below** the x-axis.

   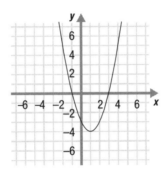

   b.

   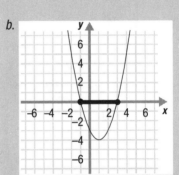

   ☐ c. Write the solution.

   c. $-1 \leq x \leq 3$

2. Solve $2x^2 - x - 36 > 0$ using the test point method.

   ☑ a. Solve the equation
      $2x^2 - x - 36 = 0$.

   $$2x^2 - x - 36 = 0$$
   $$(x + 4)(2x - 9) = 0$$
   $$x + 4 = 0 \quad \text{or} \quad 2x - 9 = 0$$
   $$x = -4 \quad \text{or} \qquad x = \frac{9}{2}$$

   ☑ b. Use the solutions to divide the number line into intervals.

   ☐ c. Substitute a test point from
      each interval into the expression
      $2x^2 - x - 36$ to determine its
      sign, for example, $x = -5$,
      $x = 0$, and $x = 6$.

   Test $x = -5$:
   $2(-5)^2 - (-5) - 36$
   $50 + 5 - 36 = 19$ is **positive**

   Test $x = 0$:

   Test $x = 6$:

   c. $2(0)^2 - (0) - 36$
   $0 - 0 - 36 = -36$ is **negative**

   $2(6)^2 - 6 - 36$
   $72 - 6 - 36 = 30$ is **positive**

   ☐ d. Find the intervals that
      satisfy the original inequality.

   d. $x < -4$ or $x > \frac{9}{2}$

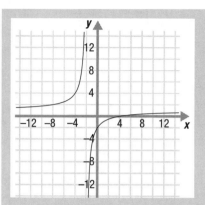

Figure 13.3.5

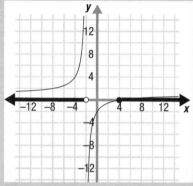

Figure 13.3.6

Notice x = –2 is not included as part of the solution. This is because the function is undefined at x = –2.

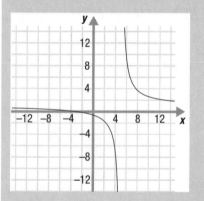

Figure 13.3.7

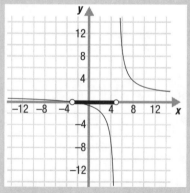

Figure 13.3.8

RATIONAL INEQUALITIES

Summary

Graphing Method

Just as you can use graphing to solve linear or quadratic inequalities, you can also use it to solve rational inequalities.

If you graph the function corresponding to...	The solutions are ...
$\frac{ax + b}{cx + d} > 0$	the x-coordinates of all the points on the graph that lie **above** the x-axis.
$\frac{ax + b}{cx + d} \geq 0$	the x-coordinates of all the points on the graph that lie **on and above** the x-axis.
$\frac{ax + b}{cx + d} < 0$	the x-coordinates of all the points on the graph that lie **below** the x-axis.
$\frac{ax + b}{cx + d} \leq 0$	the x-coordinates of all the points on the graph that lie **on and below** the x-axis.

To find the solutions of a rational inequality by graphing:

1. Graph the corresponding function, $y = \frac{ax + b}{cx + d}$.

2. Find the x-coordinates of the appropriate points on the graph, as indicated in the table above.

For example, to solve the rational inequality $\frac{x - 4}{x + 2} \geq 0$ by graphing:

1. Graph the function $y = \frac{x - 4}{x + 2}$. See Figure 13.3.5.

2. Find the x-coordinates of the points on the graph that lie **on and above** the x-axis. See Figure 13.3.6.

So $\frac{x - 4}{x + 2} \geq 0$ when $x < -2$ or $x \geq 4$.

As another example, to solve the rational inequality $\frac{x + 3}{x - 5} < 0$ by graphing:

1. Graph the function $y = \frac{x + 3}{x - 5}$. See Figure 13.3.7.

2. Find the x-coordinates of the points on the graph that lie **below** the x-axis. See Figure 13.3.8.

So $\frac{x + 3}{x - 5} < 0$ when $-3 < x < 5$.

Test Point Method

Instead of using graphing to find the solutions of a rational inequality, you can look at the value of the corresponding function at different intervals on the x-axis.

The dividing points of a rational function occur when either the numerator or denominator of the function equals 0. When the numerator of the function is 0, the value of the function is 0. When the denominator of the function is 0, the function has a break since division by 0 is undefined.

Since the value of a rational function is always positive or negative in an interval between dividing points, you can evaluate the function at any "test point" in the interval to find out whether all the values for points in that interval are positive or negative.

To find the solutions of a rational inequality without graphing:

1. Find where the numerator of the function equals 0 and where the denominator of the function equals 0.

2. Use these values to divide the number line into intervals.

3. Substitute a test point from each interval into $\frac{ax + b}{cx + d}$ to determine its sign.

4. Find the intervals that satisfy the original inequality.

For example, to solve the rational inequality $\frac{x + 3}{2x - 5} \geq 0$ without graphing:

1. Find where the numerator equals 0. Find where the denominator equals 0.

    $$x + 3 = 0 \quad \text{or} \quad 2x - 5 = 0$$
    $$x = -3 \quad \text{or} \qquad x = \frac{5}{2}$$

2. Divide the number line into intervals, using $x = -3$ and $x = \frac{5}{2}$ as the dividing points.

3. Substitute a test point from each interval into $\frac{x + 3}{2x - 5}$ to determine its sign; for example, $x = -10$, $x = 0$, and $x = 10$.

    Test $x = -10$:

    $$\frac{-10 + 3}{2(-10) - 5}$$
    $$= \frac{-7}{-25}$$
    $$= \frac{7}{25} \text{ is \textbf{positive}}$$

    Test $x = 0$:

    $$\frac{0 + 3}{2(0) - 5}$$
    $$= \frac{3}{-5} \text{ is \textbf{negative}}$$

Test $x = 10$:

$$\frac{10 + 3}{2(10) - 5}$$

$$= \frac{13}{15} \text{ is } \textbf{positive}$$

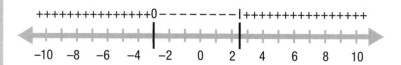

4. Find the intervals where $\frac{x + 3}{2x - 5} \geq 0$.

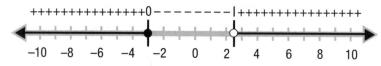

So $\frac{x + 3}{2x - 5} \geq 0$ when $x \leq -3$ or $x > \frac{5}{2}$.

If the expression in the numerator is not linear, you can still solve the rational inequality without graphing.

For example, to solve $\frac{3x^2 + 2x - 8}{x - 4} < 0$ without graphing:

1. Find where the numerator equals 0. Find where the denominator equals 0.

   $3x^2 + 2x - 8 = 0$ or $x - 4 = 0$

   $(x + 2)(3x - 4) = 0$ or $x = 4$

   $x + 2 = 0$ or $3x - 4 = 0$

   $x = -2$ or $x = \frac{4}{3}$ or $x = 4$

2. Divide the number line into intervals, using $x = -2$, $x = \frac{4}{3}$, and $x = 4$ as the dividing points.

3. Substitute a test point from each interval into $\frac{3x^2 + 2x - 8}{x - 4}$ to determine its sign; for example $x = -10$, $x = 0$, $x = 3$, and $x = 10$.

   Test $x = -10$:

   $$\frac{3(-10)^2 + 2(-10) - 8}{-10 - 4}$$

   $$= \frac{300 - 20 - 8}{-14}$$

   $$= \frac{272}{-14} \text{ is } \textbf{negative}$$

   Test $x = 0$:

   $$\frac{3(0)^2 + 2(0) - 8}{0 - 4}$$

   $$= \frac{-8}{-4}$$

   $$= 2 \text{ is } \textbf{positive}$$

*You want the value of the function to be **greater than or equal to 0**, that is, **nonnegative**.*

*The point $x = \frac{5}{2}$ is **not** included in the solution because $\frac{x + 3}{2x - 5}$ is undefined at this point.*

Test $x = 3$:

$$\frac{3(3)^2 + 2(3) - 8}{3 - 4}$$

$$= \frac{27 + 6 - 8}{-1}$$

$$= \frac{25}{-1} \text{ is \textbf{negative}}$$

Test $x = 10$:

$$\frac{3(10)^2 + 2(10) - 8}{10 - 4}$$

$$= \frac{300 + 20 - 8}{6}$$

$$= \frac{312}{6} \text{ is \textbf{positive}}$$

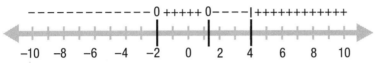

4. Find the intervals where

$$\frac{3x^2 + 2x - 8}{x - 4} < 0.$$

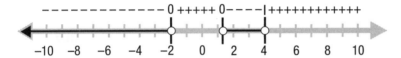

So $\frac{3x^2 + 2x - 8}{x - 4} < 0$ when $x < -2$ or $\frac{4}{3} < x < 4$.

When working with a rational inequality, make sure the right side of the inequality is 0. Otherwise you can't use this technique. If the right side is not 0, first rewrite the inequality so that the right side is 0, then follow the steps above.

For example, to solve $\frac{2x + 5}{x - 3} > 1$ without graphing:

1. Rewrite the inequality so the right side is 0.

$$\frac{2x + 5}{x - 3} - 1 > 1 - 1$$

$$\frac{2x + 5}{x - 3} - \frac{x - 3}{x - 3} > 0$$

$$\frac{2x + 5 - (x - 3)}{x - 3} > 0$$

$$\frac{x + 8}{x - 3} > 0$$

2. Find where the numerator equals 0. Find where the denominator equals 0.

$$x + 8 = 0 \quad \text{or} \quad x - 3 = 0$$
$$x = -8 \quad \text{or} \quad x = 3$$

3. Divide the number line into intervals, using $x = -8$ and $x = 3$ as the dividing points.

4. Substitute a test point from each interval into $\frac{x + 8}{x - 3}$ to determine its sign; for example $x = -10$, $x = 0$, and $x = 10$.

Test $x = -10$:

$$\frac{-10 + 8}{-10 - 3}$$

$$= \frac{-2}{-13}$$

$$= \frac{2}{13} \text{ is } \textbf{positive}$$

Test $x = 0$:

$$\frac{0 + 8}{0 - 3}$$

$$= \frac{8}{-3} \text{ is } \textbf{negative}$$

Test $x = 10$:

$$\frac{10 + 8}{10 - 3}$$

$$= \frac{18}{7} \text{ is } \textbf{positive}$$

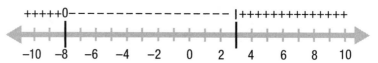

5. Find the intervals where $\frac{x + 8}{x - 3} > 0$.

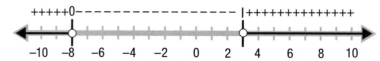

Since $\frac{x + 8}{x - 3} > 0$ is just $\frac{2x + 5}{x - 3} > 1$ rewritten, $\frac{2x + 5}{x - 3} > 1$ when $x < -8$ or $x > 3$.

So $\frac{x + 8}{x - 3} > 0$ when $x < -8$ or $x > 3$.

Sample Problems

1. Solve $\frac{x - 4}{x + 2} > 0$ by graphing.

   ☑ a. Graph the corresponding function, $y = \frac{x - 4}{x + 2}$.

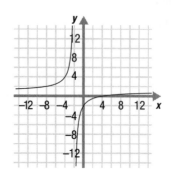

☐ b. Find the *x*-coordinates of the points on the graph that lie above the *x*-axis.

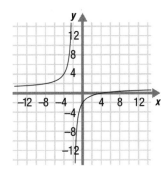

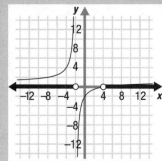

☐ c. Write the solution.

2. Solve $\dfrac{3x + 2}{x - 5} \leq 0$ without graphing.

☐ a. Find where the numerator equals 0. Find where the denominator equals 0.

$3x + 2 = 0$ or $x - 5 = 0$

$x =$ _____ or $x =$ _____

☐ b. Use these values to divide the number line into intervals.

☐ c. Substitute a test point from each interval into $\dfrac{3x + 2}{x - 5}$ to determine its sign, for example, $x = -10$, $x = 0$, and $x = 10$.

Test $x = -10$:

$\dfrac{3(-10) + 2}{-10 - 5}$

$= \dfrac{-28}{-15}$

$= \dfrac{28}{15}$ is **positive**

Test $x = 0$:

Test $x = 10$:

☐ d. Determine which intervals satisfy the original inequality.

3. Solve $\dfrac{5x + 9}{x - 1} \geq 2$ without graphing.

   ✓ a. Rewrite the inequality so the
   right side is 0.

$$\dfrac{5x + 9}{x - 1} - 2 \geq 2 - 2$$

$$\dfrac{5x + 9}{x - 1} - \dfrac{2(x - 1)}{x - 1} \geq 0$$

$$\dfrac{5x + 9}{x - 1} - \dfrac{2x - 2}{x - 1} \geq 0$$

$$\dfrac{5x - 2x + 9 + 2}{x - 1} \geq 0$$

$$\dfrac{3x + 11}{x - 1} \geq 0$$

   ☐ b. Find where the numerator
   equals 0. Find where the
   denominator equals 0.

$3x + 11 = 0$ or $x - 1 = 0$

b. $-\dfrac{11}{3}, 1$

$x = \underline{\hspace{1cm}}$ or $x = \underline{\hspace{1cm}}$

   ☐ c. Use these values to divide
   the number line into intervals.

c.

   ☐ d. Substitute a test point from
   each interval into $\dfrac{3x + 11}{x - 1}$
   to determine its sign, for
   example, $x = -10$, $x = 0$,
   and $x = 10$.

Test $x = -10$:

$$\dfrac{3(-10) + 11}{-10 - 1}$$

$$= \dfrac{-19}{-11}$$

$$= \dfrac{19}{11} \text{ is } \textbf{positive}$$

Test $x = 0$:

d. $\dfrac{3(0) + 11}{0 - 1}$

$= \dfrac{11}{-1}$ is **negative**

$\dfrac{3(10) + 11}{10 - 1}$

$= \dfrac{41}{9}$ is **positive**

Test $x = 10$:

e. $x \leq -\dfrac{11}{3}$ or $x > 1$

   ☐ e. Determine which intervals
   satisfy the original inequality.

4. Solve $\dfrac{x^2 + 4x - 21}{x + 5} > 0$ without graphing.

☐ a. Find where the numerator $x =$ ___ or $x =$ ___ or $x =$ ___
equals 0. Find where the
denominator equals 0.

☐ b. Use these values to divide
the number line into intervals.

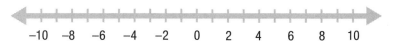

−10 −8   −6   −4   −2   0   2   4   6   8   10

b.

☐ c. Substitute a test point from   Test $x = -10$:

each interval into $\dfrac{x^2 + 4x - 21}{x + 5}$ $\dfrac{(-10)^2 + 4(-10) - 21}{-10 + 5}$

to determine its sign; for $= \dfrac{100 - 40 - 21}{-5}$

example, $x = -10$, $x = -6$, $= \dfrac{39}{-5}$ is **negative**

$x = 0$, and $x = 10$.

               Test $x = -6$:

c. $\dfrac{(-6)^2 + 4(-6) - 21}{-6 + 5}$

$= \dfrac{36 - 24 - 21}{-1} = \dfrac{-9}{-1}$

$= 9$ is **positive**

               Test $x = 0$:

$\dfrac{(0)^2 + 4(0) - 21}{0 + 5}$

$= \dfrac{-21}{5}$ is **negative**

               Test $x = 10$:

$\dfrac{(10)^2 + 4(10) - 21}{10 + 5}$

$= \dfrac{100 + 40 - 21}{15}$

$= \dfrac{119}{15}$ is **positive**

☐ d. Determine which intervals
satisfy the original inequality.

d. $-7 < x < -5$ or $x > 3$

Sample Problems

On the computer you used the Grapher to solve some inequalities by investigating the graphs of their corresponding functions.

1. The graph of $y = x^2 - 4x - 5$ is shown below. Use the graph to find the solutions of $x^2 - 4x - 5 \geq 0$.

☑ a. Find where the graph lies on or above the *x*-axis.

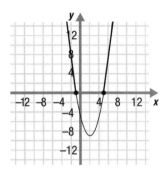

☐ b. Find the *x*-coordinates of the points on the graph that make the inequality $x^2 - 4x - 5 \geq 0$ true.

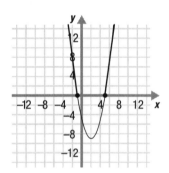

b.

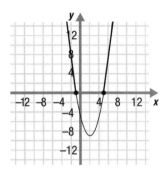

☐ c. Write the solution.

c $x \leq -1$ or $x \geq 5$

2. Given the function $y = \frac{x^2 + 1}{x - 2}$, for each value of x below, determine whether y is positive, negative, 0, or undefined.

   a. -5

   b. 0

   c. 2

   d. 7

   ☑ a. Substitute $x = -5$ into $y = \frac{x^2 + 1}{x - 2}$.

   $$y = \frac{(-5)^2 + 1}{-5 - 2}$$
   $$= \frac{26}{-7} \text{ is } \textbf{negative}$$

   ☐ b. Substitute $x = 0$ into $y = \frac{x^2 + 1}{x - 2}$.

   ☐ c. Substitute $x = 2$ into $y = \frac{x^2 + 1}{x - 2}$.

   ☐ d. Substitute $x = 7$ into $y = \frac{x^2 + 1}{x - 2}$.

HOMEWORK

Homework Problems

Circle the homework problems assigned to you by the computer, then complete them below.

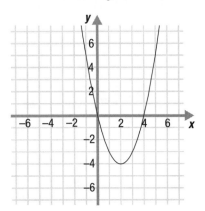

☼ Explain

Quadratic Inequalities

1. Use the equation $y = x^2 - 4x$ to find the solution of $x^2 - 4x < 0$. See Figure 13.3.9.

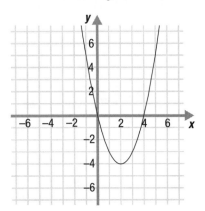

Figure 13.3.9

2. Solve $x^2 + 3x - 10 \geq 0$ using the test point method.

3. Solve $x^2 + 4x + 4 > 0$ using the test point method.

4. Use the equation $y = x^2 + 6x + 8$ to find the solution of $x^2 + 6x + 8 < 0$. See Figure 13.3.10.

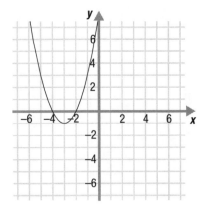

Figure 13.3.10

5. Solve $x^2 - 11x + 28 > 0$ using the test point method.

6. Solve $x^2 - 16 \leq 0$ using the test point method.

7. Use the equation $y = x^2 - 2x + 1$ to find the solution of $x^2 - 2x + 1 \leq 0$. See Figure 13.3.11.

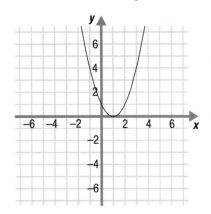

Figure 13.3.11

8. Solve $x^2 - 7x - 18 > 0$ using the test point method.

9. The profit, y, that a widget company can earn by selling x widgets can be expressed by the equation $y = -0.01x^2 + 5x - 400$. How many widgets does the company need to sell to make a positive profit?

10. The velocity, V, in feet per second, of a given object at time t seconds can be expressed by the equation $V = 3t^2 - 39t + 108$. Find the times when the object has a positive velocity.

11. Solve $x^3 + 2x^2 - 11x - 12 < 0$ using the test point method.

    Hint: $x^3 + 2x^2 - 11x - 12 = (x + 1)(x + 4)(x - 3)$

12. Solve $x^3 - 9x \geq 0$ using the test point method.

Rational Inequalities

13. The graph of the function $y = \dfrac{x}{x+3}$ is shown on the grid in Figure 13.3.13.

    Use this graph to solve the inequality $\dfrac{x}{x+3} < 0$.

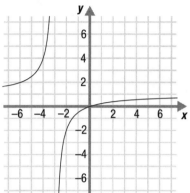

Figure 13.3.13

14. Solve $\dfrac{x+7}{x-4} > 0$ using the test point method.

15. The graph of the function $y = \dfrac{x-3}{x+1}$ is shown on the grid in Figure 13.3.14.

    Use this graph to solve the inequality $\dfrac{x-3}{x+1} < 0$.

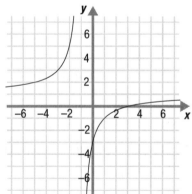

Figure 13.3.14

16. Solve $\dfrac{6-x}{6+x} \leq 0$ using the test point method.

17. The graph of the function $y = \dfrac{3x+2}{x+4}$ is shown on the grid in Figure 13.3.15.

    Use this graph to solve the inequality $\dfrac{3x+2}{x+4} > 0$.

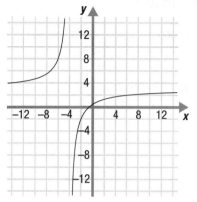

Figure 13.3.15

18. Solve $\dfrac{4x+2}{x-7} \geq 3$ using the test point method.

19. The graph of the function $y = \dfrac{x+1}{x-2}$ is shown on the grid in Figure 13.3.16.

    Use this graph to solve the inequality $\dfrac{x+1}{x-2} \leq 0$.

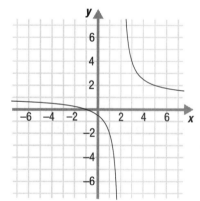

Figure 13.3.16

20. Solve $\dfrac{3x^2 + 13x + 4}{x+6} > 0$ using the test point method.

21. In a geometric sequence, 1, r, r^2, r^3, ... (where each term is the product of the previous term and a given number, r), you can find the sum of the first n terms using the formula $S_n = \dfrac{1-r^n}{1-r}$. For what value(s) of r can you not use this formula?

22. Given a cube with sides of length e, when is the surface area of the cube larger than the volume of the cube?

Hint: If x is the length of a side of a cube, surface area $= 6x^2$, and volume $= x^3$.

23. The graph of the function $y = \frac{x-3}{2x+1}$ is shown in Figure 13.3.17.

Use this graph to solve the inequality $\frac{x-3}{2x+1} \geq 0$.

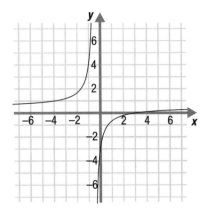

Figure 13.3.17

24. Solve $\frac{2x^2 + 30x + 91}{x + 5} \leq 7$ using the test point method.

 Explore

25. The graph of the function $y = x^2 + 8x + 15$ is shown on the grid in Figure 13.3.18. Use the graph of the function to find the solutions of $x^2 + 8x + 15 < 0$.

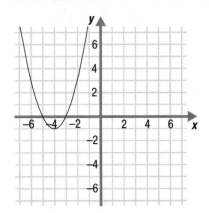

Figure 13.3.18

26. The graph of the function $y = \frac{x}{x-1}$ is shown on the grid in Figure 13.3.19. Use the graph of the function to find the solutions of $\frac{x}{x-1} > 0$.

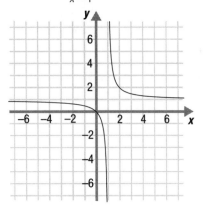

Figure 13.3.19

27. Given the function $y = 3x^4 - 7x^3 - 39x^2 + 55x + 84$, determine for each value of x below whether y is positive, negative, 0, or undefined.

   a. -10

   b. -2

   c. -1

   d. 0

   e. 4

   f. 17

28. The graph of the function $y = x^3 - x^2 - 6x$ is shown on the grid in Figure 13.3.20. Use the graph of the function to find the solutions of $x^3 - x^2 - 6x > 0$.

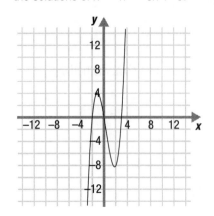

Figure 13.3.20

29. The graph of the function $y = \dfrac{x^2 + 2x - 8}{2x}$ is shown on the grid in Figure 13.3.21. Use the graph of the function to find the solutions of $\dfrac{x^2 + 2x - 8}{2x} \leq 0$.

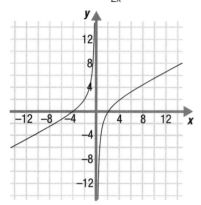

Figure 13.3.21

30. Given the function $y = \dfrac{0.2x^3 - 3x^2 + 11x - 6}{4x}$, determine for each value of x below whether y is positive, negative, 0, or undefined.

a. −10

b. −1

c. 0

d. 2

e. 7

f. 15

Practice Problems

Here are some additional practice problems for you to try.

1. The graph of the function $y = x^2 + 3x$ is shown on the grid below. Use this graph to solve the inequality $x^2 + 3x < 0$.

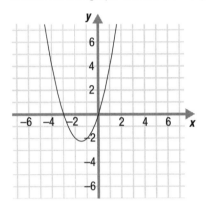

2. The graph of the function $y = x^2 + 4x + 5$ is shown on the grid below. Use this graph to solve the inequality $x^2 + 4x + 5 > 0$.

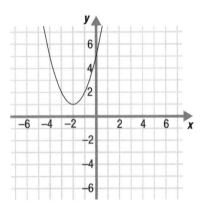

3. The graph of the function $y = x^2 - 6x + 5$ is shown on the grid below. Use this graph to solve the inequality $x^2 - 6x + 5 \geq 0$.

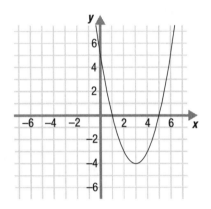

4. The graph of the function $y = -x^2 - 5x$ is shown on the grid below. Use this graph to solve the inequality $-x^2 - 5x < 0$.

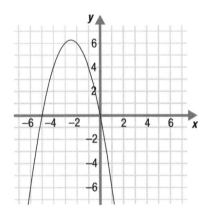

5. The graph of $y = -x^2 + 6x - 11$ is shown below. Use this graph to decide whether each statement is true or false.

    a. The solution of the inequality $-x^2 + 6x - 11 > 0$ is all real numbers.

    b. The solution of the inequality $-x^2 + 6x - 11 < 0$ is all real numbers.

    c. The solution of the equation $-x^2 + 6x - 11 = 0$ is all real numbers.

    d. The inequality $-x^2 + 6x - 11 > 0$ has no solutions.

    e. The inequality $-x^2 + 6x - 11 < 0$ has no solutions.

f. The equation $-x^2 + 6x - 11 = 0$ has no real solutions.

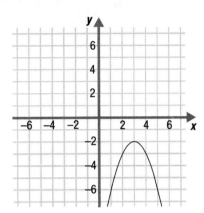

6. Solve the inequality $x^2 + 2x - 15 \geq 0$ using the test point method.

7. Solve the inequality $x^2 - 10x < -24$ using the test point method.

8. Solve the inequality $x^2 - 10 \leq -3x$ using the test point method.

9. Solve the inequality $x^2 - 8x < -16$ using the test point method.

10. Solve the inequality $x^2 - 49 \geq 0$ using the test point method.

11. The graph of the function $y = \dfrac{x}{x + 1}$ is shown on the grid below.

    Use this graph to solve the inequality $\dfrac{x}{x + 1} < 0$.

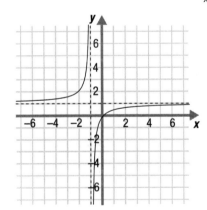

12. The graph of the function $y = \dfrac{2x - 8}{x + 1}$ is shown on the grid below.

    Use this graph to solve the inequality $\dfrac{2x - 8}{x + 1} > 0$.

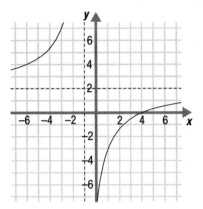

13. The graph of the function $y = \dfrac{x^2 - 1}{x}$ is shown on the grid below.

    Use this graph to solve the inequality $\dfrac{x^2 - 1}{x} \leq 0$.

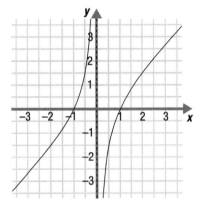

14. The graph of the function $y = \dfrac{x^2 - 4}{x^2 - 1}$ is shown on the grid below.

    Use this graph to solve the inequality $\dfrac{x^2 - 4}{x^2 - 1} \geq 0$.

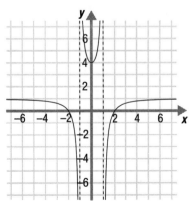

15. Solve the rational inequality $\frac{x+1}{x-3} \geq 0$ using the test point method.

16. Solve the rational inequality $\frac{3x-9}{x+1} \leq 0$ using the test point method.

17. Solve the rational inequality $\frac{(x+1)(x-2)}{x-1} \geq 0$ using the test point method.

18. Solve the rational inequality $\frac{x+2}{x^2+2x+1} < 0$ using the test point method.

19. Solve the rational inequality $\frac{2x+4}{x-8} > 0$.

20. Solve the rational inequality $\frac{x^2-7x+10}{x+3} > 0$.

EVALUATE

Practice Test

Take this practice test to be sure that you are prepared for the final quiz in Evaluate.

1. The graph of the function $y = \frac{1}{2}x^2 + \frac{5}{2}x - 3$ is shown on the grid in Figure 13.3.22. Use this graph to solve the inequality $\frac{1}{2}x^2 + \frac{5}{2}x - 3 < 0$.

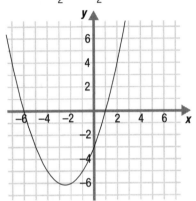

Figure 13.3.22

2. Solve the inequality $x^2 - 10 \le 3x$ using the test point method.

3. Solve the inequality $x^2 - 8x \ge -16$ using the test point method.

4. The graph of $y = x^2 + 5x + 8$ is shown on the grid in Figure 13.3.23. Use this graph to decide whether each statement below is true or false.

    a. The solution of the inequality $x^2 + 5x + 8 > 0$ is all real numbers.

    b. The solution of the inequality $x^2 + 5x + 8 < 0$ is all real numbers.

    c. The solution of the equation $x^2 + 5x + 8 = 0$ is all real numbers.

    d. The inequality $x^2 + 5x + 8 > 0$ has no solutions.

    e. The inequality $x^2 + 5x + 8 < 0$ has no solutions.

    f. The equation $x^2 + 5x + 8 = 0$ has no real solutions.

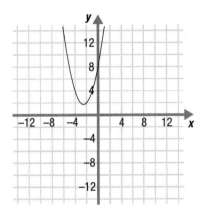

Figure 13.3.23

5. Solve the rational inequality $\frac{x-2}{x+3} > 0$ using the test point method.

6. Solve the rational inequality $\frac{1-2x}{x+9} \le 0$ using the test point method.

7. Solve the rational inequality $\frac{x^2-4}{x+5} < 0$.

8. The graph of the function $y = \frac{x^2+2}{x+1}$ is shown on the grid in Figure 13.3.24.

    Find the solution of the inequality $\frac{x^2+2}{x+1} \ge 0$.

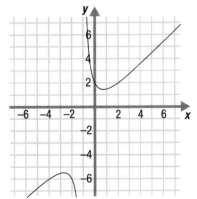

Figure 13.3.24

9. The graphs of $y = x^2 - 2x - 8$ and $y = 2x^2 - 4x - 16$ are shown on the grid in Figure 13.3.25. Find the values of x that satisfy both $x^2 - 2x - 8 \geq 0$ and $2x^2 - 4x - 16 \geq 0$.

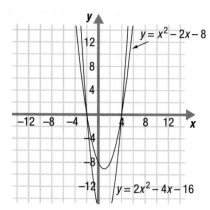

Figure 13.3.25

10. The graph of $y = \frac{1}{64}x^4 - \frac{5}{16}x^2 + 1$ is shown on the grid in Figure 13.3.26.

Find where $\frac{1}{64}x^4 - \frac{5}{16}x^2 + 1 < 0$.

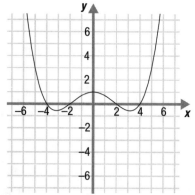

Figure 13.3.26

11. The graph of $y = \frac{x - 1}{x^2 + x - 20}$ is shown on the grid in Figure 13.3.27. Use the graph of the function to find the solution of the inequality $\frac{x - 1}{x^2 + x - 20} \geq 0$.

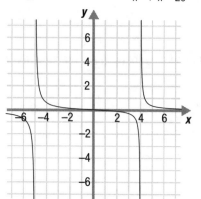

Figure 13.3.27

12. Given the function $y = \frac{x^2 - 9}{x - 5}$, for each value of x below, determine whether y is positive, negative, 0, or undefined.

a. -10

b. -3

c. 0

d. 5

e. 14

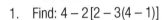

TOPIC 13 CUMULATIVE ACTIVITIES

CUMULATIVE REVIEW PROBLEMS

These problems combine all of the material you have covered so far in this course. You may want to test your understanding of this material before you move on to the next topic. Or you may wish to do these problems to review for a test.

1. Find: $4 - 2[2 - 3(4 - 1)]$

2. Find the $x-$ and $y-$intercepts of each of the functions below.

    a. $y = x^2$

    b. $y = x^2 + 2x - 8$

    c. $y = x^2 + 9$

3. Solve for x: $\sqrt{x^2 + 5} + x = 1$

4. Find the slope intercept form of the equation of the line that passes through the point $(-3, 7)$ and is parallel to the line $y = -2x + 9$.

5. Solve for x: $\log_x 16 = 2$

6. Solve for y: $\frac{1}{2}y + 3 = \frac{1}{4}y - 8$

7. Given $f(x) = x^2 + 2$ and $g(x) = 3x - 4$, find:

    a. $(f \circ g)(x)$

    b. $(f \circ g)(-2)$

8. Solve for t: $\sqrt{t + 6} - \sqrt{t + 1} = 3$

9. Rewrite in logarithmic form: $6^x = 20$

10. Simplify: $\frac{x + 3}{5xy} \cdot \frac{7yz}{x^2 + 5x + 6} \div \frac{4}{x + 2}$

11. Find: $\log_5 119$
    Approximate your answer to two decimal places.

12. The equation $y = -x^2 + 5x$ is graphed in Figure 13.1. Use the graph to find the solution of $-x^2 + 5x < 0$.

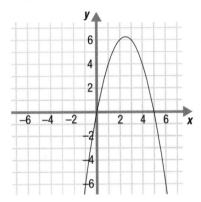

Figure 13.1

13. Solve for x: $|x + 3| - 4 = 10$

14. Given $g(x) = 3x + 1$, graph the function $y = g$ and its inverse, $y = g^{-1}$.

15. Solve for x: $|-3x + 6| > 9$

16. Factor: $5x^2y - 40xy + 35y$

17. Solve for x: $x^2 - 3x - 10 \le 0$

18. Graph the function $y = \left(\frac{1}{2}\right)^x$.

19. Find: $(2\sqrt{7} + 1)(2\sqrt{7} - 1)$

20. Solve for b: $b^3 + 2b^2 = 16b + 32$

21. Solve for x: $(x - 2)^2 = 196$

22. Simplify: $4\sqrt{3y} - 2\sqrt{75y} + 5$

23. Solve for x: $x^2 + 16 = 0$
    (Here, x can be an imaginary number.)

24. For the function $f(x) = 3x^2 + x - 1$, calculate:

   a. $f(3)$

   b. $f(0)$

   c. $f(-2)$

   d. $f(5)$

   e. $f(-10)$

25. Use the graphs in Figure 13.2 to find the number of solutions to this nonlinear system:

$$y = -x^2$$

$$x^2 + y^2 = 16$$

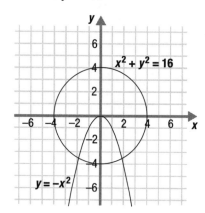

Figure 13.2

26. Solve for x: $\frac{4x}{5} - 2 = \frac{4}{3}$

27. The graph of the function $y = |x|$ is shown on the grid in Figure 13.3.
Determine the equations of the other lines shown.

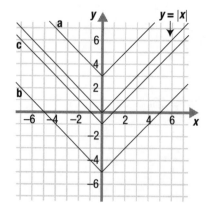

Figure 13.3

28. Solve $x^2 + 7x = 0$ by factoring.

29. Given $f(x) = 10x - 3$, find f^{-1}.

30. Solve for x: $\frac{x-2}{x+5} < 0$

31. Solve for t: $(3^2)^{t-2} = (3^3)^{4t}$

32. Solve for x: $x^2 + x - 9 = 0$

33. Simplify: $\sqrt[7]{b} \cdot \sqrt{b}$

34. Find the distance, d, between the point $(2, -2)$ and the point $(-1, 2)$.

35. Find: $(a + 5b)(2a - 3b + c)$

36. Solve for x: $2e^{x-1} = 8$

37. Find the domain and the range of each of the functions below.

   a. $f(x) = \frac{x}{x-2}$

   b. $f(x) = |x|$

   c. $f(x) = |3x| - 2$

38. Find: $\log_9 729$

39. Solve for x: $2x^2 + 6x = 3$

40. Solve for x: $\log_4(x + 1) = 1 + \log_4 x$

41. Simplify: $\sqrt{\frac{25a^4b^2}{48c^3}}$

42. Rewrite as a single logarithm: $\log 6 + 3[\log(x + 5)]$

43. Solve for k: $\left(\frac{k+2}{k}\right)^2 - 8\left(\frac{k+2}{k}\right) + 15 = 0$

44. Solve for x: $\frac{2}{x^2 + 3x} = \frac{5}{x^2 - x} + \frac{3}{x^2 + 2x - 3}$

45. Rewrite as a single logarithm: $5\log_3 x - \log_3 25$

46. Given $f(x) = x^3 + 2$ and $g(x) = x^3 - 2$, find:

   a. $(f + g)(x)$

   b. $(f - g)(3)$

   c. $(f \cdot g)(x)$

   d. $\left(\frac{f}{g}\right)(-1)$

47. Simplify using the properties of exponents:

$$\left[\frac{(3a)^2}{(2b)^3}\right]^2 \cdot a^5b^2$$

48. Solve this nonlinear system:

$$y = x^2 + 2x - 8$$

$$y + 2x + 8 = 0$$

49. Solve for b: $\frac{a-b}{c} + d = 0$

50. The equation $y = \frac{x}{x+2}$ is graphed in Figure 13.4. Use the graph to find the solution of $\frac{x}{x+2} \geq 0$.

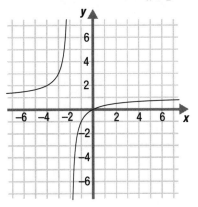

Figure 13.4

 ANSWERS

Lesson EII.A Real Numbers and Exponents
Homework

1. $<, \leq, \neq$ **2.** -5 **3.** $8 + 17$ **4.** $-\frac{1}{2}$ **5.** -4

6. $(5 \cdot 11) \cdot 23$ **7.** 9 **8.** 5

9. additive inverse $= -7$, multiplicative inverse $= \frac{1}{7}$

10. $>, \geq, \neq$ **11.** 4 **12.** -26 **13.** base: 4; exponent: 3; 64

14. 4^{19} **15.** 1 **16.** 7^{12} **17.** $7^5 \cdot 12^4$ **18.** 40 **19.** y^6

20. $x^{12} \cdot y^{24}$ **21.** $8 \cdot 9$ or 72 **22.** $3^2 y^5 z^4$ or $9 y^5 z^4$

23. $x^5 y^3$ **24.** $8x^{13}$

Practice Problems

1. $2 \leq 2, 3 \geq 0, 2 \neq 8$ **2.** $4 \neq 6, 6 \geq 0, 7 \leq 7$

3a. 3.5 **b.** 36 **c.** 0.005 **d.** 14.33 **e.** 36.1

4a. 0.26 **b.** 28 **c.** 0.13 **d.** 15.22 **e.** 2.69

5. 96 **6.** 105 **7.** 42 **8.** $5.2, -\sqrt{8}, \frac{7}{8}$ **9.** $\sqrt{7}, -\frac{1}{3}$

10. $-21, \sqrt{100}, \frac{5}{1}$ **11.** $-8; \frac{1}{8}$ **12.** $-5, \frac{1}{5}$

13. $-\frac{2}{3}; \frac{3}{2}$ **14.** $\pi, \sqrt{7}$ **15.** $\sqrt{19}$ **16.** $4, \frac{3}{5}$ **17.** 1

18. 18 **19.** 1 **20.** 11 **21.** 5 **22.** -1 **23.** -1 **24.** 4

25. -2 **26.** 0 **27.** 0 **28.** 0 **29.** base: 3; exponent: 5; 243

30. base: 2; exponent: 5; 32 **31.** 1 **32.** -1 **33.** $m^6 n^4$

34. $x^4 y^3 z$ **35.** $a^3 b^6 c^2$ **36.** 73 **37.** 24 **38.** 17

39. $7^5 - 7^{10}$ **40.** $8^8 + 8^7$ **41.** $5^7 + 5^4$ **42.** $x^4 y^2 z^4$

43. $m^6 n^{10}$ **44.** $a^3 b^{13} c$ **45.** $a^{20} b^{10}$ **46.** $x^{21} y^{35}$

47. $x^{45} y^{63} z^{72}$ **48.** $a^7 b^6$ **49.** $x^8 y^2$ **50.** $b^6 c^2$

51. $648 a^8 b^4$ **52.** $\frac{32}{9} x^6 y^5$ **53.** $12 a^5 b$ **54.** $3a^5$

55. $\frac{2}{x^2}$ or $2x^{-2}$ **56.** $\frac{1}{6b^8}$

Practice Test

1. 7 **2.** 15 **3.** $-13, \frac{1}{13}$ **4.** 20

5. $y^3 z^5$ **6.** $x^{35} \cdot y^{14}$ **7.** 144 **8.** $\frac{64 x^{11} y^4}{81}$

Lesson EII.B Polynomials
Homework

1. 6 **2.** $16xz$ **3.** $10y^2 - 7y - 12$ **4.** 10

5. $-2x^3 - 2x^2 + 14x - 23$ **6.** $4a^2 b^2 - 20abc^2 + 25c^4$

7. 15 **8.** $5x^5 - 20x^3 + 40x^2$ **9.** $8y^4 - 4y^2$ **10.** -12

11. $2x^3 - 15x^2 + 30x - 7$ **12.** $3x + 10 + \frac{13}{x-2}$

13. $6x^4 z^2 (2x^3 - 3z^2 - 4xz^4)$ **14.** $(x^2 + 6)(x^2 - 6)$

15. $(x + 2)(2 + y)$ **16.** $2xy^2 (2x - 3y)(2x + 3y)$

17. $(3 + a^2)(9 - 3a^2 + a^4)$ **18.** $2b(3 + a)(9 - 3a + a^2)$

19. $(x + 7)(x - 3)$ **20.** $(2x + 3y)(2x + 3y)$ or $(2x + 3y)^2$

21. $2x(x - 2)(x - 8)$ **22.** $(4y - 3)(7y + 2)$

23. $(2p - q^3)(4p^2 + 2pq^3 + q^6)$

24. $(x + y)(x - y)(x^2 + xy + y^2)(x^2 - xy + y^2)$

Practice Problems

1. 8 **2.** 8 **3.** 12 **4.** 15 **5.** $12x^4 + 10x^3 - 8x + 2$

6. $3x^3 + 8x^2 - 16x + 5$ **7.** $10x^3 - 4x^2 + 6x + 8$

8. $2a^3 b - 9a^2 b^2 - 9ab^3$

9. $5xy^3 - 17x^2 y + 8xy + 16x^3 y - 7x$

10. $x^3 y^2 z - 10x^2 y + 14x^2 z + 3xy^2 - 25xz^2$

11. $8x^2 + 30x + 27$ **12.** $6x^2 - 23x + 20$

13. $12y^2 + 13y - 14$ **14.** $a^4 + 4a^2b + 4b^2$

15. $16x^2y^2 - 16xyz^2 + 4z^4$

16. $9m^2n^4 - 25n^6$ **17.** $18a^5 + 24a^3 - 30a^2$

18. $5x^7 - 30x^4 + 55x^3$

19. $-21x^4y^4 + 30x^3y^3 + 27x^2y^2$ **20.** 43 **21.** -51

22. -4 **23.** $2x^3 - x + 5$ **24.** $\dfrac{4x^3}{y^2} + \dfrac{5}{4y} + \dfrac{3}{x^2}$

25. $\dfrac{3a^3}{b^3} - 7 - \dfrac{5b}{a^3}$ **26.** $3x^3 + 13x^2 + 20x + 12$

27. $2x^3 - 11x^2 + 19x - 12$ **28.** $x^3 + 8y^3$

29. $(m - 8)(m + 8)$ **30.** $(w - 7)(w + 7)$

31. $(2a + 3)(2a - 3)$ **32.** $7x^2y(5x^3y^2 - 2xy + 3)$

33. $8a^4b^2(3b^4 + a)$ **34.** $(3a + 4b)(a - 2)$

35. $(y - 3)(5x + 2)$ **36.** $6a^2(2a - 5)(a + 3)$

37. $3x^3(x + 7)(2x - 3)$ **38.** $2w(3w + 1)(w - 4)$

39. $(a + 7)^2$ **40.** $(x - 6)^2$ **41.** $(2m - 5)^2$ **42.** $(4a - 9)^2$

43. $(3x + 4)^2$ **44.** $(7m + 5n)^2$

45. $(a - 4)(a^2 + 4a + 16)$ **46.** $(2m + 1)(4m^2 - 2m + 1)$

47. $(x + 3)(x^2 - 3x + 9)$ **48.** $(x^2 + 3y)(x^4 - 3x^2y + 9y^2)$

49. $(4p - q^3)(16p^2 + 4pq^3 + q^6)$

50. $(5a^3 + 4b^3)(25a^6 - 20a^3b^3 + 16b^6)$

51. $(5m^2 + 4n)(5m^2 - 4n)$

52. $(2x - 5y^2)(2x + 5y^2)(4x^2 + 25y^4)$

53. $(2a - 3b)(2a + 3b)(4a^2 + 9b^2)$ **54.** $(x + 12)(x - 3)$

55. $(a + 2)(a - 8)$ **56.** $(4m - 3)(m + 12)$

Practice Test

1. 5 **2.** $8y^2 - 3y^4 + 1$ **3.** -30

4. $2x^3 - 13x^2y + 20xy^2 + 6x - 15y$

5. $6x^3z^4(2x^4 - 5z^5 - 7x^2z^2)$

6. $(3x - 5)(x + 6)$ **7.** $(2x - 5y)^2$ **8.** $3x(x - 2)(x - 8)$

Lesson EII.C Equations and Inequalities
Homework

1. $x = 5$ **2.** $m = \dfrac{F}{a}$ **3.** $x \geq 9$ **4.** no solution

5. $x = -\dfrac{11}{y} + 2$ **6.** $x < -\dfrac{23}{5}$ **7.** $x = 1$ **8.** $2 > y > -5$

9. $-3 < z \leq 1$ **10.** $x = 3$ **11.** $x = \dfrac{10}{17}$

12. $2 > y \geq \dfrac{1}{2}$ or $\dfrac{1}{2} \leq y < 2$

Practice Problems

1. $r = \dfrac{C}{2\pi}$ **2.** $m = \dfrac{2Fd}{v^2}$

3. $y = \dfrac{-3x + 15}{-5}$ or $y = \dfrac{3x - 15}{5}$ or $y = \dfrac{3}{5}x - 3$

4. $x = \dfrac{c - b}{a}$ **5.** $m = \dfrac{y - b}{x}$ **6.** $x < 3$ **7.** $x \geq -20$

8. $y \geq -6$ **9.** $y = -\dfrac{8}{x} + 5$ **10.** $y = 4 - \dfrac{8}{x}$

11. $y < 10$ **12.** $x < -\dfrac{44}{5}$ **13.** $y \leq -23$ **14.** $y = 1$

15. $y = -1$ **16.** $y = 3$ **17.** $2 < x < 7$

18. $-7 \leq x \leq -2$ **19.** $1 \leq x \leq 3$ **20.** $y = \dfrac{28}{9}$ **21.** $x = -9$

22. $x = -\dfrac{17}{7}$ **23.** $x = \dfrac{28y + 29}{9}$ or $x = \dfrac{28y}{9} + \dfrac{29}{9}$

24. $y = \dfrac{8x - 75}{15}$ or $y = \dfrac{8x}{15} - 5$

25. $y = \dfrac{3x + 50}{10}$ or $y = \dfrac{3}{10}x + 5$ **26.** $x = -\dfrac{68}{5}$

27. $y = -12$ **28.** $x = -14$

Practice Test

1. $x = 8$ **2.** $y = 11$ **3.** $z = -2$

4. $= -4$ **5.** $x = 10$ **6.** $x = \dfrac{z + 5y}{3}$

7. $y = -1$ and $y = 1$

8. Here is this solution graphed on the number line:

Lesson EII.D Rational Expressions
Homework

1. $\dfrac{1}{13^5}$ **2.** $\dfrac{16y^8}{x^2}$ **3.** $\dfrac{6x-2}{23xy^5}$ **4.** $\dfrac{x+5}{x-3}$ **5.** $\dfrac{x^6}{49y^4}$

6. $\dfrac{10y+2xy-3x}{15xy}$ **7.** $\dfrac{4y^2}{x^{10}}$ **8.** $\dfrac{3(x+3)}{2(x-4)}$ or $\dfrac{3x+9}{2x-8}$

9. $\dfrac{2(x^3+12)}{3x^2}$ **10.** $\dfrac{2}{5a}$ **11.** $\dfrac{6b-2a-13}{a^2+a-2}$ **12.** $\dfrac{18x-3}{12x+2}$

13. 12 **14.** $x=-\dfrac{15}{2}$ **15.** $y=\dfrac{15}{4}$ **16.** $24x^8$ **17.** $x=\dfrac{3}{2}$

18. no solution   **19.** $12(a-2)(a+2)$ **20.** $x=-7$

21. $x=-\dfrac{28}{3}$ **22.** $2y^2-y-3$ or $(2y-3)(y+1)$

23. $y=\dfrac{13}{3}$ **24.** no solution

Practice Problems

1. $\dfrac{1}{3^{10}}$ **2.** $\dfrac{1}{7^{11}}$ **3.** $\dfrac{1}{y^5}$ **4.** $\dfrac{3}{x^4}$ **5.** $\dfrac{8}{y^7}$ **6.** $\dfrac{5n^7}{m^2}$ **7.** $\dfrac{a^5}{2b^4}$

8. $\dfrac{3a^2}{b^4}$ **9.** $\dfrac{x-4}{x+2}$ **10.** $\dfrac{x+1}{x+3}$ **11.** $\dfrac{x-2}{x+1}$ **12.** $\dfrac{x+4}{x-4}$

13. $\dfrac{x(x+5)}{x-4}$ or $\dfrac{x^2+5x}{x-4}$ **14.** $\dfrac{z^2(z^2+3)}{10}$ **15.** $\dfrac{y-2}{2y}$

16. $\dfrac{y(y^2+2)}{9}$ **17.** $\dfrac{18b+2}{12b-1}$ or $\dfrac{2(9b+1)}{12b-1}$ **18.** 2

19. $\dfrac{20a-1}{12a+2}$ or $\dfrac{20a-1}{2(6a+1)}$ **20.** $\dfrac{9x}{2y^3}$ **21.** $\dfrac{-(x+5)}{x-8}$ or $\dfrac{-x-5}{x-8}$

22. $\dfrac{3}{2y^3}$ **23.** $\dfrac{y-3}{x^3}$ **24.** $\dfrac{2}{x^3}$ **25.** $\dfrac{-xy+3x+6y}{6xy}$

26. $\dfrac{2y-3x+15}{x^2+2x-35}$ or $\dfrac{2y-3x+15}{(x-5)(x+7)}$ **27.** $\dfrac{-xz+9x-6z}{6xz}$

28. $\dfrac{9-a}{a^2+5a-24}$ or $\dfrac{9-a}{(a+8)(a-3)}$ **29.** $14x^5$ **30.** $6x^3$

31. $10x^4$ **32.** $6y^2+18y$ or $6y(y+3)$

33. a^2-25 or $(a+5)(a-5)$

34. $8x(x-3)(x+3)$ or $8x^3-72x$

35. $2b^2-b-10$ or $(b+2)(2b-5)$

36. $x^3+3x^2-9x-27$ or $(x-3)(x+3)^2$

37. $2a^2+3a-35$ or $(2a-7)(a+5)$

38. $x=\dfrac{7}{2}$ or $3\dfrac{1}{2}$ **39.** $y=-\dfrac{7}{4}$ **40.** $x=-\dfrac{3}{10}$ **41.** $a=\dfrac{7}{13}$

42. $b=\dfrac{5}{7}$ **43.** $y=\dfrac{5}{11}$ **44.** $x=-\dfrac{17}{35}$

45. $y=\dfrac{17}{16}$ or $1\dfrac{1}{16}$ **46.** $x=\dfrac{2}{3}$ **47.** $x=\dfrac{2}{7}$

48. $y=\dfrac{10}{3}$ or $3\dfrac{1}{3}$ **49.** $x=-1$ **50.** $a=-1$ **51.** $b=4$

52. $a=-2$ **53.** $x=-\dfrac{7}{2}$ or $-3\dfrac{1}{2}$ **54.** $z=2$ **55.** $r=-5$

56. $q=7$

Practice Test

1. $\dfrac{7y^3}{x^{11}}$ **2.** $\dfrac{3c}{7a}$ **3.** $\dfrac{8a+5}{2a^2-5a-3}$ **4.** $\dfrac{6x^2-3x}{12x^2+2}$

5. $15(a-3)(a+4)$ **6.** $x=\dfrac{34}{9}$

7. The equation has no solution   **8.** $x=\dfrac{7}{3}$

Lesson EII.E Graphing Lines
Homework

1. $P(2,2)$, $Q(-4,1)$, $R(-2,-3)$ **2.** 7

3.

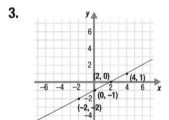

4. The x-intercept is $(9,0)$ **5.** 6

6.   **7.** $m=\dfrac{3}{7}$ **8.** 5

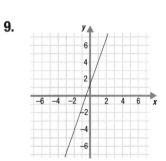

9.

10. The x-intercept is $\left(\frac{35}{16}, 0\right)$. The y-intercept is (0, −35).

11. The distance from A to B is $\sqrt{26}$, the distance from B to C is $\sqrt{10}$, the distance from A to C is $\sqrt{8} = 2\sqrt{2}$. So points A and C are closest.

12.

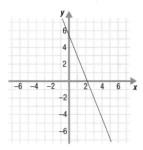

13. $y - 2 = 5(x - 1)$ **14.** $16x + y = 9$ **15.** $y = 12$

16. $y - 3 = -\frac{1}{3}(x + 7)$ or $y - 2 = -\frac{1}{3}(x + 4)$

17. $-2x + y = 4$ **18.** $x = -3$ **19.** $y = 18x + 9$

20. $-x + y = 9$ **21.** $y = \frac{1}{8}$ **22.** $y + 6 = -2(x - 4)$

23. $\frac{1}{6}x + y = \frac{5}{6}$ **24.** $y = 8x - \frac{1}{4}$

Practice Problems

1. $P(3, -2)$, $Q(-1, 5)$, $R(4, 4)$ **2.** $S(5, -5)$, $T(4, 3)$, $U(-5, 0)$

3. $(-3, 4)$ **4.** $(-3, -2)$ **5.** 5   **6.** 18   **7.** 9   **8.** 10   **9.** 13

10. 5

11.

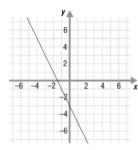

12.

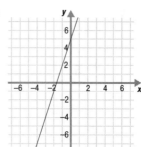

13.

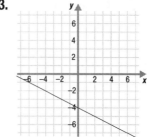

14. $m = \frac{3}{2}$ **15.** $m = -\frac{1}{4}$ **16.** $m = 0$

17. x-intercept: (5, 0); y-intercept: (0, −10)

18. x-intercept: (−5, 0); y-intercept: (0, 3)

19. x-intercept: $\left(\frac{21}{4}, 0\right)$; y-intercept: (0, 7)

20.

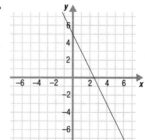

21.

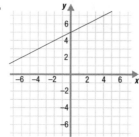

22.

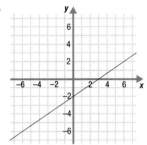

23. $y - 2x = 15$ **24.** $3x + y = 9$

25. $2x + 2y = 14$; $y = -x + 2$

26.

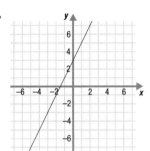

27.

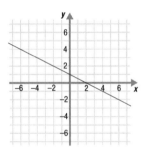

28.

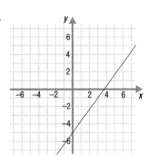

29. $y - 5 = 2(x + 3)$ **30.** $y + 4 = \frac{1}{4}(x - 2)$

31. horizontal line: $y = 6$; vertical line: $x = -2$

32. horizontal line: $y = 8$; vertical line: $x = 0$

33. horizontal line: $y = 0$; vertical line: $x = -3$

34. $x = 5$ **35.** $y = -4$ **36.** $y = -5x + 16$

37. $y = 4x + 10$ **38.** $y = \frac{1}{2}x + 8$ **39.** $y = -3x + 7$

40. $y = 2x - 9$ **41.** $y = -\frac{1}{3}x + 1$ **42.** (4,0)   **43.** (0, 7)

44. (0,1)   **45.** $x + y = -2$ **46.** $-2x + y = 1$ or $2x - y = -1$

47. $x + 3y = 18$ or $\frac{1}{3}x + y = 6$ **48.** $y = -\frac{1}{2}x + 4$

49. $y = -3x + 13$ **50.** $y = \frac{5}{2}x + 1$ **51.** $y = -3x - 4$

52. $y = \frac{2}{3}x + 3$ **53.** $y = -\frac{1}{4}x - 1$ **54.** $y = 2x - 8$

55. $y = 3x + 6$ **56.** $y = -\frac{2}{3}x - 4$

Practice Test

1. To determine in which quadrant each point lies, see the figure below.

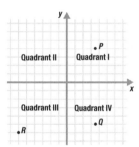

2. 15   **3.** –3   **4.** 3   **5.** $y + 7 = 9(x - 20)$

6. The y-intercept is (0, 2).

7. $x = -11$
$y = 17$

8. $-8x + y = 5$

Lesson EII.F Absolute Value
Homework

1. $x = -100$ or $x = 100$ **2.** $x = -17$ or $x = 17$

3. no solution   **4.** $x = 0$ **5.** $x = -22$ or $x = 32$

6. $x = \frac{28}{3}$ or $x = -\frac{14}{3}$ **7.** $x = 1$ or $x = \frac{7}{5}$

8. $x = -\frac{16}{3}$ or $x = 0$

9. $|x - 457| = 15$; highest guess: 472; lowest guess: 442

10. $a = 0.1716$ or $a = 0.1404$ **11.** $x = 4$

12. $x = 12$ or $x = -\frac{2}{5}$ **13.** $x \le -24$ or $x \ge 24$

14. $-8.27 < x < 8.27$ **15.** $-1 \le x \le 1$

16. $x < -\frac{15}{2}$ or $x > \frac{15}{2}$ **17.** $-11.5 < x < 25.9$

18. $-13 \le x \le 8$ **19.** $x \le -8$ or $x \ge 24$

20. $-21.9 \le x \le 18.3$ **21.** $x > \frac{28}{3}$ or $x < -6$

22. 7.96 ft. $\le a \le 8.04$ ft. **23.** $-1.1 < x < -0.1$

24. $x < -\frac{35}{6}$ or $x > \frac{49}{6}$

Practice Problems

1. $y = -128$ or $y = 128$ **2.** no solution **3.** $x = -250$ or $x = 250$

4. $y = -14$ or $y = 14$ **5.** $x = -9$ or $x = 9$

6. $x = -11$ or $x = 11$ **7.** $x = -21$ or $x = 21$

8. no solution **9.** $x = -15$ or $x = 15$ **10.** $y = -41$ or $y = 27$

11. $x = -9$ or $x = 5$ **12.** $x = -42$ or $x = 54$

13. $y = -4$ or $y = 9$ **14.** $x = -4$ or $x = 6$ **15.** no solution

16. $y = 0$ or $y = 6$ **17.** $x = -5$ or $x = 1$ **18.** $x = -6$ or $x = 1$

19. $y = -11$ or $y = 13$ **20.** $x = -33$ or $x = 27$

21. $x = -22$ or $x = 26$ **22.** $x = 4$ **23.** $y = 6$

24. $y = -11$ or $y = -1$ **25.** $x = 0$ or $x = 1$

26. $x = -4$ or $x = 1$ **27.** $y = -5$ or $y = 7$ **28.** $x = -9$ or $x = 6$

29. $-7 < y < 7$ **30.** $-3 \le x \le 3$ **31.** $-4 < x < 4$

32. $y \le -23$ or $y \ge 23$ **33.** $x < -1$ or $x > 1$

34. $x < -19$ or $x > 19$ **35.** $-9 \le y \le 9$ **36.** $x \le -15$ or $x \ge 15$

37. $x < -3$ or $x > 3$ **38.** $|y| - 18 < -18$

39. $|x| + 9 < 9$, $|x + 2| < -3$, and $-|2x + 9| > 0$

40. $y \le -4$ or $y \ge 7$ **41.** $x < -1$ or $x > 2$

42. $x < -1$ or $x > 5$ **43.** $-6 \le y \le 18$

44. $-9 \le x \le 1$ **45.** $-5 < x < 1$

46. $-6 \le y \le 10$ **47.** $-2 < x < 6$ **48.** $-2 < x < 6$

49. $y < 0$ or $y > 16$ **50.** $y \le -2$ or $y \ge 7$ **51.** $x < 2$ or $x > 4$

52. $x < -3$ **53.** $x \ge 4$ **54.** $x > -3$ **55.** $-2 < x < 5$

56. $-1 \le x \le 6$

Practice Test

1a. $x = -4$ or $x = 14$ **b.** $x = -3$ or $x = 3$

2. $x = -30$ or $x = 24$

3. $x = 2$ or $x = -4$

4. The equation has no solution.

5. The graph that corresponds to the solution of $|x - 5| \le 7$ is the first graph:

6. $x < -3$ or $x > 6$

7. $|x| > 4$

8. $3 > x > -5$

Topic EII

Cumulative Review Problems

1. $\sqrt{(-3) \cdot 5}$ **2.** $(3x - 1)(x + 7y)$ **3.** $30x^4y^2$ **4.** $x = 61$

5. $-3 < x < -\frac{8}{9}$ **6.** $x = \frac{5}{3}$ or $x = -\frac{19}{3}$ **7.** -39

8. $y = -\frac{1}{4}x - \frac{15}{4}$ **9.** $\frac{2x^2+4y^3}{3 \cdot (x+y)}$ or $\frac{2x^2+4y^3}{3x+3y}$

10. $m = -\frac{5}{7}$ **11.** $P(2, 1)$; $Q(-2, -1)$; $R(-3, 4)$

12. no solution **13.** $x = -\frac{5}{24}$ **14.** $-11 < x < 11$

15. 9 **16.** $y - 2 = 3(x + 1)$ **17.** $\frac{4096x^{17}}{729}$ **18.** $\frac{5}{x^8yz^5}$

19. $(y^2 + 7)(y^2 - 7)$ **20.** $4x^6 - \frac{1}{4}x$

21. $x = \frac{y-b}{m}$ or $\frac{y}{m} - \frac{b}{m}$ **22.** $\frac{x^3 + x^2 + 4}{32xyz^2}$ **23.** $x = \frac{3}{5}$ or $x = \frac{1}{6}$

24. $4x^4y^2(7x - 3xy^2 + 5y)$ **25.** $z < -\frac{9}{11}$ **26.** x^2

27. $\frac{3x^5 + 13y^{11}}{y^2x^2}$ **28.** $x \le -2$ or $x \ge 6$ **29.** $\frac{2y-5}{y+8}$

30. $6a^2 - 4b^2 + 4a^2b + 6ab^2 + 2ab + 14$

31. $3 < z \le 10$ **32.** $\frac{3(x-3)}{4(x+2)}$ or $\frac{3x-9}{4x+8}$ **33.** $(4x + 7)(2x - 3)$

34. $y = -\frac{12}{13}$ **35.** $d = \sqrt{41}$

36.

37. $\frac{x^2}{2}$ **38.** 0 **39.** $7 \cdot 12 + 7 \cdot 33 = 315$

40. $5x(9x - 7)(2x + 1)$ **41.** $\dfrac{2(-3x + 1)}{-4x - 1}$ or $\dfrac{-6x + 2}{-4x - 1}$

42. $\dfrac{8b + a + 31}{2a^2 + 5a - 3}$ or $\dfrac{8b + a + 31}{(a + 3)(2a - 1)}$ **43.** $d = 23$

44.

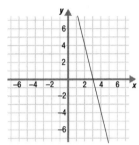

45. $x = -\dfrac{65}{6}$ **46.** $(4a + b^4)(4a - b^4)$

47. The vertical line is $x = 7$. The horizontal line is $y = 1$.

48. $x = 99$ or $x = -99$ **49.** $18x^2 + 3x - 45$

50. The x-intercept is $\left(-\dfrac{7}{5}, 0\right)$. The y-intercept is $(0, 7)$.

Lesson 6.1   Exponents
Homework

1a. 3^7 **b.** 5^7 **c.** 7^7 **3a.** 7^6 **b.** 7^6

5a. x^8 **b.** a^8 **c.** 1 **d.** $\dfrac{x^3}{4}$ **7a.** b^{18} **b.** y^{11} **c.** $\dfrac{b^3}{a^7}$

9. $\dfrac{1}{x}$ **11a.** 1 **b.** $\dfrac{1}{y^3}$ **c.** $\dfrac{1}{b^4}$ **d.** 2

Practice Problems

1. 7^8 **3.** b^{15} **5.** a^{11} **7.** 9^6 **9.** n^5

11. 5^{12} **13.** 13^{30} **15.** z^{48} **17.** $81a^4$

19. $8y^3$ **21.** $\dfrac{m^4}{n^6}$ **23.** 1 **25.** 1 **27.** 3

Practice Test

1a. 11^4 **b.** $3^2 y^5$ **c.** 5^{43} **d.** $x^{21} y^{26}$ **e.** $7^{18} b^{14}$

2a. 2^3 **b.** b^6 **c.** $\dfrac{3^3}{x^9}$ **d.** $\dfrac{1}{y^4}$

3. $\dfrac{x^6 y^2}{x^3 y^7}$ and $\dfrac{x^7 y}{x^4 y^6}$

4. $(31x^8)^0 \cdot 5y$, $\dfrac{5y^2}{y}$ and $\dfrac{(5y)^2}{5y}$

5a. b^{48} **b.** $3^{10} a^{12}$ **c.** $2^{99} x^{44} y^{66}$

6a. $\dfrac{5^4 y^{40}}{3^4 x^{32}}$ **b.** $\dfrac{7^6 a^6 b^{24}}{5^6}$

7a. -1 **b.** 1 **c.** -3 **d.** 1

8a. a^{35} **b.** $\dfrac{1}{a^{35}}$

Lesson 6.2   Polynomial Operations I
Homework

1. $3\frac{1}{4} y^3 + 3y^2 - \sqrt{5}$ **3a.** $-4y^5 - 2y^3 + 3y + 2$

b. $5, 3, 1, 0$ **c.** 5 **5.** $-4v^7 + v^3 + 6v^2 - 5v + 5$

7. $-7s^3 t^3 + 7st^2 - s^2 t + 2st - 13t + 9$

9. $2x^2 y + 10xy^2 + 4y^3 + 3$

11. $4w^2 yz + 3w^3 - 4wyz^2 + 6wyz - 4wy^2 z + 3$

13. $x^3 y^3 z^3$ **15.** $-3t^4 u^4 v^{15}$ **17.** $10p^3 r^4 + 5p^4 r^5$

19. $\dfrac{3xw^5}{y}$ **21.** $\dfrac{3a^3 d}{2b^5 c^3}$ **23.** $\dfrac{4xy^2}{3} + \dfrac{5x^2 y^3}{3}$ or $\dfrac{xy^2}{3}(4 + 5xy)$

Practice Problems

1. $2xy + 5xz; 9y^2 + 13yz - 8z^2$

3a. binomial   **b.** binomial   **c.** trinomial   **d.** monomial
  e. trinomial

5. 8 **7.** 9 **9.** 7 **11.** 6 **13.** 84 **15.** $6x^2 + 11x - 8$

17. $15m^2 n^3 + 2m^2 n^2 - 7mn$ **19.** $15a^3 b^2 + 4a^2 b - ab^3$

21. $20xy^2 z^3 - 30x^2 yz^2 + 10x^3 y^3 z$ **23.** $4x^3 + 7x - 8$

25. $y^2 + 6xy + 4y$ **27.** $11a^5 b^3 - 4a^4 b - 9b$ **29.** $15y^5$

31. $-45a^9$ **33.** $28x^4 y^8$ **35.** $-6w^2 x^5 y^3 z^3$

37. $-6a^7 b^7 + 10a^4 b^5 - 12a^4 b^2$

39. $20a^4 b^2 + 10a^4 b^3 - 35a^3 b^4 - 15a^2 b^3$

41. $12x^6 y^3 - 28x^4 y^5 + 8x^4 y^4 - 4x^3 y^4$

43. $5a^2 b^5$ **45.** $\dfrac{8a^2 b^3}{3c}$ **47.** $\dfrac{3x^3 y^2 z^5}{2w}$ **49.** $\dfrac{3n^5 p^3}{2mq}$

51. $4a + 3a^3$ **53.** $\dfrac{7}{y} + 4x^3 y$ **55.** $\dfrac{2}{x} - \dfrac{x^2 z^3}{2y^2}$

Practice Test

1. $t^2 - s + 5$, $m^5 n^4 o^3 p^2 r$, and $\frac{5}{7}c^{15} + \frac{3}{14}c^{11} - 3\pi$

2. $w^5 x^4$ is a monomial

   $2x^2 - 36$ is a binomial

   $\frac{1}{3}x^{17} + \frac{2}{3}x^{12} - \frac{1}{3}$ is a trinomial

   27 is a monomial

   $27x^3 - 2x^2 y^3$ is a binomial

   $x^2 + 3xy - \frac{2}{3}y^2$ is a trinomial

3. $8w^8 + 7w^5 + 3w^3 - 13w^2 - 2$

4a. $3x^3 y - 8x^2 y^2 - 5y^3 + xy + y^2 + 19$

   b. $7x^3 y - 8x^2 y^2 + 3y^3 + 5xy - y^2 + 7$

5. $x^8 y^3 w^5$

6. $3n^3 p^3 + 2n^5 p^5 - 35n^2 p^7$

7. $\frac{3x^4 yz^6}{2}$

8. $\frac{3t^2 u}{2v} - \frac{1}{2}t^4$

Lesson 6.3 Polynomial Operations II
Homework

1. First terms: $2p$ and p

   Outer terms: $2p$ and $-p^2$

   Inner terms: 3 and p

   Last terms: 3 and $-p^2$

3. $2s^3, 5$ 5. $12x^2 + 24x - 6yx - 12y$

7a. $9x^4 - 4$ b. $9x^4 - 12x^2 + 4$ c. $9x^4 + 12x^2 + 4$

9. $169s^2 - 4h^2$ 11. $91x^4 y^4 - 148x^5 y^2 + 60x^6$

13. $3x^2 + 4x^2 y + 7x + 8xy + 2$

15. $x^2 + x - y^2 - y$ or $x(x+1) - y(y+1)$ 17. $4x^2 - 8x$

19. $48x^5 - 48x^4 - \frac{76}{3}x^3 + 24\,x^2 + \frac{7}{3}x + 7$

21. $3x - 7 + \frac{20}{x+3}$ 23. $5x^3 - 3x^2 + 2x + \frac{1}{3} - \frac{\frac{2}{3x} + 2}{3x^2 + 2x}$

25. $9a^2 - 1$ 27. $a^2 - b^2$, $4x^2 - 9y^2$

29. $a^2 - 2ab + b^2$, $4t^6 - 16u^2 t^3 + 16u^4$

Practice Problems

1. $a^2 + 7a + 10$ 3. $x^2 - 15x + 44$ 5. $5y^2 + 7y - 24$

7. $8a^2 + 26ab + 15b^2$ 9. $18y^2 + 9xy - 5x^2$

11. $x^2 + 6x + 9$ 13. $25q^2 + 30q + 9$ 15. $z^2 - 10z + 25$

17. $t^2 - 12t + 36$ 19. $16a^2 - 56ac + 49c^2$ 21. $25m^2 - n^2$

23. $4x^2 - y^2$ 25. $25x^2 - 9$ 27. $4a^2 - 49b^2$

29. $8a^2 - 34ab + 21b^2$ 31. $18m^2 + 9mn - 20n^2$

33. $14xy + 21x - 8y - 12$

35. $3m^3 n + 11m^2 n - 4mn - 9mn^2 + 3n^2$

37. $21a^3 b + 16a^2 b + 9ab^2 - 16ab + 12b^2$

39. $10x^3 y - 30xy^2 + 19x^2 y - 12y^2 + 6xy$

41. $20m^5 n - 15m^4 n^2 + 40m^3 n^3 -$
   $15m^2 n^3 + 12m^3 n - 9m^2 n^2 + 24mn^3 - 9n^3$

43. $x^2 + 3x - 7$ 45. $x^2 + 7x + 2$

47. $x^2 + 5x - 1$ 49. $2x^2 - 3x + 1$

51. $x^2 + 2x - 3$ remainder 3 or $x^2 + 2x - 3 + \frac{3}{4x-1}$

53. $x^2 + 4x + 1$ remainder -10 or $x^2 + 4x + 1 - \frac{10}{3x+2}$

55. $2x^2 - 6x + 1$

Practice Test

1. $(2x^2)(3x^3 y) + (2x^2)(-2) + (3xy)(3x^3 y) + (3xy)(-2)$

2. $4x^2 - 12xy + 9y^2$ 3. $4x^2 + 12xy + 9y^2$

4. $4x^2 - 9y^2$ 5. $15x^3 + 14x^2 - 22x + 4$

6. $12r^8 - 15p^2 r^4 - 18p^4 - 7r^4 + 36p^2 - 10$

7. $3t^2 + t - 2 + \frac{3}{2t+1}$ 8. $2x^2 - x + 2 + \frac{-6}{4x+2}$

9a. $a^4 + a^5 - a^6 - a^7$

   b. The degree of the resulting polynomial is 7.

10. $15y^6 - 5y^5 + 4y^4 + 5y^3 - 5y^2 + 2y$

11.

	$2x^3$	$-3x$	7
$5x^4$	$10x^7$	$-15x^5$	$35x^4$
8	$16x^3$	$-24x$	56

$10x^7 - 15x^5 + 35x^4 + 16x^3 - 24x + 56.$

12.

	$5x^4$	$-7x^3$	$7x^2$	$-8x$
x^2	$5x^6$	$-7x^5$	$7x^4$	$-8x^{33}$
1	$5x^4$	$-7x^3$	$7x^2$	$-8x$

Topic 6
Cumulative Review Problems

1a. 216 **b.** x^7 **c.** $a^{20}b^8$ **3a.** $y - 7 = -\dfrac{2}{7}(x - 3)$

b. $y + -\dfrac{2}{7}x + \dfrac{55}{7}$ **c.** $2x + 7y = 55$

5. The numbers are 14 and 46. **7a.** $y + 9 = -\dfrac{8}{5}(x - 20)$

b. $y = -\dfrac{8}{5}x + 23$ **c.** $8x + 5y = 115$ **9a.** 1 **b.** -1 **c.** 1

11. $\dfrac{3x^2y^4}{2z^6}$ **13.** slope $= 4$, y–intercept $= (0, -7)$

15. $-5 < x < -3$

17. $\dfrac{26}{117}$, The LCM of 72 and 108 is 36, The GCF of 72 and 108 is 36

19.

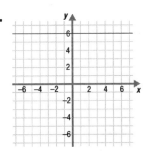

21. $\left(\dfrac{3}{2}, -3\right)$ **23.** $y = 3x - 4$ **25.** $-\dfrac{2}{13}$

27. No solution for y **29.** A and B **31.** A and C **33.** -261

35.

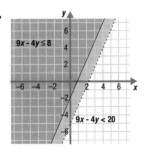

37. Manuel split his money by putting \$984 in his checking account and \$1581 in his saving account.

39a. binomial **b.** trinomial **c.** monomial **d.** binomial
e. monomial **f.** trinomial

41. $-4a^2b + 10a - 7b$

Lesson 7.1 Factoring Polynomials I
Homework

1. x^3yz^2, x **3.** $6xy$ **5.** $y(x^2 + 6y)$

7. $4(a^2 - b^2) = 4(a + b)(a - b)$ **9.** $2xy(3y^2 - 2xy + 1)$

11. $17x^2y^2z(z + 4x^8y^{30} + 9x^7y^2z^{11})$ **13.** $x^5 + y$

15. $(xy + yz)$ **17.** $(a^2 + b^2)(a - b)$

19. $(x^4y + z)(x + y + z)$ **21.** $(x^2 + y^2)(z + 3)$

23. $(x + y)(3 + z)$

Practice Problems

1. 23 and $8m^3n$ **3.** $4ab$ **5.** $5xy^2$ **7.** $3ac^2$

9. $8m(2n^4 + 1)$ **11.** $2xy(3x^3y^2 + 7)$ **13.** $2ab(4a^2b - 5)$

15. $9y^3z^5(4y^4z^3 - 5)$ **17.** $2mn(2 + 5n^2 - 9m^3)$

19. $4ab(2a^2b^3 - 3 + 5a^2)$ **21.** $8p^5q^3r(4p^2r^3 - 5q^2)$

23. $3xy^2z^3(3 - 5x^2y^3z + 7x^3z^2)$

25. $4a^2bc^2(5ab^4 + 3a^2bc - 2c)$

27. $3a^2b^2c^2(2ab^3 - 3a^2b^2c + 6b - 7a^4c)$

29. $(x + y)(z + 3)$ **31.** $(a + 9)(3b - 4)$

33. $(8m + 17)(3n^3 - 4)$ **35.** $(7x - 11)(2x^2 + 3)$

37. $(m - 3n)(5m + 2n)$ **39.** $(x + y)(w + z)$

41. $(a - b)(c + d)$ **43.** $(2a - 7)(2a + 1)$

45. $(6a + 5b)(2a + 3)$ **47.** $(5x + 2y)(3x + 7)$

49. $(2z + w)(4z - 1)$ **51.** $(4a - 5b)(3a + 2)$

53. $(6x - 5y)(3x + 2)$ **55.** $(3pr - 4s)(4r + 5)$

Practice Test

1. GCF $= x$ **2.** GCF $= xyz$ **3.** $3x^2y - 3xy^2 = 3xy(x - y)$

4. $3xy^3 - 6xy^2 + 3x^3y^4 = 3xy^2(y - 2 + x^2y^2)$

5. $13(x^2 + 4) + 6y(x^2 + 4)$
$= (x^2 + 4)(13 + 6y)$

6. $17x^2(3xyz + 4z) - 3yz(3xyz + 4z)$
$= (3xyz + 4z)(17x^2 - 3yz)$

7. $39rs - 13s + 9r - 3$
$= (3r - 1)(13s + 3)$

8. $12wz - 44z + 18w - 66$
$= 2(3w - 11)(2z + 3)$

Lesson 7.2 Factoring Polynomials II
Homework

1. $(x + 4)(x + 3)$ **3.** $(x + 5)(x + 7)$ **5.** $(x + 3)(x - 8)$

7. $(x - 3)(x + 2)$ **9.** $(x - 7)(x + 3)$

11. $(x + 36)(x - 1)$ **13.** $(x + 5)(2x + 1)$ **15.** $(2y + 3)(2y - 7)$

17. $15(a - 1)(a - 1)$ **19.** $x = \frac{1}{5}$ or $x = -\frac{2}{5}$

21. $(13x + 11)(x + 2)$ **23.** $(x + y)(x + y)$ **25.** xy

27. $\frac{y}{2}(x^2 - 1)$ or $\frac{1}{2}y(x^2 - 1)$

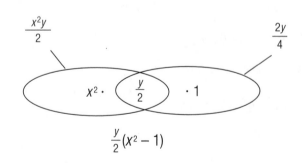

$$\frac{y}{2}(x^2 - 1)$$

29. GCF $= x$ A $= 2xy^2z$ B $= 52x^2y$

yz   2y   2x
x   13

C $= 13x$

Practice Problems

1. $(x + 1)(x + 4)$ **3.** $(x + 1)(x + 14)$ **5.** $(x + 3)(x + 5)$

7. $(x + 4)(x + 3)$ **9.** $(x - 2)(x - 6)$ **11.** $(x - 11)(x - 4)$

13. $(x - 7)(x - 3)$ **15.** $(x + 3)(x - 10)$ **17.** $(x + 7)(x - 3)$

19. $(x + 9)(x - 4)$ **21.** $(x - 9)(x + 2)$ **23.** $(x - 7)(x + 3)$

25. $(x - 9)(x + 7)$ **27.** $(x - 12)(x + 5)$ **29.** $(2x + 5)(x + 1)$

31. $(3x + 2)(x - 7)$ **33.** $(2x + 7)(x - 4)$ **35.** $(2x - 3)(x + 4)$

37. $(2x + 3)(x + 5)$ **39.** $(3x + 2)(x + 3)$ **41.** $(5x - 2)(2x - 1)$

43. $(3x + 2)(2x - 5)$ **45.** $(2x + 1)(4x - 3)$

47. $(3x - 5)(3x + 4)$ **49.** $(4x + 1)(9x + 1)$

51. $(5x - y)(x + 3y)$ **53.** $(3x + y)(x - 2y)$

55. $(3x - 2y)(3x + y)$

Practice Test

1. $x^2 - 10x + 24 = (x - 4)(x - 6)$

2. The polynomial $x^2 + 2x - 1$ cannot be factored using integers.

3. $t^2 - 16t - 17 = (t + 1)(t - 17)$

4. $r^2 + 10rt + 25t^2 = (r + 5t)(r + 5t)$

5. $5x^2 + 8x - 4 = (x + 2)(5x - 2)$

6. $27v^2 - 57v + 28 = (9v - 7)(3v - 4)$

7. $4x^2 + 57x + 108 = (x + 12)(4x + 9)$

8. $x = \frac{12}{7}$ or $x = -1$

9. The two true statements are:
  • Two factors of C are z and 2.
  • The GCF of A, B, and C is $4z$.

10. $15u^2 + 20uv$
$9uv + 12v^2$

11. $7y - 3x$

12.

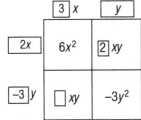

Lesson 7.3 Factoring By Patterns
Homework

1. $(x + 7)^2$ **3.** $(x + 5)(x^2 - 5x + 25)$ **5.** $x(3y - 1)(3y + 1)$

7. $(x + 4w^2)^2$ **9.** $(7y - 2x)^2$ **11.** $2x(x + 3)^2$

Practice Problems

1. $(a + 9)(a + 9)$ or $(a + 9)^2$

3. $(3x + 7)(3x + 7)$ or $(3x + 7)^2$

5. $(2a + 5)(2a + 5)$ or $(2a + 5)^2$

7. $(z - 11)(z - 11)$ or $(z - 11)^2$

9. $(4a - 5)(4a - 5)$ or $(4a - 5)^2$

11. $(3x - 2)(3x - 2)$ or $(3x - 2)^2$

13. $(x + 6)(x - 6)$

15. $25(a + 5b)(a - 5b)$

17. $(a - 6)(a^2 + 6a + 36)$

19. $(x - 5)(x^2 + 5x + 25)$

21. $(3z - 7)(9z^2 + 21z + 49)$

23. $(c + 4)(c^2 - 4c + 16)$

25. $(y + 3)(y^2 - 3y + 9)$

27. $2mn(5m + 8n)(5m - 8n)$

Practice Test

1. $9x^2 + 12x + 4$

$\quad 0.25x^2 + 8x + 64$

$\quad x^2 - 2x + 1$

2a. $x^2 - 10x + 25 = (x - 5)(x - 5)$

b. $49y^2 + 28y + 4 = (7y + 2)(7y + 2)$

c. $16x^2 - 1 = (4x + 1)(4x - 1)$

d. $9y^2 - 36 = (3y + 6)(3y - 6)$

3. $x^2 - 1000$ cannot be factored any further using integers. $9m^2 - 24mn - 16n^2$ cannot be factored any further using integers.

4. $12x^3 - 60x^2 + 75x = 3x(2x - 5)(2x - 5)$

5. $x^2 + 8x + 16$, $4x^2 - 12x + 9$

6a. $4x^2 - 24x + 36 = (2x - 6)(2x - 6)$

b. $64z^2 + 16z + 1 = (8z + 1)(8z + 1)$

c. $4w^2 - 49 = (2w + 7)(2w - 7)$

d. $9m^2 - n^2 = (3m + n)(3m - n)$

7a. $x^3 + 1000 = (x + 10)(x^2 - 10x + 100)$

b. $216y^3 - 1 = (6y - 1)(36y^2 + 6y + 1)$

c. $343x^3 + 8y^3 = (7x + 2y)(49x^2 - 14xy + 4y^2)$

8. $27w^3 + 90w^2 + 75w = 3w(3w + 5)(3w + 5)$

Topic 7
Cumulative Review Problems

1. $a^3 - 2a + 5$ **3.** $7x^2 + 5y + 2$

5. Alfredo should use 130 ml of the 15% solution and 120 ml of the 40% solution.

7. $(x - 3)^2$ **9.** a, d, e

11.

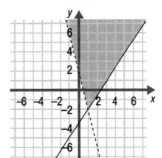

13. $(a^2 + 2)^2$

15. $6 \leq x < 10$

17a. $y - 12 = -2(x + 7)$

b. $y = -2x - 2$

c. $2x + y = -2$

19. $x = 7$ or $x = -2$

21a. $y + 3 = -\frac{8}{5}(x - 4)$

b. $y = -\frac{8}{5}x + \frac{17}{5}$

c. $8x + 5y = 17$

23. $\frac{8}{33}$ **25.** No solutions   **27.** $(5x + 7)(x - 1)$ **29.** 128

31. $a^2 + 2a - 8$

33. Jerome owed $1,190 on the credit card that charged 16% interest and $630 on the credit card that charged 14% interest.

35a. $y - 3 = \frac{5}{6}(x - 5)$

b. $y = \frac{5}{6}x - \frac{7}{6}$

c. $-5x + 6y = -7$ or $5x - 6y = 7$

37. 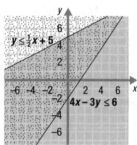   **39.**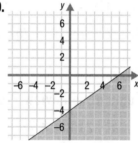

41. $(6b + 5)^2$ **43.** $4x^3 + 7xy^2 + 7$ **45.** $(-2, 5)$

47. $5x^2 + x^2y + 2xy + 25x - 15y$

49. The width of the rectangle is 17 feet and the length is 51 feet.

Lesson 8.1 Rational Expressions I
Homework

1. $x = 14$ or $x = -2$ **3.** $\frac{a}{9b^2}$ **5.** $\frac{3y^2}{7}$ **7.** $\frac{3w}{y^2}$ **9.** $\frac{5}{9a}$

11. $\frac{x(x + 1)}{x - 1}$ or $\frac{x^2 + x}{x - 1}$ **13.** $\frac{21x}{5y}$ **15.** $\frac{x + 4}{x^2 - 4}$ or $\frac{x + 4}{(x + 2)(x - 2)}$

17. $\frac{5}{x - 7}$ **19.** $\frac{7}{15x}$ **21.** $\frac{8\pi r^3}{3}$ **23.** $\frac{1}{3}$

Practice Problems

1. -5 **3.** $2, -2$ **5.** $\frac{4m^4}{3n^3}$ **7.** $\frac{4b^3}{7a^5}$ **9.** $\frac{x + 5}{x - 2}$ **11.** $\frac{14}{a^2 b^2 c^2}$

13. $\frac{14m^7}{9n^4}$ **15.** $\frac{2m^4}{3n^3 p}$ **17.** $\frac{3a}{2bc^2}$ **19.** $\frac{yz}{x}$ **21.** $\frac{3m^2 np}{2q}$

23. $\frac{2m}{n^2}$ **25.** $\frac{4a}{b^2}$ **27.** $\frac{x^2 - 3x}{2x + 14}$ or $\frac{x(x - 3)}{2(x + 7)}$ **29.** $\frac{5a}{7b}$

31. $\frac{3b + 5}{2b + 1}$ **33.** $\frac{7x + 2}{5x - 1}$ **35.** $\frac{2b - 4}{2b + 9}$ **37.** $\frac{11 + 5n}{7n}$

39. $\frac{3}{x + 2}$ **41.** $\frac{2}{x - 3}$ **43.** $\frac{4}{x - 2}$ **45.** $\frac{x + 1}{x + 3}$ **47.** $\frac{x + 5}{x + 6}$

49. $\frac{3}{2}$ **51.** $\frac{1}{x + 5}$ **53.** $\frac{1}{x - 1}$ **55.** $\frac{3}{5}$

Practice Test

1. $x = -3$ or $x = 2$ **2.** $\frac{x - 7}{x + 6}$ **3a.** $\frac{5}{9z}$ **b.** $\frac{8x}{z^2}$

4a. $\frac{6x}{7}$ **b.** $\frac{xw}{15y}$ **5a.** $\frac{7x}{13y}$ **b.** $\frac{3w + 14}{z - 8}$

6. $\frac{9y - 3}{7x}$ **7.** $\frac{4}{x - 6}$ **8.** $\frac{6y + 4}{y - 7}$

Lesson 8.2 Rational Expressions II
Homework

1. $\frac{1}{125}$ or $\frac{1}{5^3}$ **3.** 1,000   **5.** $\frac{1}{8m^3 n^3}$ **7.** $\frac{2t^{13}}{s^9}$

9. 0.00000001 cm, 0.00000005 cm   **11.** $\frac{z^{10}}{9x^8 y^2}$

13. -1 **15.** $-x$ **17.** $\frac{2x}{x + 1}$ **19.** $\frac{y + 8}{2y - 1}$

21. $\frac{z + 3}{(z - 4)(z + 5)}$ **23.** $\frac{2}{3z}$ **25.** $(x + 5)(x + 5)(x - 8)$

27. $\frac{5x + 14y}{x^2 y^3}$ **29.** $\frac{4z - 3y}{xy^2 z^2}$ **31.** $\frac{1}{y - 2}$

33. $\frac{1}{n - 1} - \frac{1}{n(n - 1)}$

$= \frac{n}{n(n - 1)} - \frac{1}{n(n - 1)}$

$= \frac{n - 1}{n(n - 1)}$

$= \frac{1}{n}$

The two fractions are $\frac{1}{4}$ and $\frac{1}{20}$:

$\frac{1}{4} - \frac{1}{20} = \frac{1}{5}$

35. $R = \dfrac{R_1 R_2}{R_1 + R_2}$; $R = \dfrac{12}{7}$ ohms

Practice Problems

1. $\dfrac{1}{8}$ **3.** $\dfrac{1}{81}$ **5.** 81 **7.** $\dfrac{1}{16}$ or 4^{-2} **9.** $\dfrac{1}{125}$ or 5^{-3} **11.** $\dfrac{4800}{89}$

13. $\dfrac{1}{a^4 b^6}$ **15.** $\dfrac{1}{m^{24} n^{12} p^4}$ **17.** $\dfrac{b^{28}}{a^{12}}$ **19.** $\dfrac{a^4 d^{10}}{c^8}$ **21.** $\dfrac{m^{16}}{n^{20} p^{12}}$

23. $\dfrac{4a^{25} b^5}{c^6}$ **25.** 5.7×10^{-5} **27.** 4,300,000 **29.** -1

31. $\dfrac{x-7}{x-5}$ **33.** $-\dfrac{x-1}{7+x}$ **35.** $-\dfrac{x+1}{9+x}$ **37.** $\dfrac{x-4}{x+3}$

39. $-\dfrac{3x(x-1)}{x+1}$ **41.** $\dfrac{-2x(x+5)}{x-5}$ or $\dfrac{2x(x+5)}{5-x}$ **43.** $\dfrac{x+5}{x-2}$

45. $\dfrac{x-5}{x+3}$ **47.** $\dfrac{x+1}{x-7}$ **49.** $\dfrac{x+11}{3}$ **51.** $\dfrac{3x}{x+7}$ **53.** $\dfrac{5x}{x-3}$

55. $\dfrac{(x-3)(x-3)}{5x(x-7)}$ or $\dfrac{x^2 - 6x + 9}{5x^2 - 35x}$ **57.** $(x+3)(x+4)(x-7)$

59. $x(x+4)(x-1)(x-1)$ **61.** $\dfrac{4b+6a}{9ab}$ **63.** $\dfrac{2y+x}{3xy}$

65. $\dfrac{2x-5}{(x+10)(x-5)}$ **67.** $\dfrac{2x^2 + x + 4}{(x+1)(x-4)}$ **69.** $\dfrac{6x^2 + 8x - 3}{(x-3)(x+2)}$

71. $\dfrac{2x^2 - x - 29}{(x+2)(x+3)(x-5)}$ **73.** $\dfrac{4n-m}{m^2 n^2}$ **75.** $\dfrac{3x - 2yz}{x^2 yz}$

77. $\dfrac{3x^2 - 2x + 88}{(x-8)(x+3)}$ **79.** $\dfrac{-3x-5}{(x+5)(x+1)(x-1)}$

81. $\dfrac{4x+10}{(x+2)(x+3)(x-1)}$ **83.** $\dfrac{3x+2}{x+4}$

Practice Test

1. 8 **2.** 7.3901, 10^{-5}, 208.1, and 0.00009019

3. $\dfrac{b^{15}}{a^6}$ **4.** $\dfrac{x^3}{64y^8}$ **5.** $-x-2$ **6a.** $\dfrac{x+3}{x+2}$ **b.** $\dfrac{2(x-1)^2}{x^2}$

7a. -3 **b.** $\dfrac{y+4}{(y+6)(y+1)}$ **8.** $\dfrac{2(x-1)}{5(x-3)}$

9. $LCM = x(x+5)(x-5)(x+7)$

10. $\dfrac{7b+17}{(b+1)(b+2)(b+3)}$ **11.** $\dfrac{y}{(y+4)(y+5)}$ **12.** $\dfrac{3x}{5(1+3x)}$

Lesson 8.3 Equations With Fractions
Homework

1. $x = 3$ **3.** $x = 6$ or $x = 1$ **5.** $x = 6$ **7.** $y = \dfrac{3}{2}$

9. 60.8 lbs. **11.** $x = 3$ or $x = -1$

12. $x = 2$, the solution is extraneous

Practice Problems

1. $x = 7$ **3.** $x = 2$ **5.** $x = -9$ **7.** $x = -8$ **9.** $x = -3$

11. $x = 4$ **13.** $x = 2$ **15.** $x = 1$ or $x = -\dfrac{4}{5}$

17. $x = 2$ or $x = \dfrac{3}{2}$ **19.** $x = 3$ or $x = -6$

21. $x = 5$ or $x = -1$ **23.** $x = 14$ or $x = -5$

25. $x = 4$ or $x = 8$ **27.** No solution

Practice Test

1. $-5 = x$

The solution is not extraneous.

2. $y = 3$

The solution is not extraneous.

3. $\dfrac{3V}{\pi r^2} = h$

4. $x = 4$

5. $4 = y$

The solution is not extraneous.

6. $x = 5$

The solution is not extraneous.

7. $\dfrac{S}{2\pi r} - r = h$

8. $y = 2$

Lesson 8.4 Problem Solving
Homework

1. 16.8 hr **3.** 357 jellybeans

5. $\dfrac{105}{22}$ hours (approximately 4.8 hours or 4 hours and 46 minutes)

7. 4 inches by 12 inches

9. $\dfrac{77}{18}$ hours (approximately 4.3 hours or 4 hours and 17 minutes)

11. 5,778 fish

Practice Problems

1. $\frac{20}{9}$ hours or approximately 2 hours 13 minutes

3. 45 minutes **5.** $\frac{21}{4}$ hours or 5 hours and 15 minutes

7. 30 minutes **9.** 24 caramels, 18 nougats

11. 35 puppies, 20 adult dogs **13.** 65 units

15. Jayme: 30 miles; Terry: 45 miles

17. Sasha: 2 miles; Leroy: 3 miles

19. Ranji: 45 miles, Paula: 30 miles

21. 68 miles per hour **23.** 180 pounds

25. 135 square inches **27.** $V = \frac{kT}{P}$

Practice Test

1. It will take them $\frac{12}{7}$ hours, or about 1 hour and 43 minutes, to wash the dishes working together.

2. Trish lives 4 miles from the park.

3. There are 252 peanuts in the bag of mix.

4. The area of the kite is 52 inches^2.

5. It will take them 36 minutes working together.

6. The hummingbird can fly at 50 km per hour and the eagle can fly at 85 km per hour.

7. The florist received 189 roses.

8. The speed of the wave is 8 feet per second.

Topic 8
Cumulative Review Problems

1. $-2 \le x < \frac{7}{5}$ **3.** $(a + b)(a - b)$ **5.** 13 units

7. $y = 3x + 11$ **9.** $x = -1$

11. $m = -\frac{9}{5}$, y-intercept $= (0, \frac{11}{5})$

13. $t = \frac{20}{13}$ hours $\approx$ 1 hour and 32 minutes

15. $\frac{4x - 11}{(x - 2)(x - 3)}$ or $\frac{4x - 11}{x^2 - 5x + 6}$ **17.** $(x^2 + y^2)(x + y)(x - y)$

19. no solution **21.** 61 units

23. $\frac{3a}{c^3} - \frac{4c^2}{a^3}$ or $\frac{3a^4 - 4c^5}{a^3c^3}$ **25.** A and B

27.

29. $m = \frac{9}{2}$ **31.** $6ab^2(a^2 - 4ab + 4)$ **33.** $\frac{x}{x + 5}$

35. $x = -9$ **37a.** $y - 6 = \frac{4}{3}(x + 2)$ **b.** $y = \frac{4}{3}x + \frac{26}{3}$

c. $-4x + 3y = 26$

39a. 3 **b.** 1 **c.** $\frac{1}{a^3b^6c^3}$ or $a^{-3}b^{-6}c^{-3}$

41. radius: 5; center: (2, –6)

43. $(7x - 1)(7x - 1)$ or $(7x - 1)^2$ **45.** infinite solutions

47. $y = -\frac{1}{2}x + 2$ **49.** $(x - 3)(x - 1)$

Lesson 9.1 Rational Exponents
Homework

1. $\sqrt[3]{8} = 2$ **3.** $2^{\frac{13}{20}}$ **5.** 4 **7.** not a real number

9. 1024 cells **11.** $\frac{x^{\frac{1}{7}}}{y^5}$ **13.** $\frac{3}{5}$ **15.** $\sqrt{x}$ **17.** $-3x^2y$

19. $2x^3y\sqrt[3]{2yz^2}$ **21.** $\sqrt[3]{2}m$ **23.** $2x^2y^3z\sqrt{xy}$

25. $7\sqrt[4]{360}$, $\frac{\sqrt[4]{360}}{72}$, $\frac{-11}{4}\sqrt[4]{360}$

27. -450 **29.** $16\sqrt[3]{3} + \frac{1}{4}\sqrt[3]{2} + 1$

31. $\sqrt[3]{24}, 2\sqrt[3]{3}, \sqrt[3]{8} \cdot \sqrt[3]{3}, \frac{\sqrt[3]{3}}{27}, 3^{\frac{1}{3}}$ **33.** $\sqrt{2}\pi$

35. $\frac{\sqrt[3]{98}}{7}$

Practice Problems

1. $\sqrt{9^5}$ or $\left(\sqrt{9}\right)^5$; 243   **3.** $\sqrt[3]{27^2}$ or $\left(\sqrt[3]{27}\right)^2$; 9

5. $\sqrt[4]{81^3} = \left(\sqrt[4]{81}\right)^3 = 27$ **7.** 6   **9.** –3   **11.** –6

13. $312^{\frac{4}{5}}$ **15.** $200^{\frac{5}{7}}$ **17.** $y^{\frac{11}{12}}$ **19.** $x^{\frac{11}{30}}$ **21.** x

23. $\frac{16b^8}{81a^3}$ **25.** $\frac{64}{27x^2y^3}$ **27.**

29. $\frac{11}{8}$ **31.** $\frac{13}{24}$ **33.** $-\frac{4}{5}$ **35.** $-\frac{2}{3}$ **37.** $\frac{3}{2}$

39. 56   **41.** –47   **43.** $\frac{\sqrt[4]{7}}{x}$ **45.** $6ab^3$ **47.** $8x^2y^3z^5$

49. $6m^2n^4\sqrt{3mn}$ **51.** $4abc^3\sqrt[3]{3b^2}$ **53.** $2np^2\sqrt[5]{5m^2n^2p^2}$

55. $2m^2\sqrt[3]{4n^2}$ **57.** $\sqrt{5}, \frac{1}{3}\sqrt{5}, \frac{7\sqrt{5}}{8}$ **59.** $-3\sqrt{5}$

61. $-2\sqrt{3}$ **63.** $20\sqrt{2} - 5\sqrt{3}$

65. $6x\sqrt[3]{2x^2} - 13\sqrt[3]{2x^2}$ or $(6x - 13)\sqrt[3]{2x^2}$

67. $4x\sqrt[3]{2x} - 11\sqrt[3]{2x}$ or $(4x - 11)\sqrt[3]{2x}$ **69.** $30 - 14\sqrt[3]{45}$

71. $42y\sqrt{5} + 12\sqrt{6y}$ **73.** $20z - 14\sqrt[3]{44z^2}$

75. $\sqrt{10} - 5\sqrt{3} - 2\sqrt{5} + 5\sqrt{6}$ **77.** $45z - 6$

79. $50y - 3x$ **81.** $\frac{\sqrt{2x}}{x}$ **83.** $\frac{5x\sqrt{2} + 10}{x^2 - 2}$

Practice Test

1. $x^{\frac{7}{10}}$ **2.** $243^{\frac{3}{5}}$ **3.** $\sqrt[3]{-125}$ **4.** $\frac{64}{343x^2y}$

5. $\frac{13}{15}$ **6.** 29   **7.** $\frac{\sqrt[3]{3}}{2}$ **8.** $\frac{9xy\sqrt{z}}{11z}$

9. $21\sqrt{5x} - 3$ **10.** –19   **11.** $-6 - 12\sqrt{2}$ **12.** $y^{\frac{1}{6}}$

Topic 9
Cumulative Review Problems

1. $x^3 + 3x^2y^2 + 36xy^2 + 13x^2 + 12x$ **3.** $y = 1$ or $y = -3$

5. Loan = \$625   **7.** $x = -4$ or $x = -9$ **9.** no solution

11. $11\frac{1}{9}$ minutes

13. point-slope form: $-3 = \frac{5}{3}(x + 7)$

slope intercept form: $y = \frac{5}{3}x + \frac{44}{3}$

standard form: $3y - 5x = 44$

15. $2a^2b^2 + 5ab + a - b$

17a. $\frac{25}{2}$ **b.** $-27a^6$ **c.** $x^{24}y^{16}z^4$ **19.** $x = 14$

21. $(-4y - 3)(4y + 3)$ or $(-4y + 3)(4y - 3)$ or $-(4y - 3)^2$

23. True: The GCF of 52 and 100 is 4. The LCM of 30 and 36 is 180.

25. $(x + 13)(x - 10)$ **27.** slope $= -\frac{3}{4}$, y-intercept $= (0, -\frac{9}{2})$

29. $\frac{3x\sqrt{2x}}{y}$ **31.** $\frac{24 - 8\sqrt{11}}{-2}$ or $-12 + 4\sqrt{11}$

33. $x^2 + 11x - 8$ **35.** $7y^2(x + 1)^2$

37. $1 \le y < 5$ **39a.** $6\sqrt{5}$ **b.** $6x + 3x\sqrt{2}$ **c.** $a^2 - b$

41. 9 pins, 19 balls   **43.** $\sqrt{x} + \sqrt{y}$

45. $(2x - 1)(4x^2 + 2x + 1)$ **47.** $a = 2$ **49.** $\frac{4y^6}{9x^8}$

Lesson 10.1 Quadratic Equations I
Homework

1. $2x^2 - 3x - 5 = 0$ **3.** $y = 0, y = -4$

5. $-11x^2 = 0, 2a(a + 5) = 4, 6x - 9x^2 = 8$ **7.** $z = \pm 5$

9. The dimensions are 2m by 13m by 20m.

11. $x = 1, x = \frac{-5}{3}$ **13.** $a = \pm 10$ **15.** $x = \pm 6$

17. $x = \pm 9$ **19.** $x = 1 \pm 5\sqrt{3}$

21. The tree was approximately 11 feet tall $\left(\frac{20\sqrt{3}}{3} \text{ feet}\right)$.

23. $x = \frac{9 \pm \sqrt{5}}{2}$

Practice Problems

1. $x = -7$ or $x = 0$ **3.** $x = 4$ or $x = 0$ **5.** $x = -5$ or $x = 0$

7. $x = 6$ or $x = 0$ **9.** $x = 1$ or $x = 6$ **11.** $x = -5$ or $x = -7$

13. $x = -3$ or $x = -6$ **15.** $x = -2$ or $x = 9$

15 19. $x = 5$ or $x = -2$

10 23. $x = -6$ or $x = 9$

25. $x = 6$ or $x = -5$ **27.** $x = -12$ or $x = 15$

29. $x = -10$ or $x = 10$ **31.** $x = -16$ or $x = 16$

33. $x = -4\sqrt{3}$ or $x = 4\sqrt{3}$ **35.** $x = -4\sqrt{2}$ or $x = 4\sqrt{2}$

37. $x = -9$ or $x = 9$ **39.** $x = -3\sqrt{2}$ or $x = 3\sqrt{2}$

41. $x = -2\sqrt{3}$ or $x = 2\sqrt{3}$ **43.** $x = -9$ or $x = 9$

45. $x = -12$ or $x = 2$ **47.** $x = -18$ or $x = 0$

49. $x = 3 - \sqrt{13}$ or $x = 3 + \sqrt{13}$ **51.** $x = -11$ or $x = 5$

53. $x = -9$ or $x = 5$ **55.** $x = \frac{4}{5} - \sqrt{3}$ or $x = \frac{4}{5} + \sqrt{3}$

Practice Test

1. $2x^2 - 17x - 2 = 0$
So $a = 2$, $b = -17$, and $c = -2$.

2. $x = 0$ or $x = 4$

3. $2 = (x - 3)^2$, $x(x + 9) = 4$, and $x^2 - 9 = 7x + 2$

4. $x = -\frac{5}{2}$ **5.** $\sqrt{(-8)^2}$, $\frac{\sqrt{256}}{\sqrt{4}}$, and $\sqrt{\frac{64}{5}} \cdot \sqrt{5}$

6. $x = +7\sqrt{7}$ and $x = -7\sqrt{7}$

7. $\frac{\sqrt{5}}{\sqrt{2}}$ or $\frac{\sqrt{10}}{2}$ **8.** $x = 5 + 2\sqrt{41}$ and $x = 5 - 2\sqrt{41}$

Lesson 10.2 Quadratic Equations II
Homework

1. $x^2 + 13x + \frac{169}{4}$, $\left(x + \frac{13}{2}\right)^2$

3. $x = 2 \pm \sqrt{5}$ **5.** $x = 18 \pm 2\sqrt{91}$

7. $x = \frac{-3}{2} \pm \frac{\sqrt{13}}{2}$

9. Seana's speed before lunch was approximately 16.5 mph and after lunch it was approximately 14.5 mph.

11. $x = \frac{5}{6} \pm \frac{\sqrt{17}}{2}$ **13.** $x = -5$ **15.** $x = -1$, $x = -\frac{1}{2}$

17. $-4 \pm \sqrt{13}$ **19.** $\frac{15 \pm \sqrt{465}}{6}$

21. The formulas yield the same child's dosage at approximately 1.2 and 9.8 years.

23. $\frac{5 \pm \sqrt{97}}{4}$ **25.** $3x^2 - 11x - 20 = 0$

27. $c = -15$, $x = -3$ and $x = \frac{5}{3}$ **29.** $18x^2 - 3x - 10 = 0$

Practice Problems

1. $x = -3$ or $x = 9$ **3.** $x = -5$ or $x = 9$ **5.** $x = -10$ or $x = 2$

7. $x = 1$ or $x = 7$ **9.** $x = -3 - \sqrt{21}$ or $x = -3 + \sqrt{21}$

11. $x = 3$ or $x = -7$ **13.** $x = 9 - 2\sqrt{6}$ or $x = 9 + 2\sqrt{6}$

15. $x = \frac{-3 \pm \sqrt{73}}{2}$ **17.** $x = \frac{-7 \pm \sqrt{85}}{2}$ **19.** $x = \frac{1 \pm \sqrt{57}}{2}$

21. $x = -7$ or $x = 3$ **23.** $x = -6$ or $x = 2$ **25.** $x = -8$ or $x = 18$

27. $x = \frac{5 \pm \sqrt{109}}{6}$ **29.** $x = \frac{1}{4}$ or $x = 1$ **31.** $x = 2 \pm \sqrt{3}$

33. $x = 3 \pm \sqrt{5}$ **35.** $x = \frac{-3 \pm \sqrt{89}}{10}$ **37.** $x = 1 \pm 2\sqrt{2}$

39. $x = \frac{-3 \pm \sqrt{41}}{2}$ **41.** $x = \frac{6}{5}$ or $x = 8$ **43.** $x = \frac{3 \pm \sqrt{37}}{2}$

45. $x = \frac{1 \pm \sqrt{5}}{2}$ **47.** $x = \frac{9 \pm \sqrt{33}}{8}$

49. The equation has two unequal real solutions.

51. The equation has one real solution.

53. The equation has one real solution.

55. The equation has no real solutions.

Practice Test

1. $x^2 + 9x + \frac{81}{4}$; the perfect sqare is $x^2 + 9x + \left(\frac{9}{2}\right)^2$

2. $x = -1 \pm \sqrt{39}$ **3.** $x^2 + 8x = -14$

4. $x = 1$ or $x = \frac{1}{4}$ **5.** $2x^2 + 4x - 7 = 0$

6. $x = \frac{-3 \pm \sqrt{14}}{5}$

7. $x^2 + 4x + 11 = 0$, $x^2 - x + 1 = 0$, and $x^2 + 2x + 5 = 0$

8. $+36$ and -36 **9.** $x = \frac{7 \pm 3\sqrt{5}}{2}$

10. $2x^2 - 3x + 6 = 0$ **11.** $5x^2 + 14x - 3 = 0$

12. $c = \frac{49}{8}$

Lesson 10.3 Complex Numbers
Homework

1. $-4i + 7$; $6 + 1 - 4i$; $7 - 3i - i$

3a. $4i$ **b.** $5i$ **c.** $9i$ **d.** -20 **e.** 1 **f.** $-i$

5a. $17 + 2i$ **b.** $3 + 7i$ **c.** $12 + 33i$

7a. 157 **b.** $\dfrac{15 + 136i}{97}$ or $\dfrac{15}{97} + \dfrac{136}{97}i$ **c.** $x = -2 \pm 2i$

9a. $34 + 8i$ **b.** $\dfrac{-3 + 4i}{5}$ or $-\dfrac{3}{5} + \dfrac{4}{5}i$

11a. $37 - 50i$ **b.** $\dfrac{-14 + 29i}{17}$ or $-\dfrac{14}{17} + \dfrac{29}{17}i$

  c. $x = -1 \pm 2i\sqrt{2}$

Practice Problems

1. $13i$ **3.** $4 - 5i$ **5.** i **7.** 1 **9.** $-5\sqrt{2}$ **11.** $19 - 3i$

13. $13 - 5i$ **15.** $18 - 6i$ **17.** $-15 + 23i$ **19.** $32 - 22i$

21. 73 **23.** $\dfrac{43}{53} - \dfrac{18}{53}i$ **25.** $\dfrac{7}{25} - \dfrac{26}{25}i$ **27.** $x = -4 \pm \sqrt{2}i$

Practice Test

1a. $11 + 8i$ **b.** $3 - 4i$ **2.** $-8i^2 - 4i^3$ **3.** $5 + 7i$

4. $2 + 26i$ **5.** 34 **6.** $\dfrac{8 - 14i}{65}$ or $\dfrac{8}{65} - \dfrac{14}{65}i$

7. i^{37} and $-i^3$

8a. $x = \dfrac{-3 \pm i\sqrt{19}}{2}$ **b.** $x = \dfrac{5 \pm i\sqrt{11}}{2}$ **c.** $x = \dfrac{-1 \pm i\sqrt{2}}{2}$

Topic 10
Cumulative Review Problems

1. $(3b + 5)(2x + 1)$ **3.** $5xy^3 - 6x^2y + 14xy - 8$

5.

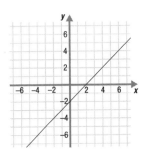

7. $x = 4$ or $x = -1$ **9.** $y = 2i$ or $y = -2i$

11a. $-12 \le x \le 2$ **13a.** $27 - 31i$ **b.** 40

15a. $y + 4 = -\dfrac{6}{5}(x - 11)$ **b.** $y = -\dfrac{6}{5}x + \dfrac{46}{5}$

  c. $6x + 5y = 46$

17. 58 **19.** $y = 7$ or $y = \dfrac{3}{2}$ **21.** $\dfrac{16}{3}$

23. $5w^2xy - 6w^2x + y$ **25.** $\dfrac{16}{47}$ **27.** $x = 6$ or $x = -\dfrac{20}{7}$

29. $x = \dfrac{5}{2}$ or $x = -\dfrac{19}{2}$ **31.** $2x^3 - \dfrac{239}{15}x^2 + 7x$

33. $x = -1$ or $x = -\dfrac{4}{3}$ **35.** $(5y - 3)(5y - 3)$ or $(5y - 3)^2$

37. $|11 - 5| = 6$
The GCF of 64 and 81 is 1
If $S = \{1, 2, 3, 4, 5, 6, 7, 8, 9, 10\}$, then $16 \notin S$.

39. $4a^8 - 25b^4$

41. $x = \dfrac{2 \pm \sqrt{3}}{4}$ **43.** $-4 \le y < \dfrac{5}{3}$

45. x-intercept: $\left(\dfrac{5}{4}, 0\right)$, y-intercept: $\left(0, -\dfrac{5}{3}\right)$

47. $(x^2 + y)(x^2 - y)$ **49.** $x = \dfrac{11}{3}$ or $x = -2$

Lesson 11.1 Functions
Homework

1.

x	y
-5	2
-4	3
-3	4
-2	5
-1	6
0	7
1	8
2	9

3a. domain: all real numbers; range: all real numbers

  b. domain: all real numbers; range: all real numbers

  c. domain: all real numbers; range: real numbers ≥ 0

  d. domain: all real numbers; range: real numbers ≥ 1

  e. domain: all real numbers except 0; range: all real numbers except 0

5a. -5 **b.** 11 **c.** 44 **d.** -1 **e.** 31

7a. 9 **b.** 0 **c.** 1 **d.** −26 **e.** 2

9a. 259,200 people **b.** 1,814,400 people
 c. 94,608,000 people

11a. domain: all real numbers except 3; range: all real
 numbers except 1

 b. domain: all real numbers except −1 and −2; range: all
 real numbers except those which are greater than −8
 and less than or equal to 0.

 c. domain: all real numbers ≥ −2; range: real numbers ≥ 0

 d. domain: all real numbers except −4, 8, and 11;
 range: all real numbers except those which are greater
 than $-\frac{4}{9}$ and less than or equal to 0.

 e. domain: all real numbers greater than or equal to −2
 and less than or equal to 2; range: all real numbers
 greater than or equal to 0 and less than or equal to 2.

13. $y = 2x + 1$; $y = 5 - 8x$

15a. domain: all real numbers; range: all real numbers
 b. domain: all real numbers; range: real numbers ≥ 0
 c. domain: all real numbers; range: all real numbers
 d. domain: all real numbers; range: real numbers ≥ 0

17a. domain: all real numbers; range: all real numbers
 b. domain: all real numbers; range: real numbers ≥ −5
 c. domain: all real numbers; range: real numbers ≥ 0

19. a: $y = |x + 1| + 4$; b: $y = |x + 1| - 1$; c: $y = |x + 1| - 3$

21. domain: $0 \le x \le 4$; range: $0 \le y \le 2$

23.

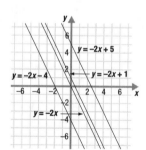

25. $y = 4 - 2x + 6x^2$; $y = 8x^2 + 9x - 1$

27a. domain: all real numbers; range: real numbers ≥ 3
 b. domain: all real numbers; range: real numbers ≥ −5

 c. domain: all real numbers; range: real numbers ≥ 0
 d. domain: all real numbers; range: real numbers ≥ 4

29a. x-intercept: (0, 0); y-intercept: (0, 0)
 b. x-intercepts: (−3, 0) and (−1, 0); y-intercept: (0, 3)
 c. x-intercept: (−4, 0); y-intercept: (0, 16)

31. a: $y = \frac{1}{2}x^2 + 5$; b: $y = \frac{1}{2}x^2 + 1$; c: $y = \frac{1}{2}x^2 - 4$

33. domain: 0 ≤ time ≤ 6; range: 0 ≤ height ≤ 40

35.

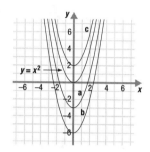

37. graphs a, b, c, and e

39. Reservoir 1

41a. 0 **b.** 5 **c.** 3 **d.** 10 **e.** 2

Practice Problems

1. $(1, 9)$, $\left(-\frac{5}{4}, 0\right)$, and $(2, 13)$

3. $(1, 4)$, $\left(\frac{1}{3}, 0\right)$, and $(2, 10)$

5. $(1, -3)$, $(-1, -9)$, $(2, -3)$

7. Tables may vary.

x	y
0	−2
1	1
2	4

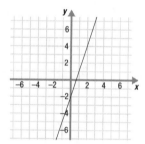

9. Tables may vary.

x	y
0	−4
1	−2
2	0

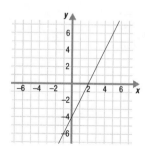

11. Tables may vary.

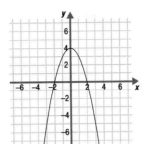

x	y
0	4
1	3
2	0

13a. 14 **b.** 5 **c.** 0 **d.** −1 **e.** 2

15a. −2 **b.** −3 **c.** 0 **d.** 7 **e.** 18

17a. 2 **b.** $\sqrt{2}$ **c.** 0 **d.** $2\sqrt{2}$ **e.** $2\sqrt{3}$

19. domain: all real numbers; range: all real numbers ≥ −16

21. domain: all real numbers; range: real numbers ≥ 9

23. domain: all real numbers ≥ 0; range: all real numbers ≥ 2

25. domain: all real numbers; range: all real numbers ≤ 3

27. domain: all reals except $x = -\frac{1}{2}$; range: all reals except $y = \frac{3}{2}$

29. $f(x) = 3x + 5; f(x) = \frac{x}{3} + 1$

31. $f(x) = 2x - 1; f(x) = \frac{x}{2}$

33a. $y = -\frac{1}{3}x - 3$ **b.** $y = -\frac{1}{3}x + 1$

35a. $y = |x + 2| - 3$ **b.** $y = |x + 2| + 4$

37.

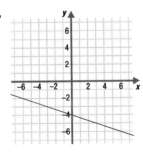

39.

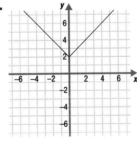

41.

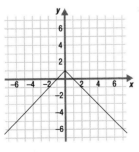

43.

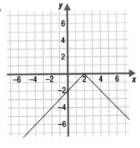

45. domain: all real numbers; range: all real numbers

47. domain: all real numbers; range: all real numbers

49. domain: all real numbers; range: all real numbers ≤ 0

51.

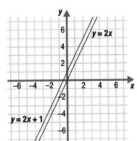

53.

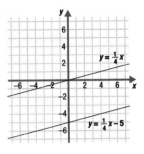

55.

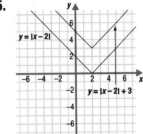

57. $y = x^2 - 5, y = -x^2, y = 9 - x^2$

59. $y = x^2 + 4; y = -5x^2 - 9; y = 3 - 2x - x^2$

61a. up **b.** up **c.** down **d.** up **e.** down

63. domain: all real numbers; range: all real numbers ≥ −6

65. domain: all real numbers; range: real numbers ≥ 9

67a. $y = -x^2 + 5$ **b.** $y = -x^2 - 3$

69a. $y = (x + 2)^2 - 2$ **b.** $y = (x + 2)^2 + 4$

71.

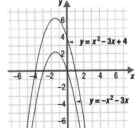

73. x-intercepts: (−5, 0) and (2, 0); y-intercept: (0, −10)

75. x-intercepts: (−5, 0) and (3, 0); y-intercept: (0, −15)

77a. true **b.** true **c.** false **d.** false

79. (3, –5) **81.** (4, –1) **83.** $y = x^2 + 4x + 6$

Practice Test

1. (0, 1), (–2, 13), (1, 4)

2a. $f(0) = 0$ **b.** $f(3) = 18$ **c.** $f(-2) = -2$ **d.** $f(-5) = 10$

3. The domain of this function is all x.
The range of this function is $y \geq -2$.

4a. $f(7) = 23$ **b.** $g(7) = -20$ **c.** $f(-2) = -13$ **d.** $g(-2) = 7$

5a. $y = 2x + 3$ **b.** $y = 2x - 4$

6. The domain of this function is all real numbers.
The range of this function is $y \geq 0$.

7. The true statements are:
   Both functions have the same domain.
   The point (0, 3) satisfies both equations.

8. The only points that satisfy the equation $y = x - 4$ are
   (5, 1), (2, –2), and (0, –4).

9. The graph crosses the x-axis at the points (–1, 0) and
   (5, 0). These are the x-intercepts of the function.

10. The true statements are:
   Both functions have the same domain.
   Both graphs have the same x-intercepts.

11. The only functions that are not quadratic are $y = 4x + 1$
   and $f(x) = -3x + 7$.

12. (–2, 2) **13a.** $y = 0$ **b.** $y = -2$ **c.** $y = -4$

14. The true statements are:
   The domain of $y = 2|x - 1| + 4$ is all real numbers.
   The range of $y = 2|x - 1| + 4$ is all real numbers ≥ 4.

15a. This is the graph of a function.
   b. This is the graph of a function.
   c. This is not the graph of a function.
   d. This is the graph of a function.

16a. The domain is all real numbers.
   b. The range is all real numbers ≤ 9.
   c. The vertex of the function is the point (–2, 9).

Lesson 11.2 The Algebra of Functions
Homework

1. $(f + g)(x) = -x + 1$ **3.** $(f \circ h)(x) = 3x + 1$

5. $(f - g)(-3) = -14$ **7.** $(f \cdot g)(-1) = -35$

9. $h[g(F)] = \dfrac{961 + F}{3}$ **11.** $\left(\dfrac{f}{g}\right)(2) = -3$ **13.** $f^{-1}(x) = \dfrac{x + 1}{6}$

15.
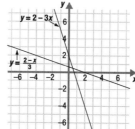
$f^{-1}(x) = \dfrac{2 - x}{3}$

17.
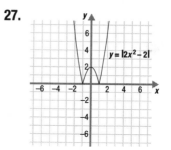
The function is not one-to-one.

19. $g^{-1}(x) = \dfrac{2}{x - 1}$

21. $f^{-1}(x) = \dfrac{10000 - x}{20}$ or $f^{-1}(x) = -\dfrac{x - 10000}{20}$

23. g has an inverse; $g^{-1}(x) = \dfrac{3}{x + 1}$

25.

x	r(x)	c(x)	p(x)
5	20	600	–580
11	44	720	–676
20	80	900	–820

27.
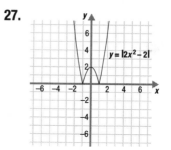

29.
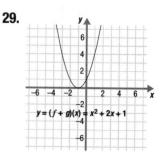

Practice Problems

1. $(g + h)(x) = x^2 - 4x + 5$ **3.** $(f + g)(x) = 2x^2 - 3x + 7$

5. $(g - h)(x) = -2x^2 - 2x + 10$ **7.** $(f \cdot g)(x) = 40x^2 + 90x$

9. $(g \cdot h)(x) = 4x^3 + 7x^2 - 20x - 35$ **11.** $(g \cdot h)(0) = 0$

13. $\left(\dfrac{f}{g}\right)(x) = \dfrac{2x}{x + 4}$ **15.** $\left(\dfrac{f}{g}\right)(x) \dfrac{x - 3}{4x^2 - 5}$

17. $(g \circ h)(x) = 6x - 5$ **19.** $(f \circ g)(x) = x^2 + 4x + 4$

21. $(f \circ h)(x) = x^2 - 3x + 2$ **23.** $(g \circ h)(3) = -3$

25. $(g \circ h)(4) - (h \circ g)(4) = 8$

27. $f(x) = 2x$ and $g(x) = \dfrac{x}{2}; f(x) = 4x + 7$ and $g(x) = \dfrac{x - 7}{4}$

29. $f^{-1}(x) = x - 8$ **31.** $f^{-1}(x) = \dfrac{x + 1}{4}$ **33.** $f^{-1}(x) = \dfrac{1}{x - 2}$

35. $f^{-1}(x) = \dfrac{2}{3(x - 1)}$

37. The function is one-to-one.

39. The function is not one-to-one.

41. The function is one-to-one. **43.** $(f^{-1} \circ f)(x) = x$

45.

x	$y = f^{-1}(x)$
−4	−1
−3	0
5	2

47.

x	$y = f^{-1}(x)$
1	−2
2	1
3	6

49.

x	$y = f^{-1}(x)$
−3	−3
−1	0
1	3

51. 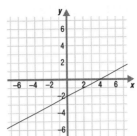 **53.**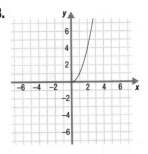

55. x-intercept: $(12, 0)$; y-intercept: $(0, -4)$

Practice Test

1. $(f - h)(-3) = -42$

2. $\left(\dfrac{f}{g}\right)(x) - (f \cdot g)(x) = 2x - 18x^7$ or $2x(1 - 9x^6)$

3. The only pair of functions for which $(f \circ g)(x) = x$ is:

$$f(x) = 2x$$
$$g(x) = \frac{x}{2}$$

4. $(f \circ g)(4) = 113$

5. Here is the completed table of some of the ordered pairs that satisfy $f^{-1}(x)$:

$f^{-1}(x)$

x	y
−5	−2
−1	0
3	2

6. $h^{-1}(x) = \dfrac{5x - 9}{2}$

7. $(g \circ f)(x) = x$

8. (b), (c), and (d)

9. $(f \cdot g)(5) = 33$

10. $(f - g)(2) = -9$

11. $(f \circ g)(2) = -3$

12. If $f(x) = 2x + 8$, the statements that are true are:

$$f^{-1}[f(x)] = x$$

$$(f \circ f)(x) = 4x + 24$$

$$(f \cdot f)(x) = (2x + 8)^2$$

Topic 11
Cumulative Review Problems

1. $\frac{15i}{8}$ **3.** $x = \frac{4}{3}$ or $x = 1$

5. $f^{-1}(x) = \frac{2x - 12}{x - 7}$ or $f^{-1}(x) = \frac{2}{x - 7} + 2$

7. 6 **9.** $93 + 27\sqrt{2}$ **11.** $(f + g)(x) = -5x + 6$

13. $-\frac{51}{8}$ **15.** $2 + 9i$ and $2 - 9i$; $7 + 3i$ and $-3i + 7$

17. $x = \frac{7}{3}$ or $x = -1$

19. x-intercept is (2, 0); y-intercept is (0, −4)

21. −54 **23.** $7^{\frac{7}{3}}$ **25.** $f^{-1}(x) = \frac{x - 4}{7}$

27. No, f does not have an inverse.

29. $x = -\frac{3}{2}\left(1 \pm \sqrt{5}\right)$ or $x = \frac{-3 \pm 3\sqrt{5}}{2}$

31. $x = 1$ or $x = -\frac{5}{2}$

33a. x-intercepts: $\left(-\frac{3}{2}, 0\right)$ and (3, 0); y-intercept: (0, −9)

b. x-intercepts: (0, 0) and $\left(\frac{1}{9}, 0\right)$; y-intercept: (0, 0)

c. x-intercepts: (2, 0) and (−2, 0); y-intercept: (0, 4)

35. $\frac{a^{\frac{7}{2}}}{b^3}$ or $a^{\frac{7}{2}}b^{-3}$ **37.** $(f^{-1} \circ f)(x) = x$ **39.** $a = \pm 8$

41a. domain: all real numbers; range: all real numbers

b. domain: all real numbers; range: real numbers ≥ 6

c. domain: all real numbers; range: real numbers ≥ 0

43. $x = \frac{3}{4} \pm \frac{i\sqrt{31}}{4}$ or $x = \frac{3 \pm i\sqrt{31}}{4}$

45a. −4 **b.** −2 **c.** 4 **d.** 14 **e.** −2

47. $x = \frac{7}{5}$ or $x = -\frac{11}{5}$ **49.** $2\sqrt{313} + 4\sqrt{47} - 6$

Lesson 12.1 Exponential Functions
Homework

1.

x	y
3	8
2	4
1	2
0	1
−1	$\frac{1}{2}$
−2	$\frac{1}{4}$
−3	$\frac{1}{8}$

3. 0.07

5.

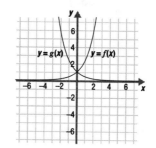

7. The base is 10.

9. Approximately 82 bacteria will be present after 2 days.

11. The base is 8.

Practice Problems

1. (0, 1), (−1, 4) **3.** (0, 1) and (−1, 1.5) **5.** $e^{1.2} \approx 3.32$

11. $f(2) = \frac{1}{4}$ **13.** $f\left(\frac{1}{2}\right) = 2$ **15.** $b = \frac{1}{3}$ **17.** $x = \frac{3}{2}$

19. $x = -\frac{2}{3}$ **21.** $t = -\frac{1}{2}$ **23.** $x = 1$ **25.** $x = 2$

27. $b = \frac{1}{7}$

Practice Test

1. (1, 2), (0, 1), and $\left(-2, \frac{1}{4}\right)$

2. Here is the completed table:

x	y
2	$\frac{1}{4}$
1.5	.3536
1	$\frac{1}{2}$
0	1
–1	2
–1.5	2.8284
–2	4

3. The only statement that is true is the fourth one. The function f is one-to-one

4. $e^3 \approx 20.09$, $e^{-1} \approx 0.37$ **5.** approximately $206.06

6. $\frac{1}{8}N_0$ **7.** $x = 1$ **8.** $x = -\frac{1}{2}$

Lesson 12.2 Logs and Their Properties
Homework

1. $\log_{10}10000 = 4$ **3.** $y = 7^x$ **5.** $\log_2 \frac{1}{64} = -6$

7. $\log_7 y = 12$

9.

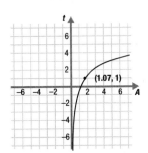

11. $\log_{\frac{1}{4}}16 = -2$ **13.** y **15.** $\log_{13}12^{14}$

17. $\log_2 3ab - \log_2 7cd$ **19.** 1 **21.** $R = \log_{10}I - \log_{10}M$

23. $\log_x \frac{1}{x} = -1$

25a. is $y = \log_3 x$ **b.** is $y = \log_7 x$ **c.** is $y = \log_{15}x$

27. Answers will vary. One answer is shown for each.

   a. $\log_4 64 = \log_4 32 + \log_4 2$

   b. $\log_4 64 = \log_4 128 - \log_4 2$

   c. $\log_4 64 = 2 \cdot \log_4 8$

29. Both graphs include the point (1,0).

Practice Problems

1. $\log_2 8 = 3$ **3.** $\log_5 625 = 4$ **5.** $\log_3 a = x$ **7.** 6 **9.** 4

11. –3 **13.** $f^{-1}(x) = 12^x$ **15.** $f^{-1}(x) = 11^x$ **17.** $f^{-1}(x) = b^x$

19. $5^3 = 125$ **21.** $8^x = 512$ **23.** $6^5 = x$

25.

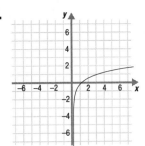

27.

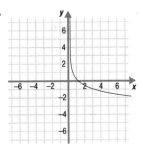

29. $4\log_2 33$ **31.** $\log_3 25^{17}$ **33.** $\log_b 7x - \log_b 8y$

35. $\log_3 5 + \log_3 x$ **37.** $\log_b 21 + \log_b x + \log_b y$

39. $\log_b 12$ **41.** $\log_5 2x^6$ **43.** $\log_7 \frac{1}{2}$ **45.** $\log_x \frac{a^3b^5}{c}$

47. 1 **49.** –3 **51.** –2

53. $\log_5 z - 4\log_5 y$ **55.** $\log_3 x - \log_3 2 - \log_3 y$

Practice Test

1. $og_{18}yz = 12x$ **2.** –3

3.

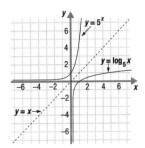

4. $\log_e \frac{P}{I} = kt$ **5.** xyz **6.** $\log_{29}19 + \log_{29}A + \log_{29}B$

7. $\log_w \frac{11}{xy}$ **8.** $\log_b 17 + 5\log_b x - 12\log_b y$

9.

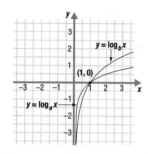

10. Graph c; Graph b **11.** This statement is true. **12.** 15.6

Lesson 12.3 Application of Logs
Homework

1. 4 **3.** $\log 2x^3$ **5.** 3.43 **7.** 0.0003

9. $\log\left[2x^3(x+1)^2(x+2)^2 \cdot 10^4\right]$ **11.** 3.731 **13.** $x=64$

15. $x=5$ **17.** $x=1$ **19.** $x=49$ **21.** $t\approx 0.17$

23. $x\approx 1.11$

Practice Problems

1. 1 **3.** -3 **5.** 7 **7.** $\log x^8$ **9.** $5\log x^{13}$ or $\log x^{65}$

11. $\ln(2x^8 + 3x^7)$ **13.** $\ln \dfrac{x^5}{4x-1}$ **15.** 1.39 **17.** 4.50

19. 4.01 **21.** 1.62 **23.** 133.35 **25.** 23.71 **27.** 0.29

29. $x=16$ **31.** $x=243$ **33.** $x=5$ **35.** $x=64$

37. $x=\sqrt{6}$ **39.** $x=2$ **41.** $x=3$ **43.** $x=3$ **45.** $x=7$

47. $x=8$ **49.** $x=\dfrac{\ln 19}{\ln 5}\approx 1.83$ **51.** $x=2+\dfrac{\ln 22}{\ln 3}\approx 4.81$

53. $x=\dfrac{\ln 6}{\ln 7}+3\approx 3.92$ **55.** $x=\dfrac{\ln 5 + 1}{4}\approx 0.65$

Practice Test

1a. ≈ 2.05 **b.** ≈ 2.19 **2.** $\log[10x(x-2)^2]$

3. $\ln 1 = 0$; $\log_e 5 = \ln 5$; $2\log x = \log x + \log x$

4. ≈ 2.12 **5.** $x=6$

6a. $x=\dfrac{1}{9}$ **b.** $x=\dfrac{3}{2}$

7. $x=13$ **8.** ≈ 1.73

Topic 12
Cumulative Review Problems

1. $t=-\dfrac{2}{11}$ **3.** $\dfrac{26+57i}{157}$

5a. $6\log_7 4$ **b.** $-3\log_2 x$ **c.** x **d.** $\log_5 10$ **e.** $\dfrac{1}{2}\log_{10} 31$

7a. $y=-2x^2+6$ **b.** $y=-2x^2+3$ **c.** $y=-2x^2-1$

9a. 1 **b.** 0 **c.** 14

11. $y=\dfrac{3}{4}x-1$

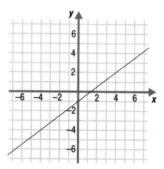

13. $6\sqrt{13}-4\sqrt{41}+4$

15a. domain: all real numbers; range: real numbers ≥ -2

 b. domain: all real numbers; range: all real numbers

 c. domain: real numbers $\neq \dfrac{1}{2}$ or -7; range: all real numbers

 d. domain: real numbers ≥ 2 or ≤ -2; range: real numbers ≥ 0

17. $2\ln x - \ln(x+2)$ **19.** $x=9$ or $x=-9$

21. $y=\log_3 x$

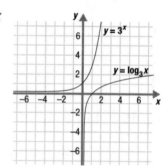

23. Yes. $y = f^{-1}(x) = \sqrt[3]{x}$

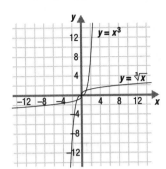

25. $y = 1$ or $y = \frac{2}{5}$ **27.** $x = \dfrac{\ln\frac{9}{4} + 6}{3}$ or $x = 2.27$

29. $A = \$515.17$ **31a.** $9i$ **b.** 5 **c.** $45i$ **d.** -45 **e.** $-i$ **f.** -1

33. $q = -1$ or $q = \dfrac{7}{2}$

35. As x becomes smaller and smaller, y approaches 1.

x	y
0	2
−1	1
1	4
−2	$\frac{1}{2}$
2	8
−3	$\frac{1}{4}$

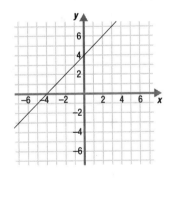

37. $(2x + y)(x - 3)$ **39.** $d = -20$ **41.** $f^{-1}(x) = \dfrac{x + 18}{33}$

43.

x	y
−6	−2
−4	0
−2	2
0	4
1	5
3	7

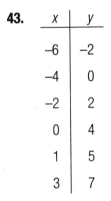

45. $m \le -\dfrac{7}{5}$ **47.** $-12 + 39\sqrt{30}$

49. $(f \circ g)(x) = x^2 + 4x - 5$

Lesson 13.1 Nonlinear Equations

Homework

1. $x = \pm\sqrt{3}$ or $x = \pm 2$ **3.** $w = \dfrac{1}{2}$ or $w = \pm i\sqrt{3}$

5. $a = 2$ or $a = 4$ **7.** $b = 0$, $b = 3$, or $b = 5$

9. $x = -1$ or $x = \dfrac{3}{2}$ **11.** $p = \dfrac{1}{3}$ or $p = 6$ **13.** $x = -2$

15. $x = -5$ or $x = -6$ **17.** $x = 7$ **19.** $x = 217$

21. $x = -2$ or $x = 10$ **23.** $x = \pm 2\sqrt{3}$

25. $x = -1$ or $x = -2$

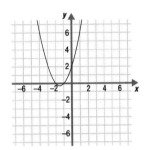

27. $x = -2$ or $x = -3$

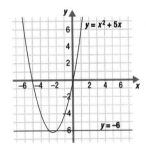

29. All values of c less than or equal to zero.

Practice Problems

1. $w = \pm 4$ or $w = \pm i$ **3.** $w = -\dfrac{1}{9}$ or $w = \dfrac{1}{4}$

5. $= 5$ or $x = \pm 3$ **7.** $z = \dfrac{3}{5}$ or $z = -2$

9. $x = 0$ or $x = -1$ or $x = 9$ **11.** $x = -3$ **13.** $x = 3$

15. $x = 5$ **17.** $x = -1$ or $x = 5$ **19.** $x = -1$

Practice Test

1. $x = -6$ or $x = \dfrac{2}{3}$ **2.** $u = x^{-4}$ **3.** $y = -7$ or $y = -2$ or $y = 2$

4. $-4 = w$ or $-2 = w$ **5.** $x = -\frac{1}{2}$ **6.** $y = -\frac{7}{16}$ **7.** $x = 3$

8. $w = 3$ or $w = -13$

9. $x = -1$ and $x = 2$

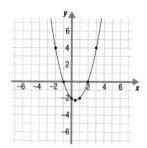

10. $x = 1$ and $x = 3$

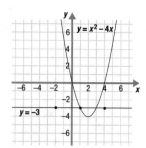

11. $\frac{1}{2}x^2 - \frac{5}{2}x - 3 = 0$ and $x^2 - \frac{8}{3}x - 1 = 0$

12. To graph the function $y = x^2 + 2x - 3$, start by making a table of ordered pairs:

x	y
-4	5
-3	0
-2	-3
-1	-4
0	-3
1	0
2	5

Then plot the ordered pairs and graph the parabola:

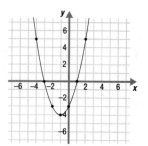

To graph the function $y = x^2 + 2x + 1$, start by making a table of ordered pairs:

x	y
-3	4
-2	1
-1	0
0	1
1	4

Then plot the ordered pairs and graph the parabola:

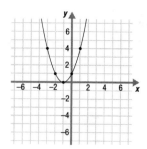

To graph the function $y = x^2 + 2x + 4$, start by making a table of ordered pairs:

x	y
-3	7
-2	4
-1	3
0	4
1	7

Then plot the ordered pairs and graph the parabola:

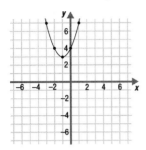

$c = 4$ and $c > 1$

Lesson 13.2 Nonlinear Systems Homework

1.

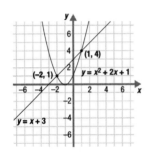

$(x, y) = (-2, 1)$ or $(x, y) = (1, 4)$

3. $(x, y) = (2, 1)$ or $(x, y) = (-2, 1)$ or $(x, y) = (2, -1)$
or $(x, y) = (-2, -1)$

5. $(x, y) = (2, 3)$ or $(x, y) = (3, 2)$

7.

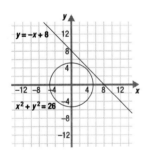

no real solutions

9. $(x, y) = (500, 25)$; 500 jackets are sold for \$25 each.

11. $(x, y) = (0, 4)$ or $(x, y) = (4, 0)$

13.

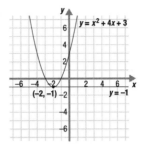

$y = -1$

15. $x^2 + y^2 = 25$ and $y = -x^2 + 4x + 5$
     $y = 0$ $y = -x + 5$

17.

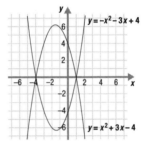

$(x, y) = (-4, 0)$ or $(x, y) = (1, 0)$

Practice Problems

1. $(x, y) = (1, 1)$ or $(x, y) = (3, 1)$

3. $(x, y) = (-2, -3)$ or $(x, y) = (1, 0)$

5. $(x, y) = (2, 10)$ or $(x, y) = (-2, 10)$

7. $(x, y) = (1, 2)$ or $(x, y) = (-1, 2)$ or $(x, y) = (1, -2)$ or
$(x, y) = (-1, -2)$

9. zero

Practice Test

1. There are four solutions.

2. $(x, y) = (-1, 4)$ or $(x, y) = (2, 1)$

3. $(x, y) = (-6, 7)$ or $(x, y) = (4, 2)$

4. There are no real solutions to this system.

5. $y = -x$

6. $(x, y) = (0, 5)$ or $(x, y) = (5, 0)$

7. $y = -2x - 2$

8. $y = x^2 - 5$, $y = -2x + 3$, $x^2 + y^2 = 5$, and $y = -x^2 + 3$

Lesson 13.3 Inequalities
Homework

1. $0 < x < 4$

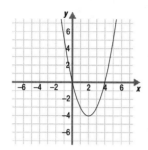

3. all real numbers except -2 **5.** $x < 4$ or $x > 7$

7. $x = 1$

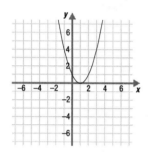

9. $100 < x < 400$ **11.** $x < -4$ or $-1 < x < 3$

13. $-3 < x < 0$

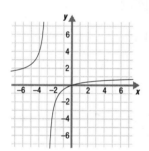

15. $-1 < x < 3$

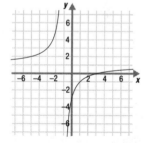

17. $x < -4$ or $x > -\frac{2}{3}$

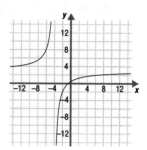

19. $-1 \le x < 2$

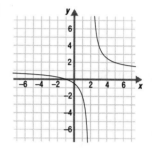

21. $r = 1$

23. $x < -\frac{1}{2}$ or $x \ge 3$

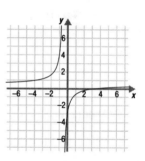

25. $-5 < x < -3$

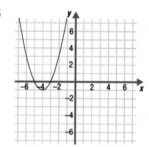

27a. positive **b.** negative **c.** 0 **d.** positive **e.** 0 **f.** positive

29. $x \le -4$ or $0 < x \le 2$

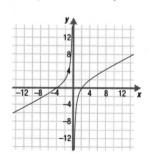

Practice Problems

1. $-3 < x < 0$ **3.** $x \le 1$ or $x \ge 5$

5a. False **b.** True **c.** False **d.** True **e.** False **f.** True

7. $4 < x < 6$ **9.** No solution **11.** $-1 < x < 0$

13. $x \le -1$ or $0 < x \le 1$ **15.** $x \le -1$ or $x > 3$

17. $-1 \le x < 1$ or $x \ge 2$ **19.** $x < -2$ or $x > 8$

Practice Test

1. $-6 < x < 1$

2. $-2 \le x \le 5$

3. Pick a test point in each interval to determine the sign of $x^2 - 8x + 16$ in that interval; for example $x = 0$ and $x = 10$:

Test $x = 0$:

$$0^2 - 8(0) + 16$$

$$= 0 - 0 + 16$$

$$= 16 \text{ is } \textbf{positive}$$

Test $x = 10$:

$$(10)^2 - 8(10) + 16$$

$$= 100 - 80 + 16$$

$$= 36 \text{ is } \textbf{positive}$$

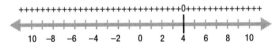

Since you are solving $x^2 - 8x + 16 \ge 0$, the solution is the intervals where the expression is positive or 0. That is, the solution is all real numbers, x.

4a. True **b.** False **c.** False **d.** False **e.** True **f.** True

5. $x < -3$ or $x > 2$ **6.** $x < -9$ or $x \ge \frac{1}{2}$

7. $x < -5$ or $-2 < x < 2$ **8.** $x > -1$ **9.** $x \le -2$ or $x \ge 4$

10. $-4 < x < -2$ or $2 < x < 4$ **11.** $-5 < x < 1$ or $x > 4$

12a. negative **b.** 0 **c.** positive **d.** undefined **e.** positive

Topic 13
Cumulative Review Problems

1. 18 **3.** $x = -2$ **5.** $x = 4$ **7a.** $9x^2 - 24x + 18$ **b.** 102

9. $\log_6 20 = x$ **11.** 2.97 **13.** $x = 11$ or $x = -17$

15. $x < -1$ or $x > 5$ **17.** $-2 \le x \le 5$ **19.** 27

21. $x = -12$ or $x = 16$ **23.** $x = \pm 4i$

25. 2 **27a.** $y = |x| + 3$ **b.** $y = |x| - 5$ **c.** $y = |x| - 1$

29. $f^{-1}(x) = \frac{x + 3}{10}$ **31.** $t = -\frac{2}{5}$ **33.** $b^{\frac{9}{14}}$

35. $2a^2 - 15b^2 + 7ab + ac + 5bc$

37a. domain: all real numbers except 2; range: all real numbers
  b. domain: all real numbers; range: real numbers ≥ 0
  c. domain: all real numbers; range: real numbers ≥ -2

39. $x = \frac{-3 \pm \sqrt{15}}{2}$ **41.** $\frac{5a^2 b \sqrt{3c}}{12c^2}$ **43.** $k = 1$ or $k = \frac{1}{2}$

45. $\log\left(\frac{x^5}{25}\right)$ **47.** $\frac{81a^6}{64b^4}$ **49.** $b = a + dc$

INDEX

Absolute value
 defined 4
 solving equations containing
 absolute values 91
 solving inequalities containing
 absolute values 97

Absolute value functions
 definition of 330
 domain of 330
 graphing of 331
 range of 330

Addition of polynomials 123

Algebraic fractions 195

Binomial(s) 121
 product of two (FOIL) 135
 square of 136

Cartesian coordinate system 71

Change of Base Formula 430

Common logarithms 425

Comparison symbols 4

Completing the square
 to solve a quadratic equation . .289

Complex conjugate 310

Complex fractions . .56, 197, 216, 222

Complex numbers 308
 adding and subtracting 310
 conjugates of 310
 dividing 312
 multiplying 311

Composition of functions 365

Conjugate 262
 of a complex number 310
 of a radical expression 262

Coordinates of a point 72

Cross multiplication 234

Degree
 of a polynomial 122
 of a term 122

Degree of a polynomial 18

Difference
 of two cubes 184
 of two squares 184

Difference of two squares 136

Discriminants 296

Distance formula 72

Division
 dividend & divisor 138
 long division 138
 of monomials 126
 of polynomials 126
 property of exponents 112

Domain of a function 324

Equations
 quadratic in one variable 172

Eratosthenes 194

Expanded form 213

Exponential equations 399, 436

Exponential functions 391
 applications 395
  compound interest 397
  growth and decay 393
  learning curves 395
 graphing 407
 with Base e 394

Exponential Growth and Decay . . .393

Exponents 255
 division 111
 multiplication 111
 power of a power 112
 power of a product 112
 power of a quotient 113
 properties 255
 properties of 111
 rational 255
 zero power 113

Exponents and their properties
 Division Property 51
 Multiplication Property 51
 Power of a Product Property . . .51

Power of a Quotient Property . . .51
Power of a Power Property 51
Zero Power Property 52

Extraneous solutions 233, 454

Factoring
 by finding the GCF 26
 by grouping 27
 by using patterns
  difference of two cubes .29, 184
  difference of two squares 29, 184
  perfect square trinomials 28, 183
  sum of two cubes 29, 184
  trinomials 27, 170
 finding the greatest
 common factor 151
 grouping 153
 to solve a quadratic
 equation 172, 277, 313

Finding the equation of a line
 horizontal lines 81
 point-slope form 78
 slope-Intercept form 80
 standard form 79
 vertical lines 81

FOIL method 21, 135

Functions
 absolute value 330
 composition of 365
 definition of 323
 domain of 324
 graphing 325
 inverse 369
 linear 329
 product and quotient of 363
 quadratic 333
 range of 324
 sum and difference of 361

Graphing linear equations
 finding the slope of a line 75
 finding x- and y-intercepts 74
 making a table of ordered pairs .73

Greatest Common Factor (GCF) . . .26

Grouping symbols 5

Horizontal line test 371

i 307
  powers of 308

Imaginary number 307

Imaginary solutions
  of nonlinear systems 470

Index .254
  of a radical 254

Inverse functions
  finding the equation of 369
  graphing 371

Least common denominator (LCD)
  . . . . .40, 219, 233

Least common multiple (LCM) . . .219

Like Terms 261
  radicals 261

Linear functions
  definition of 329
  domain of 329
  graphing 329
  range of 329

Linear Inequalities 42

Logarithmic equations 432
  natural
    finding by using a
      calculator 428
  solving 432

Logarithmic functions 407
  graphing 409

Logarithms
  common 425
    finding by using a calculator .426
    properties of 427, 429
  natural 427
    change of base formula 430
    properties of 429

Lowest terms 195

Monomial(s) 121
  division of 126
  multiplication of 125

Multiplication
  FOIL method 135
  of monomials 125
  of polynomials 138
  property of exponents 111

Natural logarithms 427

Negative exponents 209
  properties of 210
  simplifying 211

Negative integer exponents 51

Nonlinear systems of equations . .465

One-to-one function 371

Order of operations 7

Parabola 333
  vertex 333

Parallel lines 82

Pascal's triangle 142

Perfect square trinomial(s) . .136, 183

Perpendicular lines 82

Plotting points 71

Point-slope form of a line 78

Polynomial(s)
  addition of 123
  degree of 122
  division of 126
  multiplication of 138
  subtraction of 123

Polynomial expressions
  binomial 18
  degree of a polynomial 18
  monomial 18
  trinomial 18

Polynomial operations
  adding 19
  dividing 23
  evaluating 19
  multiplying 20
  multiplying binomials
    FOIL method 21
    using patterns 22
  subtracting 19

Powers
  of a power property 112
  of a product property 112
  of a quotient property 113

Properties of exponents . .9, 111, 255

Properties of real numbers 6

Proportions 234

Pure imaginary number 309

Quadrants in the *xy*-plane 71

Quadratic equations
  in one variable 172
  standard form 172
  square root property to solve . .291

Quadratic formula 294

Quadratic functions
  definition of 333
  domain of 334, 336
  graphing 333
  intercepts
    x-intercept(s) 335
    y-intercepts 335
  range of 334, 336

Radicals 257
  additional and subtraction of . .258
  division of 257
  multiplication of 257
  simplifying 258

Radicand 254

Range of a function 324

Rational exponents 254

Rational expressions 195
  addition 54, 200, 220
  complex fractions 56
  division 53, 197, 216
  multiplication 53, 196, 216
  subtraction 54, 200, 221
  writing in lowest terms 53

Rise 75

Roots .253
   cube 253
   *n*th 254
   square 253

Run 75

Scientific notation 212

Signed numbers and operations 5

Slope-intercept form of a line 80

Slope of a line
   defined 75
   of a horizontal line 81
   of a vertical line 81

Solution
   of a quadratic equation
      no real solution 292
   of an equation with rational
      expressions 233
      solution of multiplicity two 277

Solving compound linear
inequalities 43

Solving exponential
equations 399, 436

Solving linear equations
   by isolating a variable 39
   containing fractions 40

Solving linear inequalities 42

Solving logarithmic equations 432

Solving nonlinear equations
   by factoring 451
   by graphing 458
   by substitution 452
   containing radicals 454
   containing rational exponents .455

Solving nonlinear systems
   by elimination 469
   by graphing 465
   by substitution 466

Solving quadratic equations
   by completing the square 291
   by factoring 172, 277
   by square roots 281
   using the quadratic formula . . .294

Solving quadratic inequalities
   by graphing 485
   by the test point method 486

Solving rational inequalities
   by graphing 490
   by the test point method 491

Square root
   of negative numbers . . . .
   of positive numbers 280

Square root property 281

Standard form of a line 79

Standard form of a quadratic
equation 172

Subtraction of polynomials 123

Sum
   of two cubes 184

Term .122

Test point 486

Trinomial(s) 121
   factoring 183
   perfect square 183

Vertex of a parabola 333

Vertical line test 326

Word Problems 241

x-intercept of a line 74

y-intercept of a line 74

Zero Power Property 113

Zero Product Property 172, 277